Electrocatalysis - Theory and Experiment at the Interface

University of Southampton, UK

07–09 July 2008

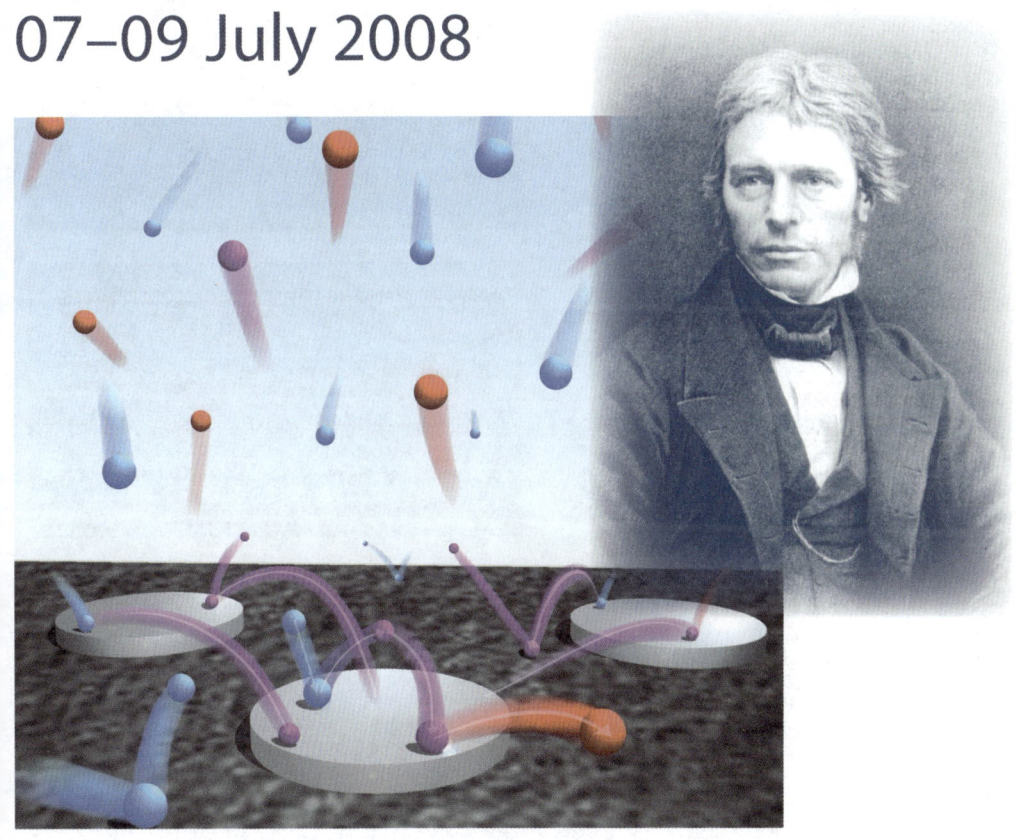

FARADAY DISCUSSIONS
Volume 140, 2008

RSC Publishing

The Faraday Division of the Royal Society of Chemistry, previously the Faraday Society, founded in 1903 to promote the study of sciences lying between Chemistry, Physics and Biology.

EDITORIAL STAFF

Editor
Philip Earis

Assistant editor
Nicola Nugent

Publishing assistant
Claire Springett

Team leader, serials production
Joanna Pugh

Technical editor
Edward Morgan

Publisher
Janet Dean

Faraday Discussions (Print ISSN 1359-6640, Electronic ISSN 1364-5498) is published 3 times a year by the Royal Society of Chemistry, Thomas Graham House, Science Park, Milton Road, Cambridge, UK CB4 0WF. Volume 140 ISBN-13: 978 0 85404 1237

2008 annual subscription price: print+electronic £519, US $1,033; electronic only £467, US $929. Customers in Canada will be subject to a surcharge to cover GST. Customers in the EU subscribing to the electronic version only will be charged VAT. All orders, with cheques made payable to the Royal Society of Chemistry, should be sent to RSC Distribution Services, c/o Portland Customer Services, Commerce Way, Colchester, Essex, UK CO2 8HP.
Tel +44 (0) 1206 226050;
E-mail sales@rscdistribution.org

If you take an institutional subscription to any RSC journal you are entitled to free, site-wide web access to that journal. You can arrange access via Internet Protocol (IP) address at www.rsc.org/ip. Customers should make payments by cheque in sterling payable on a UK clearing bank or in US dollars payable on a US clearing bank. Periodicals postage is paid at Rahway, NJ and at additional mailing offices. Airfreight and mailing in the USA by Mercury Airfreight International Ltd., 365 Blair Road, Avenel, NJ 07001, USA.

US Postmaster: send address changes to *Faraday Discussions*, c/o Mercury Airfreight International Ltd., 365 Blair Road, Avenel, NJ 07001. All despatches outside the UK by Consolidated Airfreight.

PRINTED IN THE UK

Faraday Discussions documents a long-established series of *Faraday Discussion* meetings which provide a unique international forum for the exchange of views and newly acquired results in developing areas of physical chemistry, biophysical chemistry and chemical physics.

ORGANISING COMMITTEE, Volume 140

Chairperson
A Russell (Southampton, UK)

Editor
A Russell (Southampton, UK)

E Ahlberg (Göteborg, Sweden)
C Korzeniewski (Texas, USA
E Savinova (Strasbourg, France)
P Unwin (Warwick, UK)

FARADAY STANDING COMMITTEE ON CONFERENCES

Chair
C D Bain (Durham, UK)

H M Colquhoun (Reading, UK)
G Jackson (Imperial, UK)
A J Orr-Ewing (Bristol, UK)
A Rodger (Warwick, UK)

© The Royal Society of Chemistry 2008. Apart from fair dealing for the purposes of research or private study, or criticism or review, as permitted under the Copyright, Designs and Patents Act 1988 and Related Rights Regulations 2003, this publication may only be reproduced, stored or transmitted, in any form or by any means, with the prior permission in writing of the Publishers or in the case of reprographic reproduction in accordance with the terms of licences issued by the Copyright Licensing Agency in the UK. US copyright law applicable to users in the USA. The Royal Society of Chemistry takes reasonable care in the preparation of this publication but does not accept liability for the consequences of any errors or omissions.

Royal Society of Chemistry: Registered Charity No. 207890.

⊖The paper used in this publication meets the requirements of ANSI/NISO Z39.48-1992 (Permanence of Paper).

Electrocatalysis- Theory and Experiment at the Interface

Faraday Discussions
www.rsc.org/faraday_d

A General Discussion on Electrocatalysis – Theory and Experiment at the Interface was held at the University of Southampton, UK on 7th, 8th and 9th July 2008.

RSC Publishing is a not-for-profit publisher and a division of the Royal Society of Chemistry. Any surplus made is used to support charitable activities aimed at advancing the chemical sciences. Full details are available from www.rsc.org

CONTENTS

ISSN 1359-6640; ISBN 0-85404-123-7
ISBN-978-0-85404-123-7

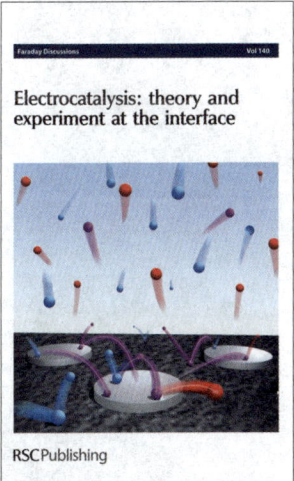

Cover
See Y. E. Seidel, A. Schneider, Z. Jusys, B. Wickman, B. Kasemo and R. J. Behm, *Faraday Discuss.*, 2008, **140**, 167–184. Transport processes in an electrocatalytic reaction on an electrode surface with active metal structures (grey) on an inert substrate (black): with arriving reactants (blue), reactive intermediates (purple) desorbing into the diffusion layer, which can re-adsorb and react further, and final products (red) leaving into the bulk electrolyte.

Image reproduced by permission of Professor R. J. Behm, from *Faraday Discuss.*, 2008, **140**, 167.

9   **Preface**
         Andrea E. Russell

INTRODUCTORY LECTURE

11   **Electrocatalysis: theory and experiment at the interface**
         Marc T. M. Koper

PAPERS AND DISCUSSIONS

25   **The role of anions in surface electrochemistry**
         D. V. Tripkovic, D. Strmcnik, D. van der Vliet, V. Stamenkovic and N. M. Markovic

Faraday Discussions

Unique international discussion meetings which focus on rapidly developing areas of physical chemistry

142: Cold and Ultracold Molecules
15-17 April 2009
Durham University, UK
www.rsc.org/FD142

143: Soft Nanotechnology
15-17 June 2009
Royal Society, London, UK
www.rsc.org/FD143

144: Multi-scale Modelling of Soft Matter
20-22 July 2009
Groningen, The Netherlands
www.rsc.org/FD144

145: Frontiers in Physical Organic Chemistry
2-4 September 2009
UWIC, Cardiff, UK
www.rsc.org/FD145

Timeline
Call for papers
11 months prior to meeting
Submission of full papers
5 months prior to meeting
Early booking and poster abstract
2 months prior to meeting
Standard registration
1 month prior to meeting

For more information, telephone: +44 (0)1223 423354/432380
or email: conferences@rsc.org

www.rsc.org/farada
Registered Charity Number 207

41 **From ultra-high vacuum to the electrochemical interface: X-ray scattering studies of model electrocatalysts**
Christopher A. Lucas, Michael Cormack, Mark E. Gallagher, Alexander Brownrigg, Paul Thompson, Ben Fowler, Yvonne Gründer, Jerome Roy, Vojislav Stamenković and Nenad M. Marković

59 **Surface dynamics at well-defined single crystal microfacetted Pt(111) electrodes:** *in situ* **optical studies**
Iosif Fromondi and Daniel Scherson

69 **Bridging the gap between nanoparticles and single crystal surfaces**
Payam Kaghazchi, Felice C. Simeone, Khaled A. Soliman, Ludwig A. Kibler and Timo Jacob

81 **Nanoparticle catalysts with high energy surfaces and enhanced activity synthesized by electrochemical method**
Zhi-You Zhou, Na Tian, Zhi-Zhong Huang, De-Jun Chen and Shi-Gang Sun

93 **General discussion**

113 **Differential reactivity of Cu(111) and Cu(100) during nitrate reduction in acid electrolyte**
Sang-Eun Bae and Andrew A. Gewirth

125 **Molecular structure at electrode/electrolyte solution interfaces related to electrocatalysis**
Hidenori Noguchi, Tsubasa Okada and Kohei Uosaki

139 **A comparative** *in situ* **^{195}Pt electrochemical-NMR investigation of PtRu nanoparticles supported on diverse carbon nanomaterials**
Fatang Tan, Bingchen Du, Aaron L. Danberry, In-Su Park, Yung-Eun Sung and YuYe Tong

155 **Spectroelectrochemical flow cell with temperature control for investigation of electrocatalytic systems with surface-enhanced Raman spectroscopy**
Bin Ren, Xiao-Bing Lian, Jian-Feng Li, Ping-Ping Fang, Qun-Ping Lai and Zhong-Qun Tian

167 **Mesoscopic mass transport effects in electrocatalytic processes**
Y. E. Seidel, A. Schneider, Z. Jusys, B. Wickman, B. Kasemo and R. J. Behm

185 **General Discussion**

209 **On the catalysis of the hydrogen oxidation**
E. Santos, Kay Pötting and W. Schmickler

219 **Hydrogen evolution on nano-particulate transition metal sulfides**
Jacob Bonde, Poul G. Moses, Thomas F. Jaramillo, Jens K. Nørskov and Ib Chorkendorff

233 **Influence of water on elementary reaction steps in electrocatalysis**
Yoshihiro Gohda, Sebastian Schnur and Axel Groß

245 **Co-adsorbtion of Cu and Keggin type polytungstates on polycrystalline Pt: interplay of atomic and molecular UPD**
Galina Tsirlina, Elena Mishina, Elena Timofeeva, Nobuko Tanimura, Nataliya Sherstyuk, Marina Borzenko, Seiichiro Nakabayashi and Oleg Petrii

269 **Aqueous-based synthesis of ruthenium–selenium catalyst for oxygen reduction reaction**
Cyril Delacôte, Arman Bonakdarpour, Christina M. Johnston, Piotr Zelenay and Andrzej Wieckowski

283 **Size and composition distribution dynamics of alloy nanoparticle electrocatalysts probed by anomalous small angle X-ray scattering (ASAXS)**
Chengfei Yu, Shirlaine Koh, Jennifer E. Leisch, Michael F. Toney and Peter Strasser

297 **General Discussion**

319 **Efficient electrocatalytic oxygen reduction by the 'blue' copper oxidase, laccase, directly attached to chemically modified carbons**
Christopher F. Blanford, Carina E. Foster, Rachel S. Heath and Fraser A. Armstrong

337 **Steady state oxygen reduction and cyclic voltammetry**
Jan Rossmeisl, Gustav S. Karlberg, Thomas Jaramillo and Jens K. Nørskov

347 **Intrinsic kinetic equation for oxygen reduction reaction in acidic media: the double Tafel slope and fuel cell applications**
Jia X. Wang, Francisco A. Uribe, Thomas E. Springer, Junliang Zhang and Radoslav R. Adzic

363 **A first principles comparison of the mechanism and site requirements for the electrocatalytic oxidation of methanol and formic acid over Pt**
Matthew Neurock, Michael Janik and Andrzej Wieckowski

379 **Surface structure effects on the electrochemical oxidation of ethanol on platinum single crystal electrodes**
Flavio Colmati, Germano Tremiliosi-Filho, Ernesto R. Gonzalez, Antonio Berná, Enrique Herrero and Juan M. Feliu

399 **Electro-oxidation of ethanol and acetaldehyde on platinum single-crystal electrodes**
Stanley C. S. Lai and Marc T. M. Koper

417 **General Discussion**

CONCLUDING REMARKS

439 **All dressed up, but where to go? Concluding remarks for FD 140**
David J. Schiffrin

ADDITIONAL INFORMATION

445 **Poster titles**
449 **List of Participants**
453 **Index of Contributors**

Preface

Andrea E. Russell*

DOI: 10.1039/b814058h

The need to develop cleaner/greener methods of both energy production and chemical synthesis has been driving renewed interest in electrocatalysis. Experimental advances in the application of spectroscopic methods, such as IR, INS, NMR, and XAS, and structural probes such as STM, AFM, high resolution TEM, and XRD are providing a wealth of data that enable structure/property relationships in electrocatalysis to be investigated. Similarly, developments in theoretical methods (MD simulations, DFT calculations, and Monte Carlo simulations combined with *ab initio* methodologies) are providing new insights regarding old catalysts and promise to provide direction in the search for new catalysts. The advent of high throughput catalyst preparation methods means that many more electrocatalyst formulations are being screened for an ever-wider variety of reactions. Directing this effort will require the combined efforts of theoretical models and the development of new experimental techniques. Such was the motivation behind the proposal for Faraday Discussion 140, which was originally proposed by David Schiffrin of the University of Liverpool whilst attending a workshop on Computational Electrochemistry in Santorini, Greece in September 2004. It was a bit of a surprise to me to find that I had been selected to put the proposal together and to chair the meeting, given that I wasn't attending the workshop. However, David can be very persuasive and it has been my privilege to organise the Discussion described in this text.

The organisation began properly at a second workshop meeting hosted by Marc Koper of Leiden University in Leiden, the Lorentz Workshop on Fuel Cell Electrocatalysis, in Oct 2006. There Elisabet Ahlberg (University of Göteborg), Carol Korzeniewski (Texas Tech Univeristy), and Elena Savinova (then dividing her time between the Boreskov Institute of Catalysis, Novosibirsk and the University of Strasbourg) and I met over dinner with Marc to draw up a plan of whom we'd like to have as invited speakers and to discuss the various themes we would want to feature in the meeting. Of particular importance to us were papers from theoreticians that would reach out to experimentalists, experimental/spectroscopy papers that were pushing the boundaries of the techniques to provide new information, discussion of the state-of-the-art catalysts for both the hydrogen and oxygen reactions, and a desire to see that electrocatalysis is more than just fuel cells. To our delight all of the speakers we approached on our most wanted list accepted the invitation and, more importantly for a Faraday Discussion, they all produced their manuscripts for discussion at the meeting and inclusion in the volume. Together with the contributed papers, where there were four times as many abstracts submitted than we could accept as full papers in the meeting, I believe that the meeting lived up to our aims.

To address the problem of demand for a contributed paper at the meeting, I organised a special issue of *Physical Chemistry Chemical Physics*, which the editorial and production teams of the RSC managed to deliver on time to be handed out to each participant at the meeting. This special issue, also entitled *"Electrocatalysis: Theory and Experiment at the Interface"* is available as volume 10, issue 25, of *Physical Chemistry Chemical Physics*. It contains 22 further papers that I wholeheartedly recommend to readers of this Discussion.

School of Chemistry, University of Southampton, Highfield, Southampton, SO17 1BJ, UK. E-mail: a.e.russell@soton.ac.uk

The meeting itself was comprised of four sessions, intended to represent those aims that Elisabeth, Carol, Elena and I discussed over dinner almost two years previously. Marc Koper gave an excellent opening lecture, setting this Discussion in the context of previous Faraday Discussions on electrochemistry related topics and then going on to highlight the power of combining theoretical and experimental methods in exploring the structure and properties of electrocatalysts. His talk enthused and energised the delegates, who went on to vigorously discuss the other papers presented for the meeting. The concluding remarks were expertly presented by David Schiffrin, who did an excellent job of both summarising the meeting and pointing out where we should be thinking of going next.

The meeting also included a packed poster session with 37 posters, which were also enthusiastically discussed. I do not believe that any student or postdoc attending the meeting ever felt that their poster was ignored and many were able to present and discuss their work at the highest level. The Organising Committee awarded the Skinner prize (for best student poster) to M. J. T. C. van der Niet (Leiden University) for her work on "Interactions between H_2O and preadsorbed H or O on Pt(533)".

On behalf of the Organising Committee, I would like to thank all those who participated in Faraday Discussion 140. I am sure that those unable to attend who will read this volume will absorb some of the excitement that was generated at the meeting, and I hope that many of the points continue to be vigorously discussed when we meet again or in the literature. We would also like to thank the RSC staff who made the organisation so much easier than I (Andrea) anticipated. In particular, we would like to thank Morwenna Gilbert and Nicola Nugent for all of their hard work and emails. We would like to thank all of those whose papers appear in this volume, for meeting the deadlines (well, almost) and keeping to the 5 minutes allowed for presentation. Finally we thank Piotr Kleszyk, Gaël Chouchelamane, Jon Speed and Rob Johnson, from my group, for their excellent skills in keeping delegates to time, managing the University's data projector systems, and otherwise keeping me calm. Finally, we acknowledge the International Society of Electrochemistry and the Electrochemical Society for their co-sponsorship of the meeting, and the University of Southampton and the Army Research Office for their generous financial support of the meeting.

Andrea E. Russell (*Chair, Editor*)

Introductory Lecture
Electrocatalysis: theory and experiment at the interface

Marc T. M. Koper

Received 25th July 2008, Accepted 25th July 2008
First published as an Advance Article on the web 22nd August 2008
DOI: 10.1039/b812859f

Introduction

Browsing through the history of Faraday Discussions typically serves as a good indicator of the development of concepts and challenges in physical chemistry. The first Royal Society Discussion Meeting held under the name "Faraday Discussions", at the University of Manchester in 1947, was on the topic of "Electrode Processes". Among the list of contributors were illustrious names such as Randles, Levich, Frumkin, Bockris, Butler, Eley, and Heyrovsky†. This was the time that electrochemical measuring techniques, such as electrochemical impedance spectroscopy, were developed, and electrochemists were still very much in the dark about the molecular nature and theoretical description of electrode reactions, in particular hydrogen evolution, which was a prominent discussion topic. One remarkable statement was that made by Eley,[1] who seriously questioned whether it would ever be possible to measure on a clean platinum surface in an electrolyte solution.

By 1968, at a Faraday Discussion on "Electrode Reactions of Organic Compounds" at the University of Newcastle-upon-Tyne, some remarkable leaps in development had taken place. Most prominently, there was by then a theory of electron transfer reactions, introduced at the meeting by Marcus. The contributions by Parsons and Conway were essentially concerned with the kinetic modeling of electrocatalytic reactions, whereas Hush used semi-empirical quantum-chemical calculations (!) to study bond breaking. Breiter had a contribution on methanol oxidation at platinum, and there was an interesting paper by Brummer and Cahill on the interaction between adsorbates, such as chemisorbed hydrogen and carbon monoxide. The relation with some of the relevant issues still raised at this meeting is remarkable. However, there was the nagging problem of specific adsorption, as put forward by Parsons[2]: "The greatest uncertainty in the adsorption behaviour arises from the lack of knowledge about the way the relative adsorption of different species depends on the nature of the metal." Thirsk, in his concluding remarks, also mentioned the "resistance to theoretical attack" of adsorption as one of the main problems of electrochemical science.

Five years later, at a meeting at Oxford University called "Intermediates in Electrochemical Reactions", one of the founding fathers of electrochemical surface science, Heinz Gerischer, was quite optimistic about the possibilities of new

---

Leiden Institute of Chemistry Leiden University, PO Box 9502, 2300 RA Leiden, The Netherlands. E-mail: m.koper@chem.leidenuniv.nl

† At the meeting it was pointed out by Professor Roger Parsons, who was also present at the Faraday Discussion 1 in 1947 as a graduate student of Professor Bockris, that the contributors from the Soviet Union and Eastern Europe (Frumkin, Levich, Ershler, Heyrovsky) were not able to be present at the meeting at Manchester. This severely limited discussion of their important work and significantly delayed the impact of their contributions. This clearly illustrates the importance of discussion in person which characterises the Faraday Discussions.

surface-sensitive techniques to probe intermediates in electrode reactions.[3] Indeed, this was the meeting where the field witnessed the introduction of new non-electrochemical methods, such as reflectance spectroscopy (Kolb and MacIntyre) and online mass spectrometry (Bruckenstein). However, only few participants of that meeting would have foreseen the explosion of new techniques that was going to transform interfacial electrochemistry in the next 10–20 years. In fact, Randles, in his closing address, seemed rather reserved about the future of electrochemistry.[4] He felt that, somehow, "chemistry has lost its glamour", though "electrochemistry still has vitality and a potential for continuing usefulness". However, "we should refrain from barren mathematising, we should use the techniques we have where they are most appropriate and think actively about possible new ones".

The lesson to be learnt? Just browse through the next Faraday Discussion volume 94 on electrochemistry "The Liquid/Solid Interface at High Resolution", held in Newcastle-upon-Tyne in 1992. There is no lack of glamour in the papers that were contributed to that meeting! Scanning tunneling microscopy and other scanning probe techniques had revolutionized the field of surface science and electrochemists such as Kolb, Itaya, and Bard, amongst many others, were actively studying the unprecedented potential of these new techniques. Also spectroscopic techniques such as infrared spectroscopy, Raman spectroscopy and non-linear methods, such as second harmonic generation, had become part of the modern electrochemist's toolbox. In the flood of STM papers, theory remained somewhat underexposed (apart from one isolated theory paper by Nagy, Heinzinger and Spohr on the modeling of water at platinum[5]).

The most recent Faraday Discussion meeting on electrochemistry was in 2002 in Berlin, entitled "The Dynamic Electrode Surface". 2007 Nobel Laureate Gerhard Ertl, Heinz Gerischer's successor at the Fritz-Haber-Institute at Berlin, pointed out in his Introductory Lecture the importance of looking at surface reactions at different time- and length scales, and the different experimental and theoretical tools needed for their proper description.[6] Theory and computational chemistry, in particular quantum-chemical calculations based on density functional theory (DFT) and (kinetic) Monte Carlo simulations of surface reactions, were beginning to play an important role in the interpretation of experimental results, as exemplified in the papers by Rikvold,[7] Weaver,[8] and myself.[9] In the electrocatalysis talks, the role of surface diffusion, especially Pt-bonded CO, was a matter of intense debate.

Since 2002, theory and computational chemistry have played an increasingly important role in fundamental electrocatalysis work. Reactions such as hydrogen oxidation and evolution, oxygen reduction, methanol oxidation and carbon monoxide oxidation have all been studied using modern computational techniques, such as first-principles DFT calculations and Monte Carlo simulations, which both complement and give input for the more widespread kinetic modeling approaches. The interaction with experiment has been become very fruitful, and it is this synergy that provided the motivation for the present Faraday Discussion.

My aim in this paper is to illustrate the ideas and concepts on which I believe this interaction or "interface" of theory and experiment should be based, and then to illustrate this on two topics of my own current research focus. Since I am primarily interested in moving electrocatalysis forward as a science, as opposed to "technology", I will only discuss the general fundamental challenges that we face, as opposed to the specific challenges related to *e.g.* fuel cell catalysis and the development of better and more efficient low-temperature fuel cells (which is undoubtedly one of the main drivers of the field).

Theoretical and computational electrochemistry

A famous statement made by Dirac in 1929[10] claims that "all of chemistry" follows from the laws of quantum mechanics, and the main difficulty lies in the fact that the equations are "too complex to be solved". However, in his statement Dirac did not foresee (or did not include) the invention of the computer, ultimately made possible by the development of quantum mechanics itself. Much of modern theoretical chemistry is concerned with developing clever ways of using the computing power of computers to solve the complex equations. The success of theoretical chemistry has been immense, as illustrated by the award of the 1999 Nobel Prize to Pople and Kohn, two founding fathers of modern computational quantum chemistry. However, modern quantum chemistry still suffers—to some extent—from Dirac's demon: the number of atoms that can be included in a full-blown "first principles" computation, especially when coupled to dynamics, is limited to at best a few hundred. The impressive work of Neurock and coworkers, as exemplified by their paper in this volume, is a good example of what is currently achievable. The extrapolation to real sizes and realistic time scales is by no means trivial, and involves approximation schemes, the accuracy of which is often unclear and difficult to assess. The accuracy and/or reliability of model calculations (and therefore of the inherent approximations) may be evaluated on the basis on a comparison to experimental data, but this is a practice which, at least in the author's opinion, should be carried out with great care, no matter how successful and gratifying such a comparison may often appear.

So what would be an ideal but still practical way to model an electrocatalytic system? In other words: how to go from quantum mechanics to, say, a real voltammogram? First of all, it is imperative that the atomic structure of the electrode surface is known accurately. Experimentally, this implies working with well-defined single crystals, and knowing how the exact structure of the single-crystalline or nanoparticulate surface and possible defects influence reactivity. This aspect becomes less important if one, for instance, compares the activity of a series of metals for a certain reaction, but at the expense of losing quite a bit of detail. Next, one has to set up a hierarchy of calculations that will finally lead to the prediction of the experimental outcome, where typically one level of calculation delivers the input for the next level. For instance, one may carry out a series of first-principles DFT calculations to estimate interaction energies between adsorbates on an electrode surface, and rate constants for reactions between adsorbates (based on transition state theory). These numbers may be input for a lattice-gas kinetic Monte Carlo (KMC) simulation of an extended surface, say 1000×1000 lattice points. The KMC simulation may directly give the desired macroscopic variable, such as for instance the electric current, as well as much more, such as the time-dependent distribution of adsorbates on the surface. Essentially, the KMC simulation serves as a way to estimate the system's partition function, *i.e.* as a simulation method to sample the long-time statistics of the system in a way that would be impossible by quantum chemistry alone. On the other hand, one often approximates the statistics of the system by assuming a perfect mixing, also known as the "mean-field approximation", which is the basis of most kinetic modeling approaches and of, for instance, the well-known Frumkin isotherm.[11] In this case, a KMC simulation should in principle still be carried out in order to confirm the validity of the mean-field approximation. Note, however, that the complexity of the system often forces one to make many more (implicit) assumptions: the lattice-gas approximation, the assumption of the additivity of interaction potentials (sometimes three-particle interactions may be included but even this is an approximation), entropic or other solvent effects are often neglected as first-principles free energy calculations are extremely expensive, assumptions about the exact structure of the surface and role of defects, assumptions about which reactions to include and which not, assumptions about the reaction mechanism (sometimes reactions may be included the rates of which are difficult to calculate and therefore they are estimated or guessed), *etc.* Although there may be examples where many of these approximations may be (or may appear to be) reasonable for the particular system under consideration, it will still be necessary to compare carefully to experimental data as well as to more accurate simulation data to finally assess the reasonableness of a model.

A key challenge in first-principles simulations of electrode systems is the introduction of the electrode potential. Since all existing quantum chemistry codes work with a fixed number of electrons in the simulation box, rather than with a fixed electrochemical potential of the electrons, applying an electrode potential to the simulation of a half cell is usually approximated by adjusting the charge distribution within the simulation cell. This can be achieved by applying an electric field to the cell,[8,12,13] adding or removing electrons to the metal slab representing the electrode,[14,15] or adding electron-drawing or withdrawing adsorbates to the metal surface.[16] The first two methods should also screen the electrostatic field or charge by an artificial background. The corresponding electrode potential is determined at the end of the fully converged calculation, typically by referring the Fermi level of the metal electrons to a field-free reference somewhere in the simulation cell, in combination with the known relation between the vacuum scale of the metal work function and the normal hydrogen electrode.[14–16] Practically all current first-principles calculations including the effect of the electrode potential are carried out in this fashion. The fundamental limitation of these methods is that they are essentially *coulostatic* methods, and cannot readily be applied to mapping out the reaction path of a charge transfer reaction, as most electrode reactions are. During a charge transfer reaction under coulostatic conditions, the electrode potential will change during the reaction. The preferred way to circumvent this problem is to switch to a grand-canonical simulation method,[17] in which the electrochemical potential of the electrons in the half cell is held fixed rather than their number. The computational limitation of this method is that it requires an additional iterative loop in the already very time-consuming simulation, such that in practice the method is still hardly used.

An interesting approach that allows a rapid assessment of the influence of the electrode potential was suggested by Nørskov *et al.*,[18] and is based on the very simple assumption that the only effect of the electrode potential is to change the energy of the electrons. For the adsorption reaction of hydrogen:

$$H^+ + e^- \leftrightharpoons H_{ads} \quad (1)$$

the reaction energy is written as:

$$\Delta G_{reaction}(E) = \Delta G_{ads}(H_{ads};E) - \Delta G_{ads}(H^+ + e^-;E) \quad (2)$$

Nørskov *et al.* make the assumption that the first term on the right-hand side of eqn (2) does not depend on potential E, whereas the second term is simply equal to $e_0 E$, with the potential referred to the NHE (normal hydrogen electrode), as the energy of $[H^+ + e^-]$ is 0 at the potential of the NHE by convention. In mathematical terms, the first assumption implies that

$$d\Delta G_{ads}(H_{ads})/dE = 0 \quad (3)$$

In so far as the electrode potential is proportional to the interfacial field F, this is equivalent to stating that the adsorbate (in this case H_{ads}) forms an apolar bond with the surface, or, more accurately, that its static surface dipole moment is zero. As a result, the quantity $(1/e_0)d\Delta G_{reaction}(E)/dE$, which is known as the electrosorption valency,[19] is equal to the (integer) number of electrons transferred in the adsorption reaction. Though this assumption seems reasonable for reaction (1), it remains to be seen how valid it is in general, and it should certainly be tested for every adsorbate considered. Note that the relationship between the electrosorption valency and the surface dipole moment is not a trivial one, and depends on the structure of the double layer.[20]

Electrocatalytic oxidation of carbon monoxide

The oxidation of carbon monoxide is not only one of the favorite model reactions in electrocatalysis, but it is also a hugely important reaction in the development of more efficient low-temperature fuel cells. We have studied the CO electro-oxidation on various stepped single crystals of platinum and rhodium,[21–24] and have demonstrated the importance of low-coordination step and defect sites in the oxidation mechanism. However, there remain quite a few fundamental issues that I believe are not yet fully understood. One of them is the importance of CO_{ads} surface mobility in the oxidative stripping of pre-adsorbed CO, an issue that was quite intensely debated 6 years ago at the Faraday Discussion in Berlin. The other is the nature of the oxygen-donating species. I will discuss these two issues in this and in the next session.

From the DFT point-of-view, CO seems to be a somewhat problematic molecule, as DFT has difficulty in correctly predicting the preferred adsorption site on a Pt(111) surface. Whereas experimentally CO prefers atop coordination on Pt(111) (in UHV at low CO coverage),[25] most DFT calculations predict the threefold hollow site to be most stable.[26] This is typically explained by the tendency of DFT to overestimate bonding interactions, which are stronger for higher coordination. Olsen, Philipsen and Baerends[27] have recently shown that using a localized basis set and taking care of achieving full convergence of the DFT calculations, the atop site is found to be the preferred site on Pt(111). Nevertheless, their binding energies seem somewhat low compared to experiment. These observations clearly underpin the statement by Feibelman *et al.*[26] that "DFT calculations cannot be used as black-box simulation tool." Typically, qualitative or relative predictions are more reliable than quantitative or absolute predictions, having error bars of *ca.* 0.1 and 0.3–0.5 eV, respectively. At any rate, these calculations suggest the *qualitative* conclusion that the corrugation potential for CO on Pt(111) should be rather flat, in agreement with the experimental observation that CO mobility on Pt(111) is high.

The oxidation of carbon monoxide, under electrochemical conditions, is believed to follow the Langmuir–Hinshelwood-type mechanism originally suggested by Gilman:[28]

$$H_2O + * \leftrightharpoons OH_{ads} + H^+ + e^- \qquad (4)$$

$$CO_{ads} + OH_{ads} \rightarrow COOH_{ads} \rightarrow CO_2 + 2* + H^+ + e^- \qquad (5)$$

The second step in this mechanism, the CO + OH combination reaction, is believed to be rate-determining, primarily because the Tafel slope observed for CO monolayer oxidation is *ca.* 70–80 mV dec^{-1},[29,30] close to the theoretical value of 60 mV dec^{-1} expected for an EC mechanism. DFT calculations indeed show that there is sizeable barrier for the CO + OH reaction: Shubina *et al.*[31] report *ca.* 0.6 eV on Pt(111) in the absence of water, whereas Janik and Neurock[32] have obtained values of 0.5–0.3 eV in the presence of water, depending on whether the surface was charged or not. Note that these values apply to $T = 0$ K, hence these are not free activation energies.

By carrying out extensive chronoamperometry measurements on a series on stepped Pt surfaces in sulfuric acid, we have shown that the rate for CO monolayer stripping is proportional to the step density, and that there is no evidence for CO slowly reaching the active step sites. This strongly suggests that OH_{ads} formation takes place preferentially on the step sites, and the CO diffusion on the terrace is rapid, in agreement with the flat corrugation potential predicted by DFT.

More recently, we have performed a similar series of experiments, but in alkaline media.[33] Alkaline media typically show higher catalytic activities than acidic media, even if the potential scale is converted to the reversible hydrogen electrode to correct for trivial pH effects.[34,35] I believe that these observations are not well understood, and cannot be explained simply by referring to the higher affinity of OH for step sites in alkaline media, as this effect must have been accounted for by the RHE scale.

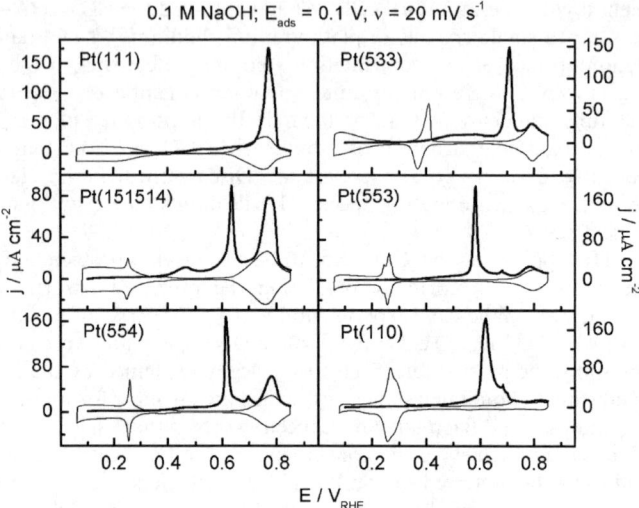

Fig. 1 CO stripping (thick solid line) and the subsequent cyclic voltammogram (thin solid line) for Pt(111), Pt(15 15 14), Pt(554), Pt(533), Pt(553) and Pt(110) in 0.1 M NaOH, sweep rate 20 mV s^{-1}, CO adsorbed at a potential of 0.1 V, no CO in solution during stripping. Reproduced with permission from ref. 33.

Fig. 1 shows the CO stripping voltammetry on a number Pt single crystals surfaces in 0.1 M NaOH. The remarkable observation here is that the CO voltammetry exhibits as many as 4 features. We can take the CO stripping voltammetry on Pt(554) as an example. By comparing to Pt(111), Pt(15 15 14), and Pt(553), we can conclude that the high-potential stripping peak between 0.72 and 0.80 V is due to the oxidation of CO on (111) terraces. Inspection of the surfaces with (110) and (100) step sites, we conclude that the stripping peak at around 0.6 V is CO oxidation at (110) sites, and that peak at *ca.* 0.70 V is due to CO oxidation at (100) sites. Note the small feature at 0.7 V in the stripping curve for Pt(554) (and Pt(553) and Pt(110) as well), which we ascribe to CO oxidation at a small amount of defects of (100) orientation. Finally, a broad low-potential potential feature, which can start at a potential as low as 0.35 V, is observed on all surfaces. This feature was also observed by Spendelow and Wieckowski[36] in their studies of CO adlayer oxidation on lightly disordered Pt(111) in alkaline media. By combining voltammetry with scanning tunneling microscopy, they suggested that this feature is due to CO oxidation on small monoatomically high islands on the Pt(111) surface, which present low-coordination sites (essentially kink sites) that are particularly active for CO oxidation. Following their assignment, we suggest that this low-potential feature is CO oxidation on "kink"-type sites, or defects in the steps. This leads to the remarkable observation that a single voltammogram such as that shown for Pt(554) reveals as many as 4 different active oxidation sites for CO: kink sites, (110) step sites, (100) sites, and (111) terrace sites, in decreasing order of activity. Such an observation is only possible if the mobility of CO on the surface is low, as in the case of high CO mobility most if not all CO would oxidize at the most active oxidation sites.

A simple way to probe the role of low CO mobility is to study the scan rate dependence of the CO stripping voltammetry, as shown in Fig. 2 for Pt(554). It is observed that at high scan rates (500 mV s^{-1}) there is a significant amount of CO oxidizing on the terraces. However, as the scan is lowered, the charge corresponding to CO oxidizing at the terraces decreases until finally at 5 mV s^{-1} it is almost negligible. This clearly suggests that at low scan rates, CO has more time to diffuse to the

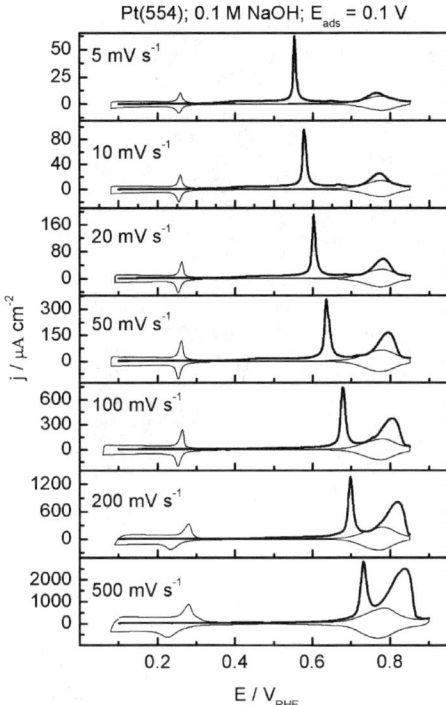

Fig. 2 CO stripping (thick solid line) and the subsequent cyclic voltammogram (thin solid line) for Pt(554) in 0.1 M NaOH at different sweep rates, $E_{ads} = 0.1$ V. Reproduced with permission from ref. 33.

step sites and react there. When the amount of CO reacted in the steps or on the terraces, as estimated from the corresponding stripping charges, is plotted as a function of the square root of the scan rate, linear relationships are observed. This "Cottrell-like" behavior is another strong indicator for the important role of slow CO surface diffusion on Pt(111) terraces in alkaline media. The slow diffusion of CO on the (111) terrace is also manifested in the chronoamperometric transients. Kinetic Monte Carlo simulations of a model for CO oxidation on stepped surfaces[37] show that in such a case, the transient is composed of two parts: an initial exponential current decay corresponding to a one-dimensional instantaneous nucleation and growth along the step, followed by a peak corresponding to an instantaneous nucleation and growth onto the terrace (see Fig. 3). Experimental transients for CO oxidation on stepped Pt in alkaline media indeed display the same characteristics.[38]

The reason for the significantly reduced mobility of terrace-bound CO on stepped Pt electrodes in alkaline media as compared to acidic media has not been fully clarified yet. In acidic media, as well as on Pt(111) in UHV, the high mobility of CO is ascribed to the conclusion made above, namely that CO does not have a strong preference for a specific adsorption site on Pt(111) (as confirmed by DFT, although not in all its details) and therefore it should be able to move over the Pt(111) surface almost barrierless. In alkaline media, the electrode potential is effectively more negative than in acidic media, however, even if the actual potential scale used is the reversible hydrogen electrode (RHE). This implies that the Fermi energy of the Pt in alkaline media (say pH = 13) is about 0.7 eV higher than in acidic media at pH = 1. If the free energy of the adsorbate under consideration is not dependent on pH, such as with CO, this may have a significant influence on the way it binds to the Pt surface. From DFT calculations on both Pt(111) clusters and slabs,[39–41]

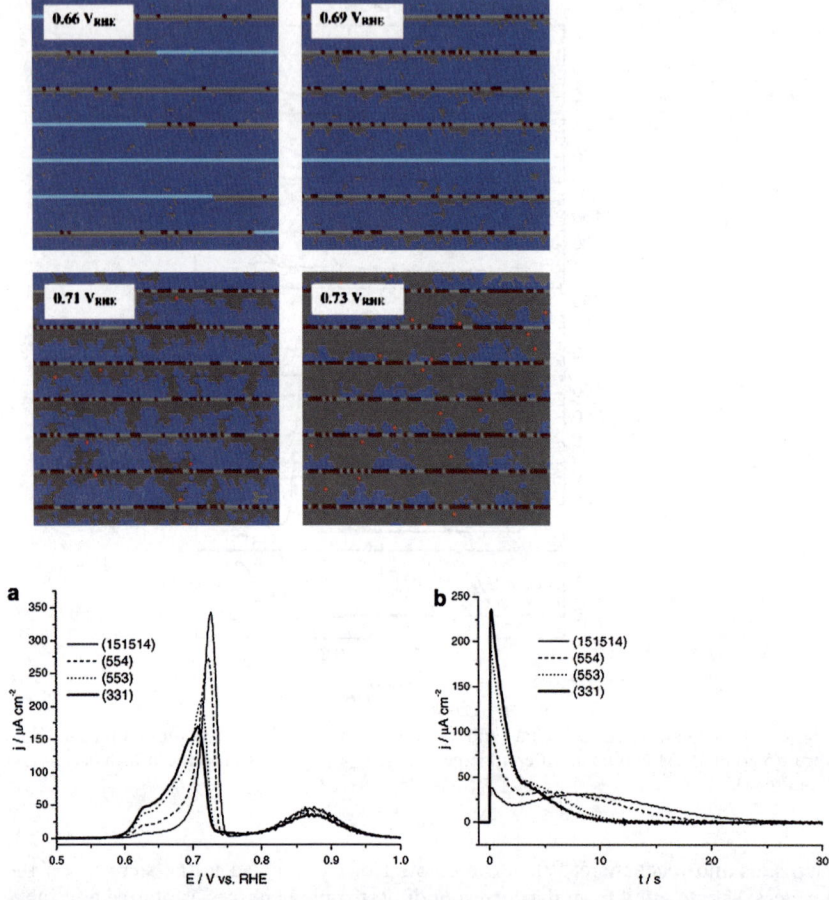

Fig. 3 Results of Kinetic Monte Carlo simulations of CO stripping on a stepped surface in the absence of diffusion of CO on the terrace. The stripping voltammetry (a) exhibits three features: CO oxidation at steps (*ca.* 0.62 V), CO oxidation at terraces (*ca.* 0.72 V), and OH adsorption on terraces (*ca.* 0.87 V). The snapshots correspond to a (554) surface, where light (dark) blue is CO on steps (terraces), light (dark) grey is water on steps (terraces) and dark (light) red is OH on steps (terraces). Figure b shows the chronoamperometry at 0.68 V, and clearly displays exponential decay first (oxidation of CO at steps) and then a peak corresponding to CO oxidation on terraces. Reproduced with permission from ref. 37.

it has been found that CO binds more strongly to the surface at negative potentials, especially to multifold coordination sites such as bridge sites and hollow sites. This preference for multifold coordination at higher Fermi level or more negative potential can be explained qualitatively by the Blyholder model.[42] At more negative potential the influence of back donation of metal electrons into the CO $2\pi^*$ orbital becomes more prominent, and since the interaction between the $2\pi^*$ orbital and the Pt d band is a bonding interaction, and therefore prefers to interact with as many surface atoms as possible, a stronger preference for multifold coordination may be expected. Recent FTIR experiments in our group[43] on CO at Pt(111) in alkaline media confirm the difference in the electronic interaction with acidic media, both through a significant change in C–O stretching frequency (which roughly corresponds to 0.7 V times the Stark tuning slope in cm^{-1} V^{-1}) as well as a clearly increased band intensity corresponding to bridge-bonded CO. However, a significantly more corrugated binding energy surface for CO on Pt(111) in alkaline media is not straightforwardly implied by these data. On the other hand, the FTIR data seem to suggest that the product of CO oxidation in alkaline media, carbonate, remains adsorbed on the surface until quite positive potentials. Strongly adsorbed carbonate may have a negative effect on the CO surface mobility, similarly to what we concluded for CO oxidation on rhodium single crystals in sulfuric acid,[44] where strongly adsorbed sulfate severely hampers CO surface diffusion.

Formation of OH$_{ads}$ on platinum

All our experiments, as well as those by various other authors, suggest that CO is oxidized by OH that is absorbed in a step or defect on the Pt surface. It is therefore all the more disconcerting that step-bonded OH has remained invisible in both spectroscopic and voltammetric experiments. Its apparent voltammetric invisibility is illustrated in Fig. 4, which compares the voltammetry of Pt(111) and a stepped Pt surface, Pt(15 15 14), in 0.1 M NaOH. Before we will attempt to explain the "anomalous" features of the stepped surface, let us discuss how the Pt(111) voltammogram may be modeled using a combination of DFT and statistical mechanics.

The Pt(111) voltammogram displays the well-known reversible features corresponding to H adsorption on the terrace (<0.35 V_{RHE}) and OH adsorption on the terrace (>0.6 V_{RHE}). These regions are reasonably well predicted by DFT calculations. Employing eqn (2) above and the assumptions stipulated there, Rossmeisl et al.[45] have calculated from DFT the equilibrium potentials for the reactions:

$$H^+ + e^- \leftrightarrows H_{ads} \quad (6)$$

and

$$H_2O \leftrightarrows OH_{ads} + H^+ + e^- \quad (7)$$

to be 0.09 and 0.81 V_{RHE} respectively. This implies that at $T = 0$ K, water is stable at Pt(111) between 0.09 and 0.81 V_{RHE}, in reasonable agreement with the experiments at room temperature. Rossmeisl et al. also calculated the field dependence of the adsorption energy of H, O and OH to check if indeed eqn (3) is satisfied for these adsorbates. All three adsorbates indeed show a variation of less than 0.1 eV within a field range of -0.3 V Å^{-1} to 0.3 V Å^{-1}, i.e. ca. -1 to 1 V (if the double layer thickness would be ca. 3 Å).

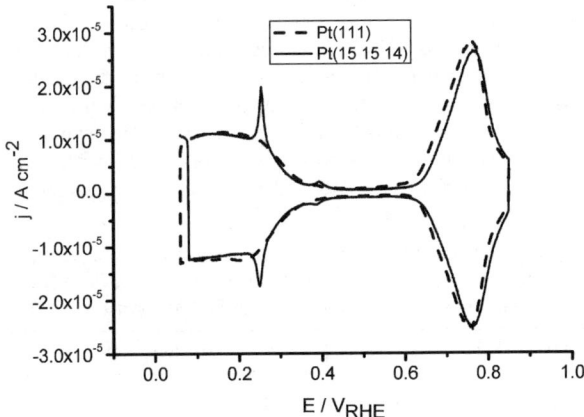

Fig. 4 Blank voltammetry of Pt(111) and Pt(15 15 14) in 0.1 M NaOH. For explanation, see text.

At room temperature, it is well known that the coverage θ of H "upd" (underpotential deposition) adsorbates on Pt(111) follow a Frumkin isotherm to a good approximation:

$$\frac{\theta}{1-\theta} = c_{H^+} \exp\left(\frac{\Delta G_H(E)}{RT}\right) \exp\left(\frac{z\varepsilon_{HH}\theta}{RT}\right) \quad (8)$$

where z is the surface coordination number of a surface atom ($z = 6$ for Pt(111)), and

$$\Delta G_H(E) = \Delta G_{H,ads} + e_0 E \quad (9)$$

From Jerkiewicz's temperature-dependent experiments[46] and our Monte Carlo simulations,[47] values for the H upd adsorption energy $\Delta G_H(E)$ and the nearest-neighbor interaction energy ε_{HH} may be determined, as summarized in the second column of Table 1. Fig. 5 shows the "hydrogen upd region" predicted by this model, compared to the exact Monte Carlo simulations. It is seen that the Monte Carlo simulation show a bit more structure, due to the relatively strong interactions, ca. 0.047 eV per pair of neighboring H adsorbates. Nevertheless, the mean-field approximation or Frumkin isotherm is a reasonable approximation. In the third column of Table 1, we give the values for the adsorption energy and the nearest-neighbor interaction as estimated from the DFT calculations by Karlberg et al.[48] The nearest-neighbor interaction energy is a bit smaller than the experimental estimate, where we note that the DFT calculations were performed without water on the surface. The Monte Carlo isotherm predicted by the DFT values is very close to the mean-field prediction,[48] as the lateral interaction is very weak.

The "OH adsorption region" on Pt(111) is more difficult to model. A simple model suggested by Rossmeisl et al.[49] elsewhere in this volume models the OH adsorption on Pt(111) by a Langmuir isotherm with a maximum OH coverage of 1/3 ML. Using the DFT value for $\Delta G_{OH,ads}$ mentioned above (0.81 eV), a reasonable agreement with experiment is obtained although a few important details are not reproduced or explained. The final coverage of OH on Pt(111) is ca. 0.4 ML in acidic media but in fact depends on pH, being slightly higher in alkaline media. The voltammogram (i.e. the derivative of the isotherm) displays a sharp peak in acidic media (but not in alkaline media), which has been explained as an order–disorder phase transition in the OH adlayer,[47] or as caused by the adsorption of two different kinds of OH.[50]

Fig. 4 compares the voltammetry of Pt(111) in 0.1 M NaOH with that of a Pt(15 15 14) surface, which has 30-atom wide (111) terraces separated by steps of (110) orientation. Whereas on the Pt(111) terrace H and OH adsorption lead to two separate features, introduction of step sites leads to only one additional feature in the voltammetry at ca. 0.25 V (note a small feature at ca. 0.4 V in Fig. 4 which corresponds to step sites of (100) orientation). Furthermore, the feature is sharp instead of broad, implying attractive lateral interactions, which is at least unexpected. The charge corresponding to this peak is ca. 1 electron per step atom in acidic media,[51] the reason why traditionally it has been attributed to hydrogen adsorption on the step site. However, if that were true, where is the feature

Table 1 Hydrogen UPD adsorption energy and nearest-neighbor interaction energy as estimated from experiment and/or fit from DFT calculations, on a Pt(111) electrodes, on the (110) step site of a Pt[n(111)×(110)] electrode, and on Pt(100) electrode

	exp./fit (111)	DFT (111)	exp./fit (110) step	exp./fit (100)	DFT (100)
$\Delta G_{H,ads}$/eV	−0.21	−0.16	−0.025	−0.45	−0.27
ε_{HH}/eV	0.047	0.019	−0.20	0.014	0.0066
References	46,47	48	53	56	48

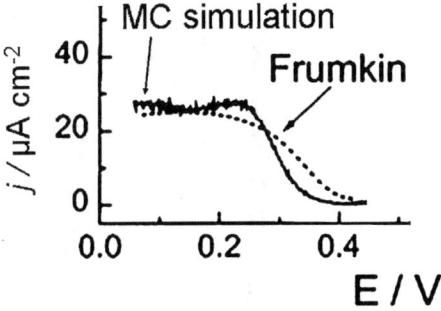

Fig. 5 Monte Carlo simulation and mean-field "Frumkin" approximation of the hydrogen region on Pt(111) using the parameters in the second column of Table 1.

corresponding to OH adsorption on the step site? Finally, whereas the features corresponding to H and OH adsorption on the (111) terrace show no significant pH dependence on the RHE scale, the step-related feature is observed at a more positive potential in alkaline media, shifting by ca. 10 mV pH^{-1} on an RHE scale.[52]

Table 1, fourth column, gives the values for $\Delta G_{H,ads}$ and ε_{HH} estimated from the peak corresponding to the (110) step site in perchloric acid solution,[53] if it is assumed that only H adsorbs on the step with a maximum step coverage of 1. Note that these numbers would imply that adsorbed H not only experiences attractive interactions on step sites, but also that H has a lower affinity to step sites than to terrace sites. This is in disagreement with ultra-high vacuum results, where it has been found that H has a higher affinity for step sites.[54] By studying the co-adsorption of H and water on a stepped Pt surface in UHV, we have recently shown that these results can also not be explained by the interaction with water. In fact, a Pt surface covered with H, be it on terraces or in steps, tends to be hydrophobic.[55]

A more general theory for sharp voltammetric peaks was formulated recently in relation to a model for the voltammetry of Pt(100) in bromide containing solution.[56] The adsorption of bromide on Pt(100) is accompanied by a sharp peak in which adsorbed H is quickly replaced by adsorbed bromide (see Fig. 6). We have modeled this voltammetric peak using a simple lateral interaction model that was solved using Monte Carlo simulations. First, we estimated the adsorption energy and interaction of upd H on Pt(100) by fitting the blank voltammetry of Pt(100) in perchloric acid solution, the results of which are also given in Table 1. Note that, as expected, H adsorbs more strongly on Pt(100) than on Pt(111), but that the interactions between the adsorbed H are weaker than on Pt(111). These results are in good qualitative agreement with DFT calculations,[48] as given in the last column of Table 1.

Introducing the potential dependent co-adsorption of Br into the model, a sharp peak is obtained, giving a good fit of the experiment (dashed line in Fig. 5), if we assume $\Delta G_{Br,ads} = -0.27$ eV and $\varepsilon_{HBr} = 0.055$ for a pair of H and Br sitting on neighboring sites, and an infinite repulsion between two Br on neighboring sites. The latter assumption leads to a maximum coverage of 0.5 ML of Br, in a c(2 × 2) adlayer, in agreement with experiment.[57] Because the interaction between H and Br is stronger than between H, and the interaction between two adsorbed Br at next-nearest neighbor sites is small, the interparticle repulsion exceeds the sum of the intraparticle repulsion and the effective interaction for competing adsorbates may be negative (i.e. attractive). For a mean-field based derivation of this condition, we refer the reader to the original paper.[56] This condition is typically satisfied if a small and a large adsorbate compete for surface sites.

A similar explanation may be applied to the sharp peak observed in the "hydrogen region" of stepped Pt surfaces. If we assume that the actual reaction corresponding to that peak is:

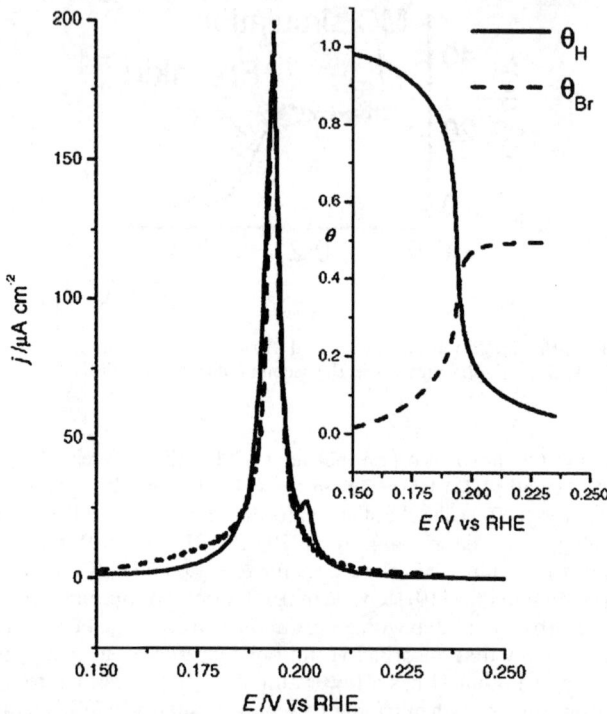

Fig. 6 Modeling hydrogen and bromine competitive adsorption from the voltammogram of Pt(100) in HClO$_4$ 0.1 M + KBr 10^{-2} M (solid line, positive scan), by using Monte Carlo simulations using the interaction energies mentioned in the text (dashed line). The inset shows the corresponding coverages of hydrogen (solid line) and bromine (dashed line) as a function of the potential. Reproduced with permission from ref. 56.

$$H_{ads} + xH_2O \leftrightarrows xOH_{ads} + (1 + x)H^+ + (1 + x)e^- \qquad (10)$$

the sharp peak may be explained by the competition between H and OH. Also, reaction 10 would explain why only a single peak is observed, and not two. On the other hand, the peak charge of 1 electron per step atom would be more difficult to explain with this model, and reaction 10 would also not explain the anomalous pH dependence.

Since it is very difficult to see adsorbed OH on Pt in a spectroscopic experiment, we have recently tried to adsorb OH in a step site of a stepped Pt(533) surface in UHV. Our tactic was similar to a method employed previously for Pt(111): by pre-adsorbing atomic oxygen and subsequently dosing water, O will react with H$_2$O to form chemisorbed OH on the Pt(111) surface.[58] In a temperature-programmed desorption experiment, this manifests as a water desorption peak that appears at higher temperature than without pre-adsorbed O. This apparent hydrophilicity is due to the stabilization of water by the exothermic reaction of water with O to OH. A similar experiment with atomic oxygen pre-adsorbed in the steps of a Pt(533) surface, without O on the terraces, does not yield a clear apparent stabilization of a monolayer of water when subsequently dosed on the Pt(533)–O$_{step}$ surface.[59] This would suggest that the reaction to OH does not take place to a significant extent in the step sites, presumably because the relative stability of atomic oxygen is higher in the step than on the surface. This obviously raises the question what the product of water dissociation is under electrochemical conditions. I believe that this is a crucial question for which at this moment we do not have a consistent answer.

Conclusions

The papers in this volume amply illustrate the glamour and vitality of modern electrocatalysis and interfacial electrochemistry, and the immense impact that modern spectroscopic techniques and modern computational chemistry, and especially their combination, are having on our understanding of the electrochemical interface at the molecular level. Heinz Gerischer, in his Introductory Lecture to the Faraday Discussion 56 in 1973,[3] quoted Julius Tafel from his 1904 paper in which he introduced the famous Tafel equation:[60] "The problem of electrode polarisation in electrolysis has been studied scientifically for about one hundred years. It is, therefore, scarcely possible to find some new aspects which previously have not been touched already, either in experiment or speculations." Now, another one hundred years later, I would conclude almost the opposite: there are many aspects that we have not yet explored and many observations that we still do not fully understand. However, the potential of the modern tools that we have at our disposal to tackle these issues is formidable and there is no question in my mind that this will lead to major advances in both the understanding and the applications of electrochemistry.

These are indeed exciting times to be an electrochemist. Not only is the unprecedented power of modern experimental and computational tools an excellent enabler for innovative fundamental research work, with the ever louder cry for alternative and more sustainable energy sources and devices, electrochemistry has every reason to put itself at the center of attention. Even prominent non-electrochemists admit that our future energy technology will have electrochemistry as one of its cornerstones. In a 2007 Science paper, Whitesides and Crabtree[61] have identified long-term research areas that should not be forgotten, and many of them are of a partial or even complete electrochemical nature. In the largely personal translation of this electrochemist:

1. The oxygen electrode (both oxygen reduction and oxygen evolution),
2. (Electro-)catalysis by design (in essence the theme of this meeting),
3. Various aspects of photoelectrocatalysis,
4. (Electro-)chemistry of carbon dioxide,
5. (Electro-)chemistry of complex systems ("emergent behavior", nonlinearity, innovative (electro-)chemical engineering),
6. Efficiency of energy use,
7. (Electro-)chemistry of small molecules (H_2O, CO, small inorganic nitrogen compounds),
8. New (but sensible) ideas.

These long-term research areas provide us with plenty of challenges to explore the potential of the interface between theory and experiment in electrocatalysis, and will continue to be discussion topics at Faraday Discussions in the decades to come.

References

1. D. D. Eley, *Discuss. Faraday Soc.*, 1947, **1**, 129.
2. R. Parsons, *Discuss. Faraday Soc.*, 1968, **45**, 40.
3. H. Gerischer, *Faraday Discuss. Chem. Soc.*, 1973, **56**, 1.
4. J. E. B. Randles, *Faraday Discuss. Chem. Soc.*, 1973, **56**, 379.
5. G. Nagy, K. Heinzinger and E. Spohr, *Faraday Discuss.*, 1992, **94**, 307.
6. G. Ertl, *Faraday Discuss.*, 2002, **121**, 1.
7. S. J. Mitchell, S. Wang and P. A. Rikvold, *Faraday Discuss.*, 2002, **121**, 53.
8. S. A. Wasileski and M. J. Weaver, *Faraday Discuss.*, 2002, **121**, 285.
9. M. T. M. Koper, N. P. Lebedeva and C. G. M. Hermse, *Faraday Discuss.*, 2002, **121**, 301.
10. P. A. M. Dirac, *Proc. R. Soc. London, Ser. A*, 1929, **A123**, 714.
11. A. N. Frumkin, *Z. Phys. Chem.*, 1926, **35**, 792.
12. P. S. Bagus, G. Pacchioni and M. R. Philpott, *J. Chem. Phys.*, 1989, **90**, 4287.
13. M. T. M. Koper, in *Modern Aspects of Electrochemistry*, ed. C. G. Vayenas, B. E. Conway and R. E. White, Kluwer Academic/Plenum Press, New York, 2003, vol. 36, p. 51–130.
14. C. D. Taylor, S. A. Wasileski, J.-S. Filhol and M. Neurock, *Phys. Rev. B: Condens. Matter Mater. Phys.*, 2006, **73**, 165402.

15 M. Otani and O. Sugino, *Phys. Rev. B: Condens. Matter Mater. Phys.*, 2006, **73**, 115407.
16 E. Skulason, G. S. Karlberg, J. Rossmeisl, T. Bligaard, J. Greeley, H. Jónsson and J. K. Nørskov, *Phys. Chem. Chem. Phys.*, 2007, **9**, 3241.
17 A. Y. Lozozvoi, A. Alavi, J. Kohanoff and R. Lynden-Bell, *J. Chem. Phys.*, 2001, **115**, 1661.
18 J. K. Nørskov, J. Rossmeisl, A. Logadottir, L. Lundqvist, J. R. Kitchin, T. Bligaard and H. Jónsson, *J. Phys. Chem. B*, 2004, **108**, 17886.
19 K. J. Vetter and J. W. Schultze, *Ber. Bunsen-Ges. Phys. Chem.*, 1972, **76**, 920.
20 W. Schmickler and R. Guidelli, *J. Electroanal. Chem.*, 1987, **235**, 387.
21 N. P. Lebedeva, M. T. M. Koper, E. Herrero, J. M. Feliu and R. A. van Santen, *J. Electroanal. Chem.*, 2000, **487**, 37.
22 N. P. Lebedeva, M. T. M. Koper, J. M. Feliu and R. A. van Santen, *J. Phys. Chem. B*, 2002, **106**, 12938.
23 T. H. M. Housmans, J. M. Feliu and M. T. M. Koper, *J. Electroanal. Chem.*, 2004, **572**, 79.
24 T. H. M. Housmans and M. T. M. Koper, *J. Electroanal. Chem.*, 2005, **575**, 39.
25 (*a*) D. F. Ogletree, M. A. Van Hove and G. A. Somorjai, *Surf. Sci.*, 1986, **173**, 351; (*b*) B. E. Hayden, K. Kretzschmar, A. M. Bradshaw and R. G. Greenler, *Surf. Sci.*, 1985, **149**, 394.
26 P. J. Feibelman, B. Hammer, J. K. Nørskov, F. Wagner, M. Scheffler, R. Stumpf, R. Watwe and J. Dumesic, *J. Phys. Chem. B*, 2001, **105**, 4801.
27 R. A. Olsen, P. H. T. Philipsen and E. J. Baerends, *J. Chem. Phys.*, 2003, **119**, 4522.
28 S. Gilman, *J. Phys. Chem.*, 1964, **68**, 70.
29 L. Palaikis, D. Zurawski, M. Hourani and A. Wieckowski, *Surf. Sci.*, 1988, **199**, 183.
30 N. P. Lebedeva, M. T. M. Koper, J. M. Feliu and R. A. van Santen, *J. Electroanal. Chem.*, 2002, **524–525**, 242.
31 T. E. Shubina, C. Hartnig and M. T. M. Koper, *Phys. Chem. Chem. Phys.*, 2004, **6**, 4125.
32 M. Janik and M. Neurock, *Electrochim. Acta*, 2007, **52**, 5517.
33 G. García and M. T. M. Koper, *Phys. Chem. Chem. Phys.*, 2008, **10**, 3802.
34 N. M. Markovic and P. N. Ross Jr., *Surf. Sci. Rep.*, 2002, **45**, 117.
35 J. S. Spendelow, J. D. Goodpaster, P. J. A. Kenis and A. Wieckowski, *J. Phys. Chem. B*, 2006, **110**, 9545.
36 J. S. Spendelow and A. Wieckowski, *Phys. Chem. Chem. Phys.*, 2007, **9**, 2654.
37 T. H. M. Housmans, C. G. M. Hermse and M. T. M. Koper, *J. Electroanal. Chem.*, 2007, **607**, 67.
38 G. García and M. T. M. Koper, in preparation.
39 M. T. M. Koper and R. A. van Santen, *J. Electroanal. Chem.*, 1999, **476**, 64.
40 M. T. M. Koper, R. A. van Santen, S. A. Wasileski and M. J. Weaver, *J. Chem. Phys.*, 2000, **113**, 4392.
41 D. Curulla Ferré and J. W. Niemantsverdriet, *Electrochim. Acta*, 2008, **53**, 2897.
42 G. Blyholder, *J. Phys. Chem.*, 1964, **68**, 2772.
43 G. García, P. Rodriguez and M. T. M. Koper, in preparation.
44 T. H. M. Housmans and M. T. M. Koper, *Electrochem. Commun.*, 2005, **7**, 581.
45 J. Rossmeisl, J. K. Nørskov, C. D. Taylor, M. J. Janik and M. Neurock, *J. Phys. Chem. B*, 2006, **110**, 21883.
46 G. Jerkiewicz, *Prog. Surf. Sci.*, 1998, **57**, 137.
47 M. T. M. Koper and J. J. Lukkien, *J. Electroanal. Chem.*, 2000, **485**, 161.
48 G. S. Karlberg, T. F. Jaramillo, E. Skulason, J. Rossmeisl, T. Bligaard and J. K. Nørskov, *Phys. Rev. Lett.*, 2007, **99**, 126101.
49 J. Rossmeisl, G. S. Karlberg, T. Jaramillo and J. K. Nørskov, *Faraday Discuss.*, 2008, **140**, DOI: 10.1039/b802129e, paper 16.
50 A. Berná, V. Climent and J. M. Feliu, *Electrochem. Commun.*, 2007, **9**, 2789.
51 J. Clavilier, K. El Achi and A. Rodes, *J. Electroanal. Chem.*, 1989, **272**, 253.
52 M. J. T. C. van der Niet, N. García-Araez, J. M. Feliu and M. T. M. Koper, in preparation.
53 M. T. M. Koper, J. J. Lukkien, N. P. Lebedeva, J. M. Feliu and R. A. van Santen, *Surf. Sci.*, 2001, **478**, L339.
54 A. T. Gee, B. E. Hayden, C. Mormiche, T. S. Nunney, *J. Chem. Phys.* 112, p. 7660.
55 M. J. T. C. van der Niet, I. Dominicus, M. T. M. Koper, L. B. F. Juurlink, *Phys. Chem. Chem. Phys*, submitted.
56 N. Garcia-Araez, J. J. Lukkien, M. T. M. Koper and J. M. Feliu, *J. Electroanal. Chem.*, 2006, **588**, 1.
57 N. Garcia-Araez, V. Climent, E. Herrero and J. M. Feliu, *Surf. Sci.*, 2004, **560**, 269.
58 C. Clay, S. Haq and A. Hodgson, *Phys. Rev. Lett.*, 2004, **92**, 046102.
59 M. J. T. C. van der Niet, I. Dominicus, O. T. Berg, L. B. F. Juurlink and M. T. M. Koper, in preparation.
60 J. Tafel, *Z. Phys. Chem.*, 1904, **50**, 641.
61 G. M. Whitesides and G. W. Crabtree, *Science*, 2007, **315**, 796.

PAPER www.rsc.org/faraday_d | Faraday Discussions

The role of anions in surface electrochemistry

D. V. Tripkovic, D. Strmcnik, D. van der Vliet, V. Stamenkovic and N. M. Markovic*

Received 3rd March 2008, Accepted 15th April 2008
First published as an Advance Article on the web 21st August 2008
DOI: 10.1039/b803714k

Some issues of the current state of understanding in the surface electrochemistry are discussed, with emphases on the role of specifically adsorbing anions in hydrogen adsorption and oxide formation, adsorption and ordering of molecular adsorbates and metal ions, metal deposition, restructuring and stability of surface atoms, and kinetics of electrochemical reactions.

1. Introduction

The adsorption of anions on metal electrodes has been one of the major topics in surface electrochemistry. Of particular interest are chemisorbed (also called contact adsorbed or specifically adsorbed) anions, whose adsorption is controlled by both electronic and chemical forces. Halides (F^-, Cl^-, Br-, I^-) and oxyanions (ClO_4^-, and SO_4^{2-}) are the most extensively studied specifically adsorbed anions.[1,2] As pointed out by Conway,[3] the degree of specific adsorption on metal surfaces increases in the following order: $F^- < ClO_4^- < SO_4^{2-} < Cl^- < Br^- < I^-$, that reflects the decreasing energy of solvation of these species, with strongly solvated F^- and ClO_4^- being nonspecifically or only weakly adsorbed. In contrast, weakly solvated SO_4^{2-}, Cl^-, Br^-, and I^- could form direct chemical bonds with the metal surface, resulting in an ionic surface concentration that exceeds one induced by pure electrostatic interaction. It has been realized for a long time that these specifically adsorbing anions have an important, and generally adverse, effect in a number of electrochemical reactions, including adsorption and ordering of adsorbates from a supporting electrolyte, structure of metal substrates, kinetics of electrochemical reactions, metal deposition, and corrosion.

The earlier studies, which were carried out on polycrystalline metal electrodes,[4] indicated that specific adsorption of anions may affect electrochemical processes in a number of ways: (i) blocking of active sites on which otherwise reactant and/or intermediates could be adsorbed; (ii) modification in adsorption energy for sites adjacent to adsorbed anions; (iii) changes in the potential distribution across the interface; and (iv) surface restructuring. Given that all of these effects may influence electrochemical reactions simultaneously, the role of anions in surface electrochemistry on metal electrodes was considered qualitatively and phenomenologically.

Most of the progress in understanding the role of anions in surface electrochemistry has come from the advent of *in situ* surface sensitive probes[5–11] and the development of the efficient methods to prepare clean single crystal surfaces that are well-characterized with respect to the geometric location and composition of surface atoms.[12–15] These well-defined surfaces have offered an ability to find the potential dependence of the surface coverage and structure by anions, the energetics of adsorption, their effects on ordering, reactivity and stability of electrochemical interfaces.

Materials Science Division, Argonne National Laboratory, University of Chicago, Argonne, Illinois, 60439, USA. E-mail: nmmarkovic@anl.gov

The objective of this Faraday Discussion report is to provide a brief discussion of the importance of specific adsorption of anions in surface electrochemistry. The many investigations concerned with the potential dependence of the phase transitions in ordered anion structures on metal surfaces have been omitted here, in preference of systems for more catalytic interest. We focus on the insight into the physical factors influencing the observed anion effects on adsorption of hydrogen, oxygenated species and CO, restructuring and stability of platinum surface atoms, kinetics of fuel cell reactions, and metal deposition of Cu. The presentation here is restricted to constrained overview; further details (including experimental procedure, *etc.*) can be found in references cited.

2.1 Adsorption of hydrogen and hydroxyl species

The adsorption of hydrogen (dubbed underpotentially deposited; $H^+ + e^- = H_{upd}$) on Pt single crystal surfaces has been the subject of intensive investigation. As early as 1965, Will recognized that the H_{upd} formation on platinum single crystals is a function of the crystallographic orientation of the surface atoms, even though the crystals he used were, by today's standards, neither well-characterized nor well-prepared.[16] In the late 1970's, structure sensitivity of adsorbed hydrogen was derived from *ex-situ* ultrahigh-vacuum (UHV) analysis of immersed surfaces[4,17–19] and/or from analyzing the results obtained by using very simple flame annealing method, the latter being introduced by Clavilier.[12] An examination of the literature from this *ex-situ* and flame annealing era revealed many conflicting reports about the number, relative size and shape of the H_{upd} peaks observed on Pt(*hkl*) in various electrolytes. Interestingly, the major difference was found between the cyclic voltammograms recorded in "non-adsorbing electrolytes, *e.g.*, HF and $HClO_4$.[13] For example, while the voltammetry curve for flame annealed Pt(100) in 0.1 M HF (Fig. 1) shows the asymmetry between the positive and negative potential sweep in the H_{upd} potential region, the H_{upd} peaks in 0.1 M $HClO_4$ are rather symmetrical. Since F^- and ClO_4^- are not thought to adsorb significantly on Pt, differences in these two electrolytes were proposed to be due residual Cl^- impurities in the HF. As seen in Fig. 1 for Pt(100), the addition of HCl to $HClO_4$ in trace amounts has substantial effect on both the H_{upd} peak position as well as the H_{upd} peak shape, *e.g.*, the

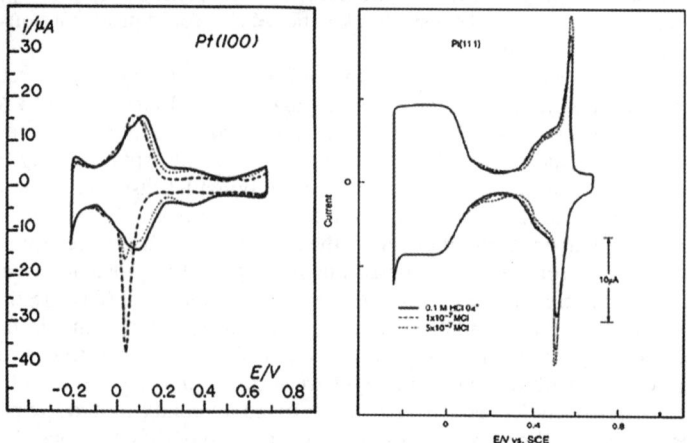

Fig. 1 Left: Cyclic voltammetry for flame-annealed Pt(100) in 0.1 M $HClO_4$ (solid line); 0.1 M $HClO_4$ + 5 × 10^{-7} M Cl^- (dotted line); 0.1 M $HClO_4$ + 5 × 10^{-6} M Cl^- (dashed line); sweep rate 50 mVs^{-1}; electrode area 0.283 cm^2. Right: Cyclic voltammetry for flame-annealed Pt(111) in 0.1 M $HClO_4$ (solid line); 0.1 M $HClO_4$ + 1 × 10^{-7} M Cl^- (dotted line); 0.1 M $HClO_4$ + 5 × 10^{-7} M Cl^- (dashed line); sweep rate 50 mVs^{-1}; electrode area 0.283 cm^2.

symmetrical voltammetric curve is changed to an asymmetric curve, becoming almost identical to the voltammogram observed in "clean" HF. The sharp peak in the hydrogen region observed in the negative sweep is probably caused by rapid desorption of Cl^-, which is accompanied by the concomitant adsorption of H_{upd}. In contrast to Pt(100), traces of Cl^- have much less effect on the hydrogen adsorption on Pt(111), see Fig. 1. It appears that even in the presence of a small concentration of Cl^-, the potential of zero charge (pzc) for Pt(111) is still quite positive to the hydrogen adsorption potential region, and thus the specific adsorption of Cl^- is negligible in this potential region.

At more positive potentials, however, the effect of trace level of Cl^- on the cyclic voltammetry is observed clearly in the "double layer" potential region and, more significantly, in the so-called "butterfly" potential region where the hydroxyl adsorption (hereafter denoted as OH_{ad}) on Pt(111) is accompanied by Cl^- desorption and the corresponding sharpening of the butterfly peak.[20] Based on this observation alone, it was proposed that Cl^- is present as an impurity (at least 10^{-7} M) even in the most meticulously prepared $HClO_4$. The level of Cl^- impurity is significantly smaller in $HClO_4$ than in HF and, therefore, all the differences observed in cyclic voltammetry of Pt(*hkl*) between these two "non-adsorbing" electrolytes are caused by different amounts of Cl anions present in these solutions.

Further successful "fishing" of trace level of Cl^- in $HClO_4$ has been achieved recently from two experimental approaches. One strategy involved utilization of the rotating disk electrode (RDE) technique,[21,22] which has allowed Cl^- concentration to be enhanced in the vicinity of the electrode surface by increasing the rate of Cl^- mass transfer from a bulk of electrolyte to the electrode surface. The second strategy has relied on the fact that the onset of adsorption of anions on metal surfaces with lower values of pzc is shifted towards more negative potentials and, if the geometry of surface atoms is preserved, it would be possible to "catch" a trace level of anions even on a surface with the (111) geometry. A notable published example of these two tactics concerns Cl^- adsorption on Pt(111) modified be a pseudomoprhic Pd monolayer.[22] As can be seen in Fig. 2, the rotation of the

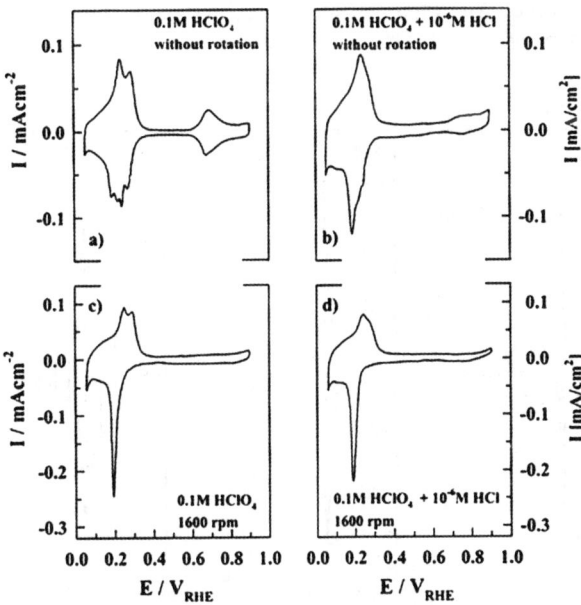

Fig. 2 Cyclic voltammograms of Pt(111)–1 ML Pd (a) in 0.1 M $HClO_4$; (b) 0.1 M $HClO_4$ + 1 × 10^{-6} M Cl^-; (c) same conditions as in (a) but rotation of electrode with 1600 rpm; (d) same conditions as in (b) but rotation of electrode with 1600 rpm; sweep rate 50 mVs^{-1}.

electrode (1600 rpm) has a significant effect on both the shape of the H_{upd} peaks and on the adsorption of OH_{ad}. In particular, the observed H_{upd} peaks in the voltammogram of the rotated electrode exhibit an asymmetry, in contrast to the relatively symmetrical H_{upd} peaks observed on a stationary electrode. Interestingly, a very similar asymmetric peak is observed for the adsorption of hydrogen on Pt(100) in $HClO_4$ containing 5×10^{-6} M Cl$^-$ (Fig. 1), suggesting that it is possible to detect a trace level of Cl$^-$ by increasing mass transport limitations. Notice that the observed voltammetric features in the presence of a small amount of Cl$^-$ are qualitatively similar to the effect induced by the rotation of the electrode in pure $HClO_4$, e.g., in the cathodic sweep direction the various peaks merge into a single peak located at 0.2 V. This is a confirmation that an enhanced mass transport of the small amount of Cl$^-$ ($\approx 10^{-7}$ M) from the bulk of pure $HClO_4$ solution to the electrode surface by forced convection has a similar effect as a small addition of Cl$^-$ to the electrolyte. Furthermore, the OH adsorption is largely suppressed by either increasing the rotation rate or by the addition of Cl$^-$, suggesting that trace amounts of Cl$^-$ and *not* the high concentration of perchlorate anions control the adsorption properties of Pt-group metals. Although the binding energy of OH_{ad} is stronger on Pd than on Pt, due to strong Pd–Cl$^-$ interactions the OH_{ad} coverage is higher on Pt(111) than on Pt(111)–1 ML Pd, a fact which will have a consequences for the interpretation of catalytic activity of Pt and Pd surfaces (see section 2.4).

In contrast to trace level of anions, with an increased amount of added Cl$^-$ to the $HClO_4$ or in pure HCl electrolyte (Fig. 3) the cyclic voltammograms of Pt single crystals in the H_{upd} potential region become very sharp and exceptionally symmetrical.[13] This is consistent with a coupling between the H_{upd} and Cl$^-$ adsorption–desorption processes, suggesting that a unique synergy of voltammetric features on metallic surfaces in acidic solutions is arising rather through the structure sensitive adsorption of anions than the adsorption of H_{upd} and OH_{ad}. An important consequence of such coupling effects is that even after four decades of comprehensive research on single crystal surfaces in acid solutions we are not able to resolve *intrinsic* structure sensitivity even for the two elementary steps in surface electrochemistry; e.g., the adsorption of hydrogen and the formation of the hydroxyl layer. Nevertheless, in the various mechanisms for accounting of anion effects on adsorption of hydrogen and hydroxyl species, the authors believe that the coupling of anion

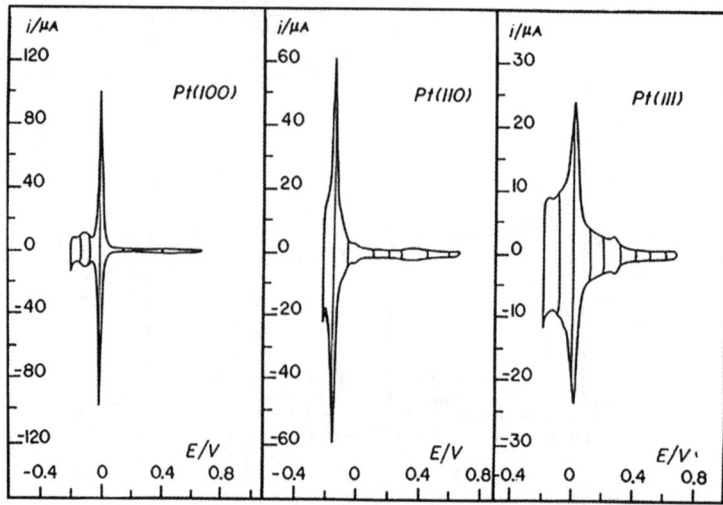

Fig. 3 Cyclic voltammetry for flame-annealed Pt single crystals in 0.1 M HCl; sweep rate 50 mVs^{-1}; electrode area 0.283 cm^2.

adsorption to the hydrogen/hydroxyl adsorption is predominant. Depending on the anion concentration in the bulk of electrolytes, the pseudocapacitive features corresponding to H_{upd} and OH_{ad} may show either an asymmetry or may be sharp and symmetrical.

2.2 Adsorption of CO and electrooxidation of CO

Beyond the question of anion effects in determining the pseudocapacitive features in cyclic volammograms, a central issue in electrochemical surface science concerns the role of anions in orchestrating the surface structures (and activity) of co-adsorbed reactants and reaction intermediates. Given its practical importance, a particularly interesting class of electrochemical adlayer systems receiving the widest attention so far is CO adsorption and electrooxidation on Pt(111).[23-31] Very recently, our group has focused on studying the role of anions on the electrocehmistry of CO_{ad} on metal surfaces for two reasons:[31] (i) CO_{ad} is suitable to be characterized by vibrational spectroscopies, *e.g.*, infrared absorption-reflection spectroscopy (IRAS), and to be probed structurally by utilizing *in-situ* methods of scanning tunneling microscopy (STM) and surface X-ray scattering (SXS); (ii) and in part because the anodic oxidation of CO_{ad} plays the role of a "test molecule" in electrocatalysis, as it does in gas-phase catalysis. For our purposes here, by summarizing the effects of anions on the potential-dependent vibrational behavior of linear CO_{ad} band frequencies (v_{CO}^l) as well as on corresponding O–C–O vibrational frequency of the dissolved CO_2, we shall demonstrate not only that anions are controlling kinetics of CO oxidation reaction through its competition with OH_{ad} but, in addition, that they are affecting CO clustering on the surface.

Representative cyclic voltammogram of Pt(111) in 0.5 M H_2SO_4 is shown in Fig. 4a. The presence of hydrogen adsorption region below 0.25 V and bisulfate/hydroxide adsorption region at higher potentials are familiar features of Pt(111) voltammetry in sulfuric acid solution. The CO-stripping curve (recorded *simultaneously* with IRAS spectra) shows that the onset of CO oxidation commences at ≈0.35 V, forming between 0.35 < E < 0.6 V what we have previously referred to as a pre-ignition potential region.[1,25] In the ignition potential region ($E < 0.6$ V) the stripping voltammetry is characterized by a sharp peak centered at 0.68 V. In both potential regions the CO oxidation reaction takes place in the Langmuir–Hinshelwood type reaction between CO and OH (CO + OH = CO_2 + H^+ + e^-).[1,25]

In all previous reports the potential dependence of v_{CO} is seen to be essentially linear, at least to the potentials as high as ≈0.45 V. Fig. 4b shows that the dependence v_{CO}^l *vs.* E is, in fact, nonlinear and that a linear slope is observed only below 0.3 V, *e.g.*, in the potential region where CO adlayer is stable (no CO_2 production in Fig. 4a). In contrast, concomitant with the development of CO_2 at 0.35 V (grey dots), the v_{CO}^l frequencies (solid line) first slightly increase with respect to the expected linear relation (dotted trace in Fig. 4b) then, over the potential range 0.5 < E < 0.62 V, v_{CO}^l frequencies downshift substantially and, finally, the marked *increase* in v_{CO}^l frequencies above 0.62 V is mirrored by the rapid CO oxidation (CO_2 production in Fig. 4a). Clearly, in addition to the effect of electric field, other factors, such as CO adlayer compression/dissipation should be taken into account to explain the anomalous deviations from a linear dependence. The phenomenon of CO adlayer compression/dissipation is observed in segregated systems where the presence of one species causes the other species to segregate into patches in which the local density is higher/lower than that observed when an equivalent amount of particular species is adsorbed alone. Extending this phenomenon to Fig. 4b, we suggest that the v_{CO}^l frequency deviations arise from different distribution of surface coverage by reactive (CO and OH) and inactive (HSO_4) adsorbates and the mutual interaction among them. The most plausible explanation for v_{CO}^l blueshift deviation between 0.35 < E < 0.5 V is a relatively simple model, where a small yet clearly discernable CO oxidation (initiated by OH adsorption on defect sites[25]) is followed

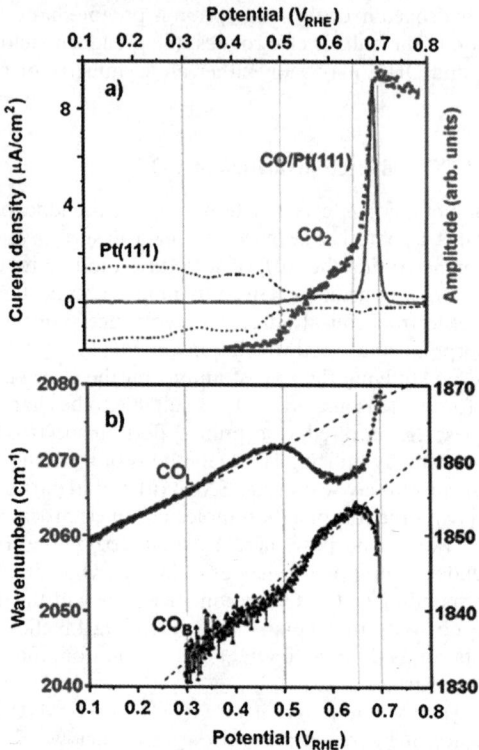

Fig. 4 (a) Cyclic voltammogram of Pt(111) in 0.5 M H_2SO_4 (dashed line) 2 mV s^{-1}; CO stripping current (solid line, 0.5 mV s^{-1}) and corresponding CO_2 production (grey dots); (b) ν_{CO} vs. E plots recorded *simultaneously* with CO stripping.

by bisulfate anion adsorption on Pt sites, which below the onset of CO oxidation were occupied by CO_{ad}. Given that the CO–HSO_4^- interaction is most likely repulsive, the higher ν_{CO}^l frequency than expected most plausibly reflects mild compression of CO_{ad} islands (the enhanced dipole–dipole coupling) engendered by HSO_4^- co-adsorption. On the other hand, marked ν_{CO}^l redshift observed between $0.4 < E < 0.55$ V reflects the reduced dipole–dipole coupling and temporal dissipation of the CO adlayer into less compressed clusters. Such behavior is symptomatic of increased CO oxidation, induced by an enhanced OH adsorption and hindered HSO_4^- adsorption. In the potential region where intensive CO oxidation occurs ($E > 0.65$ V), however, the adsorption of bisulfate is more favorable than OH^- (surface provides larger ensembles of CO-free Pt sites), acting again as a driving force for CO_{ad} to cluster into small compressed islands. This, in turn, will enhance the dipole–dipole coupling, e.g., a substational blueshift in the CO_{ad} vibrational frequency.

To test further the anion-induced island compression model, CO oxidation experiments were carried out in CO saturated H_2SO_4 (Fig. 5a), $HClO_4$ (Fig. 5b) and KOH (Fig. 5c) solutions. We chose to show the data for ν_{CO} vs. E relationships in CO saturated solution to demonstrate that the results summarized in Fig. 4 are not unique to the CO stripping experiments. In particular, above 0.55 V in H_2SO_4 solution the anomalous ν_{CO}^l deviations are also observed in the presence of CO in solution. The comparison between the anion-dependent ν_{CO}^l deviations in the solution containing strongly (HSO_4^-) and weakly (ClO_4^-) adsorbing anions shows that weakly adsorbing perchloric acid anions allow formation of reactive OH (which may aid the dissipation of CO islands) without compressing the remaining CO into smaller islands. Because OH^- adsorption rather than weak anion adsorption is preferred

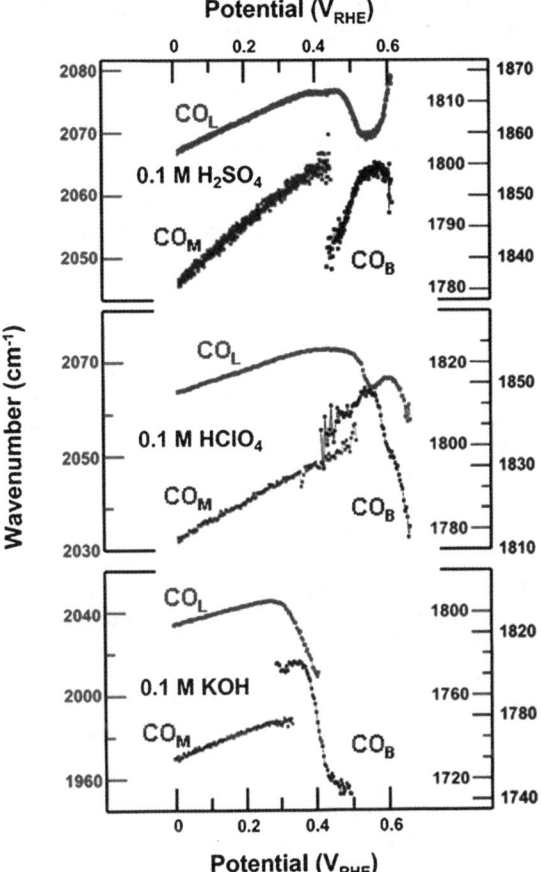

Fig. 5 ν_{CO} vs. E plots for CO_{ad} (light grey: linearly bonded, ν_{COl}; grey: multicoordenated bonded, ν_{COm}; dark grey: bridge bonded ν_{COb}) in the different electrolytes. Notice the existence of multicoordinated CO (CO_m) in the CO saturated solution; for details see ref. 31.

at higher potentials, then the ν_{CO}^l redshift at $E > 0.65$ V follows from the fast oxidative removal of CO_{ad} and, thus, the reduced dipole–dipole coupling. Consequently, instead of a "U-shaped" ν_{CO}^l vs. E dependence, characteristic for H_2SO_4, in $HClO_4$ only moderate non-linear deviations are observed in the same potential range. Not surprisingly, in the solution containing only OH^- anions, the effect of facile oxidation of CO adlayer (ν_{CO}^l redshift in Fig. 4c), induced by the adsorption of reactive OH^-, is even more pronounced. Consequently, the anomalous "Stark-tuning" behavior is completely missing in KOH solution (Fig. 5c).

A related, yet separate, issue concerns the possible role of anions in influencing the kinetics of CO oxidation reaction. Although the role of anions in electrocatalysis of a fuel cell reaction will be discussed in section 2.4 below, for our purposes here, we continue with our discussion by analyzing the role of anions in the electrooxidation of CO_{ad} on Pt(111). Fig. 5 shows that the kinetics of CO oxidation are strongly pH dependent; a notable feature is that the kinetics in alkaline solution is faster than in acid media. If the L–H mechanism is operative the higher catalytic activity in the alkaline solution (evidenced by the lower CO oxidation potential, CO_2 production) implies that the surface coverage by OH^- is larger in alkaline than in acid solution. There is a strong competition between OH^- and anions for the Pt sites in acid solution, and therefore the surface coverage by OH^- is significantly reduced with respect to alkaline solution and thus the kinetics is strongly inhibited at low pH. The explanation, for the remarkable effect of pH on the rate of CO oxidation is the pH-dependent adsorption of OH$^-$ which, in turn, is determined by competitive adsorption of anions. Therefore, as in the case of H_{upd} and OH_{ad}, it appears that the ordering of CO_{ad} in an adlayer (see ref. 25), the morphology of the adlayer during CO oxidation (Fig. 4), the structure sensitive kinetics of CO_{ad} electrooxidation (see ref. 1 and 25), and the rate of CO oxidation on the same electrode but in various electrolytes (Fig. 5) is completely determined by the structure sensitive adsorption of anions. In what follows, we demonstrate that the same applies for metallic adsorbates as well.

2.3 Underpotential deposition of Cu

Another fundamental question concerns the possible role of anions in both structural ordering of underpotentially deposited (upd) metals as well as in the kinetics of metal deposition. An essentially well studied example, which also illustrates the importance of SXS in analyzing metal–anion adsorbing structures is the formation of Cu adlayer(s) on Pt(111) in acidic media. The interpretation of processes associated with the formation of the Cu monolayer on Pt(111), and the nature of the Pt(111)–Cu structure, have been the subject of considerable controversy. Overviews of several different perspectives can be found in references.[32–34] The rotating ring disk (RRDE) method[33,34] was successfully applied to investigate the kinetics of Cu^{2+} deposition and to determine the potential dependent surface coverage by Cu_{upd} ($\Theta_{cu(upd)}$).[33,34] Fig. 6 shows that Cu deposition in 0.1 M $HClO_4$ is independent of electrode rotation, manifesting that Cu deposition is a kinetically limited reaction in the absence of specifically adsorbing anions. The effect of addition of Cl$^-$ becomes apparent in the voltammetry of Fig. 6b, with the two major distinguished features:

1. The kinetics of Cu UPD is dramatically enhanced in the presence of Cl$^-$, resulting in the diffusion-limiting transport of Cu^{2+} in the presence of Cl$^-$; *e.g.* in both the UPD and overpotential deposition (OPD) potential regions the current is proportional to $\omega^{1/2}$. The promotive effect of Cl$^-$ can be rationalized based on appreciative polarizability and highly deformable hydration shells of Cl$^-$. One might anticipate that such a Cl$^-$ anion may perturb the solvation shell of strongly hydrated Cu^{2+}, thereby promoting electron transfer kinetics in the discharge reaction of Cu^{2+} to Cu^0.

2. Cl$^-$ produces a splitting of the voltammetric Cu UPD peak.[32–34] This finding encouraged a careful evaluation of this system with SXS and RRDE methods. In combination, these studies have demonstrated that Cu_{ad} coverage and thus structure in between the two peaks (Fig. 7) is completely determined by the co-adsorbing anions (Cl$^-$ or Br$^-$). In nearly halide free supporting electrolytes, Cu appears to be deposited as metallic islands (or "patches") having the Pt lattice constant, *i.e.*, pseudomorphic adlayer.[32–34] In the presence of halides, however, a multi-step deposition occurs with the formation of ordered anion adlattices. A representation of a possible Pt(111)–Cu_{upd}-anion structural model as a function of Cu_{upd} coverage is shown in Fig. 7 (for more details see ref. 34). Clearly, Cu_{upd} is either sandwiched between the Pt surface and anions or is in contact with the anions adsorbed on the adjacent Pt sites, in Fig. 7 states 1 to 4. At higher surface coverages by Cu_{upd} (0.5 < Θ_{Cu} < 1), anions are entirely displaced from the surface by Cu_{upd} (state 3). Note that the state 3 is representative of an ordered (4 × 4) Cu_{upd}–Cl_{ad} (Br_{ad}) bilayer structure, which is formed in between the Cu UPD peaks in the cyclic voltammetry (Fig. 7), each with coverage of 0.585 ML. The final step in the Cu UPD is the filling-in of the Cu_{upd} monolayer to form a bilayer phase: *i.e.*, a pseudomorphic (1 × 1) Cu_{upd} monolayer.

The observation that anions are affecting the kinetics of metal deposition as well as ordering of metal adlayers has far-reaching implications in surface electrochemistry. While both Cu and halides are known to transfer most (or all) of their charge

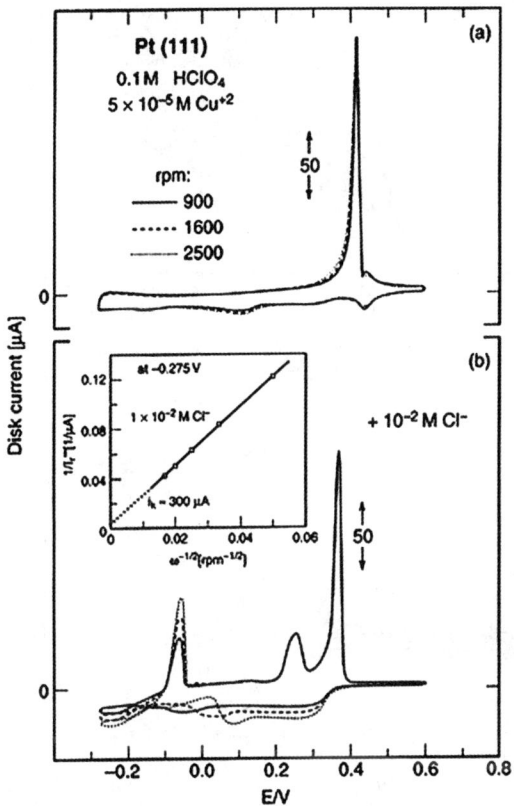

Fig. 6 Cyclic voltammogram for Cu UPD on the Pt(111) disk electrode in 0.1 M HClO$_4$; (b) Cyclic voltammogram for Cu UPD on the Pt(111) disk electrode in 0.1 M HClO$_4$ + 1 × 10^{-2} M Cl$^-$; inset; Levich plot for unshielded ring currents at −0.257 V.

upon chemisorption, the Coulombic stabilization afforded by an ionic-solid-style adlattice may well trigger their common formation at electrochemical interfaces. Understanding the factors controlling the nature and occurrence of such double layer coadsorbate adlattices is clearly a topic worthy of theoretical as well as greater experimental attention in the near future.

2.4 Electocatalysis of fuel cell reactions

The term *electrocatalysis* is commonly employed to describe the study of electrode processes where charge-transfer reactions have a strong dependence on the nature of the electrode material.[35–37] Not surprisingly, virtually every electrochemical reaction where chemical bonds are broken or formed is electrocatalytic, and the kinetics varies by many orders of magnitude for different electrode materials. This is true even for the simplest electrochemical reaction where chemical bonds are broken, the hydrogen evolution/oxidation reaction (HER/HOR), and for the more complex reactions such as the oxygen reduction reaction (ORR), and certainly for small organic molecules such as the oxidation formic acid and methanol. Although largely different in nature, the kinetics of electrochemical reactions involving either inorganic or organic compounds is governed by the same electrocatalytic law: while the reaction rate passes through a maximum for metals adsorbing reaction *intermediates* moderately, the reaction rate is very slow on metals which adsorb intermediates either strongly or weakly, *i.e.*, the Sabatier Principle. The establishment of more quantitative relationships between the energetics of intermediates and the rate of

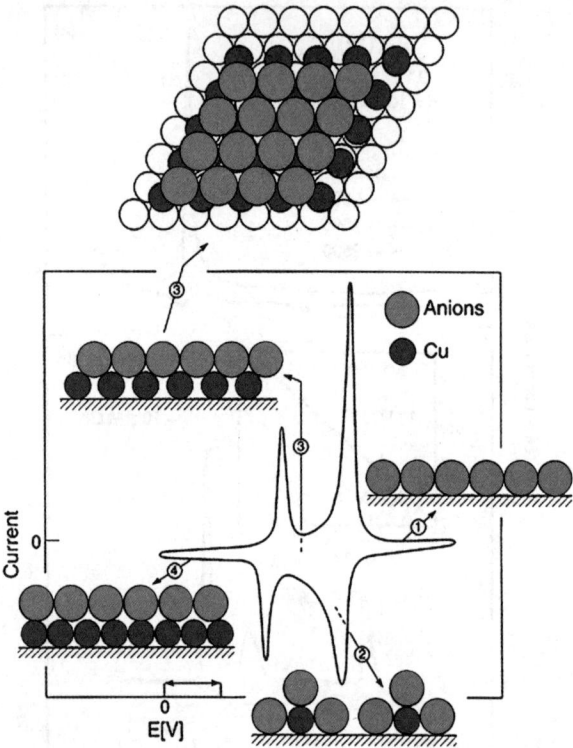

Fig. 7 The representation of the proposed Cu-halide structure on Pt(111) in which the Pt atoms are shown by open circles and halide overlayer atoms by filled circles. Bottom: proposed mechanism for Cu UPD in the presence of halide anions. For details see text.

electrochemical reaction is somewhat difficult owing in part to the absence of directly measured values for the surface–intermediate bond energy, and in part to the fact that an *electrocatalytic reaction occurs on electrode surfaces which are always modified by reaction intermediates and "spectator" species*. Therefore, for the future development of electrocatalysis as science, it is essential that these difficulties/complexities be overcome in order to progressively link the atomic/molecular-level properties of the electrochemical interface to the macroscopic kinetic process. The aim of this section is to summarize some of these relationships and to give selected examples of each, with particular emphasis on the systems that have been studied because of their inherent interest in the development of efficient energy converting systems.

2.4.1 Methanol and formic acid oxidation. Methanol and formic acid are probably the most studied of the C_1 compounds because of their potential as a logistical fuel and a feedstock for fuel cells. For both molecules, the mechanism on Pt in acid solution is reasonably well-established, *via* the so-called "dual-pathway", for details see ref. 38. Although a reaction scheme is shown for the oxidation of methanol, with some small modifications the same reaction scheme could be applicable for the oxidation of formic acid as well,[1] *e.g.*,

While numerous details remain uncertain, this reaction scheme involves the adsorption of CH_3OH (HCOOH) (k_{ad}), followed by the (*non-faradic*) dehydration of CH_3OH (HCOOH), and the formation of chemisorbed "poison" (reaction 2) in competition with the direct dehydrogenation path *via* one or more reactive intermediates. As we shall see below, the rate of this step is determined by the surface

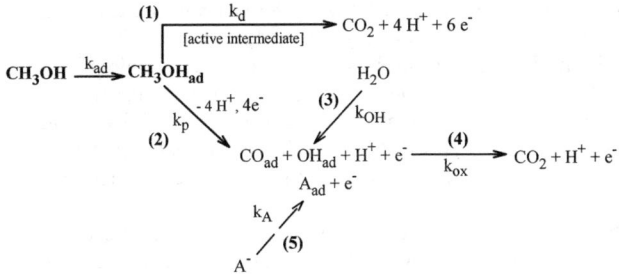

coverage of H_{upd}, anions (A_{ad}), OH_{ad}, and "poisoning" species. The major "poisoning" species was identified clearly as adsorbed CO. Besides being a "poison", CO_{ad} may also act as an intermediate, where some fraction of the CO_{ad} can be further oxidized to produce CO_2 (reaction 4). The active surface oxidant is most likely adsorbed OH_{ad}, as proposed in the section above for CO_{ad} oxidation. Following the reaction scheme for oxidative removal of CO_{ad}, the adsorption of oxygenated species is in a strong competition with anion adsorption (k_A), and consequently the rate of reaction 4 (k_{ox}) is determined by the delicate balance between the rate constants k_p, k_{OH}, and k_A. Of the various proposed rate determining steps for accounting for oxidative removal of CO,[1] the authors believe that the adsorption of OH (k_{OH}) and, thus, competition between OH^- and anion adsorption is controlling the rate of CO electrooxidation.

Important examples of the role of anions in electrooxidation of small organic molecules includes the effects of Cl^- on oxidation of methanol (Fig. 8). As anticipated, increasing the Cl^- concentration inhibits significantly the methanol electrooxidation on Pt(111) and even more on Pt(100). On Pt(100) the effect of Cl^- is generally more complicated than on Pt(111), showing a strong dependence of the pre-history of the electrode; with the inhibition becoming stronger with each successful sweep. This "ageing" effect appeared to be related to both the stronger adsorption of Cl^- on Pt(100) than on Pt(111), requiring a lower potentials to desorb Cl_{ad} from

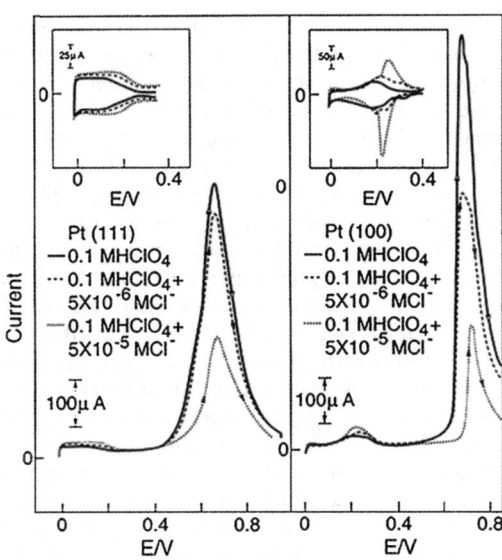

Fig. 8 Cyclic voltammetry for flame-annealed Pt(100) in 0.1 M $HClO_4$ with addition of HCl. Inset: the effect of Cl^- on hydrogen adsorption-desorption pseudocapacitance.

Pt(100) than Pt(111) and, as we discuss in the last section, due to anion structure sensitive halide-induced restructuring of surface atoms. Nevertheless, on both surfaces the inhibition of methanol oxidation is caused by blocking of surface sites by Cl_{ad}, e.g., the anion adsorption is predominant over adsorption of methanol (deactivating step 1) and effectively blocks the formation of hydroxyl adlayer and thus inhibiting oxidative removal of CO_{ad} from the surface (step 4). The same is valid for the oxidation of formic acid, see ref. 1 and references cited therein.

2.4.2 Oxygen reduction reaction. Oxygen reduction reaction is a multi-electron reaction that may include a number of elementary steps involving different intermediates. The detailed mechanism is still not known, since there are available neither *ex-situ* nor *in-situ* techniques capable of identifying all reaction intermediates which might be formed under genuine reaction conditions. Regardless of the reaction pathway, however, we suggested[1] that the rate of the reaction can be described as a simple relationship between the availability of active metal sites for the O–O bond making and the bond breaking (the $1 - \Theta_{ad}$ term) and the energetic term ($\Delta G_\theta^* = \Delta G_0^* + \gamma \Delta G_\Theta^0 \Theta_{ad}$), which is reflecting the energy of adsorption of O_2 and reaction of intermediates and how this energy is modified by changing the Θ_{ad} values:[1] e.g.,

$$i = nFkc_{O_2}(1 - y\Theta_{ad})^x \exp\left(\frac{-\beta FE}{RT}\right)\exp\left(\frac{-\gamma r\Theta_{ad}}{RT}\right) \quad (1)$$

In deriving eqn 1 it is assumed that while the $(1 - \Theta_{ad})$ term is determined by the surface coverage by blocking adsorbates (H_{upd}, OH_{ad}, anions), the reactive intermediates $(O_2^-)_{ad}$ and $(OH^-)_{ad}$ are adsorbed only at a low coverage, *i.e.* they are not a significant part of Θ_{ad}. All these factors are uniquely related to the electronic properties of electrode materials and the nature of solution used, either through the adsorption energy of reaction intermediates, which is controlled by the position of the d-band center,[39,40] or through the potential of zero charge (E_{pzc}),[41] which defines the onset potential of adsorption of spectator species. In what follows, we use this simple rate equation to discuss the role of anions in the ORR on IB group metals in acidic media. For comparison, the corresponding results for the ORR on Pt(111) are also included in the same figure.

Comparison between the cyclic voltammograms and the ORR on the IB group metals reveals that the order of activity is closely related to the fractional surface coverage by spectator anion/OH_{ad} species *at the constant potential*, which for the IB metals increases from Cu to Ag to Au. This, in turn, is related to the values of E_{pzc} of coinage metals. Evaluating E_{pzs} (*i.e.*, the potential at which the surface charge Q equals 0) of metal–aqueous electrochemical interfaces has long been recognized as a key requirement for understanding the double-layer properties of these important systems. However, the experimental evaluation and even the meaning of E_{pzc} for metal interfaces are complicated by the occurrence of potential-dependent chemisorption. As a consequence, it is very difficult, if not impossible, to evaluate the *exact* values for E_{pzc}. Because of that, a useful consideration will be to find the *trend* in E_{pzc} on the IB group metals and to explore how this trend correlates with the adsorption of spectator species, and thus with the surface activity. Following the relationship between the E_{pzc} and the work function (Φ) measured in vacuum[42] it is clear that E_{pzc} on IB group metals should increase in the order Cu(111) < Ag(111) < Au(111). Considering that the *onset* of adsorption of anions from supporting electrolytes should follow the same trend as E_{pzc}, it is not surprising that the *anion adsorption from electrolyte* is observed first on Cu than on Ag and finally on Au (see Fig. 9). Therefore, the present results indicate that the major effect of E_{pzc} is in controlling the adsorption isotherm of spectator species, *i.e.*, the $(1 - \Theta_{ad})$ term in eqn (1), and thus the availability of active metal sites at the constant overpotential for adsorption of O_2 and intermediates. Despite our conclusion about the

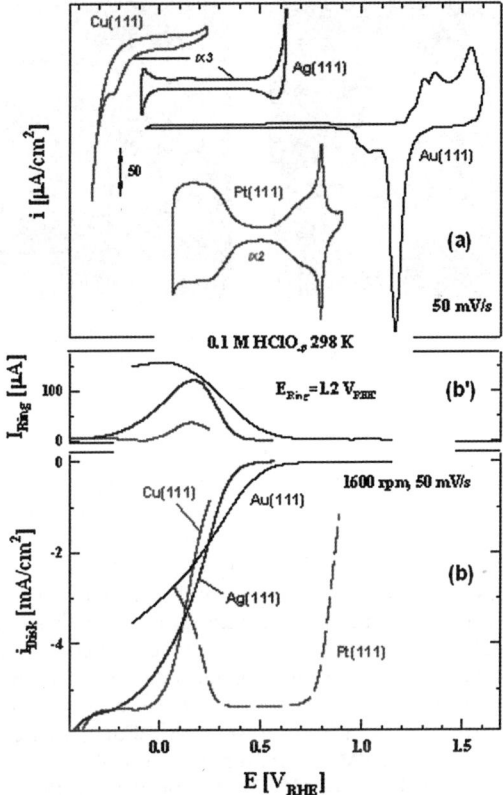

Fig. 9 (a) Cyclic voltammograms of (111) single crystal orientation of Au, Ag, Cu and Pt in 0.1 M HClO$_4$ at 298 K, and 50 mV s^{-1}. CVs are shifted along the *i*-axis from the origin for clarity. (b) ORR polarization curves in positive going sweep at 50 mV s^{-1} and 1600 rpm; (b″) corresponding currents detected on the ring electrode at 1.2 V$_{RHE}$.

importance of E_{pzc} in surface electrochemistry of the ORR, there are number of challenges that need to be met in the future in order to understand electrocatalysis on atomic and molecular levels. Of those, a fundamental issue concerns the developments of theoretical and/or computational framework that can at least rationalize, and ultimately understand, the importance of E_{pzc} in surface electrochemistry.

2.5 Surface restructuring

Finally, we now turn the discussion on the role of anions in restructuring of Pt surface atoms, where only few selected systems have been studied so far. In general, modern surface crystallographic studies have shown that on the atomic scale even the most clean metal surfaces tend to minimize their surface energy by two kinds of surface atom rearrangements, *relaxation* and *reconstruction*, which collectively may be called *restructuring*. Relaxation of metal surfaces is usually defined as a small interlayer spacing change relative to the ideal bulk lattice.[43–47] The displacements should be small compared to the near-neighbor distance, such that no bond-breaking/bond-making events take place within the substrate. Adsorbate induced relaxations occur in many varieties: as *interlayer spacing changes*, where in the near-surface region the top layer of metal atoms relaxes either inward or outward from the bulk atoms; *lateral relaxation*, in which adsorbates shift near-surface atoms parallel to the surface (frequently inducing collective rotation of substrate atoms

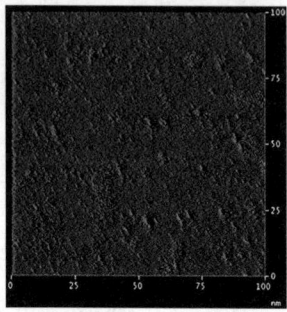 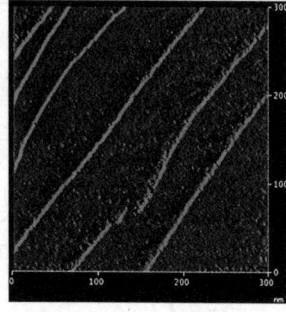

Fig. 10 STM images (100 × 100 nm size) of Pt(111) covered by CO for: (left) oxide-annealed surface (20 potential cyles) up to 0.95 V and (right) Br-annealed surface. The images (U_{tip} = 0.15 V, I_{tip} = 1 nA) illustrate the presence of islands on oxide-annealed surface and steps on Br-annealed surface.

around the adsorbate sites); and *layer buckling*, whereby a coplanar atomic layer loses its coplanaraty because of an adsorbate pulls or pushes some substrate atoms out of the plain relative to the lattice atoms. Reconstruction, on the other hand, involves large atomic displacements both perpendicular to and parallel to the surface plane, so that causes re-bonding and change in the periodicity within the substrate. In our laboratory, the examination of surface restructuring has been persuaded by means of both SXS and STM, especially for the effect of anions on lifting the surface reconstruction of Au(*hkl*),[11] halide effects on relaxation of Pt surface atoms and very recently on changing the surface morphology of Pt surface atoms by fast mass transport of adatoms from the terrace to the step sites. For the latter system, Fig. 10 shows such an example of STM images for the effect of Br⁻ on restructuring the "perfect" Pt(111) surface in 0.1 M HClO$_4$. The protocol in these experiments was the same as one described recently in ref. 48: (1st step) in an electrochemical cell twenty electrochemical oxidation–reduction cycles in Ar saturated solution were applied from 0.005 V to 0.95 V; (2nd step) following this so-called "oxide annealing" preparation method, the electrode was held at 0.05 V and covered with CO$_{ad}$; (3rd step) the CO-protected electrode was transferred into the STM cell, and (4th step) after the STM images were obtained, the electrode was returned to the experimental cell to record cyclic voltammetry. As shown in Fig. 10, STM images for the oxide-annealed surfaces are dominated by large clusters with diatomic (0.5 nm) and some triatomic (0.75 nm) heights which are *randomly* distributed over the (111) surface terraces. In the presence of Br⁻, not only is a reduction in the number and size of the islands observed after Br-annealing up to 0.95 V but, in addition, adislands completely disappear and are replaced by a series of smooth terrace-step structures. Since this transformation only occurs after Br-annealing up to 0.95 V, we conclude that it is Br⁻ coadsorption with OH⁻, as opposed to simple Br⁻ adsorption, that leads to step formation. It is plausible that Br⁻ acts as a surfactant that transports adatoms from terraces to steps. Step formation therefore appears to involve a delicate balance between the tendency of adsorbed oxygenated species to induce adisland formation at high potentials and the contrasting tendency of Br$_{ad}$ to facilitate diffusion of adatoms/adislands across the surface with concomitant incorporation into step edges. The above example shows unique ability of anions to trigger stepwise alternations in surface stability and morphology *via* electrode potential and much more can be learned regarding the anion role in controlling mobility and ordering of Pt surface atoms that is utterly inaccessible in UHV-based systems.

In summary, this article presented some facets of our current state of physicochemical understanding that anions have in surface electrochemistry. By utilizing both powerful *in-situ* techniques for structural determination and electrochemical methods for adsorption and kinetic analysis of electrochemical processes, we illustrate the remarkable insight into the effects of specifically adsorbed anions on the adsorption of hydrogen, oxygen, oxygenated species, CO, surface restructuring, stability of platinum surface atoms, kinetics of fuel cell reactions, and metal deposition of Cu.

Acknowledgements

This work was supported by the contract (DE-AC02-06CH11357) between the University of Chicago and Argonne, LLC, and the US Department of Energy.

References

1 N. M. Markovic and P. N. Ross, *Surf. Sci. Rep.*, 2002, **45**, 117.
2 O. M. Magnussen, *Chem. Rev.*, 2002, **102**, 679.
3 B. E. Conway, in *Progress in Surface Science*, ed. S. Davison, Pergamon Press, Fairview Park, NY, 1984, p. 1.
4 J. C. Huang, W. E. O'Grady and E. Yeager, *J. Electrochem. Soc.*, 1977, **124**, 1732.
5 K. Itaya, *Prog. Surf. Sci.*, 1998, **58**, 121.
6 M. G. Samant, M. F. Toney, G. L. Borges, K. F. Blurton and O. R. Melroy, *J. Phys. Chem.*, 1988, **92**, 220.
7 B. M. Ocko, J. Wang, A. Davenport and H. Isaacs, *Phys. Rev. Lett.*, 1990, **65**, 1466.
8 M. F. Toney and B. M. Ocko, *Synchrotron Radiat. News*, 1993, **6**, 28.
9 I. M. Tidswell, N. M. Markovic and P. N. Ross, *Phys. Rev. Lett.*, 1993, **71**, 1601.
10 C. Lucas, N. M. Markovic and P. N. Ross, *Surf. Sci.*, 1996, **340**, L949.
11 D. M. Kolb, *Prog. Surf. Sci.*, 1996, **51**, 109.
12 J. Clavilier, *J. Electroanal. Chem.*, 1980, **107**, 211.
13 N. Markovic, M. Hanson, G. McDougall and E. Yeager, *J. Electroanal. Chem.*, 1986, **214**, 555.
14 M. Wasberg, L. Palaikis, S. Wallen, M. Kamrath and A. Wieckowski, *J. Electroanal. Chem.*, 1988, **256**, 51.
15 L. A. Kibler, M. Cuesta, M. Kleinert and D. M. Kolb, *J. Electroanal. Chem.*, 200, **484**, 73.
16 F. Will, *J. Electrochem. Soc.*, 1965, **112**, 451.
17 A. T. Hubbard, R. M. Ishikawa and J. Katekaru, *J. Electroanal. Chem.*, 1978, **86**, 271.
18 A. S. Homa, E. Yeager and B. D. Cahan, *J. Electroanal. Chem.*, 1983, **150**, 181.
19 F. T. Wagner and P. N. Ross Jr., *J. Electroanal. Chem.*, 1983, **150**, 141.
20 N. Markovic and P. N. Ross, *J. Electroanal. Chem.*, 1992, **330**, 499.
21 N. M. Markovic, H. A. Gasteiger and P. N. Ross, *J. Phys. Chem.*, 1995, **99**, 3411.
22 M. Arenz, V. Stamenkovic, T. J. Schmidt, K. Wandelt, P. N. Ross and N. M. Markovic, *Surf. Sci.*, 2003, **523**, 199.
23 I. Villegas and M. J. Weaver, *J. Chem. Phys.*, 1994, **101**, 1648.
24 I. Villegas, X. Gao and M. J. Weaver, *Electrochim. Acta*, 1995, **40**, 1267.
25 N. M. Markovic, B. N. Grgur, C. A. Lucas and P. N. Ross, *J. Phys. Chem. B*, 1999, **103**, 487.
26 A. Wieckowski, M. Rubel and C. Gutiérrez, *J. Electroanal. Chem.*, 1995, **382**, 97.
27 H. Kita, H. Naohara, T. Nakato, S. Taguchi and A. Aramata, *J. Electroanal. Chem.*, 1995, **386**, 197.
28 A. Rodes, R. Gomez, J. M. Feliu and M. J. Weaver, *Langmuir*, 2001, **16**, 811.
29 N. M. Markovic, C. Lucas, A. Rodes, V. Stamenkovic and P. N. Ross, *Surf. Sci.*, 2002, **499**, 149.
30 N. M. Markovic, C. Lucas, B. N. Grgur and P. N. Ross, *J. Phys. Chem.*, 1999, **103**, 9616.
31 V. Stamenkovic, K. C. Chou, G. A. Somorjai, P. N. Ross and N. M. Markovic, *J. Phys. Chem. B*, 2005, **1**, 1.
32 N. Markovic and P. N. Ross, *Langmuir*, 1993, **9**, 580.
33 N. M. Markovic, H. A. Gasteiger and P. N. Ross, *Langmuir*, 1995, **11**, 4098.
34 N. M. Markovic, C. Lucas, H. A. Gasteiger and P. N. Ross, Jr., *Surf. Sci.*, 1997, **372**, 239.
35 N. Kobosev and W. Monblanova, *Acta Physiochem. URSS*, 1934, **1**, 611.
36 W. T. Grubb, *Nature*, 1963, **198**, 883.
37 J. O. M. Bockris and A. K. N. Reddy, *Modern Electrochemistry*, Plenum Press, New York, 1970.
38 T. D. Jarvi and E. M. Stuve, in *Electrocatalysis*, ed. J. Lipkowski and P. N. Ross, p. 75, Wiley-VCH, Inc., New-York, 1998.

39 J. K. Nørskov, J. Rossmeisl, A. Logadottir, L. Lindqvist, J. R. Kitchin, T. Bligaard and H. Jonsson, *J. Phys. Chem. B*, 2004, **108**, 17886–17892.
40 V. Stamenkovic, B. S. Mun, K. J. J. Mayrhofer, P. N. Ross, N. M. Markovic, J. Rossmeisl, J. Greeley and J. K. Nørskov, *Angew. Chem., Int. Ed.*, 2006, **45**, 2897.
41 B. B. Blizanac, V. Stamenkovic and N. M. Markovic, *Z. Phys. Chem.*, 2007, **221**, 1379.
42 S. Trasati, *J. Electroanal. Chem.*, 1972, **39**, 163.
43 G. A. Somorjai and M. A. Van Hove, *Prog. Surf. Sci.*, 1989, **30**, 201.
44 G. A. Somorjai, *Introduction to Surface Chemistry and Catalysis*, John Wiley & Sons, New York, 1993.
45 P. A. Thiel and P. J. Estrup, in *The Hanbook of Surface Imiging and Visualization*, ed. A. T. Hubbard, CRC Press, Boca Raton, 1995.
46 R. I. Masel, *Principles of Adsorption and Reaction on Solid Surfaces*, John Wiley & Sons, Inc., 1996.
47 M. A. Van Hove, in *Physics of Covered Solid Surfaces*, ed. Landolt-Boernstein, 1999.
48 D. Srtmcnik, P. Rebec, M. Gaberscek, D. Tripkovic, V. Stamenkovic, C. Lucas and N. M. Markovic, *J. Phys. Chem. C*, 2007, **111**, 18672.

PAPER

From ultra-high vacuum to the electrochemical interface: X-ray scattering studies of model electrocatalysts

Christopher A. Lucas,*[a] Michael Cormack,[a] Mark E. Gallagher,[b] Alexander Brownrigg,[a] Paul Thompson,[a] Ben Fowler,[a] Yvonne Gründer,[c] Jerome Roy,[c] Vojislav Stamenković[d] and Nenad M. Marković[d]

Received 3rd March 2008, Accepted 23rd April 2008
First published as an Advance Article on the web 21st August 2008
DOI: 10.1039/b803523g

In-situ surface X-ray scattering (SXS) has become a powerful probe of the atomic structure at the metal–electrolyte interface. In this paper we describe an experiment in which a Pt(111) sample is prepared under ultra-high vacuum (UHV) conditions to have a p(2 × 2) oxygen layer adsorbed on the surface. The surface is then studied using SXS under UHV conditions before successive transfer to a bulk water environment and then to the electrochemical environment (0.1 M KOH solution) under an applied electrode potential. The Pt surface structure is examined in detail using crystal truncation rod (CTR) measurements under these different conditions. Finally, some suggestions for future experiments on alloy materials, using the same methodology, are proposed and discussed in relation to previous results.

1. Introduction

Since the early days of modern surface science, the main goal in the electrochemical community has been to find correlations between the microscopic structures formed by surface atoms and adsorbates and the macroscopic kinetic rates of a particular electrochemical reaction. The establishment of such relationships, previously only developed for catalysts under ultra-high vacuum (UHV) conditions, has been broadened to embrace electrochemical interfaces. In early work, determination of the surface structures in an electrochemical environment was derived from *ex situ* UHV analysis of emersed surfaces. Although such *ex situ* tactics remain important, the relationship between the structure of the interface in electrolyte and that observed in UHV was always problematic and had to be carefully examined on a case-by-case basis. The application of *in situ* surface sensitive probes, most notably synchrotron based surface X-ray scattering (SXS)[1–6] and scanning tunneling microscopy (STM)[7,8] has overcome this "emersion gap" and provided information on potential-dependent surface structures at a level of sophistication that is on a par with (or, even, in advance of) that obtained for surfaces in UHV.[9]

[a]Oliver Lodge Laboratory, Department of Physics, University of Liverpool, Liverpool, UK L69 7ZE. E-mail: clucas@liv.ac.uk
[b]Surface Science Research Centre, University of Liverpool, Liverpool, UK L69 7ZE
[c]European Synchrotron Radiation Facility, BP220, 38043 Grenoble, France
[d]Materials Science Division, Argonne National Laboratory, Argonne, IL, 60439, USA

Central to all electrochemical reactions in water-based electrolytes and, indeed, to a wide range of physical phenomena in nature, is the structure and bonding of water at a surface. The interaction of water with solid surfaces plays a crucial role in many areas of science. The water structure at well-defined metal surfaces is of particular importance in catalysis and electrochemistry, as the activation of water is the crucial step in many surface reactions. Despite this importance, at the electrified interface numerous experimental and theoretical studies have yet to provide a detailed picture of the atomic-scale structure of water and its behavior remains poorly understood. In traditional surface science experiments, there has recently been a leap forward in understanding the structure and bonding of water on transition metal surfaces.[10] Using density-functional theory (DFT) calculations, Feibelman postulated that the stable configuration for water on Ru(0001) was a partially dissociated overlayer containing a hexagonal network and that the partial dissociation was a necessary prerequisite for wetting, at least for Ru(0001).[11] Vibrational spectroscopy results[12] suggest that a similar layer is present on Pt(111) but consisting of intact water molecules, although low energy electron diffraction (LEED) analysis is consistent with a layer consisting of dissociated OH.[13] The issue of dissociation was addressed recently by Clay et al.,[14] who studied a mixed OH/H_2O ("OH_x") layer on Pt(111) as a function of its composition to determine the role of hydrogen bonding in stabilizing the overlayer. They found that the optimal structure is a mixed ($OH + H_2O$) phase forming a hexagonal ($\sqrt{3} \times \sqrt{3}$)R30° lattice with a weak (3×3) superstructure caused by ordering of the hydrogen bonds. The mixed overlayer can accommodate a range of H_2O/OH compositions but becomes less stable as the H_2O content is reduced.

Although water structures can be observed by LEED on metal surfaces under ultra-high vacuum (UHV) conditions, the electrochemical interface is much harder to study due to the presence of the bulk electrolyte. In order to understand the role of hydration water molecules on the electrochemical double layer, it is imperative to reveal the structure under control of the electrode potential. In electrolyte solutions free of strongly adsorbing anions, *e.g.* KOH, the electrochemical interface offers the opportunity of controlling the surface coverage by hydrogen, H_2O and OH species simply by controlling the applied electrode potential. To probe, *in situ*, the structure of the water layer at the metal/electrolyte interface is a technically challenging experiment as the scattering signal from the ordered oxygen atoms is relatively small compared to the diffuse scattering from the bulk of the electrolyte solution.

In this paper, we describe preliminary results obtained using a UHV transfer system to provide unprecedented control over the electrode surface structure during transfer from UHV to the electrochemical environment. To illustrate the experimental possibilities, in section 3 we describe a single experiment in which a Pt(111) surface was prepared in UHV, dosed with oxygen to form a p(2×2) oxygen adlayer and then studied with SXS, initially under UHV conditions, then in a N_2 atmosphere before a droplet of water was contacted with the surface. Measurements of the surface structure were then performed on the surface modified by the bulk water overlayer before the water was exchanged with 0.1 M KOH electrolyte and the Pt(111) electrode put under potential control to allow potential-dependent studies of the interface structure. In section 4, results for the Pt_3Ni alloy system obtained by UHV preparation followed by external transfer to the X-ray electrochemical cell are described. In this case, it is found that the (111) surface of Pt_3Ni is extremely stable both during the transfer and under potential control in the electrochemical environment. Variations in the compositional profile at the surface are probed by resonant surface X-ray diffraction techniques. Results for the other low-index Pt_3Ni surfaces, (100) and (110), are less conclusive and indicate that the UHV-transfer system would offer a more controlled way in which to probe the surface structure and electrochemistry of these model electrocatalysts.

2. Experimental methods

Surface X-ray scattering (SXS) is now a well-established technique for probing the atomic structure at the electrochemical interface and, since the first in situ synchrotron X-ray study in 1988,[1] several groups have used the technique to probe a variety of electrochemical systems.[1,2,9,15] As in the UHV environment, the extraction of structural information, such as surface coverage, surface roughness and layer spacing (both adsorbate–substrate distances and the expansion/contraction of the substrate surface atoms themselves), at the electrified solid–liquid interface relies on measurement of the crystal truncation rods (CTR's).[16,17] By combining measurements of several symmetry-independent CTR's it is possible to build up a 3-dimensional picture of the atomic structure at the electrode surface. If the surface or adlayer adopts a different symmetry from that of the underlying bulk crystal lattice then the scattering from the surface becomes separate from that of the bulk in reciprocal space and it is possible to measure the surface scattering independently.

In this paper, we describe SXS studies of electrode surfaces that are prepared in the ultra-high vacuum (UHV) environment. This methodology has the advantage that the surface quality can be checked during preparation by standard surface science techniques such as low energy electron diffraction (LEED) and auger electron spectroscopy (AES). UHV preparation is also vital for bimetallic surfaces for which precise surface compositions are dependent on annealing temperatures. The Pt(111) electrode was prepared by cycles of sputtering and annealing in the UHV system in the surface characterization laboratory (SCL) at the ESRF.[18] After a sharp (1 × 1) LEED pattern was obtained, the sample was exposed to molecular oxygen at room temperature until a sharp p(2 × 2) LEED pattern, characteristic of the stable Pt(111)–O structure with a saturation coverage of 0.25 monolayers of atomic oxygen, was observed.[19] The sample was then transferred to a portable chamber (called TRECXI) connected with a CF38 UHV valve to a docking port of the main UHV chamber. The TRECXI chamber[20] (which has portable UHV ion pumps to maintain the UHV conditions and a large cylindrical Be window to allow the incident and scattered X-ray beams to pass through the chamber) was then detached from the main chamber and mounted on the ID32 diffractometer at the ESRF. Full details of the TRECXI chamber describing the transfer process and operation can be found in ref. 20.

SXS measurements were performed on the 6-circle surface diffractometer at the ID32 beamline at the ESRF. The horizontal sample surface was aligned with the X-ray beam for a fixed incidence angle of 0.2° and the orientation of the crystal lattice was then determined. The close-packed (111) surface has a hexagonal unit cell that is defined such that the surface normal is along the $(0, 0, L)_{hex}$ direction and the $(H, 0, 0)_{hex}$ and $(0, K, 0)_{hex}$ vectors lie in the plane of the surface and subtend 60°. The units for H, K and L are $a^* = b^* = 4\pi/\sqrt{3}a_{NN}$ and $c^* = 2\pi/\sqrt{6}a_{NN}$ where a_{NN} is the nearest-neighbor distance in the crystal ($a_{NN} = 2.78$ Å). Due to the ABC stacking along the surface normal direction, the unit cell contains three monolayers and the Bragg reflections are spaced apart by multiples of three in L. The incident X-ray beam (energy = 22.2 keV) was collimated to a spot size of 40 μm (vertically) by 0.3 mm (horizontally) and the scattered X-ray beam was detected after reflection from a graphite analyzer crystal. CTR data were obtained by performing rocking scans around the surface normal at successive L values to obtain background-subtracted integrated intensities at each L position.

Following the experiments performed under UHV conditions (section 3.1), the TRECXI chamber was filled with inert gas and selected X-ray measurements were performed to verify the stability of the Pt(111)–O surface. The electrochemical cell was then mounted above a UHV valve on top of the Be cylinder and can be lowered toward the sample once the chamber is filled with inert gas at ambient temperature and the UHV valve is open. The electrochemical cell in this setup consists of a glass tube (about 25 cm long) that has an open end for forming a droplet of electrolyte on one side and is connected to a glass cross on the other side. The glass cross contains connectors for the electrochemical inlet and outlet tubings, the counter-electrode (Pt wire) and a reference electrode (a commercial Ag/AgCl microelectrode). The cell is part of a complete electrolyte handling system including a computer-controlled pumping system and potentiostat (full details are given in ref. 20). Initially pure water was used in place of electrolyte (without potential control) and a droplet was contacted with the Pt(111)–O surface (measurements described in section 3.2). During the experiment the droplet is monitored by an endoscope and can be regularly adjusted in volume to replace evaporated liquid. Following the measurements in pure water, 0.1 M KOH was used as the electrolyte and potential contact to the sample was made at −1 V (relative to Ag/AgCl). The measurements performed in the electrolyte solution are described in section 3.3.

The Pt_3Ni samples were mechanically cut and polished into 6 mm diameter disks (2 mm thick), with the surface orientated to within 0.5° of the low index crystal planes and then prepared and characterized in a UHV system with a base pressure of 1×10^{-10} Torr. In addition, the surface was examined in a separate UHV system using synchrotron-based high resolution ultraviolet photoemission spectroscopy (UPS). After UHV characterization the sample was transferred into an electrochemical cell where cyclic voltammetry (CV) measurements were performed. Low energy ion scattering (LEIS) showed that, after a final anneal, the surface atomic layer was composed of pure Pt, the so called 'Pt-skin' surface. This is in agreement with previous LEED studies of a $Pt_{0.78}Ni_{22}(111)$ sample where a Pt-rich surface atomic layer accompanied by a damped oscillation in the Pt occupation over the next two atomic layers was observed.[21,22] Full details of the sample preparation and UHV characterization can be found in ref. 23.

Following the UHV measurements the sample was transferred to the X-ray electrochemical cell with a drop of pure water protecting the surface.[24] In all of the Pt_3Ni experiments presented, the electrolyte was 0.1 M $HClO_4$, the reference electrode was a saturated calomel electrode (SCE) and after immersion into the electrolyte the electrode potential was contacted at 0.05 V (*versus* the reversible hydrogen electrode (RHE)). The SXS measurements were carried out on beamline 7–2 at the Stanford Synchrotron Radiation Laboratory (SSRL) and beamline BM12-BESSRC at the Advanced Photon Source (APS), Argonne National Laboratory. The $Pt_3Ni(111)$ surface was indexed using the standard hexagonal unit cell notation described above, whereas the regular fcc unit cell was used for the $Pt_3Ni(100)$ surface with the L direction along [1 0 0]. The Pt_3Ni lattice parameter was measured to be 3.841 Å during the X-ray diffraction experiment, which is close to a 75 : 25 combination of the Pt and Ni fcc lattice constants. Measurement of the bulk Bragg reflections revealed that the crystal had a well-defined mosaic spread with a rocking scan showing several distinct peaks over an angular range of ∼1 degree. This mosaic spread was found for all three low-index single crystals that were cut from the as-grown Pt_3Ni crystal and reflects the difficulty in preparing single crystals of this material.

3. Results and discussion

3.1 Pt(111)–p(2 × 2)-O

The first experiment after preparation and transfer of the Pt(111)–O surface in the baby chamber to the beamline was to characterize the surface under UHV conditions. Due to the increased size of the p(2 × 2)-O unit cell, scattering is expected at the half integer positions in the surface plane (H, K) of the reciprocal lattice.[19] Fig. 1 shows a radial scan (along the $\langle H, K \rangle$ direction) and a rocking scan measured at ($\frac{1}{2}$, $\frac{1}{2}$, 0.1). The count time in this measurement was 10 seconds per point, which illustrates the weakness of the scattering signal. This is expected if the structure consists of only 0.25 monolayers of O (which is a weak scatterer due to the low atomic number). Fits of a Lorentzian lineshape to the data (shown by the solid lines

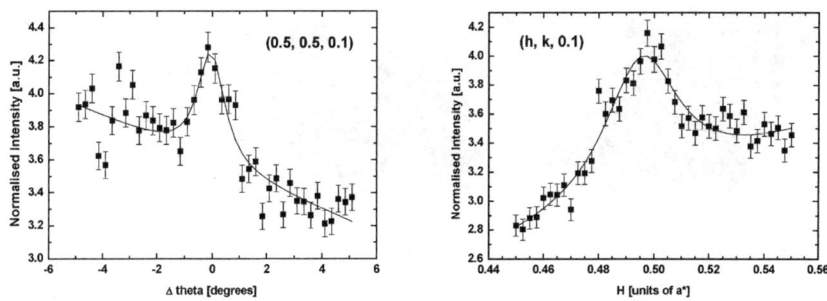

Fig. 1 A rocking scan (left) and a radial scan (right) through the (0.5, 0.5, 0.1) reciprocal lattice position from the Pt(111)–p(2 × 2)-O surface in UHV.

in Fig. 1) give a coherent domain size in the range 90–130 Å. Due to the weakness of the scattering it was impossible to collect a full set of structure factors that could then be used for structural refinement. In fact the intensity distribution implied that there was no significant Pt contribution to the scattering from the p(2 × 2) unit cell in contrast to previous SXS measurements of the structure in which relaxation of multiple Pt layers was proposed.[25]

In order to fully characterize the Pt surface, we measured 7 symmetry-independent CTR's and representative results (6 of the measured CTR's) are shown in Fig. 2. The

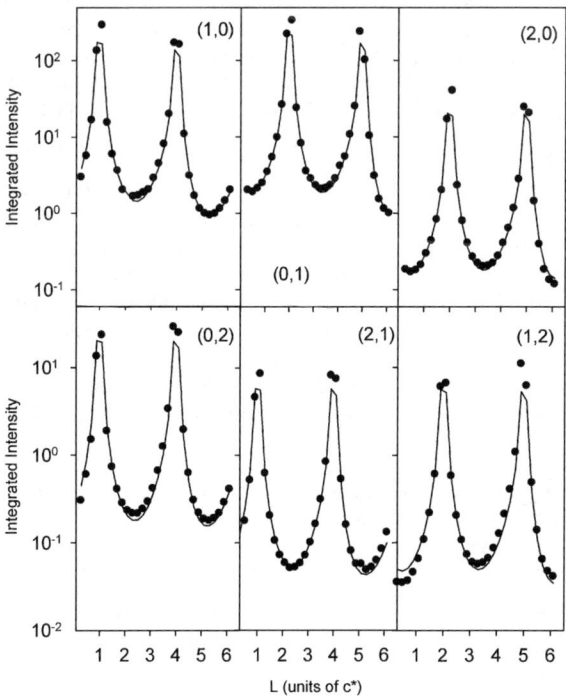

Fig. 2 Crystal truncation rod (CTR) data of the Pt(111)-p(2 × 2)-O surface measured in UHV. The circles correspond to background-subtracted integrated intensities that are corrected for the instrumental resolution. The (H, K) values for each CTR are indicated. The solid line is a fit to the data using the structural model shown in Fig. 3 according to the parameters listed in Table 1.

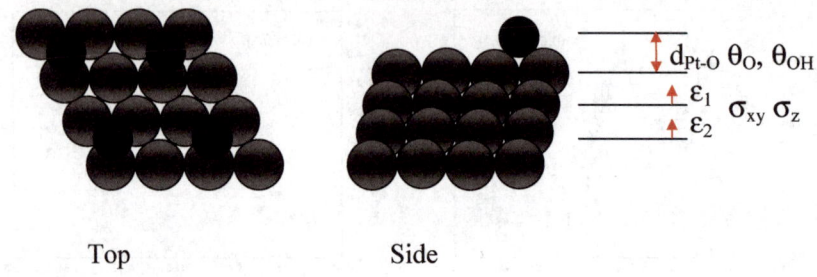

Fig. 3 A schematic of the structural model used to fit the CTR data in Fig. 2, 3, 6 and 7. The structural parameters to the fit are indicated in the Figure and the values obtained are listed in Table 1.

data has been corrected for the instrumental resolution according to the formalization developed by Vlieg.[26] Errors due to counting statistics are relatively small, but in fitting the data a 10% systematic error was associated with each data point. This error is typical of that obtained by comparison of data measured at symmetry-equivalent CTR positions.[27] The lack of any significant features in the CTR data is consistent with the weakness of the p(2 × 2) scattering (Fig. 1) which implies that there is no substantial relaxation in the surface Pt layers. The data was fitted using the simple structural model shown in Fig. 3. Inclusion of the p(2 × 2)-O unit cell made a slight improvement to the calculated fit (χ^2 improved from 4.0 to 3.5) which was obtained by least squares refinement of the structural parameters shown in Fig. 3 with the O atom adsorbed in the 3-fold hollow Pt surface site (as previously determined in UHV experiments).[19] The best fit parameters are listed in Table 1. Including buckling of the surface and sub-surface Pt atoms in the p(2 × 2) unit cell (as found in the LEED-IV study of the Pt(111)–O surface) did not improve the fit to the data. The data was therefore modelled with a Pt-(1 × 1) unit cell allowing for relaxation, according to the parameters ε_1 and ε_2, and an asymmetric static Debye–Waller factor of the form $\exp\left(-\frac{Q^2_{\text{in-plane}}\sigma^2_{xy}}{2}\right)\exp\left(-\frac{Q^2_z\sigma^2_z}{2}\right)$ to account for small distortions of the Pt lattice. This same model was then used to fit all of the CTR measurements made for the different states of the Pt surface, as described in the subsequent sections of this paper. The structural parameters imply that there is outward relaxation of the surface Pt atomic layer by 0.05 Å (compared with the previous LEED-IV study[19] that found an outward relaxation of 0.01 Å with an additional bucking of the adlayer so that 3 of the 4 Pt atoms in the unit cell were additionally expanded by 0.04 Å). The Pt–O vertical spacing was found to be 1.5 ± 0.2 Å compared to 1.24 Å found in the LEED-IV study and 1.46 Å in density functional theory (DFT) calculations.[28] The SXS results are thus generally in good agreement with

Table 1 Structural parameters for the fits to the Pt(111) CTR data. f indicates a parameter that was fixed. Uncertainties are indicated in brackets. ε_1 is also quoted as a percentage of the Pt(111) atomic layer spacing.

System	θ_O, θ_{OH}	d_{Pt-O}/Å	ε_1/Å	ε_2/Å	σ_{xy}/Å	σ_z/Å
Pt(111)–O	0.25f	1.5 (0.2)	+0.05 (0.005) (2.2%)	0.005 (0.005)	0.06 (0.01)	0.0
Pt(111)–H$_2$O	0.7 (0.3)	1.9 (0.4)	+0.03 (1.3%)	0.01	0.10	0.0
Pt(111)–KOH ($E = -1.0$ V)	0	—	+0.05 (2.2%)	0.005	0.06	0.0
Pt(111)–KOH ($E = -0.1$ V)	0.55 (0.2)	2.1 (0.4)	+0.04 (1.8%)	0.0	0.06	0.0

the previous LEED-IV study and theoretical results. As described above, buckling of the Pt surface layers in the p(2 × 2) unit cell did not improve the fit to the data and so was not included in the structural model.

Following the measurements under UHV conditions, the chamber was brought up to an N_2 atmosphere and the measurements shown in Fig. 1 and some of the CTR measurements shown in Fig. 2 were repeated. These results were identical indicating that the Pt(111)–p(2 × 2)-O surface was stable in the inert atmosphere. A droplet of water was then contacted with the surface according to the experimental procedures described in section 2.

3.2 Pt(111)–H_2O interface

One of the key aims of this experiment was to search for the existence of ordered water structures at the Pt(111)–bulk water interface. Adsorption of water onto the Pt(111)–p(2 × 2)-O surface was chosen as this helps to pin the water molecules into an ordered phase when such surfaces are prepared under UHV conditions at low temperatures.[14,29] Following contact with the water droplet, a search for scattering due to ordered structures was made in the surface plane (H,K) of reciprocal space by scans along the high symmetry directions ($\langle H, 0 \rangle$, $\langle 0, K \rangle$ and $\langle H, K \rangle$ at fixed L values, 0.1, 0.8). Additionally, rocking scans were performed at key reciprocal lattice positions, *i.e.* half-integer and 1/3 integer, where scattering from the p(2 × 2)-O structure,[19] the ($\sqrt{3} \times \sqrt{3}$)-R30° and (3 × 3) structures of water that are formed in the UHV experiments[14,29] at monolayer coverage of water would be expected. In all cases no scattering above the background level was observed. It should be noted that the background scattering due to the presence of the liquid droplet was significant and it cannot be ruled out that the background signal is too high to observe the weak scattering that would be expected from such an ordered structure (particularly given the weakness of the scattering from the p(2 × 2)-O layer shown in Fig. 1). At this stage it is impossible to determine if an ordered water adlayer is present at the interface. A future experiment would be to prepare an ordered ($\sqrt{3} \times \sqrt{3}$)-R30° water monolayer on Pt(111) under UHV conditions and contact this surface with the water droplet. This would require a development of the TRECXI chamber to facilitate cold transfer of the sample.

In order to characterize changes to the surface atomic structure, measurements of the CTR data (as shown in Fig. 2) were repeated. The CTR data measured with the surface exposed to the bulk water environment was then normalized to the data shown in Fig. 2 to obtain a ratio data set and the results are shown in Fig. 4. The ratio data set shows a clear systematic change in the intensities of the CTR's which corresponds to a change in the interface structure due to the presence of the water overlayer. The data was modeled using the same structural model shown in Fig. 3. Adsorption of the X-ray beam by the electrolyte essentially leads to a constant attenuation (compared to the UHV data) as the pathlength of the X-ray beam does not change significantly as a function of L. This was accounted for by allowing the scale factor multiplying the CTR data to vary in fitting the ratio data set. The best fit of the ratio data set is shown by the solid lines in Fig. 4 according to the structural parameters listed in Table 1. A constant error of ±0.1 was assumed for each ratio data point as indicated by the error bars in Fig. 4.

Given that no in-plane ordering at the Pt(111)–H_2O interface was detected, it is impossible to build detailed structural models in order to reproduce the CTR ratio data. However, the systematic changes observed in Fig. 4 are consistent with a structural rearrangement at the interface and so it is intuitive to explore the origin of the changes. The calculated curves (shown by the solid lines in Fig. 4) used a very simple (1 × 1) unit cell allowing only for changes in the relaxation parameters. A fit to the data without the presence of a θ_O adlayer gave a χ^2 of 1.9. The principal structural changes are a reduction in the Pt surface expansion accompanied by an increase in the in-plane component of the Debye–Waller factor consistent with some in-plane

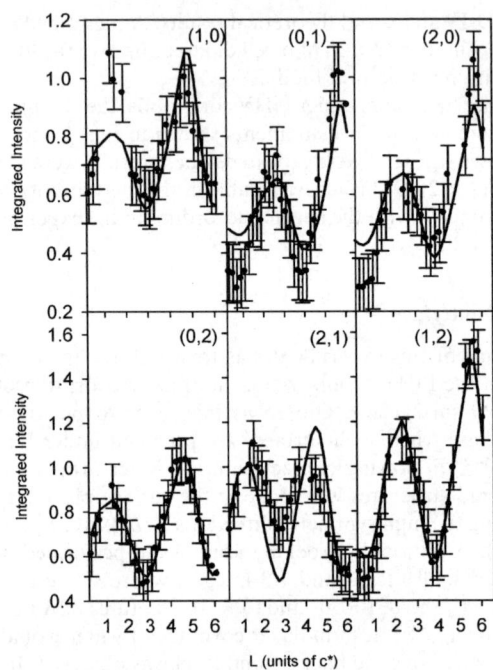

Fig. 4 Ratios of CTR data (Pt(111)–H$_2$O divided by Pt(111)–p(2 × 2)-O, i.e. data in Fig. 1). The circles correspond to the measured ratios (a constant error of ±0.1 is estimated). The (H, K) values for each CTR are indicated. The solid line is a fit to the data using the structural model shown in Fig. 3 according to the parameters listed in Table 1 as described in the text.

distortion of the Pt lattice. This result would be expected for an increased surface coverage by oxygenated species. Including an oxygen layer in the model gave the fit to the data shown in Fig. 4 ($\chi^2 = 1.3$). Different adsorption sites in the (1 × 1) unit cell were tried but there was a clear preference for the 3-fold hollow site. Models based on the ($\sqrt{3} \times \sqrt{3}$)R30° and (3 × 3) structures, previously observed for the mixed OH–H$_2$O layer in UHV,[29] did not improve the fit to the data obtained without an adsorbed O layer. In particular, it was found that a good fit could not be obtained with O adsorbed at a Pt on-top site. In UHV studies the sharpest LEED patterns are observed when water is adsorbed onto a disordered oxygen adlayer with a reduced coverage of $\theta_O \sim 0.15$[14,30] compared to the p(2 × 2) phase ($\theta_O = 0.25$). It is possible, therefore, that in our experiment the oxygen was stabilized in the 3-fold hollow site although it is clear that a reaction with bulk H$_2$O has occurred. At present it is thus difficult to reconcile the results with the UHV studies of water adsorption at low temperature. As noted in the following section, the inclusion of the oxygen adlayer to fit the data in Fig. 4 is statistically significant, as fits to the CTR data in 0.1 M KOH at −1 V, where atomic hydrogen is adsorbed onto the surface,[31] are consistent with the removal of the oxygen adlayer (see below). At this stage, however, we cannot rule out that the CTR data could be reproduced with more complex surface models.

3.3 Pt(111)–0.1 M KOH

Following the measurements performed with the water droplet, the water was replaced by 0.1 M KOH electrolyte and potential contact was established at −1 V (*versus* Ag/AgCl). A cyclic voltammogram and X-ray voltammogram (at (1, 0, 3.6), a position that is sensitive to the Pt surface expansion) were then recorded

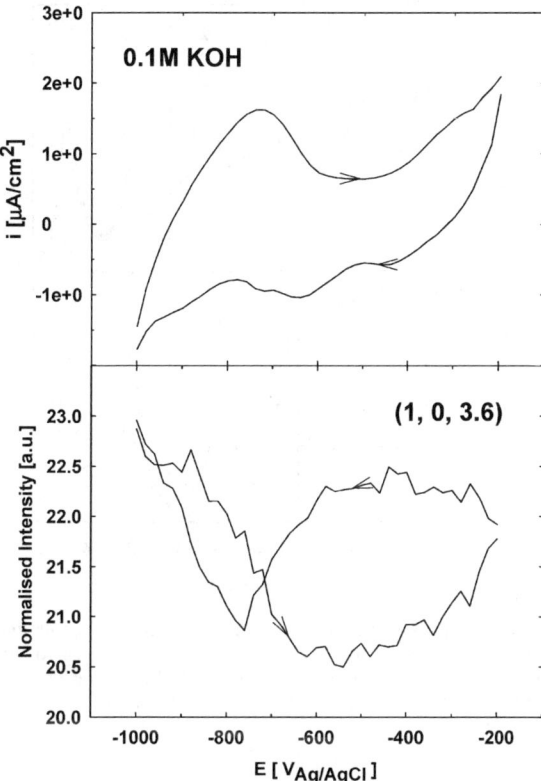

Fig. 5 (Top) The cyclic voltammogram measured in the droplet cell for the Pt(111) electrode in 0.1 M KOH. (Bottom) The simultaneously measured X-ray voltammetry (XRV) measured at (1, 0, 3.6), a reciprocal lattice position sensitive to the Pt surface expansion (the sweep rate was 2 mV s^{-1}).

and the results are shown in Fig. 5 (sweep rate 2 mV s^{-1}). Starting at -1.0 V and scanning anodically it is clear that the desorption of hydrogen from the surface leads to an inward relaxation reaching a minimum at ~-0.6 V. Adsorption of OH anions then causes an outward relaxation which is stable during the cathodic sweep until the OH desorbs. Adsorption of hydrogen then again results in the outward expansion of the surface Pt layer and the XRV is seen to be fully reversible.

Characterization of the surface structure was again obtained by measuring and modeling of the CTR data. Fig. 6 shows the CTR data measured at $E = -1.0$ V. At this potential there is adsorption of ~ 0.66 monolayers of hydrogen on the surface[31,32] and so the data was modeled without any adsorbed oxygenated species, i.e. $\theta_{OH} = 0$. The results indicate that there is an outward relaxation of the topmost Pt layer by $\sim 2.2\%$ of the bulk layer spacing which is in excellent agreement with previous measurements on this system.[33] Interestingly, this surface expansion in the presence of adsorbed hydrogen is also in excellent agreement with the local-density-functional (LDF) calculations by Feibelman for hydrogen adsorbed onto Pt(111).[34] As noted above, fits to a CTR ratio data set (not shown) for I_{KOH}/I_{H_2O} indicated that the oxygen adlayer is not present in 0.1 M KOH at -1 V, and so its presence at the Pt(111)–H$_2$O interface is statistically significant.

After potential cycles over the range shown in Fig. 5, the potential was held at -0.1 V and the SXS measurements were repeated. As for the Pt(111)–H$_2$O interface, searches in the (H, K) surface plane for scattering due to an ordered OH adlayer were unsuccessful. The data in Fig. 7 is the ratio CTR data set obtained by dividing

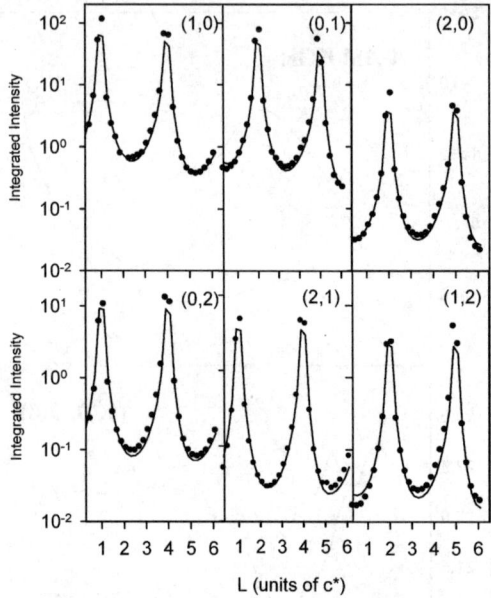

Fig. 6 CTR data of Pt(111) in 0.1 M KOH measured at an electrode potential of −1 V (*versus* Ag/AgCl). The circles correspond to background-subtracted integrated intensities that are corrected for the instrumental resolution. The (*H*, *K*) values for each CTR are indicated. The solid line is a fit to the data using the structural model shown in Fig. 3 according to the parameters listed in Table 1.

the CTR data measured at −0.1 V by the data measured at −1 V (Fig. 6). Systematic changes in the measured intensities are again observed, although it should be noted that the changes are significantly smaller than those observed in Fig. 4. The same (1 × 1) structural model was used to fit the data in Fig. 7 giving the structural parameters listed in Table 1. As for Pt(111)–H$_2$O, the best fit was obtained with OH species adsorbed into the Pt 3-fold hollow sites although the Pt–O spacing seems rather large in this case. The fact that there are features in the ratio data set that are not reproduced in the calculation implies that the model is too simple. Further analysis and measurement is required to get more insight into the Pt(111)–OH$_{ad}$ interface.

4. Future directions: alloy surfaces

As with all PtNi alloys, Pt$_3$Ni forms an fcc solid solution with random occupation of the lattice sites by Pt and Ni. An important initial step is the preparation and characterization of Pt$_3$Ni(*hkl*) alloy surfaces in ultra-high vacuum (UHV) (Fig. 8). The surface sensitive techniques employed included LEED, AES, low energy ion scattering (LEIS) and synchrotron based high-resolution ultraviolet photoemission spectroscopy (UPS). Each of these methods has certain unique advantages, and they yield complementary information. For example, the atomic structure of the surface is obtained from LEED analysis, which shows that whereas the Pt$_3$Ni(111) surface exhibits a (1 × 1) pattern (Fig. 8), the atomically less dense Pt$_3$Ni(100) surface shows a clear (1 × 5) reconstruction pattern (the so-called "hex" phase) in both the [011] and [0-11] directions (Fig. 8). Analysis of the Pt$_3$Ni(110) LEED data (Fig. 8) indicates that this surface may exhibit a mixture of (1 × 1) and (1 × 2) periodicities, the latter being known as the (1 × 2) missing row structure.[35] The exact surface composition of the outermost atomic layer is obtained by utilizing

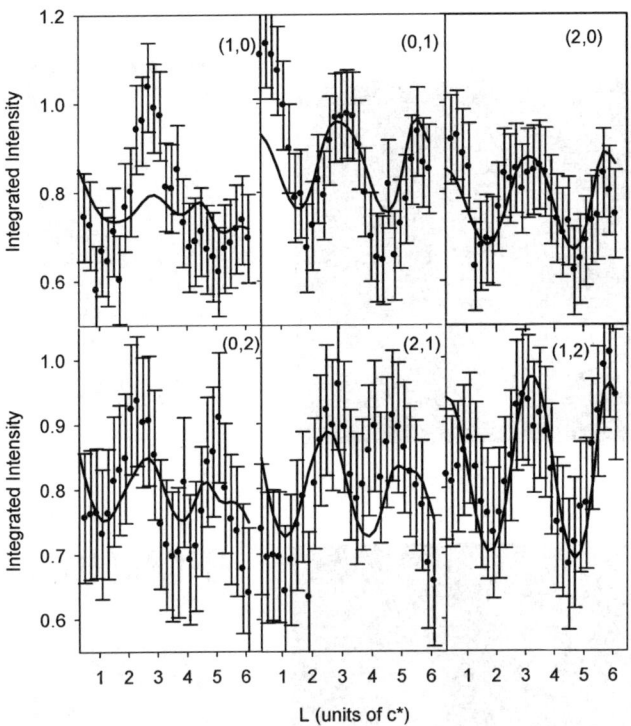

Fig. 7 Ratios of CTR data for Pt(111) in 0.1 M KOH (data measured at $E = -0.1$ V divided by data measured at $E = -1.0$ V, *i.e.* data in Fig. 6). The circles correspond to the measured ratios (a constant error of ±0.1 is estimated). The (H, K) values for each CTR are indicated. The solid line is a fit to the data using the structural model shown in Fig. 3 according to the parameters listed in Table 1 as described in the text.

LEIS, as previously shown for Pt$_3$Ni polycrystalline alloys.[36,37] The LEIS spectra showed that, after a final anneal, the surface atomic layer of all three Pt$_3$Ni(hkl) crystals is pure Pt, *i.e.* they all form the so-called Pt-skin structure.[38,39] Earlier reports from similar studies suggested that this surface enrichment of Pt in the first layer is counterbalanced by its depletion in the next two to three atomic layers, resulting in a concentration profile that oscillates around the bulk value.[40,41] These types of near-surface compositional changes account for the unique electronic properties of alloys.

While the phenomenon of surface segregation at bimetallic surfaces has been well studied under UHV conditions, the stability of such surfaces in the electrochemical environment is relatively unknown, and yet this is crucially important to their potential application as electrocatalysts. After immersion of the crystal at 0.05 V into the electrolyte the Pt$_3$Ni(111) surface structure was determined by measurement and analysis of the crystal truncation rods (CTR's).[17] Fig. 9a shows the (0, 0, L) and (0, 1, L) CTR data measured at 0.05 V. The data points, indicated by the error bars, represent background-subtracted integrated intensities obtained from rocking scans over the full mosaic range which were taken at sequential L values along the CTR. The CTR data can be modeled using kinematical scattering theory. The model assumes the crystal to be a perfectly random alloy with an fcc lattice, each atom having an average atomic form factor of $0.75f_{Pt} + 0.25f_{Ni}$.[42] In order to provide uniqueness in the modeling of the CTR data, energy dependent measurements at two reciprocal lattice positions were performed. The data points in Fig. 9b correspond to background-subtracted integrated intensities obtained from rocking scans

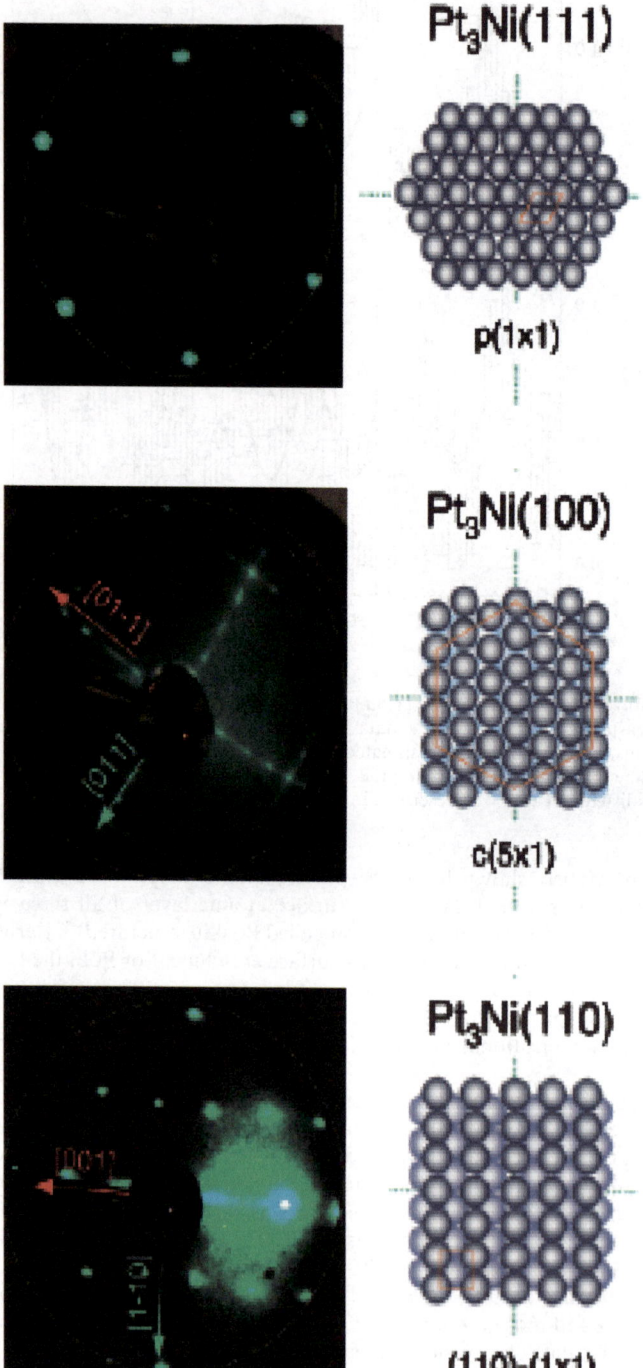

Fig. 8 Surface characterization of the Pt₃Ni single crystals in UHV. LEED patterns and the corresponding ball and stick models for the three low-index surfaces.

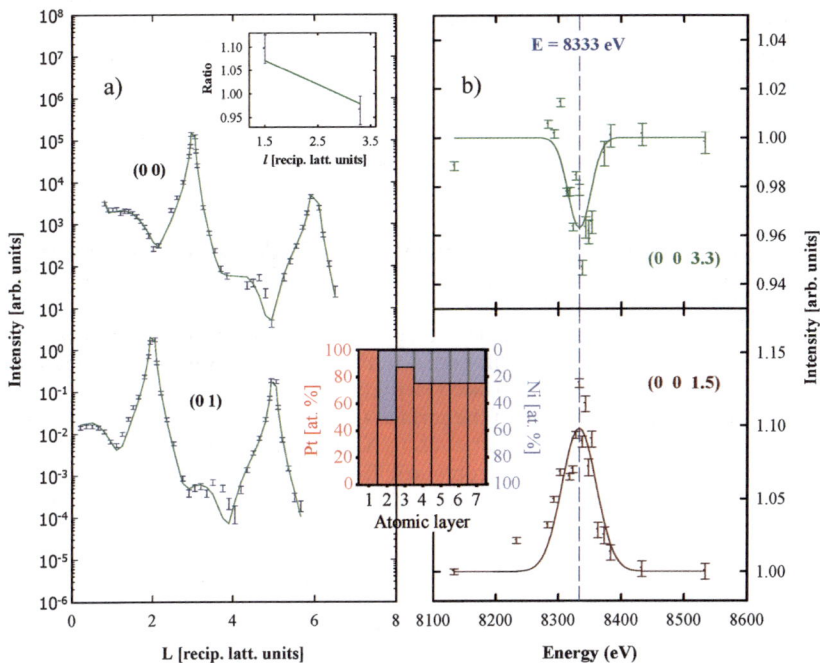

Fig. 9 CTR data of the Pt$_3$Ni(111) crystal surface in 0.1 M HClO$_4$ with an applied potential of 0.05 V (*vs.* RHE). (a) The error bars are the data points and the solid line is the best fit to the data. (Inset) Energy ratio data as described in the text. The data points are denoted by circles and the best fit to the data is a solid line. (b) Measurements at the (0, 0, 1.5) and (0, 0, 3.3) reciprocal lattice positions as a function of the incident X-ray energy. The vertical dashed line is at 8333 eV, the Ni K adsorption edge. The segregation profile obtained from the SXS measurements is shown in the center of the figure.

at (0, 0, 3.3) and (0, 0, 1.5) as a function of the incident X-ray energy. The energy range of the measurement was 200 eV either side of the Ni K adsorption edge at 8333 eV. The resultant spectra were approximately corrected for detector efficiency and adsorption by normalizing to a linear change in the background signal. Gaussian line shapes with a fixed centre at 8333 eV were fitted to both of the normalized spectra and are shown as the solid lines in Fig. 9b. From this data, an intensity ratio at each CTR position, I_{8323eV}/I_{8133eV}, was calculated and the results are plotted in the inset to Fig. 9a. Over the energy range 8133–8323 eV, *i.e.* below the Ni K adsorption edge, the change in the anomalous dispersion corrections to the Ni form factor is due to only to the real part, f', and can be calculated exactly.[43] By simultaneously fitting this ratio data, which constrains the fit to the full CTR data, sensitivity to the elemental concentration profile at the surface, *i.e.* separation from surface roughness effects in the modeling of the CTR data, is obtained.

In order to fit the model to the data, a least squares minimization was performed in which the variable structural parameters of the three outermost atomic layers were; the fractional Pt occupation assuming fully occupied atomic layers (θ_{Pt}), relaxation of the surface layer (ε) and a static enhanced Debye–Waller factor (σ). A β factor according to the model developed by Robinson[16,17] was used to model surface roughness effects. The best fit to both CTR data and the ratio data is shown by the solid line in Fig. 9a. A list of the structural fit parameters is given in Table 2. The surface atomic layer is thus determined to be 100% Pt, the second atomic layer to be 48% Pt, the third to be 87% and beyond that the bulk value of 75%. It is clear from these results that the Pt-rich segregated surface (as determined by LEIS in the

Table 2 Structural parameters for the calculated fits to the Pt$_3$Ni(111) data in Fig. 9

$\theta_{Pt1} = 1.00 \pm 0.11$	$\varepsilon_1 = -0.003 \pm 0.008$ Å	$\sigma_1 = 0.288 \pm 0.035$ Å
$\theta_{Pt2} = 0.48 \pm 0.11$	$\varepsilon_2 = 0.012 \pm 0.005$ Å	$\sigma_2 = 0$
$\theta_{Pt3} = 0.87 \pm 0.11$	$\varepsilon_3 = 0$	$\sigma_3 = 0$
Bulk D-W = 0.13 ± 0.04Å		$\beta = 0.20 \pm 0.06$

UHV chamber) is stable both during transfer from UHV and, subsequently, in the electrochemical environment. Furthermore, the CTR measurements also provide the composition of the sub-surface atomic layers and indicate that the second atomic layer is Ni-rich compared to the bulk alloy composition. This is key to understanding the modified electronic properties of the surface Pt layer [compared to bulk Pt(111)] which determines the surface reactivity.[23]

Following the determination of the surface atomic structure, the potential dependence of the surface was investigated using X-ray voltammetry (XRV) and cyclic voltammetry (CV). Results were obtained for both the Pt$_3$Ni(111) sample and a Pt(111) electrode for comparison, and representative results are shown in Fig. 10. The potential response of the two surfaces (Fig. 10b) can be separated

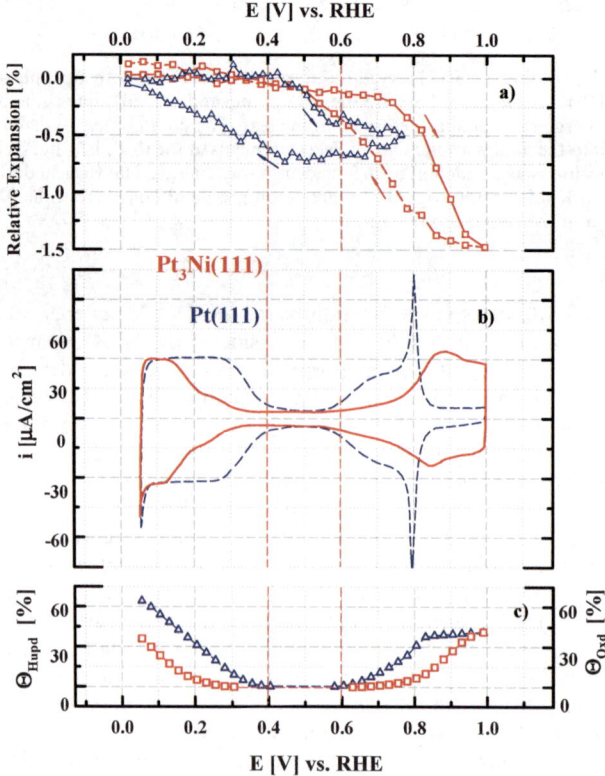

Fig. 10 Potential-dependent measurements of the Pt$_3$Ni(111) and Pt(111) crystal surfaces. (a) XRV measurements for Pt$_3$Ni(111) at the (0, 0, 2.7) (red line) and Pt(111) at (1, 0, 3.6) (blue line) with 2 mV s^{-1} sweep rate. (b) Cyclic voltammetry recorded in 0.1 M HClO$_4$ with 50 mV s^{-1} sweep rate. (c) Surface coverage by underpotentially deposited hydrogen (H$_{upd}$) and hydroxyl species (OH$_{ad}$) calculated from the cyclic voltammograms of Pt$_3$Ni(111) (red curve) and Pt(111) (blue curve).

into three regions; the first corresponds to the underpotential deposition of atomic hydrogen on the surface (H_{upd}), the intermediate region to the charging of the double layer and the positive potential region to the reversible adsorption of OH (OH_{ad}). Integration of the charge in the CV gives the surface coverages by H_{upd} and OH_{ad} and these are shown in Fig. 10b. There is a negative potential shift of ~0.15 V in H_{upd} and a positive potential shift of ~0.1 V in OH_{ad} on the Pt-skin surface relative to the Pt(111) surface. Fig. 10a shows the XRV's for the $Pt_3Ni(111)$ surface measured at (0, 0, 2.7) and for the Pt(111) surface at (1, 0, 3.6). Both of these reciprocal lattice positions, at l values just below a bulk Bragg reflection, are sensitive to surface relaxation as confirmed by the corresponding measurements at l values above the Bragg reflections, (0, 0, 3.3) and (1, 0, 4.4), which showed 'mirror-like' behavior. For this reason the intensities are converted to surface expansion (by model calculations after fits to the full CTR data sets in each case) normalized to the values obtained at 0.05 V. It is important to note that at 0.05 V the Pt(111) surface is expanded by ~2% of the lattice spacing whereas the $Pt_3Ni(111)$ surface is essentially unrelaxed. Fig. 10a shows the change in expansion relative to the value measured at 0.0 V, *i.e.* the pure Pt(111) surface is expanded by ~2% at 0.0 V and this is reduced to ~1.5% expansion at 0.8 V, whereas the $Pt_3Ni(111)$ surface is unrelaxed at 0.0 V and undergoes a ~1.5% contraction at 1.0 V. Surface expansion is dependent on the substrate-adsorbate bondstrength, *i.e.* the chemisorption properties that depend on the surface electronic structure (the d-band center).[44] On both surfaces the adsorption of OH_{ad} causes surface contraction (the shift in the potential for OH_{ad} is clearly seen in Fig. 10) and this is the precursor to oxide formation *via* place exchange with the Pt surface atoms.[45] The hysteresis observed for Pt(111) is large and is presumably due to the fact that the relaxation not only depends on the coverage but also the ordering of the adlayer species. The structural changes are thus dependent on the kinetics of ordering on the surface, a process that can have a timescale of several minutes. The reduced hysteresis on the $Pt_3Ni(111)$ surface is likely due to the weaker metal-adsorbate interaction. The stability of the Pt-skin surface is a consequence of the electronic structure (d-band center) that decreases the amount of OH_{ad} (Fig. 10b) so that there is less tendency for place exchange and irreversible roughening. This mechanism is also responsible for the increased activity for the ORR.

The $Pt_3Ni(100)$ electrode surface was also transferred into the X-ray electrochemical cell using the same methodology as for $Pt_3Ni(111)$. As shown in Fig. 8, the LEED pattern from this surface was consistent with the presence of a hexagonal surface reconstruction, analogous to the reconstruction of Au(100) found in both UHV[35] and electrolyte.[2,15] A search of the (H, K) surface plane of reciprocal space, however, was inconclusive as to the presence of such a hexagonal reconstruction. This may be due to a loss of ordering during the transfer to, or inherent to, the electrochemical environment. The (0, 0, L) CTR was then measured at a potential of 0.05 V, *i.e.* at the same potential as the data for $Pt_3Ni(111)$ shown in Fig. 9. The (0, 0, L) CTR is only sensitive to the electron density profile perpendicular to the sample surface and doesn't require registry of the surface atoms with the underlying bulk crystal. Attempts were made to fit the data with different structural models; (i) bulk termination of the Pt_3Ni lattice and (ii) variation in the Pt surface concentration (as used for the $Pt_3Ni(111)$ surface) but it was not possible to get a good fit to the data. The data in Fig. 11 shows that the secondary maxima in the CTR data occur at values of $L \sim \sqrt{3}$ and $L \sim 2\sqrt{3}$, and this implies that there is a (111) stacking at the surface that goes beyond a single atomic layer. The solid line in Fig. 11 includes two surface atomic Pt layers that a have a density corresponding to that of a hexagonal plane, *i.e.* 1.2 times greater than the cubic (100) plane, and are separated by the (111) plane spacing. These results imply that considerable Pt enrichment has occurred at the $Pt_3Ni(100)$ surface leading to the formation of hexagonal Pt overlayers. This in turn may explain the reduced activity for the oxygen reduction reaction (ORR) on $Pt_3Ni(100)$ compared to $Pt_3Ni(111)$.[23]

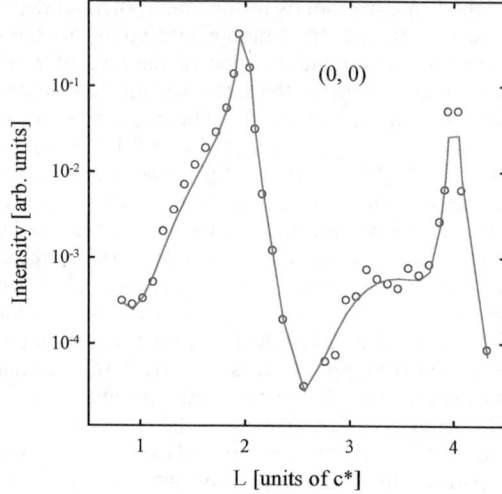

Fig. 11 Specular CTR data (0, 0, L) of the Pt$_3$Ni(100) crystal surface in 0.1 M HClO$_4$ with an applied potential of 0.05 V (vs. RHE). The solid line is a fit to the data described in the text.

For both Pt$_3$Ni(100) and Pt$_3$Ni(110), it was not possible using SXS to observe any scattering due to the ordered surface structures that were observed by LEED (Fig. 8). Furthermore, it was not possible to obtain any reliable XRV measurements (as shown for Pt$_3$Ni(111) in Fig. 10) from these surfaces, *i.e.* follow structural changes as a function of the applied potential. This implies either that the surfaces are unstable during the transfer from UHV or that they are unstable in the electrochemical environment. Gaining more insight into these more open surfaces would be feasible using the UHV-transfer system described in section 3, as it would then be possible to characterize the surfaces under UHV conditions before adding the electrolyte droplet.

5. Conclusions

In this paper we have described a new experimental methodology for the study of model electrocatalyst materials using surface X-ray scattering (SXS). As a test case we have performed a systematic study of a Pt(111) surface. Initially this was prepared in UHV to have a p(2 × 2)-O layer adsorbed onto the surface. This was characterized by SXS, principally by measurement and modeling of the CTR data. The sample was then contacted with a droplet of pure water and the SXS measurements were repeated. Unfortunately, it was not possible to observe any ordering in the water layers at the Pt–water interface, although it cannot be ruled out that this is due to the weakness of the expected signal compared to the background scattering from the liquid droplet. Analysis of the CTR data showed systematic changes in the measured intensities associated with restructuring at the interface. The data was reproduced with a simple structural model showing reduced outward relaxation in the Pt surface accompanied by increased in-plane distortion of the surface Pt layer and an increased coverage by oxygenated species. Measurements performed in 0.1 M KOH were consistent with previous SXS results and give some insight into the adsorption of hydrogen and oxygenated species.

The experiment on Pt(111) indicates the possibility of studying a range of model electrocatalysts using this methodology. For example, experiments on the low index surfaces of Pt$_3$Ni, using a less direct transfer method from UHV to the electrochemical environment, were only fully successful for the (111) surface, whereas results on Pt$_3$Ni(100) and Pt$_3$Ni(110) were more ambiguous. These systems would be ideal

future experiments using the direct transfer system. It is also envisaged that the UHV characterization methods could be developed to include direct imaging techniques, such as STM. A transfer system which enabled both transfer to and from the electrochemical environment would then offer unique opportunities to further probe the link between atomic structure and reactions at the electrochemical interface.

Acknowledgements

Tien-Lin Lee is acknowledged for his assistance with the ID32 beamline and discussion of the correction factors that are applied to the CTR data. The work at the Argonne National Laboratory was supported by the U.S. Department of Energy, Office of Science, Office of Basic Energy Sciences, under contract DE-AC02-06CH11357 and by a travel grant from the EPSRC. Research was carried out in part at SSRL, which is funded by the Division of Chemical Sciences (DCS), U.S. DOE.

References

1 M. G. Samant, M. F. Toney, G. L. Borges, K. F. Blurton and L. M. O. R. Blum, *J. Phys. Chem.*, 1988, **92**, 220.
2 B. M. Ocko, J. Wang, A. Davenport and H. Isaacs, *Phys. Rev. Lett.*, 1990, **65**, 1466.
3 M. F. Toney and B. M. Ocko, *Synchrotron Radiat. News*, 1993, **6**, 28.
4 I. M. Tidswell, N. M. Markovic and P. N. Ross, *Phys. Rev. Lett.*, 1993, **71**, 1601.
5 C. A. Lucas, N. M. Markovic and P. N. Ross, *Surf. Sci.*, 1996, **340**, L949.
6 C. A. Lucas, N. M. Markovic and P. N. Ross, *Phys. Rev. Lett.*, 1996, **77**, 4922.
7 D. M. Kolb, *Prog. Surf. Sci.*, 1996, **51**, 109.
8 K. Itaya, *Prog. Surf. Sci.*, 1998, **58**, 121.
9 C. A. Lucas and N. M. Markovic, in *9th Volume of Advances in Electrochemical Science and Engineering*, ed. R. Alkire, D. M. Kolb, P. N. Ross and J. Lipkowski, Wiley-VCH, 2006, Chapter 1.
10 D. Menzel, *Science*, 2002, **295**, 58.
11 P. J. Feibelman, *Science*, 2002, **295**, 99.
12 A. Glebov, A. P. Graham, A. Menzel and J. P. Toennies, *J. Chem. Phys.*, 1997, **106**, 9382.
13 A. P. Seitsonen, Y. J. Zhu, K. Bedürftig and H. Over, *J. Am. Chem. Soc.*, 2001, **123**, 7347.
14 C. Clay, S. Haq and A. Hodgson, *Phys. Rev. Lett.*, 2004, **92**, 046102.
15 I. M. Tidswell, N. M. Markovic, C. A. Lucas and P. N. Ross, *Phys. Rev. B: Condens. Matter Mater. Phys.*, 1993, **47**, 16542.
16 I. K. Robinson and D. J. Tweet, *Rep. Prog. Phys.*, 1992, **55**, 599.
17 I. K. Robinson, *Phys. Rev. B: Condens. Matter Mater. Phys.*, 1986, **33**, 3830.
18 http://www.esrf.eu/UsersAndScience/Experiments/SurfaceScience/ID32/SurfaceLab.
19 N. Materer, U. Starke, A. Barbieri, R. Doll, K. Heinz, M. A. Van Hove and G. A. Somorjai, *Surf. Sci.*, 1995, **325**, 207.
20 F. U. Renner, Y. Grunder and J. Zegenhagen, *Rev. Sci. Instrum.*, 2007, **78**, 033903.
21 Y. Gauthier, Y. Joly, R. Baudoing and J. Rundgren, *Phys. Rev. B: Condens. Matter Mater. Phys.*, 1985, **31**, 6216.
22 Y. Gauthier, R. Baudoing, Y. Joly, J. Rundgren, J. C. Bertolini and J. Massardier, *Surf. Sci.*, 1985, **162**, 342.
23 V. R. Stamenković, B. Fowler, B. S. Mun, G. Wang, P. N. Ross, C. A. Lucas and N. M. Marković, *Science*, 2007, **315**, 493.
24 C. A. Lucas and N. M. Marković, in *Encyclopedia of electrochemistry*, ed. E. J. Calvo, Wiley-VCH, 2003, vol. 2, section 4.1.2.1.2.
25 M. Nakamura, K. Sumitani, M. Ito, T. Takahashi and O. Sakata, *Surf. Sci.*, 2004, **563**, 199.
26 E. Vlieg, *J. Appl. Crystallogr.*, 1997, **30**, 532.
27 R. Feidenhans'l, *Surf. Sci. Rep.*, 1989, **10**, 105.
28 M. T. M. Koper and R. A. van Santen, *J. Electroanal. Chem.*, 1999, **472**, 126.
29 G. Held, C. Clay, S. D. Barrett, S. Haq and A. Hodgson, *J. Chem. Phys.*, 2005, **123**, 064711.
30 K. Bedurftig, S. Volkening, Y. Wang, J. Wintterlin, K. Jacobi and G. Ertl, *J. Chem. Phys.*, 1999, **111**, 11147.
31 N. M. Markovic and P. N. Ross, *Surf. Sci. Rep.*, 2002, **45**, 117.
32 N. Marinkovic, N. M. Markovic and R. R. Adzic, *J. Electroanal. Chem.*, 1992, **330**, 433.
33 C. A. Lucas, *Electrochim. Acta*, 2002, **47**, 3065.
34 P. J. Feibelman, *Phys. Rev. B: Condens. Matter Mater. Phys.*, 1997, **56**, 2175.

35 P. A. Thiel and P. J. Estrup, in *The Handbook of Surface Imaging and Visualization*, ed. A. T. Hubbard, CRC Press, Boca Raton, FL, USA, 1995.
36 V. Stamenkovic, T. J. Schmidt, N. M. Markovic and P. N. Ross, *J. Phys. Chem. B*, 2002, **106**, 11970.
37 V. Stamenkovic, T. J. Schmidt, P. N. Ross and N. M. Markovic, *J. Electroanal. Chem.*, 2003, **554–555**, 191.
38 V. R. Stamenković, B. S. Mun, K. J. J. Mayrhofer, P. N. Ross, N. M. Marković, J. Rossmeisl, J. Greeley and J. K. Nørskov, *Angew. Chem., Int. Ed.*, 2006, **45**, 2897.
39 V. R. Stamenković, B. S. Mun, M. Arenz, K. J. J. Mayrhofer, P. N. Ross, C. A. Lucas and N. M. Marković, *Nat. Mater.*, 2007, **6**, 241.
40 Y. Gauthier, R. Baudoing and J. Rundgren, *Phys. Rev. B: Condens. Matter Mater. Phys.*, 1985, **31**, 6216.
41 Y. Gauthier, *Surf. Rev. Lett.*, 1996, **3**, 1663.
42 B. E. Warren, *X-Ray Diffraction*, Dover Publications Inc., 1990.
43 f' changes from -4.04 electrons to -6.50 electrons, calculated from: S. Brennan and P. L. Cowen, *Rev. Sci. Instrum.*, 1992, **63**, 850.
44 B. Hammer and J. K. Norskov, in *Chemisorption and Reactivity on Supported Clusters and Thin Films*, ed. R. M. Lambert and G. Pacchioni, Kluwer Academic, 1997.
45 H. You, D. J. Zurawski, Z. Nagy and R. M. Yonco, *J. Chem. Phys.*, 1994, **100**, 4699.

Surface dynamics at well-defined single crystal microfacetted Pt(111) electrodes: *in situ* optical studies

Iosif Fromondi and Daniel Scherson*

Received 25th March 2008, Accepted 9th May 2008
First published as an Advance Article on the web 7th October 2008
DOI: 10.1039/b805040f

The assembly and electrochemical oxidation of well-defined CO adlayers on Pt(111) microfacets in aqueous CO-saturated 0.1 M H_2SO_4 were examined by a combination of *in situ, simultaneous*, time-resolved, reflectance spectroscopy (RS) and second harmonic generation (SHG), and potential step and linear scan techniques. Optical transients were collected following potential steps from a value high enough for a full monolayer of bisulfate ($\theta = 0.2$) to adsorb on the Pt(111) facet, E_{ox}, to potentials E_{ads}, at which either the c(2 × 2)-3CO or $\sqrt{19} \times \sqrt{19}R23.4°$-13CO phase is expected to form once surface saturation is achieved. Similar experiments involving subsequent steps from E_{ads} to E_{ox} provided unambiguous evidence that the rates of oxidation of c(2 × 2)-3CO on such quasi-perfect Pt(111) facet *at constant overpotential* are much slower than those of $\sqrt{19} \times \sqrt{19}R23.4°$-13CO, an effect attributed to the presence of intrinsic vacant sites within the latter, more dilute phase, which are required for oxidation of adsorbed CO to ensue. Furthermore, continuous CO adsorption–oxidation cycles were found to increase the rate of oxidation of the c(2 × 2)-3CO phase. This phenomenon was tentatively ascribed to the progressive emergence of defects along the edge of the facet (and/or within the facet itself) which serve as nucleation sites for the oxidation of adsorbed CO.

Introduction

Applications of structural and spectroscopic techniques to the study of ultrafast interfacial processes are expected to open exciting new prospects for the further understanding of the factors that govern electrocatalysis with potential impact in areas of both fundamental and technological importance. Methods that are inherently interface specific are particularly attractive, as contributions to the measured signals due, for example, to changes in the composition of the bulk solution induced by the passage of current can be effectively neglected. Indeed, both linear, *e.g.* normalized reflectance (RS)[1–3] and infrared reflection absorption spectroscopies (IRAS)[4–8] and, especially, non-linear optical techniques, including second harmonic generation (SHG),[9–12] and, more recently, sum frequency generation (SFG)[13] are beginning to provide information with unparalleled specificity and temporal and, in some cases, spatial resolution.[14] The changes in the intensity of reflected light in the case of RS, as well as the cross sections for SHG and SFG are very small. It becomes therefore essential to average the results of hundreds to thousands of replicate experiments, in order to increase the signal to noise and thereby obtain meaningful time-resolved data. This requirement demands the careful design of protocols

Department of Chemistry, Case Western Reserve University, Cleveland, OH, 44106, USA

that render the system in precisely the same state prior to the beginning of each run. Attention in our laboratory has been focused on the oxidation of CO adlayers on Pt(111) in aqueous acidic electrolytes, a canonical interfacial system that has been the subject of numerous investigations by a growing number of techniques. Our emphasis has been placed on the use of quasi-perfect Pt(111) facets prepared by the melting and cooling of fine Pt wires, a procedure found to yield surfaces displaying properties believed to be associated with the presence of a very low density of defects. The specific measurement strategy we have implemented in our laboratories involves the simultaneous use of two different laser beams for monitoring light reflected from the facet (RS) and generated at the interface, (SHG or SFG), during application of judiciously selected potential step protocols.[15] It is the main objective of this contribution to address certain issues relating to the role of defects on the oxidation of the CO adlayer, which complements information reported earlier by the groups of Stimming[16] and Koper[17] for low index Pt single crystals under otherwise similar conditions to those used in this study.

Experimental

The instrumental array involved in these measurements, as well as the methods employed to prepare facetted Pt single crystals have been described in detail in previous communications[15] and will be only briefly summarized here. As specified therein, a low power CW (633 nm HeNe) and a high power pulsed polarized laser beam (590 nm, 3 ps) were aimed parallel to one another to a single focusing lens to achieve a common focal point on a single Pt(111) facet at an angle of incidence of 45°. Additional optical components were used to direct each of the reflected beams to either a filter/monochromator/photomultiplier/photon counter system (SHG), or a silicon detector (RS). The spectroelectrochemical cell used was a 10 × 10 mm quartz fluorimeter cuvette with five transparent windows. A reversible hydrogen electrode (RHE) placed in a different compartment (to avoid CO contamination) and connected to the cell *via* Teflon tubing was used as a reference electrode. All experiments were performed at room temperature in 0.1 M H_2SO_4 (Ultrex) solutions (Barnstead ultrapure water) purged using either UHP Ar (Praxair) or CO (Mathesson Gas purity grade). The size of the (111) microfacets were determined with an optical microscope yielding diameters in the range 20 to 30 μm.

Results and discussion

Cyclic voltammetry

Plots of the intensity of the SHG signal, $I_{pp}(2\omega)$, where pp refers to p input and p output polarizations, *vs.* potential, E, in CO-saturated 0.1 M H_2SO_4 recorded during voltammetric cycles (see black line, Panel A, Fig. 1) at a scan rate, $v = 2$ V s^{-1}, were found to be remarkably similar to those reported earlier in CO-saturated 0.1 M $HClO_4$.[14] In particular, $I_{pp}(2\omega)$ decreased linearly with E in the range $0.05 < E < 0.45$ V, *i.e.* where the CO adlayer is expected to be present in the (2 × 2)-3CO phase, irrespective of the direction of the scan (see Panel A, Fig. 1). The increase in $I_{pp}(2\omega)$ observed in the scan in the positive direction at about 0.85 V is ascribed to the onset of the kinetically hindered (2 × 2)-3CO → $\sqrt{19} \times \sqrt{19}$R23.4-13CO phase transition, as originally reported by Akemann *et al.*[9] As the scan was continued, $I_{pp}(2\omega)$ decreased rather suddenly at E ca. 0.95 to a very low value, signaling the full oxidation of the CO adlayer, which remained virtually unchanged upon reversing the scan at the upper limit, ca. 1.0 V, down to ca. 0.8 V. This specific behavior may be attributed to the adsorption of bisulfate at saturation coverage, $\theta_{HSO_4^-} = 0.2$, as evidenced from experiments performed in the strict absence of CO in the electrolyte (see grey lines, in this figure). As E reached ca. 0.8 V, in the scan in the negative direction, $I_{pp}(2\omega)$ began to increase markedly to yield values ca. 80% larger than those

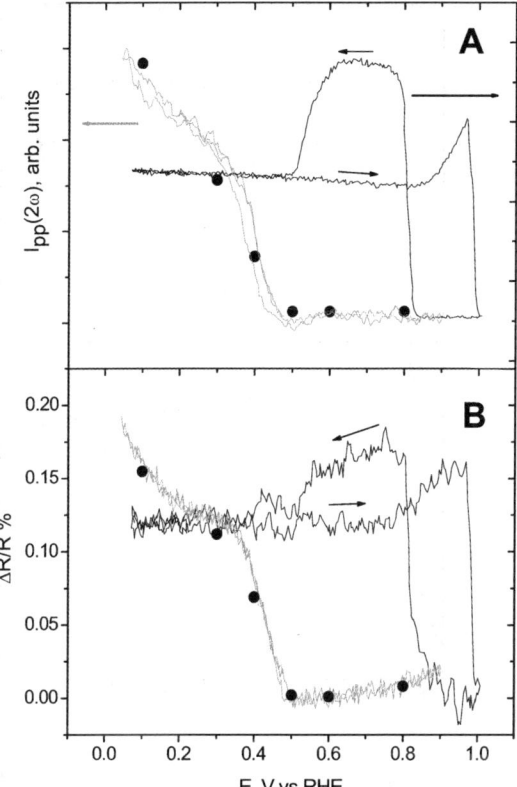

Fig. 1 Plots of $I_{pp}(2\omega)$, (Panel A) and $\Delta R/R$ ($E_{ref} = 0.1$ V), (Panel B) vs. potential (E) for a Pt(111) microfacet in neat (grey lines, $\nu = 2$ V s^{-1}) CO-saturated 0.1 M H$_2$SO$_4$ (black lines, $\nu = 0.1$ V s^{-1}). The solid black circles are values of $\Delta R/R$ and $I_{pp}(2\omega)$ observed immediately following application of E_{ads} (see text for details). Photon counter: SR400; accumulation time $T = 4$ ms per count (grey) and $T = 30$ ms (black); reading frequency, $R_f = 4$ ms.

observed for the (2 × 2)-3CO phase, which are characteristic of the √19 × √19R23.4-13CO phase.

A rather similar overall behavior, albeit with poorer signal to noise was found for $\Delta R/R = [R(E) - R(E_{ref})]/R(E_{ref})$ (see Panel B, Fig. 1), where $R(E)$ and $R(E_{ref})$ are proportional to the intensity of the reflected light at the detector for the electrode polarized at an arbitrary (E) and reference (E_{ref}) potential, respectively.

Also shown in grey lines in Panel B, Fig. 1, is the $\Delta R/R$ response obtained for the bare Pt(111) surface 0.1 M H$_2$SO$_4$ solutions devoid of CO, where the increase at E ca. 0.5 V, also found in $I_{pp}(2\omega)$ (see grey lines, Panel A) during the scan in the negative direction, is ascribed to the onset of bisulfate desorption. Quantitative aspects of bisulfate adsorption on Pt(111) bisulfate containing solutions as monitored by these two optical techniques may be found elsewhere.[4]

Potential steps

Shown in Fig. 2 are averaged (ca. 700 acquisitions, Acq.) plots of $I_{pp}(2\omega)$ (Panel A) and $\Delta R/R$ (Panel B) vs. time following a potential step from $E_{ox} = 0.98$ V, i.e. sufficiently positive for CO$_{ads}$ oxidation to ensue, to $E_{ads} = 0.10$ (curves a), 0.30 (b), 0.40 (c), 0.50 (d), 0.60 (e) and 0.80 V vs. RHE (f)) in CO-saturated 0.1 M H$_2$SO$_4$. It is important to stress that the quasi perfect single crystal sphere was annealed

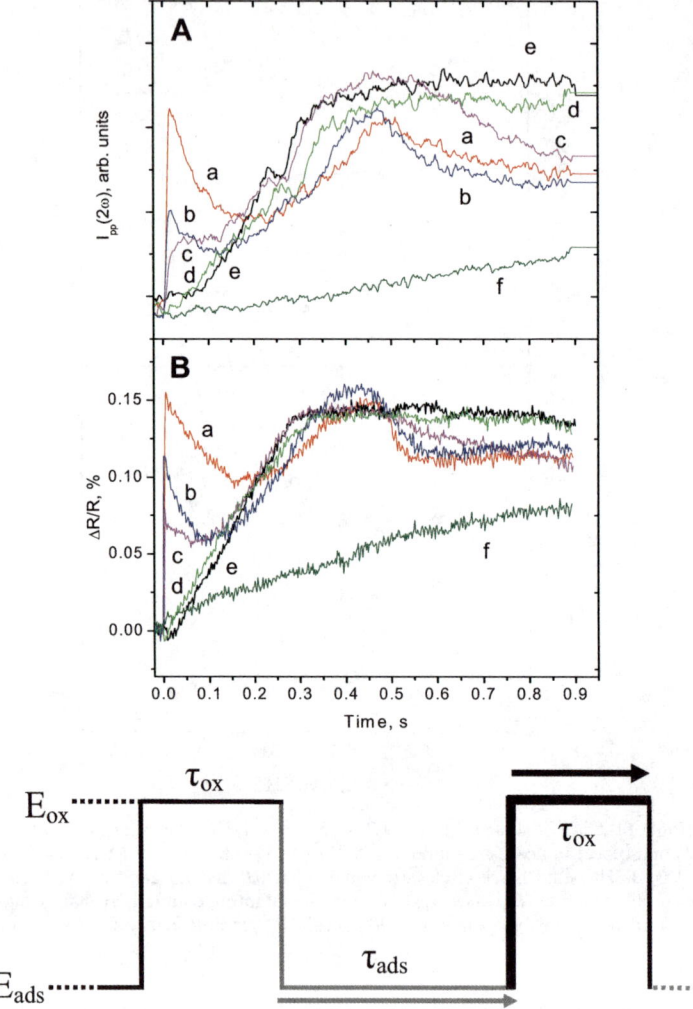

Fig. 2 Plots of $I_{pp}(2\omega)$ (Panel A) and $\Delta R/R$ ($E_{ref} = E_{ox}$, Panel B) vs. time, following application of a potential step from $E_{ox} = 0.98$ V down to $E_{ads} = 0.10$ (curves a), 0.30 (b), 0.40 (c), 0.50 (d), 0.60 (e) and 0.80 V vs. RHE (f) in CO-saturated 0.1 M H_2SO_4 for $\tau_{ads} = 1$ s and $\tau_{ox} = 0.1$ s (see grey section in the protocol in the insert). Each of the curves represents the average of ca. 700 acquisitions, Acq. Electrodes were reannealed following acquisition of a single set of averaged E_{ads} data. Photon counter: SR400, $T = 4$ ms per count, $R_f = 4$ ms.

before each set of experiments, involving a specific E_{ads}, was initiated. As indicated by the E–time protocol (see the insert, Panel B), the electrode was first held at E_{ox} for $\tau_{ox} = 0.1$ s, a period of time long enough for the CO adlayer to fully oxidize, then stepped to the desired E_{ads}, in the range between 0.1 and 0.6 V, which correspond, respectively, to values at which the CO adlayer at saturation is known to be present in the (2×2)-3CO and $\sqrt{19} \times \sqrt{19}$R23.4-13CO phases. The potential was then held at E_{ads} for $\tau_{ads} = 1$ s to allow for the full adlayers to form, and immediately thereafter stepped to E_{ox} to reinitialize the protocol.

Except for their magnitudes, the overall qualitative features of the $I_{pp}(2\omega)$ and $\Delta R/R$ transients were found to be rather similar. In fact, the values of both optical responses recorded immediately following application of the adsorption step, i.e. $t = 0$ in Fig. 2, shown as solid circles in Fig. 1, were very close to those found at slow scan rates in solutions devoid of CO (see grey lines in that same figure), regardless of the values of E_{ads}. In particular, following a step to $E_{ads} = 0.1$ V (red curve), a value at which the coverage of bisulfate would be exceedingly small, $I_{pp}(2\omega)$ and $\Delta R/R$ suddenly increased, as would be expected for the instantaneous adsorption of hydrogen (and thus not unlike that observed upon application of the same protocol in Ar-saturated solutions). The gradual decrease in both optical responses found for short times immediately thereafter is consistent with the gradual adsorption of CO, a species that displaces adsorbed hydrogen from the surface. As time elapsed, and thus the CO coverage, θ_{CO}, increases, both optical signals also increased, reaching a maximum at $ca.$ 0.4–0.5 s, and then decreased thereafter. The most likely explanation for this phenomenon is an increase in the amount of CO adsorbed on bridge (two-fold) sites, responsible for enhancements in both $I_{pp}(2\omega)^6$ and $\Delta R/R$, as shown in Fig. 1. As time elapses, θ_{CO} increases and the CO bridge migrates to on-top (one-fold), and hollow (three-fold) sites, as prescribed by the c(2 × 2)-3CO phase, causing a decreases in the magnitude of both optical signals. Evidence in support of this site displacement model may be found in the work of Weaver et al.,[6] who performed in situ IRAS measurements in the same base electrolyte containing CO at concentrations of $ca.$ 2 × 10^{-5} M and higher, to determine the site occupation of CO$_{ads}$. As shown by these authors, coadsorption of water or hydrogen and CO on Pt(111) prepared by dosing (as opposed to stripping) yielded, for small θ_{CO}, spectral features consistent with CO adsorbed in a bridge (two-fold) position. Although somewhat speculative, this increase in the bridge site occupation is also consistent with the formation of small $\sqrt{19} \times \sqrt{19}R23.4°$-13CO domains at these low θ_{CO}.

Unlike the behavior found for $E_{ads} = 0.1$ V, the transient optical responses for $E_{ads} = 0.6$ V (see black curves), reached monotonically values characteristic of the $\sqrt{19} \times \sqrt{19}R23.4°$-13CO phase. Close inspection of the $I_{pp}(2\omega)$ transient revealed the presence of a small peak at about 180 ms (see arrow). The same feature is also visible, albeit not as well-defined, for $E_{ads} = 0.3$, 0.4 and 0.5 V, a potential range in which bisulfate adsorbs on Pt(111).[18] On this basis, it seems likely that this peak is caused by the dynamic interplay between CO adsorption and bisulfate desorption. In particular, for potentials at which the bisulfate coverage is below saturation, CO can adsorb on vacant sites and/or displace bisulfate from the surface. Both of these processes would be expected to lead to an increase in $I_{pp}(2\omega)$, as the data in Fig. 2 indicates. This feature, however, was not found in the reflectance measurements, pointing to differences in the specificity of the two techniques to the interfacial states involved.

Linear potential scans

Insight into the oxidation dynamics of the adsorbed CO layer was obtained from in situ RS/SHG experiments in which the potential was scanned linearly at 0.05 V s^{-1} starting in one case from $E_{ads} = 0.1$ V, i.e. (2 × 2)-3CO, and $E_{ads} = 0.4$ V in the other, after holding the potential at the prescribed values for an adsorption time $\tau_{ads} = 2$ s (see insert, Panel B in Fig. 3). Similar conditions were employed by Lopez-Cudero et al.[8] in their strictly electrochemical studies involving massive Pt(111) crystals, providing means for a direct comparison with the results reported in this work. Once again, the single crystal sphere was annealed before each set of measurements was performed. Shown in Panel A, Fig. 3, are plots of $I_{pp}(2\omega)$ vs. time for $E_{ads} = 0.1$ V and 0.4 V. As clearly evidenced by the data, shown in expanded form in Panel B in this figure, the differences between the two curves may be regarded as statistically insignificant. This may not be surprising, as based on the $I_{pp}(2\omega)$ results in Fig. 1, the CO adlayer is present in the (2 × 2)-3CO phase in the range $0.1 < E < 0.4$ V. These results, are at variance with those reported by Lopez-Cudero et al., for which the voltammetric peak associated with CO oxidation shifted toward more positive potentials for $E_{ads} = 0.4$ V compared to $E_{ads} = 0.1$ V. Moreover, as E_{ads} was

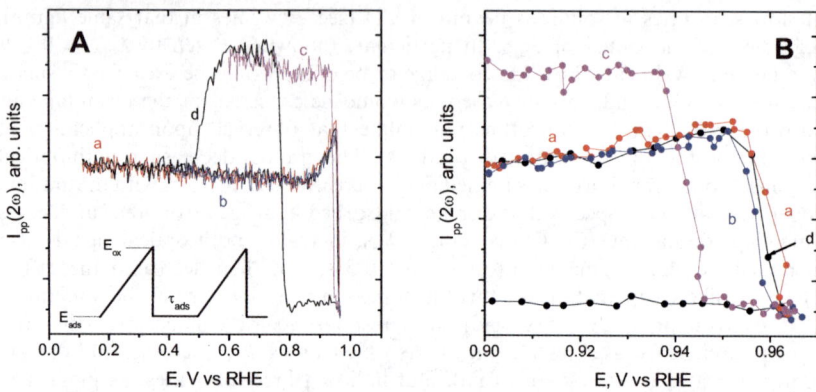

Fig. 3 A. Average (Acq. *ca.* 30) plots of $I_{pp}(2\omega)$ *vs.* E collected during linear scans at 0.05 V s^{-1} toward positive potentials following CO adsorption at either $E_{ads} = 0.1$ V (a), $E_{ads} = 0.4$ V (b) and $E_{ads} = 0.6$ V (c) for $\tau_{ads} = 2$ s, and $E_{ox} = 0.965$ V *vs.* RHE. Also, shown in curve d in this figure are the results obtained while the electrode was being cycled continuously between $E_{ads} = 0.1$ V and E_{ox}. The actual potential protocol is shown in the inset. Each of the measurements was performed with the same facet following reannealing. Photon counter: SR400, $T = 60$ ms per count, 500 pts per record. B. Expanded view of the data in A in this figure in the range $0.90 < E < 0.965$ V.

increased to 0.6 V, a potential at which the adlayer is present in the more dilute $\sqrt{19} \times \sqrt{19}R23.4°$-13CO phase, the onset of oxidation of the CO adlayer as determined by optical means was found to shift toward more negative and not more positive values as reported by those authors. Efforts to unveil the reasons underlying this discrepancy were made by intentionally damaging the Pt(111) surface.

Effect of surface microstructure

Marked changes in the optical transient profiles using the same protocol as that depicted in the inset, Fig. 3, were observed following application of repeated potential steps to $E_{ox} = 1.4$ V, a potential at which Pt would undergo substantial oxidation rendering a highly damaged surface. As evidenced from the $\Delta R/R$ transients following a potential step to $E_{ox} = 0.96$ V *vs.* RHE (see Fig. 4) the oxidation of

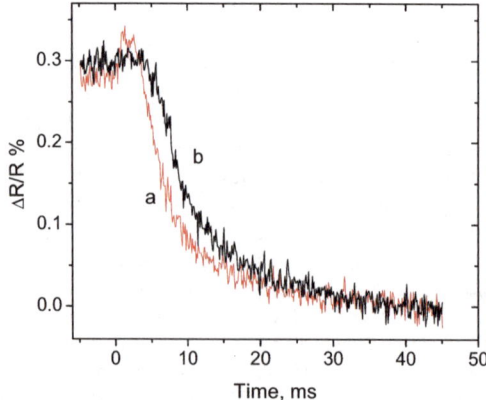

Fig. 4 Plots of $\Delta R/R$ ($E_{ref} = E_{ox}$), *vs.* t data for an intentionally disordered Pt(111) facet (see text for conditions) in CO satd. 0.1 M H$_2$SO$_4$ for $E_{ads} = 0.1$ (curve a) and $E_{ads} = 0.6$ V (b) and $E_{ox} = 0.96$ V. Other conditions are specified in the caption of Fig. 2.

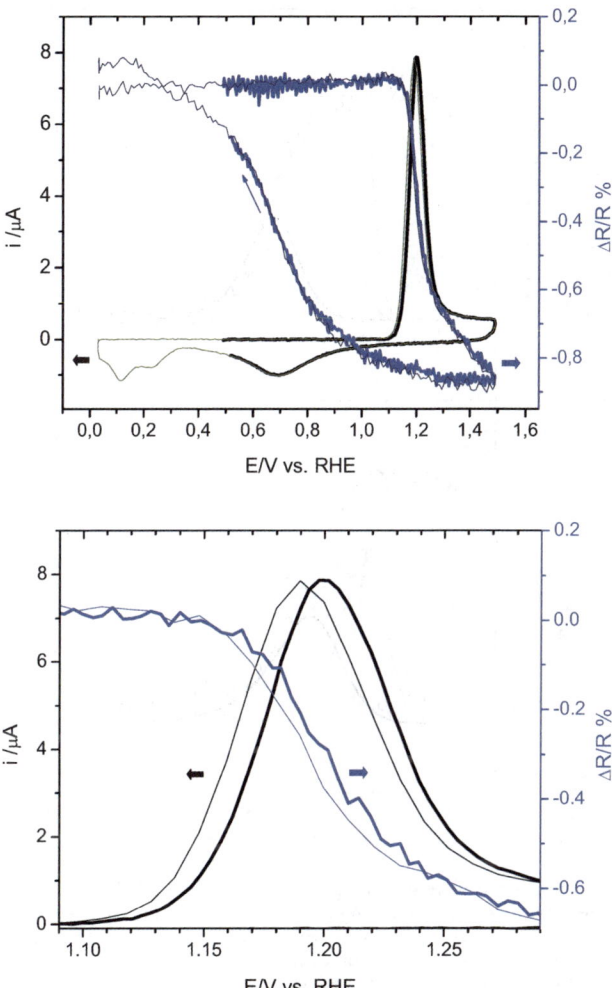

Fig. 5 Panel A. Simultaneous cyclic voltammogram (left ordinate) and $\Delta R/R$ vs. E ($E_{\text{ref}} = 0.5$ V, right ordinate) for a 25 μm diameter Pt microelectrode in CO-saturated 0.5 M H_2SO_4 at 100 V s^{-1} for CO adsorption potential $E_{\text{ads}} = 0.5$ V (thick lines, Acq = 1712) and $E_{\text{ads}} = 0.05$ V (thin lines, Acq = 2165) and $\tau_{\text{ads}} = 2$ s. Panel B. Expanded view of the data in the upper panel in this figure in the potential range of $ca.\, 1.1 < E < 1.3$ V vs. RHE.

CO adlayers formed on such intentionally damaged surfaces was faster for $E_{\text{ads}} = 0.1$ compared to $E_{\text{ads}} = 0.6$ V and thus contrary to the behavior found for the pristine Pt(111) facet. These results suggest that despite their well-behaved voltammetric response, the defect density of massive single crystal surfaces prepared by the Clavilier technique is higher than that of Pt(111) facets grown by the method herein implemented. In fact, the same trend was found earlier for polycrystalline Pt electrodes in 0.5 M H_2SO_4 using potential scans as opposed to steps (see Fig. 5 and captions for details).[19]

Influence of repetitive application of oxidation protocol

Interesting effects were noted upon collecting replicate sets of sequential measurements with the same precise facet (no annealing) under otherwise identical

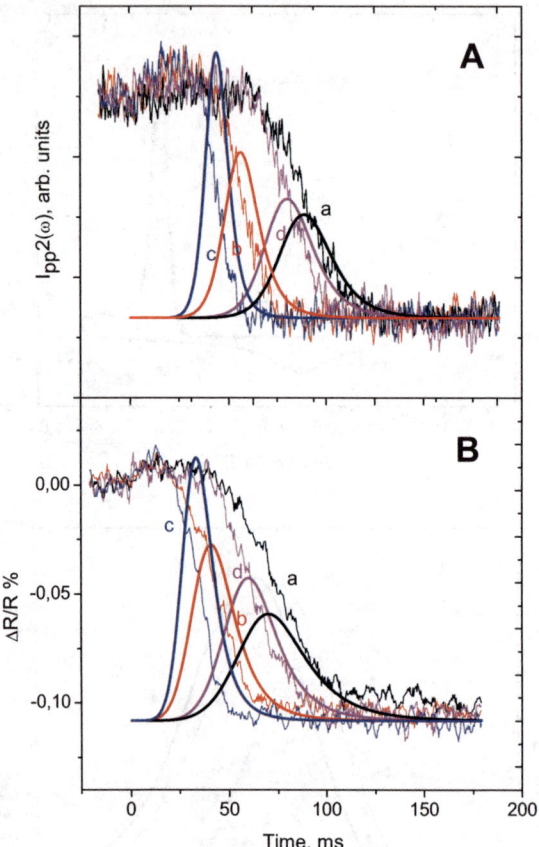

Fig. 6 Plots of three consecutive sets of $I_{pp}(2\omega)$ (Panel A) and $\Delta R/R\%$ ($E_{ref} = E_{ads}$, Panel B) vs. time measurements (curves a, b and c in sequence, with no reannealing) following a potential step from $E_{ads} = 0.1$ V, i.e. c(2 × 2), to $E_{ox} = 0.98$ V. Curve d was obtained after the same electrode was reannealed. Each of the curves represents the average of ca. 700 acquisitions and 5 pts. AA smoothing. Photon counter: SR430, bin width = 163.84 µs, 1000 pts per record. The solid curves are the derivatives of the best fits to the raw transient data.

conditions. As shown in Fig. 6 for $E_{ads} = 0.1$ V, i.e. (2 × 2)-3CO, and $E_{ox} = 0.98$ V, curves a through c collected in sequence yielded shorter τ_{ind} times. Following reannealing, however, this parameter increased approaching a value similar to that found for the original pristine facet (see curve d in this figure). Further insight into these effects may be gleaned from the derivatives to the best fits of the transients shown in solid smooth lines in the same figure. As indicated, not only does the maximum shift to shorter times in agreement with the trend of the raw data, but the width of the peaks becomes smaller as the induction time decreases. One possible explanation for this effect may be found in the emergence of additional nucleation sites at the periphery of the facet as the number of oxidation cycles is increased, which would trigger multiple oxidation fronts, thereby decreasing the time required for full oxidation of the adlayer. Although the results obtained for the same sequence of experiments involving a single specific facet (not reannealed) for $E_{ads} = 0.6$ V, i.e. $\sqrt{19} \times \sqrt{19}$R23.4°-13CO, (see Fig. 7) appear to suggest a similar trend, the quality of the data may not be sufficient to firmly support this view.

In conclusion, the data herein presented underscores the role of defects and vacant sites in promoting the rates of CO oxidation adsorbed on Pt(111) in aqueous

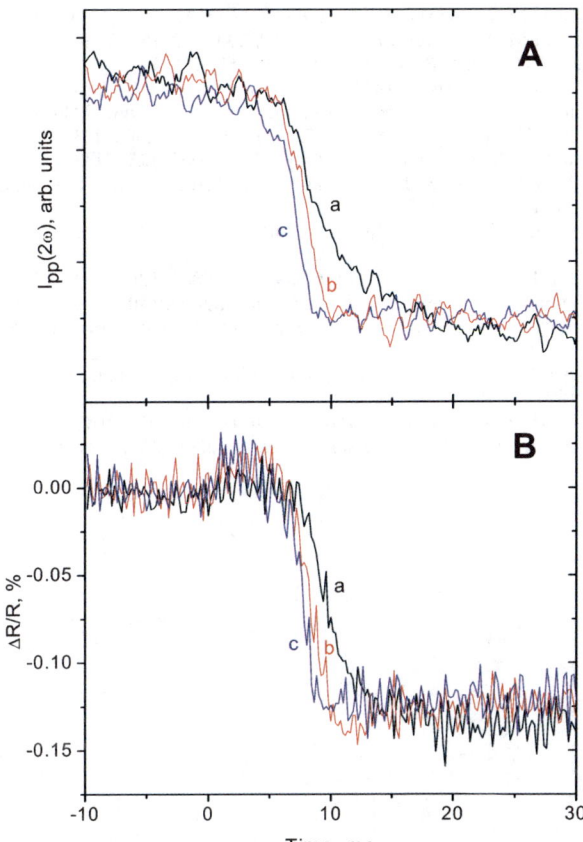

Fig. 7 Plots of three consecutive sets of and $I_{pp}(2\omega)$ (Panel A) and $\Delta R/R$ % ($E_{ref} = E_{ads}$, Panel B) vs. time measurements (curves a, b and c in sequence, with no reannealing) following a potential step from $E_{ads} = 0.6$ V, i.e. $\sqrt{19} \times \sqrt{19}R23.4°$-13CO, to $E_{ox} = 0.98$ V. Each of the curves represents the average of ca. 700 acquisitions and 5 pts. AA smoothing. Photon counter: SR430, bin width = 163.84 µs, 1000 pts per record.

electrolytes and the extreme care that must be exercised to control their density. Comparison with data reported in the literature appears to suggest that, despite its sensitivity, cyclic voltammetry may not provide means to identify defect sites nor their density which may have a profound effect in controlling the rate of thoroughly studied heterogeneous electron transfer processes, such as the oxidation of CO adlayers adsorbed on Pt(111).

Acknowledgements

This work was supported by a grant from the NSF.

References

1 F. V. Molina and R. Parsons, *J. Chim. Phys. Phys.–Chim. Biol.*, 1991, **88**, 1339–1352.
2 F. Huerta, E. Morallon, C. Quijada, J. L. Vazquez and L. E. A. Berlouis, *J. Electroanal. Chem.*, 1999, **463**, 109–115.
3 I. Fromondi, A. L. Cudero, J. Feliu and D. A. Scherson, *Electrochem. Solid-State Lett.*, 2005, **8**, E9.
4 F. Kitamura, M. Takahashi and M. Ito, *Surf. Sci.*, 1989, **233**, 493–499.

5 S. C. Chang and M. J. Weaver, *J. Chem. Phys.*, 1990, **92**, 4582–4594.
6 S. C. Chang and M. J. Weaver, *Surf. Sci.*, 1990, **238**, 142–162.
7 T. Iwasita and F. C. Nart, *Prog. Surf. Sci.*, 1997, **55**, 271–340.
8 A. Lopez-Cudero, A. Cuesta and C. Gutierrez, *J. Electroanal. Chem.*, 2005, **579**, 1–12.
9 W. Akemann, K. A. Friedrich and U. Stimming, *J. Chem. Phys.*, 2000, **113**, 6864–687.
10 B. Pozniak, Y. Mo and D. A. Scherson, *Faraday Discuss.*, 2002, **121**, 313–322.
11 B. Pozniak and D. A. Scherson, *J. Am. Chem. Soc.*, 2003, **125**, 7488–7489.
12 B. Pozniak, Y. B. Mo, I. C. Stefan, K. Mantey, M. Hartmann and D. A. Scherson, *J. Phys. Chem. B*, 2001, **105**, 7874–7877.
13 A. Lagutchev, G. Q. Lu, T. Takeshita, D. Dlott and A. Wieckowski, *J. Chem. Phys.*, 2006, **125**, 154705.
14 B. Pozniak and D. A. Scherson, *J. Am. Chem. Soc.*, 2004, **126**, 14696–14697.
15 I. Fromondi and D. A. Scherson, *J. Phys. Chem. B*, 2006, **110**, 20749–20751.
16 A. V. Petukhov, W. Akemann, K. A. Friedrich and U. Stimming, *Surf. Sci.*, 1998, **402**, 182–186.
17 N. P. Lebedeva, M. T. M. Koper, J. M. Feliu and R. A. van Santen, *J. Phys. Chem. B*, 2002, **106**, 12938–12947.
18 I. Fromondi and D. A. Scherson, *J. Phys. Chem. C*, 2007, **111**, 10154–10157.
19 P. Shi, I. Fromondi and D. A. Scherson, *Langmuir*, 2006, **22**, 10389–10398.

PAPER

Bridging the gap between nanoparticles and single crystal surfaces

Payam Kaghazchi,[a] Felice C. Simeone,[b] Khaled A. Soliman,[b] Ludwig A. Kibler[b] and Timo Jacob*[ab]

Received 20th February 2008, Accepted 28th April 2008
First published as an Advance Article on the web 7th August 2008
DOI: 10.1039/b802919a

Using density functional theory calculations and the extended *ab initio* atomistic thermodynamics approach, we studied the adsorption of oxygen on the different surface faces, which are involved in the faceting of Ir(210). Constructing the (p,T)-surface phase diagrams of the corresponding surfaces in contact with an oxygen atmosphere, we find that at high temperatures the planar surfaces are stable, while lowering the temperature stabilizes those nano-facets found experimentally. Afterwards, we constructed the $(a,T,\Delta\phi)$-phase diagram for Ir(210) in contact with an aqueous electrolyte and found that the same nano-facets should be stable under electrochemical conditions. Motivated by this prediction from theory, experiments were performed using cyclic voltammetry and *in-situ* scanning tunneling microscopy. The presence of nanofacets for Ir(210) gives rise to a characteristic current-peak in the hydrogen adsorption region for sulfuric acid solution. Furthermore, first results on the electrocatalytic behavior of nano-faceted Ir(210) are presented.

Introduction

Highly-disperse nanoparticles are often used to catalyze (electro-)chemical reactions. Unfortunately, not all nanoparticles have the same size and shape, but show a relatively large distribution. Since the electronic properties of the nanoparticles are correlated with their morphology, experimental measurements usually represent the averaged behavior of (almost) the entire ensemble of particles. This limits our understanding of the ongoing processes and makes direct comparison with theoretical studies difficult.

One way out of this dilemma is the formation of well-defined nanostructures or facets on single-crystal surfaces, which provide a reproducible basis and model systems for studying structural sensitivity in (electro-)catalytic reactions. Surface faceting can be understood as a morphology change from a flat bulk-truncated surface to a hill-and-valley structure. While clean surfaces rarely facet, adsorbate-induced faceting of surfaces, driven by the anisotropy of surface free energy, is a general phenomenon observed in many systems.[1–3] Usually the facets have more close-packed surface structures than the original surface, resulting in a minimized surface free energy although the total surface area may be increased. Therefore, in order to actively select and control a desired surface morphology, it is necessary to deepen our understanding of adsorbate-induced faceting. Furthermore, this

---

[a] *Fritz-Haber-Institut der Max-Planck-Gesellschaft, Faradayweg 4-6, D-14195 Berlin, Germany. E-mail: jacob@fhi-berlin.mpg.de; Fax: +49-(0)30-8413-4701; Tel: +49-(0)30-8413-4816*
[b] *Institut für Elektrochemie, Universität Ulm, Ulm, D-89081, Germany*

would provide model systems to study structural sensitivity in catalytic reactions[4-6] and may be used as templates to grow nanostructures.[7,8]

So far experimental studies of adsorbate-induced faceting of metal surfaces focused mainly on body-centered cubic or face-centered cubic metals, such as W(111),[1,2] Mo(111),[9,10] Ni(210),[11,12] Pt(210),[13] Ir(210),[14] Rh(553),[15] and vicinal Cu surfaces.[16-19] Although the enhancement of the anisotropy in surface free energy is the thermodynamic driving force for facet formation, in most cases this process is hindered by kinetic limitations. Therefore, not only is a critical adsorbate coverage required but also a minimum annealing temperature, allowing the system to overcome all kinetic barriers in the process of facet formation.

Recently the group of Madey found that on particular rough surfaces certain adsorbates are able to induce the formation of well-defined nanostructures after annealing the system to elevated temperatures.[14,20] Using scanning tunnelling microscopy (STM) and low-energy electron diffraction (LEED) under ultra-high vacuum (UHV) they could demonstrate that an initially planar Ir(210) surface becomes faceted when being covered with more than 0.5 ML oxygen and annealed to temperatures above 600 K. The facets that form were characterized as an array of three-sided pyramidal nanostructures having Ir(311), Ir(31−1) and Ir(110) faces. Furthermore, higher resolution STM images showed that while the (311) and (31−1) faces are always unreconstructed, some (110) faces are partially reconstructed. This *superstructure* was proposed to be a "stepped double-missing-row"-(110) surface.[21]

After facet formation, oxygen that still remains on the surface can be removed by reaction with H_2 at $T < 400$ K. During this reaction the nanopyramidal surface structure is not affected, since the kinetic barrier of facet destruction is not reached at these low temperatures. The clean nanofacets remain stable up to ~600 K, and for higher temperatures the initial planar Ir(210) surface becomes stable again.

Similar behavior could also be observed in the case of Re(11−21), where by adsorption of oxygen pyramid-like facets having each two (01−11) and (10−11) faces could be generated. However, changing the adsorbate to ammonia and annealing to 900 K, led to the formation of two-sided ridges with (13−42) and (31−42) faces.[22]

It has also been demonstrated that planar and nano-faceted Ir(210) surfaces can be prepared outside an UHV chamber by inductive heating in a $N_2 + H_2$ mixture and a nitrogen atmosphere, respectively.[23] Cooling the sample in a reducing gas atmosphere yields a planar surface according to the preparation of unreconstructed low-index planes of iridium.[24-27] The presence of trace amounts of oxygen in nitrogen gas was found to be crucial for facet formation on Ir(210).[23] Nano-faceted Ir(210) in contact with aqueous sulfuric acid is easily characterised by a sharp voltammetric current peak around −0.2 V *vs.* SCE.[23] The similarity of the nano-pyramids obtained outside an UHV chamber with those reported by Madey *et al.*[14,21] has been verified by *in-situ* scanning tunnelling microscopy (STM).[23]

By the combination of theory and experiments we will demonstrate that choosing appropriate adsorbate and potential conditions it should be possible to electrochemically generate a reproducible and well-defined basis for studying catalytic reactions on unsupported monometallic nanostructures with controllable size and shape.

In the following, we will first describe the theoretical and experimental methods that were used to investigate the faceting of Ir(210). Afterwards, a brief description on calculations for oxygen-induced facet-formation on Ir(210) is given, which provides the basis for generating the electrochemical phase diagram. The theoretical prediction that facet formation should also be possible electrochemically was then studied experimentally. In this context, cyclic voltammetry and *in-situ* STM studies on the electrochemical behaviour of Ir(210) in contact with perchloric acid solution are described, followed by electrocatalytic investigations for different simple reactions.

Methods

Theoretical calculations

The energy required to form facets can be expressed as a sum of changes in the Gibbs free energies mainly related to surface, edge, kink and strain contributions:

$$\Delta G^{\text{form}} = \Delta G^{\text{surface}} + \Delta G^{\text{edge}} + \Delta G^{\text{kink}} + \Delta G^{\text{strain}} + \ldots \quad (1)$$

As long as the facets are large enough, such that contributions from step-edges, kinks, and strain are negligible compared to surface contributions, the overall formation energy can be approximated by the surface contribution only. This condition, usually referred as Herring-condition, is comparable to the so-called Wulff-construction. On the basis of this condition, facet formation should occur when

$$\Delta G^{\text{form}} \approx \Delta G^{\text{surface}} = \sum_f A_f^{\text{final}} \gamma_f^{\text{final}} - A^{\text{initial}} \gamma^{\text{initial}} < 0 \quad (2)$$

where the initial surface is characterized by a surface free energy γ^{initial} and an overall surface area A^{initial}, and the f th-face of the facets accordingly by γ_f^{final} and A_f^{final}. Since in the present case facets showing different faces are formed on the initially planar surface after adsorption of oxygen, eqn (2) converts into the following condition, which has to be fulfilled in order to show facet formation:

$$\frac{S_{311}}{\cos\theta_{311}}\gamma_{311}(T, a_{H_2O}, \Delta\phi) + \frac{S_{110}}{\cos\theta_{110}}\gamma_{110}(T, a_{H_2O}, \Delta\phi) < \gamma_{210}(T, a_{H_2O}, \Delta\phi) \quad (3)$$

Here S_{311} and S_{110} specify the partial contributions of the different faces to each pyramidal-shaped facet, while θ_{311} and θ_{110} are the tilt angles of the faces with respect to the initial substrate, T is the temperature, a the water activity, and $\Delta\phi$ the electrode potential. Experimentally and geometrically obtained values for S_f and θ_f are summarized in Table 1. It should be noted that since Ir(311) and Ir(31−1) show the same surface morphology, both have been combined.

The interfacial free energies γ, which are relevant for eqn (3), give the stability of the corresponding electrode/electrolyte-interfaces. As described in ref. 28 and 29, an exact evaluation of the interfacial free energies is in principle possible, but requires a self-consistent modeling of the entire interfacial region, which might range up to several 100 Å. Since this is currently beyond capabilities of *ab initio* approaches, we reduce our model to the electrode and the adlayer only and assume a constant influence of the electrolyte, allowing us to neglect its presence when studying relative stabilities only. Consequently, the interfacial free energy reduces to

Table 1 Partial surface contributions (S) and tilt angles (θ) for the two types of nanopyramids, those consisting of (311) and (110) faces and those consisting of (311) and (110)-superstructure faces

Surface	S	θ^{exp}/deg	Θ^{geom}/deg
(311)/(31−1)	0.70	18.7 ± 0.7[a]	19.29
(110)	0.30	19.0 ± 0.9[a]	18.43
(311)/(31−1)	0.47	18.7 ± 0.7[a]	19.29
(110)-superstr.	0.53	7.0 ± 1.0[a]	7.13

[a] Ref. 21.

$$\gamma(T, a_{H_2O}, a_{Ir}, \Delta\phi) = \frac{1}{A}\big[G(T, a_{H_2O}, a_{Ir}) - N_{Ir}g_{Ir}^{bulk}(T, a_{Ir}) \\ - N_O\big(\mu_{H_2O}(T, a_{H_2O}) + 2e\Delta\phi\big)\big] \quad (4)$$

where the last term comes from the assumption that every oxygen that adsorbs on the surface induces the formation of facets originates from a water-splitting reaction in the bulk-electrolyte, allowing us to reference the electrode potential to the reversible hydrogen electrode (RHE).. In eqn (4), G is the Gibbs energy of the interface, which is now reduced to the electrode surface plus the adlayer and g_{Ir}^{bulk} is the Gibbs energy of Ir-bulk, which is one of the reservoirs the system should be in contact with. The properties of the second reservoir (*i.e.*, water) are determined by the chemical potential of the water being present in the electrolyte

$$\mu_{H_2O}(T, a_{H_2O}) = \bar{\mu}_{H_2O}(T, a^0) + k_B T \ln\left(\frac{a_{H_2O}}{a^0}\right) \quad (5)$$

where the first term on the right side denotes the standard chemical potential at temperature T and a water activity of one.[30]

With eqn (4) we now can evaluate (approximate) the interfacial free energies of the different surface faces of the facets, since all relevant quantities can be deduced from first principles, here density functional theory calculations. These can then be used together with eqn (3) to finally obtain the electrochemical phase diagram.

In order to calculate the total energies of different surface structures, which are required for eqn (4), we performed DFT slab calculations using the CASTEP code[31] with Vanderbilt-type ultrasoft pseudopotentials[32] and the generalized gradient approximation (GGA) exchange–correlation functional proposed by Perdew, Burke and Ernzerhof (PBE).[33] Layer-converged supercells consisting of 16-layer slabs for Ir(210), 11-layer slabs for Ir(311), 12-layer slabs for Ir(110), and 7-layer slabs for Ir(110)-superstructure were used to model oxygen adsorption with different coverages and adlayer structures. To decouple the interactions between neighboring slabs in the supercell geometry, repeated slabs were separated by a ~12 Å vacuum. For Ir(210), Ir(311) and Ir(110)-superstructure, the bottom three layers, and for Ir(110) the bottom four layers, were fixed at the calculated bulk structure, while the geometry of the remaining layers plus adsorbates were fully optimized (to <0.03 eV Å^{-1}). The Brillouin zones of the (1 × 1)-surface unitcells of Ir(210), Ir(311), Ir(110), and the superstructure were sampled with 10 × 8, 14 × 8, 14 × 10, and 4 × 4 Monkhorst-Pack k-point meshes, respectively. Finally, a plane-wave basis set with an energy cutoff 340 eV was used.

Investigating the error sources related to slab thickness, vacuum size, plane-wave cutoff and k-point mesh, we found the maximum overall error bar in the surface free energy to be <5 meV Å^{-2}, when using optimized values for each parameter.

Throughout this section oxygen binding energies are with respect to half an gas-phase oxygen molecule.

Experimental

The Ir(210) single crystal (MaTecK, Jülich, Germany) was a cylinder of 4 mm height and 4 mm diameter. The surface had been polished down to 0.03 μm and oriented to better than 1°. The electrode was heated up to 1000 °C for about 30 s by an induction coil in presence of inert (nitrogen) or reducing gas (nitrogen + hydrogen), cooled down slowly in the same atmosphere and immersed into 0.1 M H_2SO_4 solution under potential control. A conventional three electrode glass cell was used for electrochemical measurements. A platinum wire and a saturated calomel electrode (SCE) were used as counter and reference electrode, respectively. The solution was prepared from suprapure chemicals and ultrapure water (18.2 MΩ cm at

25 °C, total organic carbon < 1 ppb). After voltammetric experiments, the electrode was eventually transferred directly to an STM cell without further surface preparation. Pt wires were used in the STM cell as counter and reference electrodes. The STM images were recorded with a Digital Instruments Nanoscope III (Santa Barbara, California). For the preparation of the STM tips, a Pt/Ir wire (80/20) was etched in 4.5 M NaCN and coated with an electrophoretic paint to reduce the Faraday current at the tip to below 50 pA.

Results and discussion

Electrochemical surface phase diagram

UHV-experiments on faceting of Ir(210) performed in the group of Madey[14,21] showed that adsorption of oxygen is capable of stabilizing the faceted surface morphology. Motivated by these observations, in a previous work we described theoretical studies on facet formation on Ir(210) in contact with a gaseous atmosphere by first studying the oxygen adsorption on each relevant Ir-surface (see Fig. 1) separately and then combining the obtained energetics to an overall (p_{O_2}, T)-phase diagram.[34] There we had seen that the presence of surface oxygen enhances the anisotropy in surface free energy for the different orientations, which finally causes the faceted surface to become thermodynamically favorable. Since even for the smallest experimentally prepared nanopyramids with 4 nm base-length we obtained good agreement between theory and experiment, this can be used as upper limit nanostructure size at which the Herring-condition is applicable. However, further investigations will aim on evaluating step-edge, kink and strain contributions.

Besides the external parameters temperature and pressure, it is well-known that under electrochemical conditions even the electrode potential is able to cause surface oxidation. Therefore, in the following we will combine both concepts, the oxygen-induced surface faceting and the potential-induced electrooxidation. We will show that under electrochemical conditions, potential-induced surface faceting should be possible.

In order to generate the electrochemical phase diagram shown in Fig. 2, we made use of the assumption that oxygen binding energies as well as the electrolyte structure and properties should be potential-independent, which in turn allowed us to directly use the DFT-energies that were obtained when investigating the gas-phase system.[34] This is certainly a strong assumption and different theoretical studies have been performed on the role of surrounding water on binding energies and reaction barriers.[35-40] But in previous studies on Pt-oxide formation we were able to reproduce the experimental CV-curve on the basis of this approximation. However, so far it is not clear how the surrounding water might influence the nanostructured surface that shows a variety of lower coordinated sites (*e.g.* step-edges or kinks).

On the basis of this approach, we distinguish between clean and oxygen-covered surfaces of:

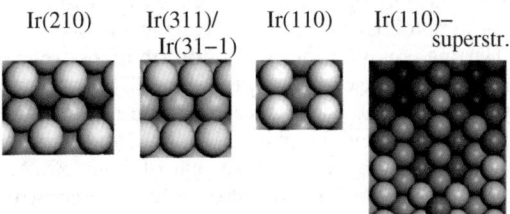

Fig. 1 Hard-sphere models of Ir(210), Ir(311), Ir(110), and Ir(110)-superstructure, which are the surfaces relevant for faceting of Ir(210).

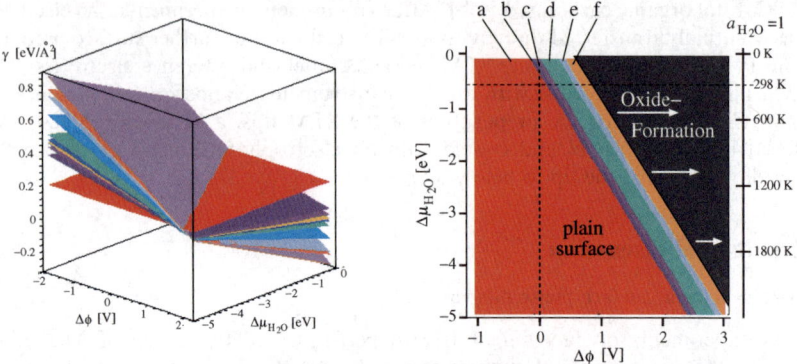

Fig. 2 $(a, T, \Delta\phi)$-phase diagram for the electrochemical faceting of Ir(210) in an aqueous electrolyte. The left figure shows the interfacial free energy γ as function of the water chemical potential and electrode potential (referenced to RHE), while the right figure shows the view to the bottom. In addition, the temperature scale, which corresponds to $a = 1$, is given on the right side of the phase diagram. The structure-labeling corresponds to the models shown in Fig. 3.

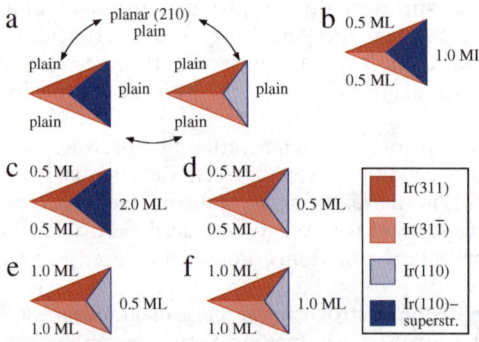

Fig. 3 Models of the different surface structures that are present on Ir(210) at specific electrode potentials (see Table 2). For all surfaces 1 ML is defined as one oxygen atom per surface unit cell.

- planar Ir(210),
- nanopyramids with (311), (31−1) and (110)-regular faces,
- and nanopyramids with (311), (31−1) and (110)-superstructure faces.

The surface free energies of the faceted surfaces were calculated using the left side of eqn (3) with the parameters for the partial surface areas and facet tilt angles summarized in Table 1. Each summand of this equation was evaluated by eqn (4), where the main temperature- and pressure-dependence is assumed to be dominated by the water chemical potential eqn (5). As already mentioned above, this approach is based on the Herring-condition, in which contributions from step-edges, kinks and surface stress or strain are considered to be small.

Although experimentally the coexistence of (110)-regular and (110)-superstructure was observed on most nanopyramids, we only consider the extremes in which the entire (110)-faces of all pyramids have one of both structures at the same time. However, in the following we will also discuss the consequences of having a mixture (coexistence).

Since all binding energies were calculated for oxygen adsorption on pure Ir, the following discussion is restricted to the part of the phase diagram where the IrO_2

bulk-oxide is not stable yet. Equivalent to eqn (4), this can be translated to the following stability condition, which restricts the electrode potential range of interest

$$+1.29 \text{ eV} - 2e\Delta\phi < \mu_{H_2O} < 0 \text{ eV}. \tag{6}$$

Fig. 2 shows the final phase diagram, where on the left side γ is plotted against the water chemical potential and the electrode potential referenced to half an oxygen molecule

$$\Delta\mu_{H_2O} = \mu_{H_2O} - \frac{1}{2}E_{O_2}^{tot} \tag{7}$$

while the bottom view to this plot is shown on the right side. Since $\Delta\mu_{H_2O}$ depends on temperature and water activity, we also added the temperature scale, which corresponds to an ideal solution/solvent with $a_{H_2O} = 1$. The phase diagram makes apparent that at room temperature (dashed horizontal line) and without externally applied electrode potential (dashed vertical line), no oxygen is adsorbed on the surface, which is either purely planar or shows a coexistence between planar and faceted Ir(210). The latter conclusion is motivated by two aspects. Although we find the lowest surface free energy for a faceted surface in which the nanopyramids consist of (311), (31−1) and (110)-superstructure faces, the second type of nanofacets, that show regular Ir(110) instead of the superstructure, as well as the planar non-faceted Ir(210) are only 3–4 meV Å^{-2} less stable, which is certainly within the accuracy of the calculations. Furthermore, so far we have neglected the presence of step-edges and kinks, which usually act as destabilization. Since these contributions, which are only relevant for the faceted surfaces, can be estimated to be in the same energy range, it is most likely that the planar Ir(210) becomes the most stable structure under these conditions. This would be in agreement with the (UHV) experimental observation of a phase transition from the clean faceted to the clean planar Ir(210) at temperatures above ∼600 K. To combine this discussion, so far we assume the coexistence of planar and faceted Ir(210) for $\Delta\phi < 0.1$ V.

Above 0.1 V adsorption of oxygen takes place, causing the formation or stabilization of the nanofaceted surface. While at lower electrode potentials (0.1 V > $\Delta\phi$ > 0.26 V) nanopyramids of the type (311)/(31−1)/(110)-superstructure are most stable, at more positive potentials (0.26 V > $\Delta\phi$ > 0.85 V) the (110)-superstructure face is replaced by regular (110)-(1 × 1). In this range, increasing the potential does not change the structure, but causes the coverage of oxygen on the facets to increase. Finally the IrO_2 bulk-oxide appears as stable phase for electrode potentials above 0.85 V. Overall, we find the surfaces phases summarized in Table 2. Interestingly, the overall phase transition behavior is qualitatively comparable to the temperature decrease in the gas-phase UHV system.

Table 2 Potential ranges at which the different phases are thermodynamically stable (for labeling, see Fig. 2)

Phase	Potential range/V
a	$\Delta\phi < 0.10$
b	$0.10 < \Delta\phi < 0.18$
c	$0.18 < \Delta\phi < 0.26$
d	$0.26 < \Delta\phi < 0.57$
e	$0.57 < \Delta\phi < 0.65$
f	$0.65 < \Delta\phi < 0.85$
oxide	$0.85 < \Delta\phi$

Finally, regarding the phase diagram, two aspects should be mentioned. While clean planar Ir(210) is most probably the stable phase at low electrode potentials (phase a), oxygen coverages of around or above 0.5 ML are required in order to stabilize the nanofacets. If one would only concentrate on coverages below this value, oxygen-covered planar Ir(210) would become stable at potentials above 0.13 V. This coverage-dependence is in agreement with experimental observations.[14,21]

Moreover, the presence of the (110)-superstructure face at potentials of 0.1 V > $\Delta\phi$ > 0.26 V is rather remarkable. On a planar Ir(110) surface this superstructure is always less favorable than regular (110), but this is different for the faceted Ir(210) surface. There, the superstructure forms on the (110)-side of the nanopyramids at lower potential, which is a consequence of the nonlinear dependency of the surface free energy on the tilt angle [see prefactors on the left side of eqn (3)] and the fact that the (110)-faces of the nanopyramids are already tilted with respect to the (210)-substrate. Again, this behavior is also confirmed experimentally.[21]

In order to evaluate the influences coming from choosing the PBE exchange–correlation functional, we recalculated the most relevant surface structures with the LDA functional and generated the equivalent surface phase diagram. Comparison shows that with the LDA functional all phase transitions are shifted toward lower potentials, without causing any changes in the ordering of the stable phases. Furthermore, the stability ranges, respectively chemical potential ranges, of the different phases are almost the same with both xc-functionals. Therefore, it is reasonable to assume that the conclusions drawn above are qualitatively independent of the xc-functional.

Electrochemical behaviour

Differences in the electrochemical behaviour of planar and faceted Ir(210) in acidic solutions are obviously manifested in the hydrogen adsorption region. While voltammograms for planar Ir(210) are almost featureless, the faceted surface is characterized by narrow peaks, the position and shape of which are strongly dependent on electrolyte anions. In the case of 0.1 M H_2SO_4, a sharp voltammetric peak located around −0.2 V can be taken as indicator for the presence of nano-facets, as proven by STM measurements.[23] While the Ir(210) surface structure was found to be unchanged in 0.1 M H_2SO_4, the stability has been tested with other electrolytes. In Fig. 4, voltammograms for planar and faceted Ir(210) in 0.1 M $HClO_4$ are shown.

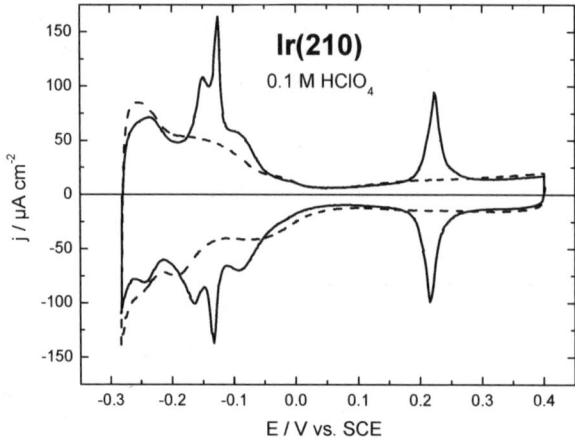

Fig. 4 Cyclic voltammograms for planar (dashed line) and nano-faceted (solid line) Ir(210) in 0.1 M $HClO_4$. Scan rate: 50 mV s^{-1}.

It is well-known that the perchlorate anion can easily be reduced by iridium to chloride. Therefore, stable curves representing stationary behaviour show slightly smaller peaks as the curves in Fig. 4 for the 1st cycles. A variety of peaks can be discerned in the hydrogen adsorption region, *i.e.*, at potentials between −0.3 and +0.1 V. These might be attributed to well-ordered domains of unreconstructed Ir(210) in the case of the planar surface or to the (110) and (311) faces for the faceted surface. However, since details on the electrochemical behaviour of Ir(110) and Ir(311) are not available yet, a direct assignment of the peaks in Fig. 4 is not possible. When going towards the faceted surface, three peaks are emerging between −0.2 and 0 V, while the peak at −0.25 V is slightly decreasing. Even more evidently, a relatively sharp reversible peak at 0.22 V is related to the presence of nano-facets on the surface (Fig. 4). This peak is probably related to the initial stages of OH or O adsorption from water and absent in the case of planar Ir(210). Restricting the positive potential limit to 0.4 V did not give rise to significant changes in the voltammograms. Thus, we conclude that simple potential variation does not lead to an electrochemical faceting of planar Ir(210) at room temperature.

The STM image shown in Fig. 5 was recorded for nano-faceted Ir(210) in 0.1 M $HClO_4$ after measuring the respective voltammogram in the electrochemical cell (see Fig. 4). The electrode was contacted with the solution in the STM cell at 0.4 V. The surface consists of pyramids with facets of different orientation corresponding to slightly different size. It can be seen in Fig. 5 that the pyramids are uniformly distributed. Moreover, there are no planar zones alternating with the pyramids. The vertical distance between two consecutive pyramidal-extremes (apex/valley) never exceeds 3 nm and the average width at the base of the pyramids is 30 nm. It should be mentioned that the size of the pyramids changes slightly for various measurements, pointing out that subtle differences in the preparation procedure (*e.g.*, different annealing times and/or temperatures) may influence the faceting process. By carefully watching the disposition of the pyramids, some order can be seen. The pyramids seem to share one edge along which they align. This could be a consequence of the faceting mechanism.

The STM measurements give also the opportunity to estimate the tilt angle that the different facets form with the horizontal plane of the STM images. The (311) facet forms a tilt angle of 18 ± 2°, a value already obtained with different techniques by Madey and co-workers.[14] For the other facet, the angle is 9 ± 2° and indicates the

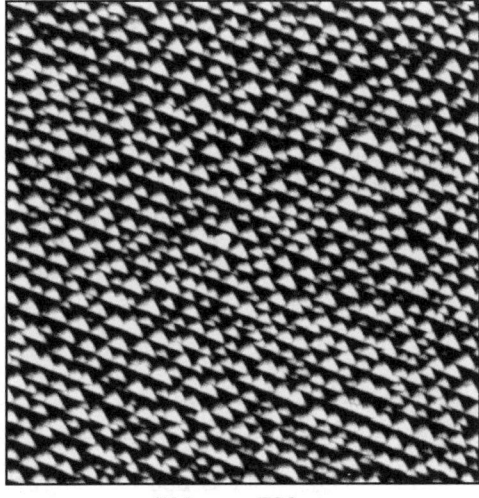

700nm x 700nm

Fig. 5 STM-image for nano-faceted Ir(210) in 0.1 M $HClO_4$ at 0.4 V.

presence of a superstructure formed by the Ir atoms of the top most plane of a reconstructed (110) facet.[21] The model proposed by Madey and co-workers,[21] which identifies the geometry of the pyramids as made by two (311) and one (110), describes correctly the experimental results reported here.

Since the adsorption of oxygen was seen to provoke facet formation, an attempt was made to apply more positive potential values, where surface oxidation is expected to take place. Such experiments were also performed with 0.1 M H_2SO_4 to rule out any disturbance by perchlorate reduction. In addition, the differences in the voltammograms for planar and faceted Ir(210) are much more pronounced for sulfuric acid compared with perchloric acid solution. Although slight changes in the voltammetric profile could be discerned after potential excursions to 0.8 V, a straightforward electrochemical method for nanofacet formation on Ir(210) as function of electrode potential has still to be explored.

Electrocatalytic behaviour

Since both the planar and the faceted Ir(210) surface represent stable and well-ordered systems of distinct structure, they are predestined for model studies in electrocatalysis. Furthermore, the faceted structure of Ir(210) constitutes a highly ordered array of (110) and (311) oriented nano-pyramids of narrow size distribution for a single component, *i.e.*, without foreign substrate as it is common for metal nanostrucures. Not comprehending the mechanistic details, for both planar and faceted Ir(210) the onset of oxidation or reduction was tested for a series of rather simple electrocatalytic reactions. The latter involved (i) adlayer oxidation of carbon monoxide, (ii) formic acid oxidation, (iii) hydrogen evolution reaction, (iv) oxygen reduction, and (v) nitrous oxide reduction. Without going into details, the main result for all of these reactions is that the reaction overpotential is larger for the faceted Ir(210) surfaces compared with the planar one in all cases. The adlayer oxidation of CO for the two types of Ir(210) is shown in Fig. 6 as representative example. The difference in the onset potential is more than 0.1 V. CO adlayer oxidation on the planar Ir(210) surface starts around 0.2 V. The presence of two distinct oxidation peaks at 0.25 V and 0.4 V suggests that diffusion of reaction partners may be involved in the mechanism. It is likely that surface defects, such as monoatomic-high steps, act as active centres for CO oxidation. These sites are absent for the faceted Ir(210) surface, which explains the higher overpotential. In essence, it can

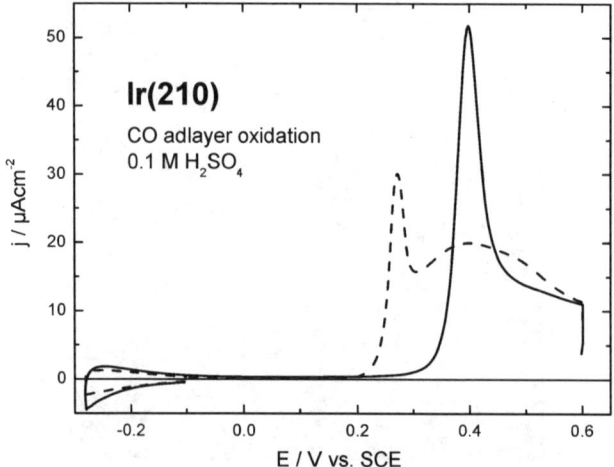

Fig. 6 Current–potential curves for CO adlayer oxidation on planar (dashed line) and on nano-faceted (solid line) Ir(210) in 0.1 M H_2SO_4. Scan rate: 10 mV s^{-1}.

be concluded that the planar surface is, in general, more active than the faceted surface. Changes in the local geometry of the Ir surfaces thus have significant impact on the interaction with adsorbates and therefore on reaction rates, which is in agreement with the findings for oxidation of CO under UHV conditions.[4]

It should be mentioned that higher overpotentials have also been reported for CO monolayer oxidation on Pt nanoparticles supported on glassy carbon electrodes.[41] Spatially confined formation of oxygen containing species at active sites and slow diffusion of CO molecules to the active sites were given as main reasons for slower kinetics compared to extended Pt surfaces.[41] While size effects have been addressed for the latter Pt systems,[42] it still remains to be a challenge to significantly vary the size of the nano-facets on Ir(210). However, the faceted Ir(210) surface can ideally serve as a model system to study structure sensitivity of electrocatalytic reactions.

Conclusion

In this paper we have shown that surface faceting of Ir(210) should not only be a phenomenon observable under gas-phase or UHV-conditions, respectively, but should also be possible under electrochemical conditions. There, the important parameter to tune the surface morphology is the electrode potential. Focusing on the former parameter, our theoretical studies showed that although the process itself might be kinetically hindered at room temperature, nanofacets become thermodynamically stable above +0.1 V. Since this potential range is easily accessible by experiments, we afterwards presented first cyclic voltammetry and *in-situ* STM measurements on planar and nanofaceted Ir(210). After the morphological characterization, we further investigated the electrocatalytic behavior of both surface structures with different simple reactions, finding an increased overpotential for faceted Ir(210). Future studies will aim to develop suitable methods, so that facet formation can be induced electrochemically by the electrode potential.

Acknowledgements

P.K. and T.J. gratefully acknowledge support by the "Fonds der Chemischen Industrie" (FCI), the "Deutscher Akademischer Austauschdienst" (DAAD), and the "Deutsche Forschungsgemeinschaft" (DFG) within the Emmy-Noether-Program.

References

1 T. E. Madey, J. Guan, C.-H. Nien, C.-Z. Dong, H.-S. Tao and R. A. Campbell, *Surf. Rev. Lett.*, 1996, **3**, 1315.
2 T. E. Madey, C.-H. Nien, K. Pelhos, J. J. Kolodziej, I. M. Abdelrehim and H.-S. Tao, *Surf. Sci.*, 1999, **438**, 191.
3 Q. Chen and N. V. Richardson, *Prog. Surf. Sci.*, 2003, **73**, 59.
4 W. Chen, I. Ermanoski, T. Jacob and T. E. Madey, *Langmuir*, 2006, **22**, 3166.
5 W. Chen, I. Ermanoski, Q. Wu, T. E. Madey, H. H. Hwu and J. G. Chen, *J. Phys. Chem. B*, 2003, **107**, 5231.
6 W. Chen, I. Ermanoski and T. E. Madey, *J. Am. Chem. Soc.*, 2005, **127**, 5014.
7 R. Bachelet, S. Cottrino, G. Nahélou, V. Coudert, A. Boulle, B. Soulestin, F. Rossignol, R. Guinebretière and A. Dauger, *Nanotechnology*, 2007, **18**, 015301.
8 H. Wang, M. Reyhan, T. E. Madey, unpublished data.
9 K.-J. Song, J. C. Lin, M. Y. Lai and Y. L. Wang, *Surf. Sci.*, 1995, **327**, 17.
10 D. B. Danko, M. Kuchowicz and J. Kolacziewicz, *Surf. Sci.*, 2004, **552**, 111.
11 R. E. Kirby, C. S. McKee and M. W. Roberts, *Surf. Sci.*, 1976, **55**, 725.
12 R. E. Kirby, C. S. McKee and L. V. Renny, *Surf. Sci.*, 1980, **97**, 457.
13 M. Sander, R. Imbihl, R. Schuster, J. V. Barth and G. Ertl, *Surf. Sci.*, 1992, **271**, 159.
14 I. Ermanoski, K. Pelhos, W. Chen, J. S. Quinton and T. E. Madey, *Surf. Sci.*, 2004, **549**, 1.
15 J. Gustafson, A. Resta, A. Mikkelsen, R. Westerström, J. N. Andersen, E. Lundgren, J. Weissenrieder, M. Schmid, P. Varga and N. Kasper, *Phys. Rev. B: Condens. Matter Mater. Phys.*, 2006, **74**, 035401.
16 P. J. Knight, S. M. Driver and D. P. Woodruff, *Surf. Sci.*, 1997, **376**, 374.

17 S. Vollmer, A. Birkner, S. Lucas, G. Witte and C. Wöll, *Appl. Phys. Lett.*, 2000, **76**, 2686.
18 N. Reinecke and E. Taglauer, *Surf. Sci.*, 2000, **454**, 94.
19 D. A. Walko and I. K. Robinson, *Phys. Rev. B: Condens. Matter Mater. Phys.*, 2001, **64**, 045412.
20 I. Ermanoski, W. Swiech and T. E. Madey, *Surf. Sci.*, 2005, **592**, L299.
21 I. Ermanoski, C. Kim, S. P. Kelty and T. E. Madey, *Surf. Sci.*, 2005, **596**, 89.
22 H. Wang, A. S. Y. Chan, P. Kaghazchi, T. Jacob and T. E. Madey, *ACS Nano*, 2007, **1**(5), 449.
23 K. A. Soliman, F. C. Simeone, L. A. Kibler, in preparation.
24 S. Motoo and N. Furuya, *J. Electroanal. Chem.*, 1984, **167**, 309.
25 S. Motoo and N. Furuya, *J. Electroanal. Chem.*, 1984, **172**, 339.
26 T. Pajkossy, L. A. Kibler and D. M. Kolb, *J. Electroanal. Chem.*, 2005, **582**, 69.
27 T. Pajkossy, L. A. Kibler and D. M. Kolb, *J. Electroanal. Chem.*, 2007, **600**, 113.
28 T. Jacob, *J. Electroanal. Chem.*, 2007, **607**, 158.
29 T. Jacob, M. Scheffler, to be submitted.
30 *JANAF Thermochemical Tables*, ed. D. R. Stull and H. Prophet, U.S. National Bureau of Standards, U.S. EPO, Washinghton, DC, 2nd edn, 1971.
31 M. D. Segall, P. L. D. Lindan, M. J. Probert, C. J. Pickard, P. J. Hasnip, S. J. Clark and M. C. Payne, *J. Phys.: Condens. Matter*, 2002, **14**, 2717.
32 D. Vanderbilt, *Phys. Rev. B: Condens. Matter Mater. Phys.*, 1990, **41**, 7892.
33 J. P. Perdew, K. Burke and M. Ernzerhof, *Phys. Rev. Lett.*, 1996, **88**, 3865.
34 P. Kaghazchi, W. Chen, H. Wang, I. Ermanoski, T. E. Madey and T. Jacob, *ACS Nano*, 2008, **2**, 1280.
35 C. Hartnig, P. Vassilev and M. T. M. Koper, *Electrochim. Acta*, 2003, **48**, 3751.
36 P. Vassilev and M. T. M. Koper, *J. Phys. Chem. C*, 2007, **111**, 2607.
37 C. D. Taylor and M. Neurock, *Curr. Opin. Solid State and Mater. Sci.*, 2005, **9**, 49.
38 Z. Gu and P. B. Balbuena, *J. Phys. Chem. A*, 2006, **110**, 9783.
39 G. S. Karlberg, T. F. Jaramillo, E. Skúlason, J. Rossmeisl, T. Bligaard and J. K. Nørskov, *Phys. Rev. Lett.*, 2007, **99**, 126101.
40 A. Roudgar and A. Gross, *Chem. Phys. Lett.*, 2005, **409**, 157.
41 O. V. Cherstiouk, P. A. Simonov, V. I. Zaikovskii and E. R. Savinova, *J. Electroanal. Chem.*, 2003, **554/555**, 241.
42 F. Maillard, E. R. Savinova and U. Stimming, *J. Electroanal. Chem.*, 2007, **599**, 221.

Nanoparticle catalysts with high energy surfaces and enhanced activity synthesized by electrochemical method†

Zhi-You Zhou, Na Tian, Zhi-Zhong Huang, De-Jun Chen and Shi-Gang Sun*

Received 4th March 2008, Accepted 9th May 2008
First published as an Advance Article on the web 14th August 2008
DOI: 10.1039/b803716g

Electrochemical shape-controlled synthesis of metal nanocrystal (NC) catalysts bounded by high-index facets with high surface energy was achieved by developing a square-wave potential route. Tetrahexahedral Pt NCs with 24 {$hk0$} facets, concave hexoctahedral Pt NCs with 48 {hkl} facets, and multiple twinned Pt nanorods with {$hk0$} facets were produced. The method was employed also to synthesize successfully trapezohedral Pd NCs with 24 {hkk} facets, and concave hexoctahedral Pd NCs with 48 {hkl} facets. It has been tested that, thanks to the high-index facets with high density of atomic steps and dangling bonds, the tetrahexahedral Pt NCs exhibit much enhanced catalytic activity for equivalent Pt surface areas for electrooxidation of small organic fuels such as ethanol. These results demonstrate that the developed square-wave potential method has surmounted the limit of conventional chemical methods that could synthesize merely metal nanocrystals with low surface energy, and opened a new prospect avenue in shape-controlled synthesis of nanoparticle catalysts with high surface energy and enhanced activity.

1. Introduction

Metal nanoparticles display often novel properties thanks to their nanosize effects, surface effects, and other unique effects.[1,2] In particular, nanoparticles of platinum group metals (PGM: Pt, Pd, Rh, Ir, Ru and Os) play a vital role as catalysts for many important reactions applied in industrial chemical processing, in petroleum reform, in motor vehicle catalytic converters that reduce exhaust pollution, in fuel cells and in sensors.[3-6] Since the price of PGM is extremely high due to the rare reserve of the PGM on the earth and the continuing increase in demand, to find and design new type PGM catalysts with higher activity and stability are therefore key issues in development of the above momentous fields.[7] At present, the commercially available Pt nanocrystal (NC) catalysts exist often as cubes, cuboctahedra, tetrahedra and octahedra, which are enclosed with low energy surfaces characterized by "low-index" facets such as {100} and {111}.[8] From the knowledge gained in studies of model catalysis using single crystal planes, it is known that the "low-index" planes do not present, in general, good catalytic properties; and instead, the "high-index" planes with high densities of atomic steps, ledges, and kinks display

State Key Laboratory of Physical Chemistry of Solid Surfaces, Department of Chemistry, College of Chemistry and Chemical Engineering, Xiamen University, Xiamen, 361005, China. E-mail: sgsun@xmu.edu.cn

† The HTML version of this article has been enhanced with colour images.

often higher catalytic activity and stability.[9–11] Unfortunately, it is rather challenging to synthesize metal nanoparticles enclosed by high-index facets. The surface energy of different crystal planes of face-centered cubic (fcc) metals such as Pt is increased in the order of $\gamma_{\{111\}} < \gamma_{\{100\}} < \gamma_{\{110\}} < \gamma_{\{hkl\}}$.[12] As a consequence, the growth rate in the direction perpendicular to a high-index facet with high surface energy is much faster than that along the normal direction of a low-index facet, which results in a rapid disappearance of high-index facets during the formation of nanoparticles,[13] and yields NCs with shapes of cube, cuboctahedron, tetrahedron and octahedron, as those produced in shape-controlled synthesis through conventional chemical ways.[14–16]

We have developed, recently, an electrochemical square-wave potential method to control the growth and the surface structure of metal nanoparticles, and synthesized successfully tetrahexahedral (THH) Pt NCs.[17,18] This new Pt NC is enclosed by 24 high-index facets of {730} and vicinal planes, and exhibits a superior catalytic activity and stability; its electrocatalytic activity per unit area can be as much as four times of the existing commercial Pt catalysts.

In the current paper, we will present our new results concerning synthesis of nanoparticle catalysts of PGM enclosed by high energy surfaces, including (1) single crystalline nanoparticles of different shapes, such as tetrahexahedron (THH) enclosed by {hk0} facets, trapezohedron by {hkk}, and concave hexoctahedron by {hkl}; (2) multiple twinned crystals. The successful synthesis of nanoparticle catalysts enclosed by high-index facets has promoted effectively the study that bridges the fundamental gained from model catalysts of single crystal planes with the design and fabrication of real catalysts of high performances.

2. Correlation between crystal planes and nanocrystal shape

For fcc metals, such as Pt, a unit stereographic triangle (Fig. 1) is usually employed to illustrate the coordinates of different crystal planes.[19] The three low-index or basal planes, *i.e.* (111), (100), and (110) locate at three vertexes. Among them, the (111) and (100) planes are flat with closely packed surface atoms, whereas the (110) plane is rough with step atoms. The coordination numbers (CNs) of top-layer atoms on (111), (100), and (110) are 9, 8, and 7, respectively. Other planes located at the sidelines and inside of the triangle are high-index planes. The three sidelines represent [01$\bar{1}$], [1$\bar{1}$0], and [001] crystallographic zones, the planes lying in these zones exhibit terrace-step structure.[20] The CNs of step atoms on the planes belonging to the [001] zone is 6,

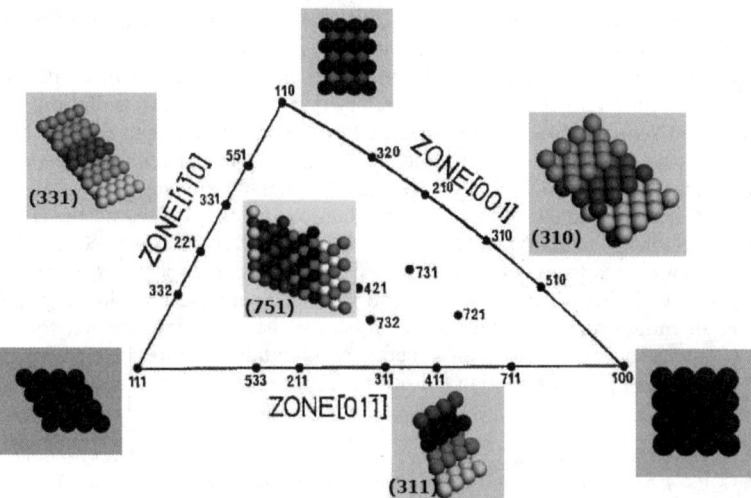

Fig. 1 Unit stereographic triangle of fcc single-crystal and models of surface atomic arrangement.

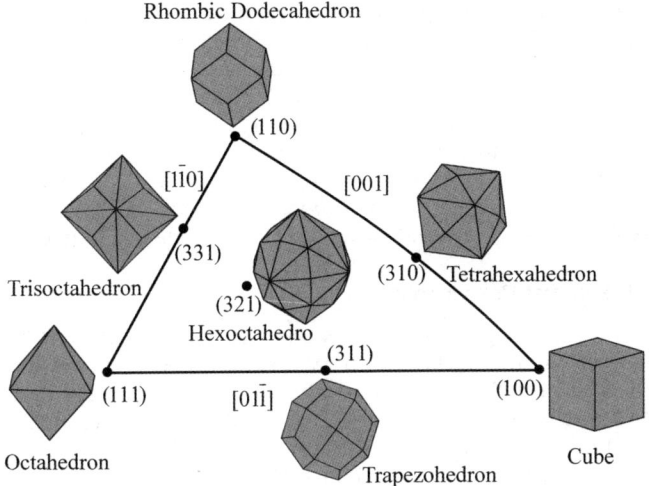

Fig. 2 Unit stereographic triangle of polyhedral nanocrystals bounded by different crystal planes.

and 7 for planes lying in the other two zones. In the [001] zone, the Pt(210) plane possesses the highest density of step atoms (5.81×10^{14} cm^{-2}). Along with decreasing the width of (100) terraces, the density of step atoms decreases to 4.11×10^{14} cm^{-2} on Pt(310), and to 2.55×10^{14} cm^{-2} on Pt(510). The planes inside the triangle are distinguished by their chirality and kink atoms with CNs of 6. These low-coordinated step and kink atoms (CNs = 6–7) intrinsically exhibit very high chemical activity.[21]

In analogue with the unit stereographic triangle, there is also an intrinsic triangle that coordinates the crystal surface index and the shape of metal NCs,[22] as shown in Fig. 2. Three vertexes represent the coordinates of polyhedral nanocrystals bounded by basal facets, *i.e.* cube covered by {100}, octahedron by {111}, and rhombic dodecahedron by {110}. The polyhedral NCs lying in the sidelines of the triangle are THH bounded by {*hk*0} facets, trapezohedra by {*hkk*} facets, and trisoctahedra by {*hhl*} facets. They all have 24 facets. The polyhedra situated inside the triangle are hexoctahedra bounded by 48 {*hkl*} (*h* > *k* > *l* > 0) facets. These polyhedra belong to Catalan solids or Archimedean duals,[22] and their shapes are complex and unconventional. To identify them quickly, the THH can be considered as a cube with each face capped by a square pyramid; the trisoctahedron can be considered as an octahedron with each face capped by a pyramid. It is worthwhile to note that along with variation of the geometric parameters of NCs, their surface structure can be changed following those planes as illustrated in the unit stereographic triangle shown in Fig. 1. Taking the THH as an example, the relationship between the facet index (*hk*0) and the nanocrystal parameters, *i.e.* the height of the square pyramid, *b*, and the side length of the cube, *a*, is expressed as follows:

$$\frac{b}{a} = \frac{k}{2h} \tag{1}$$

3. Electrochemically shape-controlled synthesis of Pt, Pd nanocrystals with high-index facets

We have developed an electrochemical route for synthesis of PGM nanocrystals bounded by high-index facets.[17] In brief, the electrochemical synthesis was carried out in a standard three-electrode cell at room temperature (about 25 °C). The working electrode was glassy carbon (GC, $\phi = 6$ mm), the counter and reference electrodes were a Pt foil and a saturated calomel electrode (SCE), respectively.

Electrode potential was controlled by a PAR 263A potentiostat/galvanostat (EG&G) *via* home-developed software that can generate arbitrary potential waveform. The electrochemical synthesis procedures include: (1) Pt nanospheres of ~750 nm in diameter were deposited on the GC substrate by pulse electrodeposition in 2 mM K_2PtCl_6 + 0.5 M H_2SO_4 solution; (2) the Pt nanospheres were then subjected to a treatment of square-wave potential at 10 Hz, with upper potential of 1.20 V and lower potential between −0.10 to −0.20 V, in a solution of 0.1 M H_2SO_4 + 30 mM ascorbic acid for 5–60 min. The Pt NCs bounded by high-index facets were grown exclusively on GC surface at the expense of Pt nanospheres.

3.1 Tetrahexahedral Pt nanocrystals

Fig. 3 shows the scanning electron microscopy (SEM) images of THH Pt NCs with different sizes by controlling the growth time. Three perfect square pyramids can be seen clearly, which is fully consistent with the THH model. The size distribution of THH Pt NCs is relatively narrow with relative standard deviation ranging from 10% to 15%, and the yield in the final products is over 90%.

The surface structure (Miller indices) of THH Pt NCs was identified to be mainly {730} facets by high-resolution transmission electron microscopy (HRTEM) and selected-area electron diffraction (SAED). Besides, some THH Pt NCs bounded

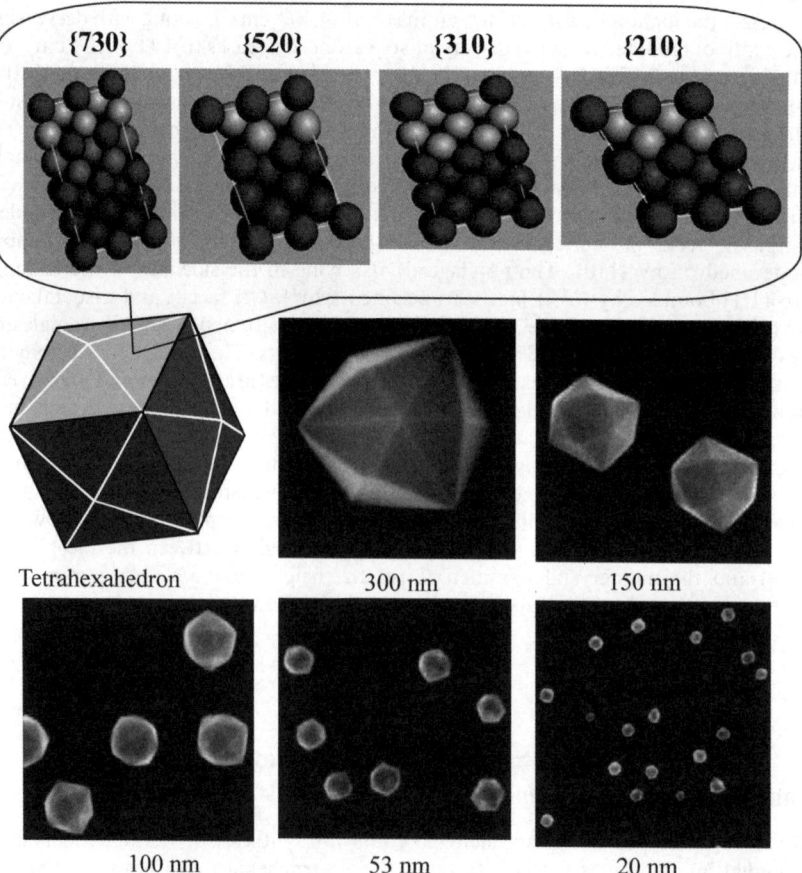

Fig. 3 Model of a tetrahexahedron (THH) and atomic arrangement structure of {$hk0$} high-index planes (above). SEM images of THH Pt NCs of different sizes prepared by an electrochemical square-wave potential method and supported on GC.

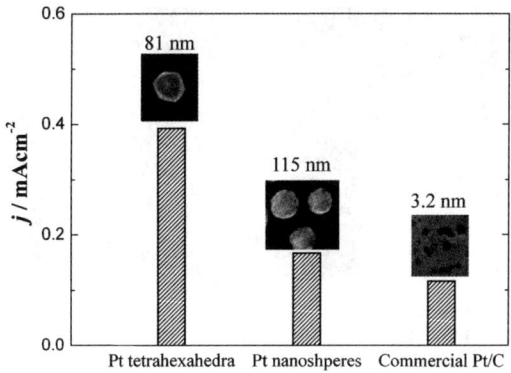

Fig. 4 Comparison of electrocatalytic activity (current density) towards ethanol oxidation at 0.25 V on THH Pt NCs (81 nm), Pt nanospheres (115 nm, the sphere is composed of primary Pt nanoparticles of ~3–5 nm), and commercial Pt/C (3.2 nm) catalysts.

by {210}, {310}, or {520} were also observed. The atomic arrangements of these surfaces are also illustrated in Fig. 3. The density of step atoms on Pt(730) is as high as 5.1×10^{14} cm^{-2}, that is, 43% of the total number of atoms on the surface. Therefore, the THH Pt NCs exhibit high electrocatalytic activity. As for the electrooxidation of ethanol, the steady-state current density recorded at 0.25 V is 0.39 mA cm^{-2} on the THH Pt NCs of 81 nm, 0.16 mA cm^{-2} on Pt nanospheres of 115 nm, and 0.12 mA cm^{-2} on commercial 3.2 nm Pt/C catalyst, as demonstrated in Fig. 4. In addition, at a fixed current density of technical interest, such as 0.20 mA cm^{-2}, the oxidation potential on the THH Pt NCs was shifted negatively about 80 mV in comparison with that of the Pt nanospheres and the commercial Pt/C catalyst.

It was found that the formation of THH Pt NCs with high-index facts is related to the oxygen adsorption/desorption generated by square-wave potential. Under this condition, low-index facets with high coordinated surface atoms, such as Pt(111) and Pt(100), are disturbed through site exchange between Pt and oxygen, whereas, high-index facets of {$hk0$} with low coordinated surface atoms, such as {730} and {210} are retained.

Our recent progresses confirmed that the electrochemical square-wave potential route can be further extended to synthesize NCs of other shapes with high-index facets besides the THH Pt NCs.

3.2 Concave hexoctahedral Pt nanocrystals

In the square-wave potential route, when the 30 mM ascorbic acid is replaced by 50 mM sodium citrate, complex concave polyhedral Pt NCs were obtained. As shown in Fig. 5, on this concave polyhedral Pt nanocrystal, six facets intersect on a point in a three-fold axis, and four facets intersect on a point in a four-fold axis. This symmetry is identical to that of the convex hexoctahedron illustrated in Fig. 2. The bottom left inset to Fig. 5 is a model of concave hexoctahedron bounded by {321} facets, whose shape is similar to the Pt NCs seen from an SEM image. The result indicates that the Pt NCs are of concave hexoctahedral shape and bounded by 48 {hkl} high-index facets. In geometry, the THH can be easily transformed into concave hexoctahedron just by shrinking along the ⟨110⟩ direction. As a result, one {$h'k'0$} facet will be split into two {hkl} facets in the transformation of a THH to concave hexoctahedron.

3.3 Trapezohedral and concave hexoctahedral Pd nanocrystals

Pd nanoparticles are also very important electrocatalysts, especially for the electrooxidation of formic acid. The electrochemical square-wave potential route can also

Fig. 5 SEM images of concave hexoctahedral Pt NCs with {hkl} facets prepared by electrochemical square-wave potential method. The inset illustrates a model of concave hexoctahedron bounded by {321} facets.

be applied to synthesize Pd NCs with high-index facets. SEM images of as-prepared Pd NCs on indium-tin-oxide (ITO) glass substrate are illustrated in Fig. 6. We can observe Pd NCs of trapezohedral shape bounded by 24 {hkk} high-index facets in Fig. 6a and b, and Pd NCs with concave hexoctahedral shape bounded by 48 {hkl} high-index facets in Fig. 6c. Although the exact surface structure, *i.e.*, Miller indices of these Pd NCs are yet unknown due to the complexity of their shape, these preliminary results demonstrate that the square-wave potential route is a powerful method to synthesize metal nanoparticles bounded by high-index facets.

4. Multiple twinned crystalline Pt nanorods with high-index facets

Great efforts have been devoted to the synthesis of metal nanorods and nanowires due to their unique physical and chemical properties, and important applications in the fabrication of nanoscale devices.[23,24] The properties of nanorods can be tuned

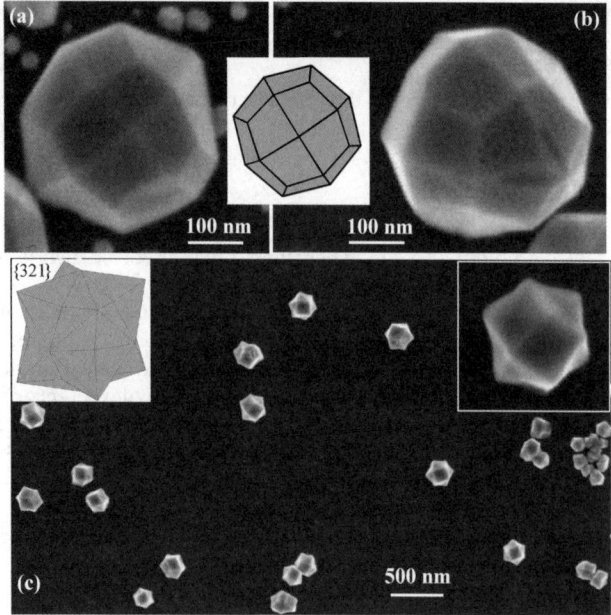

Fig. 6 SEM images of trapezohedral Pd NCs with {hkk} facets (a and b), and concave hexoctahedral Pd NCs with {hkl} facets (c) prepared by electrochemical square-wave potential method and supported on ITO substrate.

by controlling their aspect ratio and size.[25,26] Recent studies have shown that the surface structure of nanorods can also significantly affect their properties including thermal stability, chemical reactivity and surface functionalization.[24,27–29] Despite a variety of methods having been developed to synthesize metal nanorods in the last decade,[30–34] only limited cases have shown the control of surface structure. Metal nanorods grown in rigid templates usually expose a polycrystalline surface with irregular atom arrangement.[35] Although rod-shaped micelles or localized oxidative etching can be used to synthesize several kinds of metal single crystalline nanorods enclosed by a mixture of {100}, {111} and {110} facets, they are generally limited in Au and Pd nanorods.[36–39] The growth of metal nanorods with fivefold twinned structure is relatively easy, since the growth along lateral direction is greatly inhibited due to the stress originated from structure mismatch.[40] The fivefold twinned metal nanorods exhibit highly faceted structure: five {100} facets as side surfaces, and five {111} facets as end surfaces.[40] They are frequently observed in Au, Ag, Cu, and even in Pd,[41–45] but rarely in Pt.[46]

Recently, we extended the electrochemical square-wave potential route to synthesize fivefold twinned Pt nanorods with {$hk0$} high-index facets. Interestingly, the indices of the facets, i.e. the value of h and k, vary along the geometry of Pt nanorods.

4.1 Synthesis of Pt nanorods

The synthesis processes of the Pt nanorods are similar to those of the THH Pt NCs, except that the GC electrode loading Pt nanospheres was exposed in air for 3–5 h prior to the treatment of square-wave potential. It was found that the exposure could decrease the activity of the GC surface, on which new Pt nanocrystals could barely grow during the treatment of square-wave potential. Instead, Pt nanorods grew on the Pt nanospheres. It has confirmed that this step is crucial for the synthesis of the Pt nanorods. If the procedure of exposure to air was skipped, only THH Pt NCs grown on GC were obtained, as shown in Fig. 3.[17] If the exposure time was shortened to 1 h, both THH Pt NCs on GC and Pt nanorods on the Pt nanospheres were produced.

4.2 Characterization of Pt nanorods

Fig. 7a illustrates a typical SEM image of the as-prepared Pt nanorods with growth time of 30 min. The yield of Pt nanorods is about 30%. Other irregular nanostructures are the residua of Pt nanospheres. The nanorods are not uniform in diameter along the longitudinal axis; broadest at the middle and gradually tapering to both ends. The average diameter of the nanorods measured at the middle is 124 nm. The length varies from 0.7 to 1.4 μm, and the aspect ratio of Pt nanorods is about 8. Fig. 7b shows a high-magnification SEM image of a Pt nanorod. Zigzag-arranged facets can be observed on the surface of the middle part. Also, there are several ridges along the nanorod. Very interestingly, on both ends of the nanorod, five facets can be discerned clearly from the SEM image, which indicates that the ends are enclosed by ten facets, i.e. the ends are decagonal pyramids.

The structure of the Pt nanorod was analyzed by SAED and HRTEM. A TEM image of the Pt nanorod is presented in Fig. 8a. The wriggling border of the nanorod is related to the zigzag pattern on the surface. Fig. 8b and c demonstrate two typical SAED patterns obtained from the Pt nanorod. Each diffraction pattern contains two sets of diffraction from the face-centered cubic Pt. The growth direction of the Pt nanorod was determined to be along [110], according to the (220) diffraction. In Fig. 8b, the square (solid line) and rectangular (dashed line) symmetrical diffractions correspond to [001] and [$\bar{1}$12] zone axes, respectively. The SAED pattern in Fig. 8c was obtained after rotating the nanorod by 18° along the longitudinal axis. The rectangular (solid line) and rhombic (dashed line) symmetrical diffractions correspond to [$1\bar{1}$0] and [$1\bar{1}\bar{1}$] zone axes, respectively. These results are well consistent with those

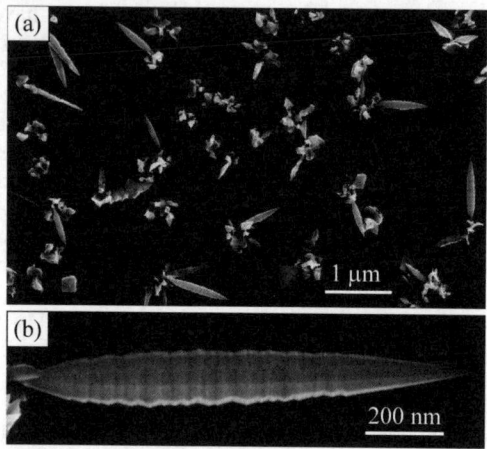

Fig. 7 SEM images of Pt nanorods. (a) Low magnification. (b) High magnification, showing fine surface facets.

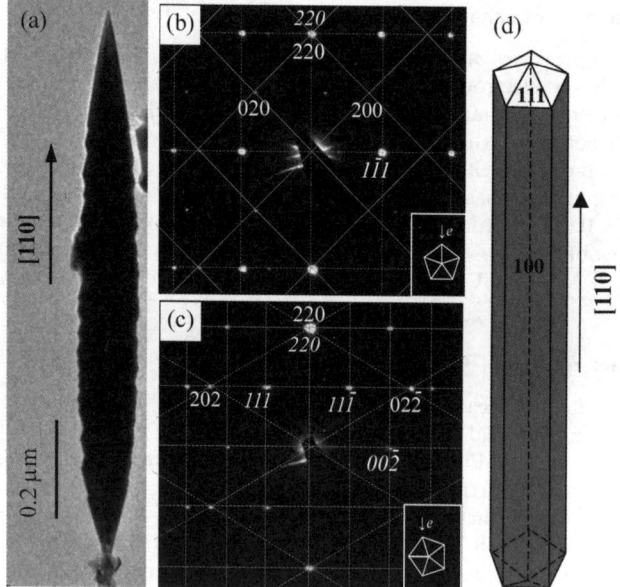

Fig. 8 (a) TEM image of a Pt nanorod. (b), (c) Two typical SAED patterns obtained from the Pt nanorod, demonstrating that the Pt nanorod is of fivefold twinned structure. (d) A model of typical fivefold twinned nanorod.

obtained from fivefold twinned nanorods of Au, Ag and Cu,[41,47,48] demonstrating that the as-prepared Pt nanorod is of fivefold twinned structure.

4.3 Surface structure of the Pt nanorods

The structure of the fivefold twinned nanorod has been well documented.[41–44,47,48] It consists of five sub-crystals twinned at (111) planes, with side surfaces bounded by five {100} facets and end surfaces by five {111} facets, as illustrated in Fig. 8d. Besides, the diameter of nanorods is uniform. Clearly, although the Pt nanorod is of fivefold twinned structure, its morphology, especially the zigzag pattern on side surfaces and the sharp ends, is quite different from that of the typical fivefold twinned nanorod (Fig. 8d). The fine facets on side surfaces of the Pt nanorod are not likely to be {100} facets, since the {100} facets should be parallel to the longitudinal [110] axis, whereas the fine facets of the Pt nanorod obviously do not satisfy this criterion. Moreover, the end surfaces of the Pt nanorod are not {111} facets, because the end of the typical fivefold twinned nanorod is a pentagonal pyramid, while the end of the as-prepared Pt nanorod is an elongated decagonal pyramid. Since the growth of the Pt nanorods are under similar conditions to those of producing the THH Pt NCs, the surface facets of Pt nanorods are very likely to be {hk0} high-index facets. We then carefully analyzed the surface structure of the ends and the middle part of the Pt nanorods, respectively.

Based on the structure characteristic of fivefold twinned nanorod, one {111} facet will evolve into two {hk0} facets if the end of nanorod is bounded by {hk0} facets. Correspondingly, the shape of the end will change from pentagonal pyramid to decagonal pyramid, as demonstrated in Fig. 9a. The enlarged SEM image of the end of the Pt nanorod is shown in Fig. 9b, from which five facets can be discerned clearly, confirming the shape of decagonal pyramid. The good consistency between the model and experimental result demonstrates that the end of the Pt nanorod is bounded by {hk0} high-index facets. The value of h and k can be determined by measuring the geometrical parameters, i.e. the ratio of height to width (H/L) and the cone angle (θ) of the decagonal pyramid:

$$\frac{H}{L} = \left(1 + \frac{2h}{k}\right) / \left(\sqrt{12} \cos 18°\right) \qquad (2)$$

$$\theta = 2\mathrm{arctg}\left(\frac{L}{2H}\right) \qquad (3)$$

Table 1 lists the theoretical values of H/L and θ for decagonal pyramids bounded by different {hk0} facets. The H/L values of the as-prepared Pt nanorods were measured between 2.7 and 2.9 for the sharp ends, and between 1.2 and 1.7 for the obtuse ends. By comparing these measured values with the theoretical ones listed in Table 1, we can identify that the sharp ends of the Pt nanorods are practically enclosed by {410} facets, and the obtuse ends are enclosed by {320}, {210} or {730} facets.

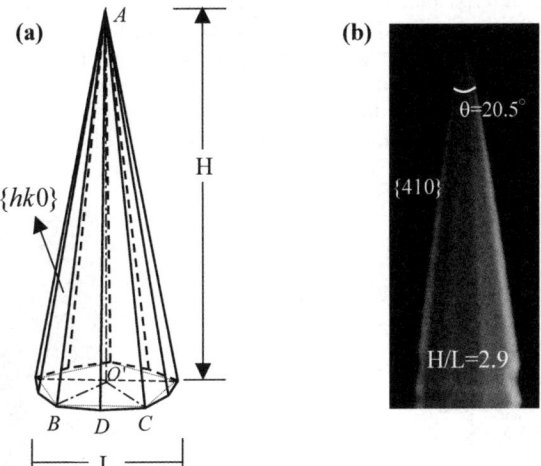

Fig. 9 (a) Schematic model of decagonal pyramid bounded by ten {hk0} facets with fivefold twinned structure. (b) SEM image of the sharp end of a Pt nanorod. Both the cone angles (θ) and H/L indicate the sharp end is enclosed by a {410} facet.

Table 1 Theoretical ratio of height to width (H/L) and cone angle (θ) of decagonal pyramids bounded by $\{hk0\}$ facets

$\{hk0\}$	$\{320\}$	$\{210\}$	$\{730\}$	$\{520\}$	$\{310\}$	$\{410\}$	$\{510\}$
H/L	1.21	1.52	1.72	1.82	2.12	2.73	3.34
θ	44.77°	36.47°	32.42°	30.70°	26.48°	20.74°	17.03°

Fig. 10a depicts the magnified SEM image of the middle part of the Pt nanorod, from which the fine facets and ridges along the nanorod can be seen more clearly. To identify the indices of the fine facets, we should first determine twin boundaries. Two kinds of ridges appearing alternatively on the nanorod can be distinguished, denoted as I and II in Fig. 10a. It can be observed that the fine facets at the two sides of the ridge I are essentially symmetrical to each other, while those at the two sides of the ridge II are asymmetrical. According to the crystal symmetry, the symmetrical facets are more likely on the same sub-crystal, so ridge II should be the boundary of two sub-crystals. Four fine facets in the boxed area marked in Fig. 10a can be considered as a surface unit. This kind of arrangement of the fine facets can be observed on a THH bouned by $\{hk0\}$ facets. Every four $\{hk0\}$ facets, i.e. $(k0h)$, $(0\bar{k}h)$, $(0kh)$, $(\bar{k}0h)$, situated in different directions, form a surface unit of the Pt nanorod, which repeatedly arranges along the growth direction of the Pt nanorod and results in the zigzag pattern on the crystal surface. The indices of these $\{hk0\}$ facets can be determined by measuring the slope angle (ϕ) between the border line and the longitudinal axis in the TEM image, as illustrated in Fig. 10c. The angle of ϕ was measured to be $15.2 \pm 0.5°$, so the corresponding θ is about 30° as $\theta = 2\phi$. By comparing these values of θ with the theoretical values listed in Table 1, the surface of the middle part of the nanorod is mainly composed of $\{520\}$ facets.

The surface structures of the Pt nanorod are summarized in Fig. 11, i.e. the sharp end is enclosed by $\{410\}$ facets, the obtuse end by $\{320\}$, $\{210\}$ or $\{730\}$ facets, the middle part mainly by the zigzag-arranged $\{520\}$ facets. The densities of stepped atoms on these facets are significantly different, among which the $\{320\}$ facet has the highest density and the $\{410\}$ facet holds the lowest one. The difference in density of stepped atoms on different parts of the Pt nanorod may correlate with the different growth rate. During the growth, the sharp end grows outward, so it can receive more Pt ions from solution, i.e. the sharp end has the highest growth rate. In contrast, the other end rooted on the Pt nanosphere has the lowest growth rate due to the interparticle diffusion coupling between the nanorod and the Pt nanosphere.[49] In this study, the $\{hk0\}$ facets were formed through the repetitive oxygen adsorption/desorption generated by square-wave potential.[17] So, slower growth rate of the obtuse end implies an intensive surface reconstruction by oxygen, resulting in $\{hk0\}$ facets with higher density of stepped atoms, such as $\{320\}$ and $\{210\}$ facets. On the other hand, the faster growth rate of the sharp end leads to

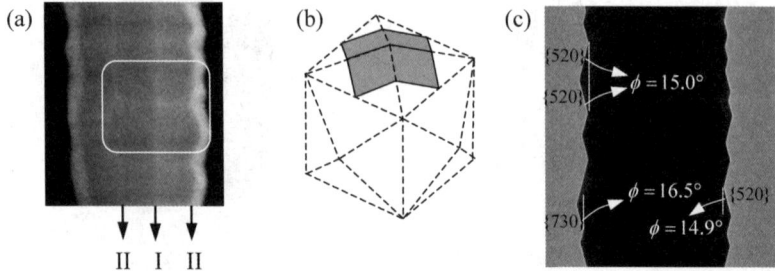

Fig. 10 (a) Magnified SEM image of the middle part of a Pt nanorod; (b) schematic illustration of the surface unit bounded by $\{hk0\}$ facets. (c) Magnified TEM image of the middle part of another nanorod, showing that the surfaces are mainly $\{520\}$ facets.

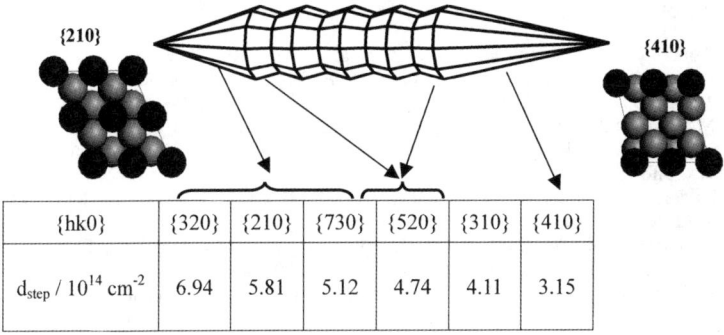

Fig. 11 The Miller indices and the densities of stepped atom of {$hk0$} facets at different parts of the Pt nanorod.

the formation of a surface with a lower density of stepped atoms, such as {410}. These results demonstrate that the indices of {$hk0$} facets of Pt NCs can be tuned by changing the growth rate during the square-wave potential treatment.

Furthermore, we have also synthesized fivefold twinned Pd nanorods with {$hk0$} and {hkk} type high-index facets, and found that the catalytic activity of Pd nanorods with {hkk} facets (near {15 1 1} facet) for the electrooxidation of ethanol in alkaline solution was about 2.5 times larger than that of a commercial Pd black catalyst.

5. Conclusions

In the current paper, we report the development of a novel square-wave potential route that has successfully overcome the difficulties in synthesis of Pt nanocrystals (NCs) bounded by high-index facets with high surface energy. It has revealed that the dynamic oxygen adsorption/desorption mediated by square-wave potential and the difficulty of site-exchange between oxygen and surface atoms of open-structure planes play key roles in the electrochemically shape-controlled synthesis of metal nanoparticle catalysts with high-index facets. By this method, tetrahexahedral (THH) Pt NCs bounded by 24 {$hk0$} high-index facets, concave hexoctahedral Pt NCs by 48 {hkl} facets, trapezohedral Pd NCs by 24 {hkk} facets, and concave hexoctahedral Pd NCs by 48 {hkl} facets have been successfully synthesized. Besides single crystalline nanoparticle catalysts, fivefold twinned Pt nanorods by high-index facets are also obtained by controlling the nucleation mode. It was determined that the sharp end of the Pt nanorods was enclosed by ten {410} facets, the obtuse end by {320}, {210} or {730} facets, and the middle part mainly by the zigzag-arranged {520} facets. Owing to the high density of atomic steps and dangling bonds on the high-index facets, the nanoparticle catalysts bounded by high-index facets display in general high catalytic reactivity and stability. Such excellent catalytic properties were confirmed on the THH Pt NCs with {$hk0$} facets and also on the Pd nanorods with {hkk} facets towards electrooxidation of small organic fuels. These results demonstrate that square-wave potential is a versatile method to synthesize metal NCs bounded by high-index facets. The study has opened a new prospect avenue in shape-controlled synthesis of nanoparticle catalysts with high performances.

Acknowledgements

This study was supported by NSFC (grant Nos. 20673091, 20503023, 2043060) and Ministry of Science and Technology of China (2002CB211804 and 2007DFA40890) and by the Natural Science Foundation of Fujian Province of China (No.2008I0025). The authors are grateful to Prof. Z. X. Xie for helpful discussion on the studies of Pt nanorods.

References

1. A. Roucoux, J. Schulz and H. Patin, *Chem. Rev.*, 2002, **102**, 3757.
2. C. Burda, X. B. Chen, R. Narayanan and M. A. El-Sayed, *Chem. Rev.*, 2005, **105**, 1025.
3. A. T. Bell, *Science*, 2003, **299**, 1688.
4. R. M. Heck and R. J. Farrauto, *Appl. Catal., A*, 2001, **221**, 443.
5. J. Larminie, A. Dicks, *Fuel Cell Systems Explained*, John Wiley & Sons, Chichester, West Sussex, 2nd edn, 2003.
6. F. Favier, E. C. Walter, M. P. Zach, T. Benter and R. M. Penner, *Science*, 2001, **293**, 2227.
7. D. J. Berger, *Science*, 1999, **286**, 49.
8. L. C. Gontard, L. Y. Chang, C. J. D. Hetherington, A. I. Kirkland, D. Ozkaya and R. E. Dunin-Borkowski, *Angew. Chem., Int. Ed.*, 2007, **46**, 3683.
9. G. A. Somorjai, *Chemistry in Two Dimensions: Surfaces*, Cornell University Press, Ithaca, 1981.
10. S. G. Sun, A. C. Chen, T. S. Huang, J. B. Li and Z. W. Tian, *J. Electroanal. Chem.*, 1992, **340**, 213.
11. G. A. Attard, *J. Phys. Chem. B*, 2001, **105**, 3158.
12. Z. L. Wang, *J. Phys. Chem. B*, 2000, **104**, 1153.
13. H. E. Buckley, *Crystal growth*, Wiley, New York, 1951.
14. T. S. Ahmadi, Z. L. Wang, T. G. Green, A. Henglein and M. A. El-Sayed, *Science*, 1996, **272**, 1924.
15. H. Song, F. Kim, S. Connor, G. A. Somorjai and P. D. Yang, *J. Phys. Chem. B*, 2005, **109**, 188.
16. J. T. Ren and R. D. Tilley, *J. Am. Chem. Soc.*, 2007, **129**, 3287.
17. N. Tian, Z. Y. Zhou, S. G. Sun, Y. Ding and Z. L. Wang, *Science*, 2007, **316**, 732.
18. Y. Ding, Y. Gao, Z. L. Wang, N. Tian, Z. Y. Zhou and S. G. Sun, *Appl. Phys. Lett.*, 2007, **91**, 121901.
19. J. F. Nicholas, *An Atlas of Models of Crystal Surfaces*, Gordon & Breach, New York, 1965.
20. M. A. Van Hove and G. A. Somorjai, *Surf. Sci.*, 1980, **92**, 489.
21. G. A. Somorjai and D. W. Blakely, *Nature*, 1975, **258**, 580.
22. A. A. Proussevitch and D. L. Sahagian, *Comput. Geosci.*, 2001, **27**, 441.
23. Y. N. Xia, P. D. Yang, Y. G. Sun, Y. Y. Wu, B. Mayers, B. Gates, Y. D. Yin, F. Kim and Y. Q. Yan, *Adv. Mater.*, 2003, **15**, 353.
24. C. J. Murphy, T. K. San, A. M. Gole, C. J. Orendorff, J. X. Gao, L. Gou, S. E. Hunyadi and T. Li, *J. Phys. Chem. B*, 2005, **109**, 13857.
25. M. A. El-Sayed, *Acc. Chem. Res.*, 2001, **34**, 257.
26. C. J. Murphy and N. R. Jana, *Adv. Mater.*, 2002, **14**, 80.
27. Y. Wang, S. Teitel and C. Dellago, *Nano Lett.*, 2005, **5**, 2174.
28. J. H. Song, F. Kim, D. Kim and P. D. Yang, *Chem.–Eur. J.*, 2005, **11**, 910.
29. J. K. Diao, K. Gall and M. L. Dunn, *Nat. Mater.*, 2003, **2**, 656.
30. Y. G. Sun, B. Gates, B. Mayers and Y. N. Xia, *Nano Lett.*, 2002, **2**, 165.
31. H. J. Choi, H. K. Seong, J. Chang, K. I. Lee, Y. J. Park, J. J. Kim, S. K. Lee, R. R. He, T. Kuykendall and P. D. Yang, *Adv. Mater.*, 2005, **17**, 1351.
32. J. C. Hulteen and C. R. Martin, *J. Mater. Chem.*, 1997, **7**, 1075.
33. J. Y. Chen, T. Herricks, M. Geissler and Y. N. Xia, *J. Am. Chem. Soc.*, 2004, **126**, 10854.
34. M. P. Zach, K. H. Ng and R. M. Penner, *Science*, 2000, **290**, 2120.
35. M. Tian, J. Wang, J. Kurta, T. E. Mallouk and M. H. W. Chan, *Nano Lett.*, 2003, **3**, 919.
36. Y. Y. Yu, S. S. Chang, C. L. Lee and C. R. C. Wang, *J. Phys. Chem. B*, 1997, **101**, 6661.
37. Z. L. Wang, M. B. Mohamed, S. Link and M. A. El-Sayed, *Surf. Sci.*, 1999, **440**, L809.
38. H. A. Keul, M. Möller and M. R. Bockstaller, *Langmuir*, 2007, **23**, 10307.
39. Y. J. Xiong, H. G. Cai, B. J. Wiley, J. G. Wang, M. J. Kim and Y. N. Xia, *J. Am. Chem. Soc.*, 2007, **129**, 3665.
40. V. G. Gryaznov, J. Heydenreich, A. M. Kaprelov, S. Nepijko, A., A. E. Romanov and J. Urban, *Cryst. Res. Technol.*, 1999, **34**, 1091.
41. I. Lisiecki, A. Filankembo, H. Sack-Kongehl, K. Weiss, M. P. Pileni and J. Urban, *Phys. Rev. B: Condens. Matter Mater. Phys.*, 2000, **61**, 4968.
42. C. J. Johnson, E. Dujardin, S. A. Davis, C. J. Murphy and S. Mann, *J. Mater. Chem.*, 2002, **12**, 1765.
43. Y. G. Sun, B. Mayers, T. Herricks and Y. N. Xia, *Nano Lett.*, 2003, **3**, 955.
44. S. H. Zhang, Z. Y. Jiang, Z. X. Xie, X. Xu, R. B. Huang and L. S. Zheng, *J. Phys. Chem. B*, 2005, **109**, 9416.
45. Y. Xiong, H. Cai, Y. Yin and Y. N. Xia, *Chem. Phys. Lett.*, 2007, **440**, 273.
46. A. J. Melmed and D. O. Hayward, *J. Chem. Phys.*, 1959, **31**, 545.
47. C. Lofton and W. Sigmund, *Adv. Funct. Mater.*, 2005, **15**, 1197.
48. H. Y. Chen, Y. Gao, H. C. Yu, H. R. Zhang, L. B. Liu, Y. G. Shi, H. F. Tian, S. S. Xie and J. Q. Li, *Micro*, 2004, **35**, 4694.
49. M. Z. Liu and P. Guyot-Sionnest, *J. Phys. Chem. B*, 2005, **109**, 22192.

General discussion

Professor Gewirth opened the discussion of the paper by Christopher A Lucas*, Michael Cormack, Mark E Gallagher, Alexander Brownrigg, Paul Thompson, Ben Fowler, Yvonne Gründer, Jerome Roy, Vojislav Stamenkovic and Nenad M. Marković:

Could the lack of structure you observe be related to an observation made on Ag(111), in that the absence of electrolyte led to only "liquid like" water as observed in SFG measurements? We saw evidence of more ordered water only with electrolyte. You didn't use electrolyte in your experiment.

Dr Lucas responded:

At this stage we cannot rule out the possibility that an ordered water structure is present at the interface for a number of reasons; (i) it may be that the X-ray scattering is just too weak and cannot be detected above the background level caused by scatter from the liquid overlayer. In Fig. 1 of the paper the signal level from the p(2 × 2) oxygen structure is very weak, and this is without a liquid overlayer. In the presence of that overlayer, the background level increased significantly. One possible experiment to overcome this would be to contact the water droplet with the surface and then withdraw it. This would presumably leave a thin water layer and would considerable reduce the background scatter.

(ii) It may be that there is no long range order in the structure. In order to detect a measureable signal the structure must be coherent over a distance of at least 50 Å. This was the smallest coherence length that was detectable by SXS in studies of ordered CO adlayers on Pt(111).[1]

(iii) It was possible that in the experiment we had a too high coverage of oxygen in the p(2 × 2) adlayer that was transferred to the liquid environment. This could frustrate the formation of the mixed (OH + H_2O) phase that has a hexagonal ($\sqrt{3} \times \sqrt{3}$)R30° lattice with a weak (3 × 3) superstructure caused by ordering of the hydrogen bonds.[2] A future experiment may be to develop cold transfer of the sample for study of the ordered water monolayer prior to forming the bulk water environment. In the paper, however, we do suggest that there is some out-of-plane ordering of the water adlayers as shown by the systematic changes in the crystal truncation rods (CTR's) (Fig. 4 of the paper). In fact, the CTR analysis is consistent with ordering both in water and electrolyte.

1. N. M. Markovic, C. A. Lucas, A. Rodes, V. Stamenkovic, and P. N. Ross, *Surf. Sci. Lett.*, 2002, **499**, L149.
2. C. Clay, S. Haq and A. Hodgson, *Phys. Rev. Lett.*, 2004, **92**, 046102

Professor Sun remarked:

Can we see the specific adsorption of anions by using x-ray scattering technique?

Dr Lucas responded:

I presume this question relates to the charge transfer process, as there have been many *in-situ* surface X-ray scattering (SXS) studies of phase transitions and ordering of specifically adsorbed anions (see ref. 1 for a review). In the case of charge transfer, it may be possible to probe it using resonant SXS techniques. These have been developed by the group of Hoydoo You at Argonne National Laboratory and used to study processes such as CO adsorption and oxide formation on Pt.[2,3] In this technique, the incident X-ray energy is tuned around an adsorption edge, usually of the metal electrode, and the elastically scattered X-ray signal (separated from the fluorescent scattering by energy analysis of the data) is measured while the diffraction conditions isolate the surface atoms. Using this technique, it may be possible to observe charge transfer during the specific adsorption of anions although recent attempts by our group on the I/Au(111) system[4] using the resonant SXS technique indicated that the measureable effects are subtle.

[1] O. M. Magnussen, *Chem. Rev.*, 2002, **102**, 679.
[2] Y. S. Chu, H. You, J. A. Tanzer, T. E. Lister and Z. Nagy, *Phys. Rev. Lett.*, 1999, **83**, 552.
[3] A. Menzel, K.-C. Chung, V. Komanicky, H. You, Y. S. Chu, Y. V. Tolmachev and J. J. Rehr, *Radiat. Phys. Chem.*, 2006, **75**, 1651.
[4] C. A. Lucas *et al.*, unpublished data.

Professor Schiffrin commented:
It is very interesting that the experimental results shown in Fig. 4 indicate a change in structure due to the presence of a water overlayer (a reduction in the Pt surface expansion). Can these results be reconciled with the calculations by Gross in Paper 15, which indicate a very weak interaction between water and the Pt surface?

Dr Lucas answered:
It should be recognized that the data in Fig. 4 represents a change in the interface structure between the Pt(111)–p(2 × 2)–O surface and the Pt(111)–H$_2$O interface. Thus, it is important to take into account the fact that the Pt surface, prior to water adsorption, is distorted by the adsorbed oxygen adlayer. The changes modelled in Fig. 4 indicate a reduction in the Pt surface expansion, which is consistent with a reduced adsorbate–substrate interaction, *i.e.* both H$_{UPD}$ (Fig. 6) and the p(2 × 2)-O adsorbates induce a larger surface expansion and so the results are not inconsistent with a weaker Pt–H$_2$O interaction (as predicted in the paper by Gross). The data do, however, indicate a larger in-plane distortion of the Pt surface atoms, although in the absence of a well-defined water adlayer (*i.e.* the ($\sqrt{3} \times \sqrt{3}$)R30° or (3 × 3) structures observed for monolayer coverages of H$_2$O on Pt(111)) it is difficult to define an unique interface structure.

Professor Behm remarked:
I would expect the 'invisibility' of the lower coverage adlayer structure is largely related to the mobility within the adlayer, both in terms of (lateral) vibrations (Debye–Waller Factor) and in terms of adsorbate hopping (temporal density fluctuations).

Dr Lucas responded:
It is certainly true that lower coverage adlayers could be 'invisible' to in-plane surface X-ray scattering due to mobility effects, such as temporal density fluctuations. That is another possible reason that structures can only be observed in the surface normal direction using CTR analysis, as there would be a fixed average bondlength in this direction even in the case of mobile species. In fact, this is what we observe for water adsorption on Pt(111) and in the case of 0.1 M KOH electrolyte at potentials where OH species are adsorbed on the surface.

Professor Wieckowski opened the discussion of the paper by D. V. Tripkovic, D. Strmcnik, D. van der Vliet, V Stamenković and N Marković*:
Generalisation of anions (halides and oxyanions) is in my judgement incorrect. Halides and oxyonions should be dealt with independently. The central issue is specificity, *e.g.* perchlorate *vs.* bisulfate. Furthermore, absorption of specifically adsorbed (bi)sulphate anions and halide ions, like chloride and bromide, on platinum or gold is reversible

Professor Markovic answered:
I respectfully disagree with this comment. Certainly there are some differences between adsorption of oxyanions and halides on metal surfaces, as we discussed

in *Surf. Sci. Rep.*, 2002, **45**, 117. However, there are many more similarities between halides and oxyanions; namely, both type of anions are controlling the availability of active sites in the ORR, HOR and oxidation of small organic molecules, with exception of Pt(111) in perchloric and sulfuric acid solutions, they are controlling the shapes of cyclic voltammograms in both the H_{upd} and oxide(s) potential regions, they are controlling the ordering and the kinetics of the UPD process, clustering of CO adlayers *etc*. Regarding the kinetics of adsorption, all the anions discussed in this paper under equilibrium conditions (slow sweep rate or steady state measurements) are reversible.

Professor Schmickler commented:
Your work reminds me of the role that anions play in the so-called underpotential deposition (upd) of metals. About 30 years ago, Kolb, Przasnyski and Gerischer[1] established a relation between the work function of metals and the upd shift based on cyclic voltammograms. We now know, that in many cases the observed process was not the formation of a pure metal layer, but of a mixed layer containing both anions and metal ions, a two-dimensional salt. The adsorption of anions depends on the potential of zero charge, which is proportional to the work function. Therefore the observed experimental correlation may be caused by the behaviour of the anions as well as by the properties of the metal.

1. D.M. Kolb, M. Przasnyski and H. Gerischer, *Surf. Sci.*, 1974, **54**, 25.

Professor Markovic replied:
I absolutely agree with your comment. The upd of metals in acidic solutions is (almost) always controlled by adsorbed anions. As in the case for the UPD of Cu in Fig. 7 the interaction is cooperative and most likely controlled by the local potential of zero charge of metal atoms surrounded by anions. Therefore, evaluation of thermodynamic parameters corresponding to the UPD of metals without including the effect of anions is (almost) impossible.

Professor Savinova said:
I fully agree with the claim that electrochemical/electrocatalytic processes are affected by the values of the potential of zero charge (pzc) of the electrodes, and by the anion adsorption on their surfaces. It is, however, obvious that the differences in the behaviours of metal electrodes cannot be fully explained by their different pzc. For example, Fig. 9b of the paper shows that the difference in the onset potentials for the ORR on Au(111) and Ag(111) is less than 0.2 V, while the difference in the pzc for these electrodes exceeds 0.9 V (pzc(Ag(111)) = −0.69 *vs*. SCE, see G. Valette, A. J. Hamelin, *Electroanal. Chem.* 1978, **45**, 301; pzc(Au(111)) = +0.23 *vs*. SCE, see D. M. Kolb and J. Schneider, *Electrochim. Acta*, 1986, **31**, 929). This example suggests that the pzc of Au and Ag have only a minor influence on the ORR.

Professor Markovic responded:
In electrochemical systems, it is impossible to use the exact values of the work function from the gas phase catalyses and apply them on the systems in which bare metal does not exist in entire potential region. Although in cyclic voltammetry of the IB group metals we do see the so-called the double layer potential region, I doubt that there is an ideal double layer, even on the gold surface. Nevertheless, one can use, as we did in our paper, trends in the values of the work function. In this case, the results for the ORR can easily be explained based on the value of the work function observed in UHV. The point I tried to make in the paper is simply that in addition to the d-band center, the work function should be considered seriously in the quest to find a true relationships between the electronic and catalytic properties of metal surfaces.

Professor Neurock commented:

The role of anions has been described here as site blocking which will inhibit rates of reaction. There are, however, a number of known catalytic reactions on which the rate is inhibited but that the selectivity for a specific channel is enhanced considerably. I would think that there are a number of electrocatalytic reactions in which you enhance the selectively as well by anion adsorption. Can you please comment?

Professor Markovic replied:

This is true. We proposed that the major effect of anions is the blocking of active sites (they are spectators). The selectivity is, in most cases, the consequence of the very same blocking effect (predominantly ensemble). For example, in contrast to the 4e$^-$ reduction (no peroxide formation) on the Pt(111) Cl system, in the presence of Br the 2e$^-$ reduction and the production of peroxide is observed in kinetically controlling potential region. The latter is produced due to lack of two atoms required to break the O–O bond.

Professor Tsirlina commented:

The potential of zero charge (pzc) as related to anions adsorption is most probably the potential of zero **free** charge (pzfc). These values still remain less certain for single crystalline platinum surfaces, in contrast to potentials of zero **total** charge (pztc). For polycrystalline platinum, iridium and rhodium at low pH pzfc is always lower than pztc, at least for the most systematically studied sulphate media (see ref. 1). This makes doubtful the location of pzc quite positive to the hydrogen adsorption region (p. 3 in the paper under discussion). When gold, silver and copper are considered in relation to OH adsorption with charge transfer, they should be also treated as perfectly (not ideally) polarisable metals. This means that the notion of pzc related straightforwardly to electronic work function would not already hold for them (the same is true for all platinum metals). Charge transfer from adsorbates contributes to the **free** charge of these electrodes, and this value can not be unambiguously estimated from more easily measured **total** charge (which only can give a formal value of electrosorption valency). For example the true charge transfer from halides is significantly below unity.[2] What is mentioned now about most or all transfered charge (p. 9) should be discussed just as chloride electrosorption valency.

These points are important for any future consideration of coadsorption of atoms and molecular species. Another important point is the 'chloride warning' declared in the paper under discussion: both HF and HClO$_4$ are dangerous if we want to escape any specifically adsorbing components. I should add that perchlorate can induce additional complication because of its reduction on platinum in the vicinity of 0.05 V RHE, with generation of chloride.

1. A. N. Frumkin and O. A. Petrii, *Electrochim. Acta*, 1975, **20**, 347.
2. V. E. Kazarinov, A. M. Foontikov and G. A. Tsirlina, *J. Electroanal. Chem.*, 1990, **282**, 253.

Dr Herrero added:

Regarding the position of the pzfc on platinum electrodes, we have published a paper in which the pzfc is estimated for the Pt(111) electrode.[1] In this case, the pzfc has the same value as the pztc since this latter potential is located in the region where only capacitative processes are found. Concerning the possible Cl$^-$ contamination of perchloric acid solutions, a recent paper shows that the Cl$^-$ levels are below 10^{-8} M in 0.1 M HClO$_4$.[2] The effects of such low concentrations in the "typical" voltammetric experiments are negligible, and only very subtle differences are observed when very detailed and careful experiments are made. Then, the chloride contamination of the perchloric acid solutions only appears when the electrode material catalyses the perchlorate reduction or in very long term experiments.

1. V. Climent, N. Garca-Araez, E. Herrero, and J. Feliu, *Russ. J. Electrochem.*, 2006, **42**, 1145.
2. A. Bern, V. Climent, J. M. Feliu, *Electrochem. Commun.*, 2007, **9**, 2789

Professor Tong said:
In your paper you have shown that the shape of hydrogen UPD and desorption depends strongly on the concentration of Cl^-. At low Cl^- concentration, the shape is asymmetric. It becomes symmetric at high Cl^- concentration. You mentioned a "coupling effect" in the paper but didn't give detailed description. Can you elaborate on what this "coupling effect" really means?

Professor Markovic answered:
The coupling effect is well established, even on polycrystalline Pt electrodes. Namely, in order to be adsorbed on the surface, H_{upd} has to "wait" for adsorbed anions to desorb from the surface. Under experimental conditions described in Fig. 1 (relatively fast sweep rate and low concentration of anions), desorption and readsorption of anions is not reversible and as a consequence the reversible adsorption of hydrogen (in anion "free" solution even at $1V\ s^{-1}$) becomes irreversible.

Dr Wang remarked:
In proposing the site-blocking kinetic equation for the ORR (line 25 on page 12), you assumed that the ORR intermediates are adsorbed only at low coverage relative to other adsorbates, which are H_{upd} and OH_{ad} formed from water if there are no specifically adsorbed anions. Why did you not include O_{ad}, which is known to be more stable than OH_{ad} near the reversible potential for the ORR? Secondly, can you elaborate on your assumption that the coverage of OH_{ad} formed from ORR is negligible at all potentials compared to the OH_{ad} coverage measured by CV in the absence of oxygen?

Professor Markovic replied:
First of all the identity of oxygenated species is still unresolved. It is, however, proposed that at high overpotentials the electrode is predominately covered with the reversible species, such as OH. However, we cannot exclude the (co)existence of atomic oxygen in the potential region where the ORR is taking place. Nevertheless, independent of the nature of oxygenated species they have blocking effect not only for the ORR but for the HER as well.

A discussion of the effect of spectator species on the ORR was discussed recently by Strmcnik *et al.* in *J. Phys. Chem. C*, 2007, **111**, 18672. Based on the fact, that cyclic voltammetry in Ar saturated solution and in the presence of O_2 are identical it has been proposed that surface coverage with reaction intermediates (included OH) must be negligible.

Dr Rossmeisl remarked:
For the ORR everything can be understood in terms of only one kind of intermediate. In the presence of O_2 the coverage is determined by the steady state, which need not to be the same as equilibrium, when the electrolyte $(1 - \theta)$ term cannot be varied without affecting the ΔG term

Professor Markovic answered:
Again, we try to emphasize that the major effect is coming through the $(1 - \theta)$ term. We are indeed aware that the energy of adsorption of reaction intermediates is dependent (decreases) on increasing the coverage of both intermediates and spectator species. However, this effect, with some exceptions, is described for the Pt(111) CO system in Fig. 4. Particular examples are structure sensitivity of the ORR, HOR and oxidation of organic molecules on the same metal but with a different geometry of surface atoms. The big effect of the ΔG term is observed if one would compare catalytic activity of the fuel cell reactions on Pt-group metals *vs.* Au.

Professor Tryk commented:

I would like to underscore the importance of anion adsorption for real fuel cells for oxygen reduction. This was recognised many years ago for the adsorption of phosphate on platinum—this slows down O_2 reduction significantly. Even now that we have the polymer electrolytes, like Nafion, the same issue can arise. Nafion initially has a very slight tendency to adsorb, but during long term operation it can break down, releasing bisulfate which can adsorb more strongly and slow down the O_2 reduction. Also, the newer polymers being developed can, in certain cases, have a stronger tendency to adsorb, so this is an issue we have to face.

Professor Markovic responded:

The question of the effect of anions in real fuel cell is of paramount importance not only in the case of phosphoric acid fuel cell but also for the PEMFC. For the latter system trace levels of Cl or sulphonic groups would significantly affect the cell performance.

Professor Schiffrin continued the discussion of the paper by Christopher A Lucas*, Michael Cormack, Mark E Gallagher, Alexander Brownrigg, Paul Thompson, Ben Fowler, Yvonne Gründer, Jerome Roy, Vojislav Stamenkovic and Nenad M. Marković:

Could you comment on why people started to investigate the Pt-Ni system? If we use the phase diagram of the alloy as criteria for the formation of stable phases, Pt and Ni would not have been a preferred choice, since these two metals form solid solutions in all the composition range and there is no evidence for the formation of an intermetallic phase of high stability.[1] Although the Ni-Pt system shows a low temperature order–disorder transition,[2,3] the thermodynamic properties of the binary alloy do not show any special features at the Pt_3Ni composition.[4] Can you discuss how to relate the interesting results that you have obtained using bulk Pt_3Ni with the properties of nanoparticles made from this alloy? These are known to retain a Pt surface layer, and although the interfacial structure does not contain just a single segregated Pt "skin", it would be interesting to be able to predict under what conditions a surface miscibility gap would appear for metals that form solid solution alloys.

1. M. A. K. Hansen, *Constitution of binary alloys*, McGraw-Hill, New York, 1958.
2. C. E. Dahmani, M. C. Cadeville, J. M. Sanchez and J. L. Morn-Lpez, *Phys. Rev. Lett.*, 1985, **55**, 1208–1211.
3. F. Lantelme and A. Salmi, *J. Phys. Chem.*, 1996, **100**, 1159–1163.
4. R.A. Walker and J. B. Darby, *Acta Metallurg.*, 1970, **18**, 1261–1266.

Dr Lucas answered:

The motivation behind the systematic study of Pt_3M alloys[1] was to try and create a surface that could form a Pt 'skin' under certain methods of preparation. It is known from UHV studies of bimetallic alloy surfaces that if the relative concentration of Pt in the alloy is too low then it is impossible to form the Pt 'skin' structure[2]. This has important consequences for the stability of the surface as the alloying element is then easily depleted in the near surface region in the hostile electrochemical environment (in the same way that the Pt 'skeleton' surfaces are formed on the polycrystalline Pt3M alloys[1]). In terms of Pt_3Ni nanoparticles, considering that the $Pt_3Ni(111)$ surface is stable over a wide potential range and is very active for the ORR,[3] the challenge would be to create stable nanoparticles with electronic and morphological properties that mimic the $Pt_3Ni(111)$ single crystal alloy. To investigate this possibility, we performed atomistic Monte Carlo (MC) simulations using modified embedded atom method (MEAM) potentials to predict whether the octahedral $Pt75Ni25$ nanoparticle, consisting of 8 (111) facets, 12(111)-(111)

step-edges and 6 vertices and the tetrahedral nanoparticle, containing 4 (111) facets, are thermodynamically stable. The results are described in a recent publication[4] and indicate that the octahedral nanoparticle is energetically stable and exhibits a surface segregation profile that is identical to the profile determined from the SXS results for the extended $Pt_3Ni(111)$ surface.

1. V. R. Stamenković, B. S. Mun, M. Arenz, K. J. J. Mayrhofer, P. N. Ross, C. A. Lucas and N. M. Marković, *Nat. Mater.*, 2007, **6**, 241.
2. M. A. Vasiliev, *J. Phys. D: Appl. Phys.*, 1997, **30**, 3037.
3. V. R. Stamenković, B. Fowler, B. S. Mun, G. Wang, P. N. Ross, C. A. Lucas and N. M. Marković, *Science*, 2007, **315**, 493.
4. B. Fowler, C. A. Lucas, A. Omar, G. Wang, V. Stamenkovic and N. M. Markovic, *Electrochim. Acta*, 2008, **53**, 6076.

Professor Markovic responded:
In the mid 70's the catalysts were put into use long before their structure and properties are clearly understood and that was certainly the case for the Pt Ni system. A rationale behind this study is unknown to me, but the approach used in this period was clearly a trial and error one. In our experiments, we relayed on the surface segregation phenomenon rather than to the formation of stable phases in intermetallic phases which, with the exception of PtSn are not effective catalysts for the fuel cell reactions. In contrast to single crystals, for nanoparticles there is no surface sensitive probe capable to explore the surface segregation of Pt. However, very recent Monte Carlo analysis has shown that for many Pt bimetallic alloys, including PtNi, the Pt would like to be on the surface, as found experimentally with single crystals. For fuel cell applications, the nanoparticles are prepared by the so-called pre-acid treatment in which the 3d element is leaching out from the surface. This methodology is producing the so-called Pt-skeleton structure which is 2 time less active than the Pt-skin structure. I think, we need much more time to learn how to synthesize the Pt-skin structures.

Professor Chorkendorff returned to the discussion of the paper by D V Tripkovic, D Strmcnik, D van der Vliet, V Stamenković and N Marković*:
You see an enhancement in the CO oxidation by going from acidic to alkaline solution, but as Marc Koper was showing in his plenary talk this could introduce carbonate, which would cause blocking. I wonder whether you also see in your IR evidence for carbonate and how the presence of this anion would influence the observed shift in the CO vibrations?

Professor Markovic responded:
In our experiments in alkaline solutions (*Surf. Sci.*, 2002, **499**, L149), rather then focusing on the adsorption of carbonate we were monitoring the position of CO bands and the oxidation of CO (production of CO_3^-). If carbonate is indeed adsorbed (I see no reason why not), then I would expect that due to the lower surface potential in alkaline solution the onset potential for adsorption should be rather positive. As a consequence, the adsorption of oxygenated species (presumably OH_{ad} on the defect sites) will be possible even in the hydrogen adsorption region, which is confirmed by oxidation of CO and the production of CO_3^-. Notice, in acid solutions these sites are blocked by specific adsorption of anions. Regarding the effects of adsorbed carbonate on the CO vibrational properties, I would expect less pronounced effect on CO clustering, *i.e.*, anomalous blue shift characteristic for the Pt(111) CO system induced by anions in acid solutions (Fig. 4), because the surface coverage by adsorbed carbonate should be significantly attenuated corresponding to anions in acidic solutions. Again due to lower surface potential the C–O stretching bands in alkaline solutions are shifted at lower wave numbers. For more details see the above reference.

Professor Sun commented:
Could we have some knowledge between the partial charge transfer of anions and the specific adsorption of anions?

Professor Markovic answered:
Although we know very little about the partial charge transfer in the process of specific adsorption of anions, it appears that, at least for halides and (bi)sulfate anions, they are highly discharged on the surface. An indirect support for this statement is that on Pt-group metals, as well as on IB group metals, compressed structures of halides is observed in STM and SXS experiments. The charge redistribution upon the adsorption is a separate issue and should be examined based on a case by case basis.

Professor Wieckowski opened the discussion of the paper by Iosif Fromondi and Daniel Scherson*:
Defects are in the metal; this should be shown by voltammetry. Voltammetry is very sensitive to defects or lack of them.

Professor Scherson replied:
Unfortunately the experimental tactic we have so far implemented for studies of this type requires for the entire sphere to be immersed in the electrolyte; hence, the voltammetry reflects areas of the surface which are not oriented along the 111 direction. We are at present developing a technique that allows voltammetry to be obtained for a surface with spatial resolution. Some aspects of this novel approach are given in Y. Chen, A. Belianinov and D. Scherson, Spatially-Resolved Interfacial Electrochemistry: Ohmic Microscopy, *J. Phys. Chem. C*, 2008, **112**(24), 8754–8758

Dr Santos opened the discussion of the paper by Payam Kaghazchi, Felice C. Simeone, Khaled A. Soliman, Ludwig A. Kibler and Timo Jacob*:
I think it is important in the light of recent developments of the technology to go further in the investigation of well known methodologies to improve their application. However, I wonder, why the previous papers of the group of Arvia have not been mentioned in this contribution. Since the pioneering work of Cerviño *et al.* in the eighties,[1] the electrochemical faceting procedure of different metal surfaces to produce preferred crystallographic orientations has been widely employed and investigated. The application of fast repetitive periodic potential pulses in different potential ranges to produce faceting in platinum, gold, palladium, rhodium is a well known procedure.[2–7] The structure and morphology of these sorts of electrodes have been studied through scanning electron and scanning tunnelling microscopy, as well as their catalytic properties[8–10] and the mechanisms of the processes involved.[11] Besides, one of the authors had already applied this method in a previous work, where the work of Arvias's group is mentioned.[12] I think the new results shown in the present contribution must be discussed in the context of the previous work.

1. R.M. Cerviño, W.E. Triaca and A. J. Arvia, *J. Electrochem. Soc.*, 1985, **132**, 266.
2. A. J. Arvia, J. C. Canullo, E. Custidiano, C. L. Perdriel and W. E. Triaca, *Electrochim. Acta*, 1986, **31**, 1359.
3. W. E. Triaca and A. J. Arvia, *J. Appl. Electrochem.*, 1990, **20**, 347.
4. C.L. Perdriel, M. Ipohorski and A. J. Arvia, *J. Electroanal. Chem.*, 1986, **215**, 317.
5. E. Custidiano, S. Piovano, A. J. Arvia, A. C. Chialvo and M. Ipohorski, *J. Electroanal. Chem.*, 1987, **221**, 229.
6. E. Custidiano, A. C. Chialvo and A. J. Arvia, *J. Electroanal. Chem.*, 1985, **196**, 423.
7. C. L. Perdriel, E. Custidiano and A. J. Arvia, *J. Electroanal. Chem.*, 1988, **246**, 165.
8. E. P. M. Leiva, E. Santos, M. C. Giordano, R. M. Cerviño and A. J. Arvia, *J. Electrochem. Soc.*, 1986, **133**, 1660.

9. B. Beden, F. Hahn, C. Lamy, J. M. Léger, N. R. de Tacconi, R. O. Lezna and A. J. Arvia, *J. Electroanal. Chem.*, 1989, **261**, 401.
10. E. P. M. Leiva, E. Santos and T. Iwasita, *J. Electroanal. Chem.*, 1986, **215**, 357.
11. S. A. Bilmes, M. C. Giordano and A. J. Arvia, *J. Electroanal. Chem.*, 1987, **227**, 183.
12. A. A. El-Shafei, R. Hoyer, L. A. Kibler and D. M. Kolb, *J. Electrochem. Soc.*, 2004, **151**, F141.

Dr Jacob and **Dr Kibler** responded:
Of course, the authors are aware of the work of Arvia *et al.* So far, however, fast repetitive periodic potential pulses have not been applied for the Ir(210) system. (It is rather paper 3, where a square-wave potential routine has been applied to synthesize faceted electrodes.) Our theoretical work explains the adsorbate-driven faceting of Ir(210). In addition, the thermodynamic stability of various surface phases as a function of electrode potential is predicted. So far, we have prepared nano-faceted Ir(210) by adsorption of oxygen during the cooling-down period after inductive heating of the electrode. We have not (yet) discovered a straightforward method for nanofacet formation on Ir(210) as a function of electrode potential, as stated in the question.

The main issues of our paper are: (i) the stability of nano-faceted Ir(210) as a function of electrode potential can be explained theoretically; (ii) planar and faceted Ir(210) surfaces reveal a clearly distinct electrochemical behavior.

It will be future work to find electrochemical strategies for facet formation on Ir(210). Fast repetitive periodic potential pulses may represent one possibility. In addition, structure–activity relations shall be established to further bridge the gap between single crystals and nanoparticles.

Professor Markovic continued the discussion of the paper by Iosif Fromondi and Daniel Scherson*:
Why are defects affected by CO oxidation reaction?

Professor Scherson replied:
In our view, the defects are originally present only on the periphery of the facet. As found in earlier *in situ* SHG studies, repeated excursions to potentials positive enough for the oxidation of adsorbed carbon monoxide to ensue increase the number and perhaps the character of these defects. This is due, in all likelihood, to the fact that the onset for the oxidation of adsorbed CO is very close to the onset for the incipient oxidation of Pt in the electrolytes in which these experiments were performed.

Professor Russell asked:
I think there is some confusion. Are you (Prof. Scherson) talking about defects in the adsorbate layer or defects on the metal surface? Please make this clear.

Professor Scherson replied:
We believe the Pt(111) facets are virtually defect free and also that the $c(2 \times 2)$ CO layer is devoid of defects.

Dr Santos commented:
About the comment about why to employ Second Harmonic Generation (SHG) to characterize this system and not other less complicated techniques. I agree with Prof. Scherson, that SHG is a very convenient and powerful method: firstly, SHG is forbidden in the bulk of centrosymmetric media. In particular, its inherent sensitivity to the interfacial region makes this technique attractive to investigate adsorption processes. In contrast to infrared spectroscopy, the signal is generated from the interface, where the symmetry is broken and no contribution of the bulk material is present. Secondly, the nonlinear response is produced by the second order polarizability, which is related to the electronic response of the surface. Therefore, *in situ*

SHG can be a useful tool for investigating the changes in the electrodynamic properties of electrode surfaces caused by the presence of adsorbates. This is an important aspect for the characterization of CO adsorption, whose bond implies mechanisms of withdrawing/back-donation of electrons between the metal substrate and CO adsorbate. For example, tuning wavelength with electronic transitions and switching resonance processes by the potential can provide valuable information about the different bonds of CO at the interface.

Finally, I should suggest the modification of the experimental set-up to work with larger single crystals in order to take advantage of the possibility to vary the direction of the polarization and to rotate the sample for obtaining the anisotropic response.

Professor Scherson answered:
We wholeheartedly agree with Prof. Santos's comments and, indeed, we are at present implementing some of the methods she suggests. The problem, however, is not related to the actual size of the facets, rather, the complication arises from the lack of easy means of rotating the facet about the normal to the surface while keeping constant both the angle of incidence of the laser beam and the position of the center of the facet relative to the beam.

Professor Gewirth remarked:
Given infinite resources wouldn't you rather use SFG or even X-ray scattering to examine CO evolution on Pt at relatively fast time scales? Then you would have chemical in addition to temporal sensitivity.

Professor Scherson answered:
Indeed, the two techniques you mention could provide very valuable complementary information to that which SHG can afford.

Professor Wieckowski added:
Correct: BB-SFG operates with a thick gap between the electrode and the window, therefore it has lower resistance (R) from equivalent FTIR system(s). This is then ideal for simultaneous IR and electrochemical experiments.

Professor Tsirlina returned to the discussion of the paper by Payam Kaghazchi, Felice C Simeone, Khaled A Soliman, Ludwig A Kibler and Timo Jacob*:
Surface geometry introduced in your calculations ignores the effects of equilibrium size-dependent surface deformation (see, e.g. in the review ref. 1). This means that the results remain very approximate for nm-size metal fragments, and it is probably better not to be so happy with the particular apparent agreement of computational and experimental results, but to look for stimulating contradictions. Have you observed any?

I would like also to support the comment of Elizabeth Santos concerning similarity of your system and what was previously and carefully studied by Arvia *et al.* for a huge number of systems. See ref. 2 and 3 as the representative examples of early and more recent studies. The difference between faceting techniques used by Arvia's group and yours (with co-authors) is not essential, as the faceting process involves very similar phenomena. What should be thoroughly compared are the experimental arguments, which support contribution of faceted regions and intermediate areas into electrochemical responses.

1. E. L.Nagaev, *Phys. Rep.*, 1992, **222**, 201–307.
2. J. Gomez, L. Vazques, A .M. Baro, N. Garcia, C. L. Perdriel, W. E. Triaca, A. J. Arvia, *Nature*, 1986, **323**(6089), 612–614.
3. F. J. Rodriguez Nieto, G. Andreasen, M. E. Martins, F .Castez, R. C. Salvarezza, A. J. Arvia, *J. Phys. Chem. B*, 2003, **107**, 11452–11466.

Dr Jacob answered:

As stated in the method section of our paper, for evaluating the phase diagram of surface faceting on Ir(210) we have employed the Herring-condition, which is the standard-procedure for these kind of studies. Although, we agree that stress- or strain-related effects become more pronounced for nano-structures below 2 nm, in the studies performed by Madey *et al.* only nano-facets larger than 4 nm were generated. Comparing our phase diagram of Ir(210)-faceting under UHV-conditions with his studies, we find deviations of less than 50 K for the temperatures of various phase transitions. Furthermore, in recent studies on O- and N-induced faceting of Re-surfaces, resulting in surface morphologies different from those observed on Ir(210), again the same accuracy is observed. Therefore, we conclude that for the structure-sizes considered here coverage effects, which are fully included in our calculations, are more important than deformation-related effects.

Regarding the second comment, we refer to the answer also given to Prof. Santos comment above, but once again would like to state that (as described in the paper) in our studies the surface facets were generated by inductive heating, which is rather different from employing fast repetitive periodic potential pulses as performed by Arvia *et al.*

Professor Schmickler commented:

It would be useful if you expanded your work to other metals, so that one could understand for which metals this facetting method works. This could also provide a theoretical basis for the older works of the Arvia[1] group.

1. *e.g.* J. Arvia *et al.*, *Electrochim. Acta*, 1986, **31**, 1359.

Dr Jacob replied:

As stated in our introduction, different groups have already observed adsorbate-induced faceting of various high-indexed transition metal surfaces experimentally. Although, it is not obvious whether the changes in the surface morphology are caused by the same driving-force, the long-term aim is certainly to explore nano-faceted structures on other metal surfaces. For instance, our group has already performed the first theoretical studies on Re-surfaces, where different strongly-adsorbing adsorbates cause the formation of different surface facets.

Professor Wieckowski asked:

Ted Madey is a surface scientist and did not do voltammetry. The work in the Timo Jacob laboratory in the domain of voltommetry should not be discussed in the context of the Ted Madey work.

Dr Jacob answered:

It has quite often been observed that the concepts of UHV-systems have an analog in electrochemistry and *vice versa*. This is also the case for the present system, where strongly-interacting adsorbates cause nano-facets to become thermodynamically more stable than the planar surface. While under UHV-conditions the adsorption of oxygen can be modified by temperature or pressure changes (as shown by Madey *et al.*), under electrochemical conditions one, in addition, has the advantage to induce oxygen or ion adsorption by changing the electrode potential. Furthermore, the facet-morphologies observed under UHV and electrochemical conditions are both three-sided nanopyramids, which somehow rationalized this comparison.

Professor Behm addressed Dr Jacob and Professor Schmickler:

The studies of Madey and co workers showed that adsorbate induced faceting occurs for metals which have an extremely strong metal–oxygen or metal–nitrogen bond. This indicates that the faceting observed for Pt after potential cycling is not directly comparable to the adsorbate reduced faceting in Madey's experiments.

Dr Jacob responded:

It might be correct that our studies and those performed by Madey *et al.* are not directly comparable to the adsorbate-reduced studies of the Arvia group. This reduces to the question of whether facet-formation that has been observed on various transition metal surfaces so far is caused by the same driving-force. It will be aim of future studies to elaborate this question.

Professor Gewirth opened the discussion of the paper by Zhi-You Zhou, Na Tian, Zhi-Zhong Huang, De-Jun Chen and Shi-Gang Sun*:

Is it possible that, at long times, the particles decompose? For example, peroxide formation at Pt. Any O_2 reduction in an MEA is known to promote Pt dissolution. Does this occur with the particles you have prepared? Have you tested an MEA composed of these particles or rods over long (say 1000 h) time scales?

Professor Sun responded:

We have not yet carried out the test of long term stability of tetrahexahedral Pt nanocrystals (THH Pt NCs). However, from our preliminary results, the stability of the THH Pt NCs is much higher than conventional Pt nanoparticles. One reason for this is that the THH Pt NCs were grown under intensive oxidation–reduction conditions, *i.e.* square-wave potential between −0.10V and 1.20 V *vs.* SCE, while conventional Pt nanoparticles can be progressively dissolved under this condition.

Professor Savinova remarked:

I agree that from the fundamental perspective it might be interesting to explore the behaviour of preferentially oriented particles. However, I cannot agree with the conclusion of the authors that "The study has opened a new prospect avenue in shape-controlled synthesis of nanoparticle catalysts with high performances". To my opinion, preferentially oriented particles are irrelevant to practical applications.

Let's compare the electrocatalytic activities towards ethanol oxidation of 81 nm preferentially oriented Pt tetrahexahedra (THH) and 3.2 nm commercial Pt/C (Fig. 4 of the paper). The specific electrocatalytic activity of the THH particles exceeds that of commercial Pt/C by a factor of *ca.* 3.3, which is, of course, interesting. However, the specific surface area of 81 nm particles is *ca.* 25 times lower than that for 3.2 nm. Hence, the mass activity (which is important for practical applications), of 81 nm particles in the ethanol oxidation will be *ca.* 7.7 times lower than that of 3.2 nm particles. The only way to overcome this limitation is to synthesise preferentially oriented particles in the size range of 2 to 5 nm. This is indeed possible, as documented in numerous publications. However, while large particles may be relatively stable in the course of electrocatalytic reactions, metal particles of smaller (few nm) size won't be able to preserve their shape. Indeed, adsorption and reaction on metal particles are known to influence their structure and morphology. The larger the particles, the higher will be the activation barrier required to change their morphology. However, for nm-sized particles this activation barrier will be rather small and they will easily change their shape in the course of electrocatalytic reactions, even if these are carried out at ambient conditions (not to speak about the fuel cell operation).

Professor Sun answered:

To improve the catalytic activity and stability of Pt nanoparticle catalysts is always vital for many energy and environmental processes, especially today we are facing to the challenge of rare Pt reserve in the earth and the continuing increase of Pt price. The control of surface structure of metal nanoparticles, *i.e.*, synthesis of preferentially oriented nanoparticles through shape-controlled methods, is an important strategy. The pioneering work was the synthesis of cubic and tetrahedral Pt nanoparticles by El-Sayed and co-workers in 1996 (El-Sayed *et al.*, Science, 1996, **272**, 1924). In the past decade, many metal nanocrystals with different shapes have been synthesized, but all these metal nanoparticles were limited to those bounded by low-index facets with low surface energy, such as {111} and {100} facets. Fundamental studies have demonstrated that the catalytic activity of these low-index surfaces is much lower than that of high-index surfaces. However, the synthesis of metal nanoparticles bounded by high-index facets is rather challenging due to their high surface energy. We have developed an electrochemical method to overcome the obstacle and synthesized successfully tetrahexahedral Pt nanocrystals (THH Pt NCs) that are bounded by {hk0} high-index facets with high catalytic activity. Besides, we have also demonstrated that nanocrystals bounded by {hkk} and {hkl} high-index facets can be also synthesized. Our study has demonstrated the possibility to prepare metal nanocrystals with diverse shape and different high-index surfaces, and to correlate their catalytic activity with surface structure as coordinated in the triangles shown in Fig. 1 and Fig. 2 in our paper. As the surface structure of metal nanocrystals in Fig. 2 may be varied along with those single crystal planes in Fig. 1, and metal nanoparticle catalysts of different surface structure exhibit peculiar activity for specified reactions, the metal nanoparticles bounded with high-index facets are therefore prospective catalysts of high performances for energy, environment and diverse applications. From this point of view, the study of developing electrochemically shape-controlled synthesis method and the success of synthesizing THH Pt NCs has opened a new prospect avenue in shape-controlled synthesis of nanoparticle catalysts with high performances.

As for the particle size, it can be controlled by varying the growing time. We have reported in a previous paper (Science, 2007, **316**, 732–735) that THH Pt NCs of *ca.* 20 nm were obtained by shortening the square-wave potential time. Taking the account of high specific electrocatalytic activity of THH Pt NCs based on active area, the mass activity of THH Pt NCs of about 10 nm will be comparable to that of commercial Pt catalysts of about 3 nm. One advantage of the catalysts of such a large size (~10 nm) is that they will have higher stability. Concerning the stability of the THH Pt NCs, theoretical calculations (see F. Ma, *et al.*, J. Phys. Chem. C, 2008, **112**, 3247–3251) have demonstrated that the surface structure of 7.8 nm THH Pt NCs can be maintained up to 860 K. Although there is still no stability test of the THH Pt NCs of 2–5 nm, relevant experiments of small preferentially oriented Pt nanoparticles were already documented. C. Wang *et al.* have reported recently that polyhedral, truncated cubic and cubic Pt nanoparticles of 3–8 nm in dimension can exhibit significant surface structure effects upon electrochemical oxygen reduction (C. Wang *et al.*, *Angew. Chem., Int. Ed.*, 2008, **47**, 3588–3591; C. Wang *et al.*, *J. Am. Chem. Soc.*, 2007, **129**, 6974), implying a good stability of these preferentially oriented Pt nanoparticles under electrochemical reaction conditions; Vidal-Iglesias *et al.* (Vidal-Iglesias *et al.*, *Electrochem. Commun.* 2004, **6**, 1080–1084) have shown that Pt cubes of 10 nm possess seven times activity higher than other shape particles towards ammonia electrooxidation. These results clearly illustrated that the preferentially oriented nanoparticles can preserve their surface structure in the course of electrocatalytic reactions.

Professor Wieckowski asked:
The voltammetry to follow stability and reactivity was missing. In other words it would be nice to see short time and long time electrochemical performance of these nano-objects.

Professor Sun responded:
In the paper, we have put emphasis upon the electrochemically shape-controlled synthesis of Pt and Pd nanoparticles bounded by high-index facets, which yields high catalytic activity and stability, as exemplified for ethanol oxidation mentioned in the paper. Recently, we have carried out the stability test of tetrahexahedral Pt nanocrystals (THH Pt NCs) using cyclic voltammetry, and found that the stability of the THH Pt NCs during a potential cycling as long as 13 h is very good.

Professor Schiffrin asked:

Can you explain why exposure to oxygen of the nanosphere seeds for 4–5 hours leads to the growth of nanorods by further cycling of the potential? This is a very unusual observation.

Professor Sun answered:

In our experiments, we have tested that the exposure of the GC electrode with Pt nanospheres is crucial for the growth of Pt nanorods. If the exposure time was shortened, both THH Pt NCs on GC and Pt nanorods on Pt nanospheres can be observed. The key reason is not the exposure of Pt nanosphere to oxygen, but the exposure of GC substrate to air. The latter may cause the GC surface to be hydrophobic or inert, on which new Pt nuclei are hardly generated; instead, Pt crystal nuclei preferentially generate on the Pt nanospheres. The reason for the formation of fivefold twinned nuclei is not clearly understood yet.

Dr Buder asked:

Agglomeration of Pt particles in fuel cells is a big issue. How is the stability of your particles under fuel cell conditions? H_2O_2 stability is also an issue for fuel cells. Are your particles stable against H_2O_2?

Professor Sun responded:

In our experiment, we did not observe agglomeration of tetrahexahedral Pt nanocrystals (THH Pt NCs) during the oxidation of ethanol when the THH Pt NCs were supported on glassy carbon substrate. At present, we have not yet used the THH Pt NCs in a fuel cell for testing. To our knowledge, the H_2O_2 will mainly cause damage to proton-exchange membranes, not to Pt catalysts. (see, *e.g.* T. Kinumoto *et al.*, J. Power Sources, 2006, **158**, 1222–1228)

Professor Strasser remarked:

Does the synthesis of high energy facet nanoparticles *via* potential cycling also work starting with very small 2–6 nm sized nanoparticles? Is the mechanism a dissolutive reprecipitation? Can you expect the formation of such structures in cathode layers of fuel cells? Can you make Pt_3Ni octahedrons *i.e.* only exposing (111) facets?

Professor Sun replied:

At the present stage, we started from a large Pt particle (>400 nm) to produce the tetrahexahedral Pt nanocrystals (THH Pt NCs). We have tried also starting from small (~3 nm) Pt particles, but we could not obtained the THH Pt NCs of *ca.* 3 nm using the same square wave potential treatment. The growth mechanism of the THH Pt NCs consists of the repetitive dissolutive re-precipitation on a same particle under intensive oxidation–reduction conditions, *i.e.* square-wave potential between −0.10 V and 1.20 V *vs.* SCE. On the one hand, at both potentials hydrogen or oxygen species may adsorb on the surface of the growing Pt particle, which leads to decreasing the surface energy of the particles, and the oxygen adsorption at 1.20 V also serves the selection of the facet on the other hand (*i.e.* the low-index facet is easily reconstructed under oxygen adsorption, while the high-index facet is stable upon oxygen adsorption as indicated by studies using metal single crystal planes). In a fuel cell, oxidation dissolution occurs on the surface of cathode catalysts, while reduction precipitation mainly occurs in the Nafion membrane layers (H_2 crossover). So we think it may be not likely to form high energy facet nanoparticles in cathode layer of fuel cells. We have not yet tried to synthesize Pt_3Ni octahedra.

Professor Neurock retuned to the discussion of the paper by Payam Kaghazchi, Felice C Simeone, Khaled A Soliman, Ludwig A Kibler and Timo Jacob*:

You have indicated that most of the restructuring is due to adsorbate induced reconstruction, which I would agree with. On the nanoparticles, electrochemical effects may also become important. I realise that it is difficult to calculate all of these effects. Can you please comment, however, on the effect of the field or potential on restructuring? They would be quite different at step edges and kink sites which could significantly influence restructuring. Can you also please comment on the influence of coverage effects?

Dr Jacob replied:

So far in our theoretical phase diagram, electrochemical effects have been taken into account simply by shifting the Fermi-level of the electrode. We certainly agree that under electrochemical conditions step edges and kinks might have a significant role. However, this would require an accurate treatment of the surface electric field on rather extended systems, which so far has not been realized, but will be the aim of future work. Regarding coverage effects, on all surfaces relevant for the faceting of Ir(210) we have studied various surface sites and coverages. Our phase diagram indicates that in order to induce faceting at least ~0.5ML oxygen would be required, which is in excellent agreement with the observations of Madey *et al.* for the UHV-system.

Professor Hayden addressed Dr Jacob and Professor Sun:

As outlined in the introduction given by Mark Koper, a key goal is to understand the increasing complexity of electro catalytic systems, and the first steps may be to advance understanding through moving from single crystal surfaces (close packed/ stepped and kinked surfaces) to nano-structured surfaces and nano-particles. The objective is to assess active sites, relative barriers/over-potentials, and surface kinetics. Is there sufficient control, characterization or homogeneity in the systems described in order to make a connection, and identify differences. In particular, is there any prospect, or have you already, studied relatively simple reactions such as CO oxidation (as outlined also by Mark Koper on single crystal surfaces) as a function of *e.g.* nano-structure, electrolyte environment *etc.* in order to make these connections?

Professor Sun answered:

As indicated in the paper by Fig. 1 and Fig. 2, the surface structure of polyhedral metal nanocrystals in the unit stereographic triangle in Fig. 2 may be varied systematically as the surface structure of metal single crystal planes in the unit stereographic triangle in Fig. 1. This provides, theoretically, a possibility to connect the catalytic activity with nanoparticles, *i.e.* the real catalysts.

However, to control the surface structure of metal nanoparticle is more difficult than to prepare a metal single crystal plane. In the latter, the orientation and cutting of a single crystal can be done precisely using well-established classical technology. The shape-control synthesis of metal nanoparticles by chemical methods was nevertheless started merely in the middle of the 1990s, and only cubes and octahedra (or tetrahedra) were frequently obtained, since these nanoparticles bounded by close-packed (100) and (111) facets, which yield low surface energy. For synthesizing metal nanoparticles with open surface structure, such as tetrahexahedra that are enclose by {730} and vicinity high-index facets, the big challenge consists of overcoming the limit of crystal growth rule, that is, crystal grows very fast along the normal direction of high-index facets, resulting in the disappearance of high-index facets. In our study we have successfully developed electrochemically shape-controlled methods for synthesizing metal nanoparticles of high surface energy, which gives a prospect to connect the catalytic activity with the surface structure of metal nanoparticles. It is worthwhile pointing out that not only to control precisely the surface structure of metal nanoparticles, but also to control the homogeneity (surface structure, size distribution, crystal shape, *etc.*) of metal nanoparticles is also challenging. Our study has just initiated the investigation to connect the catalytic activity with nanoparticles enclosed by high-index facets of high surface energy. In the present stage, this study is far from mature in comparison with the study using single crystal planes that has been started at the beginning of 1980s.

Dr Jacob replied:

This is exactly the long-term objective. According to Madey *et al.*, the size of the Ir nanofacets can be controlled under UHV conditions. A first electrochemical and *in-situ* STM characterisation of planar and faceted Ir(210) in perchloric acid solution is reported in the present work. CO adlayer oxidation for both planar and faceted Ir(210) is presented in Fig. 6. So far, we have been able to vary the extent of faceting by changing annealing and cooling conditions. For future work, it is planned to systematically vary the nano-faceted structures and deduce connections between structure and activity, *e.g.*, for CO adlayer oxidation. Such structural variations are expected to be achieved electrochemically, as theoretically predicted in the present paper.

Professor Savinova asked:

I think that comparison of the electrochemical/electrocatalytic properties of nano-faceted single crystals and metal nanoparticles is very interesting and should help in answering the question of to what extent the behavior of metal nanoparticles can be modeled by extended surfaces. I believe that extended surfaces can adequately model the properties of large metal particles, say, above 3–4 nm. However, for smaller particles such phenomena as size confinement and metal–support interactions are expected to lead to diverse properties which cannot be accounted for by extended surfaces.

As concerned with the paper, I am particularly interested in the comparison of CO oxidation on planar and on nano-faceted Ir(210) (Fig. 6). It is remarkable that the onset of CO oxidation on nano-faceted electrode is shifted very much positive. Could you comment on that?

Dr Jacob replied:

We totally agree with the comment that our studies are more relevant for larger particles. However, in case of smaller nanoparticles they might also help to distinguish between particle- and support-effects.

So far, our studies on the CO oxidation showed that there are remarkable differences between the electrochemical properties of planar and faceted Ir(210), which is in agreement with corresponding UHV-studies (*Langmuir*, 2006, **22**, 3166). Regarding the potential shift on nano-faceted Ir(210), this might be caused by site-blocking effects. While on the faceted surface CO blocks those sites favored by O, on planar Ir(210) oxygen is rather mobile without having one particular favored surface site. Presently, calculations are being performed to answer this question.

Professor Behm remarked:

I would like to question the use of the term "nanoparticle" for the very interesting nanostructures created by faceting of the Ir surface. Nanoparticles in catalysis and electrocatalysis are more or less decoupled from the support, chemically and electronically. The Ir nanostructures, in contrast, compromise of an Ir nanostructure on an Ir substrate, and the two are closely coupled, and I would expect chemical/electrochemical properties which are rather different from real Ir nanoparticles of similar size. Therefore, referring to these structures as nanoparticles could be misleading.

Dr Jacob responded:

In our paper we have not exactly used the expression "nanoparticles" for the nano-facets formed on Ir(210), but rather suggested that by exploring such nano-facets one might gain insight on the catalytic behavior of nanoparticles. However, we agree with the comment that this depends on the system and the size of the nanoparticles. While smaller nanoparticles are influenced by the support, larger particles might be decoupled.

Professor Chorkendorff continued the discussion of the paper by Zhi-You Zhou, Na Tian, Zhi-Zhong Huang, De-Jun Chen and Shi-Gang Sun*:

At this point of the discussion I feel obliged to comment that more steps and defects on the surface of nanoparticles is necessarily not good. It really depends on what kind of reaction we are looking at, *i.e.* whether we are on the right or left side of the volcano curve. If the rate limiting step is limited by the reactivity of the surface (the right side of the volcano curve) then steps and defects may improve the overall reaction since the introduction of such in general, due to their under-coordinated nature, will enhance the reactivity. However, if the rate is limited by site blocking due to too strong bonding of the various intermediates, then it naturally does not help the overall rate to introduce steps or defect as they only will lead to stronger blocking effects. This is the so called Sabbatier principle leading to volcano curves in heterogeneous catalysis and there is no reason not to adapt this principle in electrocatalysis as well.

Dr Wang asked:

Your conclusion that "Owing to high density of atomic steps and dangling bonds on the high-index facets, nanoparticle catalysts bounded by high-index facets display in general high catalytic activity and stability" is interesting and also seems questionable to me. Generally, low coordinated surface sites are more reactive. For a catalyst, this can be good or bad depending on whether the catalyzed reaction is adsorption or desorption limited. For example, specific activity of Pt decreases with decreasing particle size for the desorption-limited ORR, but not for the adsorption-limited HOR. Also, a DFT study[1] of particle size effects on ORR activity suggested two opposite trends for Pt and Au, which is rationalized by bulk Pt being too reactive and Au being too inert. Are you expecting that Pt nanoparticles with high-index facets will yield relatively high ORR activity and stability? For the oxidation of small organic molecules, dependence on the facets and particle size are likely to be complex. An example is given in Fig. 4 for ethanol oxidation in this paper. The specific activity decreases in the order of 81 nm tetrahexahedra > 115 nm nanosphere > 3.2 nm (should also be nearly spherical). The significantly higher activity of the 81 nm tetrahexahedra than that of other particles is attributed to its high step density. However, the general argument that low coordinated sites display higher activity is not supported by the lower activity of the 3.2 nm than that of the 115 nm particles. The former should have a higher percentage of low coordinated sites (edges and corners) than the latter. Maybe, steps sites differ considerably from generalized low coordinated sites? Based on the above discussion, can you clarify your conclusion and the rationale behind it?

Professor Sun answered:

We agree that low coordinated surface sites are more reactive in terms of breaking chemical bonds. However, the catalytic activity characterized by turnover frequency depends on reactions. If the rate-determining step (RDS) is breaking bonds (*e.g.*, ammonia synthesis), low coordinated surfaces will exhibit high activity; on the other hand, if the RDS is desorption processes, low coordinated surface will be easily poisoned, *i.e.*, less activity. As for the ORR, Feliu's group studies (*J. Electroanal. Chem.*, 2004, **564**, 141–150; *J. Electroanal. Chem.*, 2007, **599**, 333–343) have proved that high-index surfaces exhibit also higher activity. We have not yet tested the activity of THH Pt NCs towards ORR. The stability of the THH Pt NCs is good, since they grow under the repetitive oxidation–reduction conditions, where polycrystalline surfaces are unstable. In order to avoid any confusion, we revise the expression "…high-index facets display in general high catalytic activity and stability" in

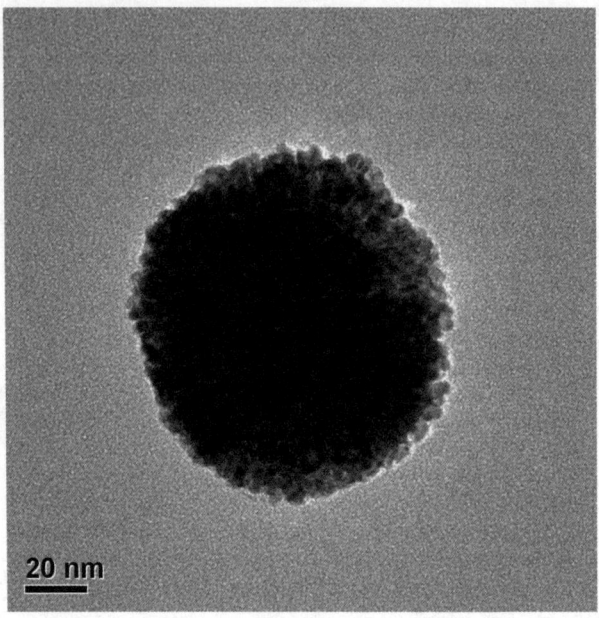

Fig. 1 SEM image of the 115 nm Pt nanosphere

our paper to "...high-index facets display in general high catalytic reactivity and stability". It should be noted that the surface of the 115 nm Pt nanospheres synthesized by electrodeposition is not smooth. In a HRTEM image (see Fig. 1), we can observe that the Pt nanosphere consists of the agglomeration of primary Pt nanoparticles of 3–5 nm, so the step density is not very low, but is comparable with commercial catalysts. Besides, the catalytic activity of the Pt agglomeration is usually higher than that of well-dispersed fine particles (*e.g.* S. Lee *et al.*, *Angew. Chem., Int. Ed.*, 2006, **45**, 7824–7828), which may be attributed to size effects. We mention now the structure of the 115 nm Pt nanospheres in the legend of Fig. 4 in the revised paper for clarification.

Mr Duca remarked:

With respect to the reactivity of the nanoparticle catalysts towards ethanol oxidation (Fig. 4, page 5), is there any possible bias due to ascorbic acid remaining on the support? In other words, what procedure is followed in order to remove possible adsorbates from the glassy carbon support prior to the electrochemical experiments? Lastly, in which conditions (ethanol concentration, concentration and nature of supporting electrolyte) are the electrochemical tests reported in section 3.1 carried out?

Professor Sun replied:

Prior to the test of electrocatalytic activity towards ethanol oxidation, tetrahexahedral Pt nanocrystals (THH Pt NCs) were cleaned in 0.1 M $HClO_4$ by potential cycling between -0.25 and 0.75 V (SCE) at 50 mV s^{-1} until a stable CV curve of THH Pt CNs was recorded. In such a way, possible adsorbates such as ascorbic acid have been removed completely by oxidation. The electrochemical tests of ethanol oxidation were carried out in 0.1 M ethanol + 0.1 M $HClO_4$, the electrode potential was varied from 0.1 V to 0.55 V (SCE) in chonoamperometry experiments.

Mr Dudzin remarked:

1. Does ascorbic acid effect the growth of the facets on the surface? Does it affect it through the preferential adsorption?

2. If these crystals are big enough isn't it only surface effects which could be studied on the particular single crystal surfaces?

Professor Sun responded:

1. We have obtained also small (<100 nm) tetrahexahedral Pt nanocrystals (THH Pt NCs) in ascorbic acid free solutions. So, in our opinion, the ascorbic acid does not play a key role in the formation of the THH Pt NCs. However, our results demonstrated that, when the solution contains ascorbic acid, the THH Pt NCs will grow slowly and reach a relatively perfect shape, especially when the THH particle size exceeds 100 nm. In this case the ascorbic acid may have preferential adsorption.

2. Yes, if the particle size is big enough, only surface effects may be studied, since nanoparticles size effects can be observed usually when the particle is smaller than 10 nm.

Professor Ren addressed Professor Sun and Professor Scherson:

In all the three talks, oxygen species are very important in the oxidation of CO or the formation of nanoparticles. Is there any direct evidence of whether the surface species is O or OH? Do you have any idea which kind of technique can be of help in providing this information?

Professor Sun answered:

In our current studies, there is no direct evidence whether the surface species is O or OH. To identify the origin of oxygen species generated on Pt surface is certainly an important issue in electrocatalysis. Different techniques have been used for this investigation, such as electrochemistry-XPS (N.H. Li *et al.*, *J. Electroanal. Chem.*, 1997, **430**, 57–67), *in situ* XFAS (Tada *et al.*, *Angew. Chem., Int. Ed.*, 2007, **46**, 4310–4315), EQCM (Jerkiewicz *et al.*, *Electrochim. Acta*, 2004, **49**, 1451–1459).

Professor Scherson responded:

Unfortunately SHG lacks the required specificity to identify the nature of the adsorbed oxygen species. In fact, the development of techniques capable of providing that type of information may be regarded as key to further advances in electrocatalysis.

Professor Schiffrin addressed Professor Hayden, Professor Scherson, Dr Jacob and Professor Sun:

Prof Hayden has raised an important issue regarding how to transpose results obtained from single crystal studies to predict the electrocatalytic properties of nanoparticles. We know at present how to synthesise a wide range of nanoparticles of different materials and geometries but the structure of the surface still represents a challenge. The problem becomes more difficult for sizes below, say, 5 nm due to the small number of surface atoms. It is possible to establish the overall geometry from TEM diffraction patterns but this does not give details of the surface structure. Structural details for functionalised nanoparticles has been recently obtained by X-ray diffraction measurements.[1] This requires, however, stabilisation of particles of a very well defined geometry and size, and more importantly, their crystallisation, something that is not easy to achieve with electrocatalytic materials. In this respect, the use of surface orientation specific reactions appears a very promising approach for assessing the structure of surfaces at the nanoscale.[2]

1. P. D. Jadzinsky, G. Calero, C. J. Ackerson, D. A. Bushnell and R. D. Kornberg, *Science*, 2007, **318**, 430–433

2. J. Solla-Gullon, P. Rodriguez, E. Herrero, A. Aldaz and J. M. Feliu, *Phys. Chem. Chem. Phys.*, 2008, **10**, 1359–1373

Professor Sun responded:

We fully agree. The surface structure of small nanoparticles may be elucidated by using surface orientation specific reactions, such as UPD of some metal ions on nanoparticles. This method is very convenient for electrochemists. Besides, we suggest also using probe molecules (e.g. adsorbed CO and NO) to characterize the surface structure of tetrahexahedral Pt nanocrystals through *in situ* FTIR spectroscopy.

Differential reactivity of Cu(111) and Cu(100) during nitrate reduction in acid electrolyte†

Sang-Eun Bae and Andrew A. Gewirth*

Received 22nd February 2008, Accepted 28th March 2008
First published as an Advance Article on the web 21st August 2008
DOI: 10.1039/b803088j

The interactions of nitrate with Cu(100) and Cu(111) in acidic solution are studied by cyclic voltammetry (CV) and *in situ* electrochemical scanning tunneling microscopy (EC-STM). CV results show that reduction of nitrate on Cu(111) commences at 0.0 V *vs.* Ag/AgCl while the corresponding potential is −0.3 V on Cu(100). EC-STM images show that the terrace of both Cu(111) and Cu(100) are atomically flat at potentials more negative than −0.7 V. The Cu(100) surface exhibits flat terraces throughout the entire cathodic potential range. Close to OCP, step edges start to corrode. In contrast to Cu(100), the first layer of Cu(111) is converted to an atomically rough and defected surface—associated with nascent surface oxidation at potentials positive of −0.7 V. This surface oxidation is correlated with nitrate reduction.

Introduction

The adsorption and reduction of nitrate at metal surfaces have long attracted considerable interest because of its environmental importance[1] and its putative role in the production of useful nitrato compounds.[2] Nitrate contamination of water must be remediated because of the deleterious effect of nitrate on human health and the environment.[1,3] Nitric acid has been long used as cleaning solution for metal contamination in the laboratory[4] and manufacturing[5] settings. The nitrate ion is also used to selectively polish on-chip copper wiring in recent CMP processes.[6] Nitrate has also been a main starting material to produce useful nitrato compounds such as ammonia, hydroxylamine, hydrazine, and azides.[2,7] Nitrate electroreduction has been studied on several metal surfaces by using conventional electrochemical techniques,[1,8–10] FTIR,[11–13] Raman,[14] and mass[7] spectroscopic methods.

In aqueous media, nitrate electroreduction proceeds by several pathways depending on electrode composition and pH of the electrolytes.[7,15,16] One of the most active electrode materials for nitrate electroreduction is Cu.[7] On copper surfaces, nitrate reduction in acidic media takes place through an eight-electron pathway, as shown in eqn (1).[8]

$$NO_3^- + 8e^- + 9H^+ \rightarrow 3H_2O + NH_3 \quad E^0 = 0.68 \text{ V } vs. \text{ Ag/AgCl} \quad (1)$$

On the basis of the observation of intermediates, it is thought that the reaction proceeds stepwise, through successively more reduced intermediates. The Tafel slopes measured for nitrate electroreduction on many transition metal electrodes are found close to 120 mV dec^{-1},[7] which suggests that the first electron transfer—involving

Department of Chemistry, University of Illinois, Urbana, IL, 61801, USA. E-mail: agewirth@uiuc.edu; Fax: +1-217-244-3186; Tel: +1-217-333-8329

† The HTML version of this article has been enhanced with colour images.

the reduction of nitrate to nitrite—is the rate determining step. However, the Tafel slope measured on the Cu surface is somewhat larger than 120 mV dec^{-1}, possibly suggesting that another mechanism might be operating.[7] Additionally, the reduction of nitrite on Cu occurs at potentials less negative than those for the nitrate reduction.[10] Recently, we observed the potential driven interconversion between nitrate and nitrite structures on a Cu(100) electrode during the course of electroreduction.[17,18] On the basis of these observations, we suggested that the nitrate to nitrite conversion is initiated at step edges. However, in contrast to Cu(100), it was impossible to observe those ordered adlattices on Cu(111). Rather, the Cu(111) surface became defected during cathodic potential excursions. The origin of the differences in reactivity between the two crystal faces is unknown.

The evolution of Cu surfaces in aqueous solution[19–23] and air[19,24,25] has long been a topic of inquiry. X-Ray diffraction results[19] for aqueous and native oxide formation on Cu(111) reveals a monolayer oxide phase at more negative potentials than that for the bulk oxide in aqueous solution. Cu single crystals are found to be oxidized at different rates depending on surface orientation.[24] The initial stages of Cu(111) oxidation are dominated by the nucleation of oxide islands at relatively low temperature and by two-dimensional oxide growth at high temperature. On the Cu(100) surface, oxidation takes place through 3D growth limited by oxygen surface diffusion.

In acidic solution, Cu oxide is typically not found, while both cupric ion and Cu metal are, according to the Pourbaix diagram.[26] In neutral and basic solution, Cu_2O, CuO, or $Cu(OH)_2$ are formed rather than cupric ion.[19,27,28] In acidic solution, the surfaces of both Cu(111) and Cu(100) cases are anodically dissolved through step flow reactions. However, at high dissolution rates, copper single crystal surfaces exhibit an isotropic etching trend.[29] If the etchant solution contains adsorbents such as chloride and sulfate ions, step flow reactions are strongly dependent on the orientations of the copper single crystal surfaces and the adsorbent dynamics.[29–32] For example, in the case of Cu(111), the step edges along {211} directions dissolve preferentially, while in the case of Cu(100), {100} directional step edges are preferentially etched.[29,31,32] Nitrate is well known not only to etch copper surfaces at open circuit potential (OCP),[33] but also to be electrochemically reduced on the copper surface in aqueous solutions.[8] However, despite the large literature examining the interaction of nitrate with copper surfaces,[34–36] the evolution of Cu single crystal surface in the presence of nitrate remains to be explored.

In this paper, we examine the orientation dependent interaction of nitrate with Cu single crystal surfaces by using cyclic voltammetry (CV) and *in situ* electrochemical scanning tunneling microscopy (EC-STM).

Experimental

Solutions used in this work were prepared from ultrapure nitric acid (J. T. Baker – Ultrex II). The electrolytes were made from ultrapure $HClO_4$ (J. T. Baker – Ultrex II) or HF (J. T. Baker – Ultrex II) by using Milli-Q water (18.2 MΩ cm, Millipore Inc.).

Working electrodes were made from 10 mm diameter Cu(100) and Cu(111) single crystals (Monocrystals Co.) that were polished to a mirror finish with 9, 3, 1 and 0.25 μm diamond suspensions, sequentially. The crystals were then electropolished in concentrated phosphoric acid,[37] rinsed with copious Milli-Q water, and immediately covered with electrolyte solution to inhibit oxidation.

Cyclic voltammetric (CV) data were obtained in a two-compartment, glass electrochemical cell, as previously reported.[38] The copper crystal was placed in the cell in a hanging meniscus configuration. The solutions were purged with Ar prior to use, and an Ar atmosphere was maintained in the cell during all electrochemical measurements. Scan rates for measurements were 50 mV s^{-1}.

EC-STM measurements were carried out using a Nanoscope III E (Digital Instrument Corp.) equipped with a fluid cell. The electropolished crystal was clamped to the bottom of the cell using an O-ring. Platinum wires were used for both the reference and counter electrodes. All potentials are reported with respect to the Ag/AgCl reference electrode for convenience. EC-STM tips were prepared by electrochemical etching of a tungsten wire (0.25 mm in diameter) in 2 M NaOH, followed by insulation with Apiezon Wax in order to reduce the area in contact with the electrolyte. This procedure minimized the Faradaic current at the tip.

III. Results

3.1 Cyclic voltammetry

Fig. 1 shows cyclic voltammograms (CV) of the Cu(100) and Cu(111) surface immersed in 0.1 M $HClO_4$ solutions with 1 mM HNO_3. In curve (a)—a CV of Cu(111) obtained in a solution containing 0.1 M $HClO_4$ + 1 mM HNO_3—the cathodic current begins at 0.0 V and increases gradually as the potential is swept negatively. The reduction current increases abruptly at −0.25 V, exhibits a peak at −0.53 V and then decreases. Cathodic current associated with the hydrogen evolution reaction (HER), begins at ca. −0.72 V. There is no corresponding anodic current seen on the positive potential sweep, which means that the reduction of nitrate on the copper surface is irreversible.

Curve (b) shows the CV obtained from a Cu(100) crystal immersed in a solution containing 0.1 M $HClO_4$ + 1 mM HNO_3. In contrast to the Cu(111) case, a featureless flat region between 0.0 and −0.3 V is observed. As the applied potential was moved to negative values, the cathodic current begins to increase around −0.3 V. The onset of the HER occurs at nearly the same potential as in the Cu(111) case. The CV shows a crossover feature at ca. −0.4 V ascribed to reduction of nitrite retained near the electrode after the negative potential excursion.[10] CVs obtained with 0.1 M HF supporting electrolyte were very similar to that in 0.1 M $HClO_4$.

3.2 *In situ* EC-STM—Cu(111)

Fig. 2 and 3 show *in situ* EC-STM images obtained from a Cu(111) crystal immersed in 0.1 M HF + 1 mM HNO_3 while the potential was swept from −0.8 V to −0.25 V. The image shown in Fig. 2a—obtained at very negative potentials—exhibits wide meandering and atomically flat terraces and steps exhibiting a height of 0.20 nm, which is the expected value for Cu(111) monoatomic steps. Fig. 2b shows an *in situ* EC-STM image obtained while moving the potential from −0.59 V to −0.47 V. The feature in the center of the figure shows that this image was obtained from the same location as that in Fig. 2a. At these intermediate potentials, the terraces are roughened and the surface is covered with a number of small white particles

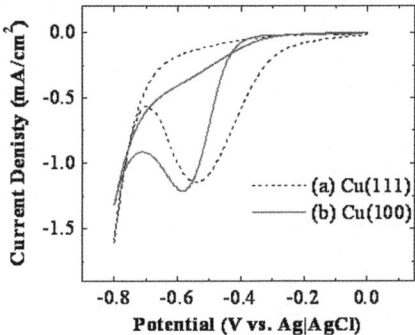

Fig. 1 Cyclic voltammograms obtained from Cu(111) (a) and Cu(100) (b) surfaces in 0.1 M $HClO_4$ + 1 mM HNO_3.

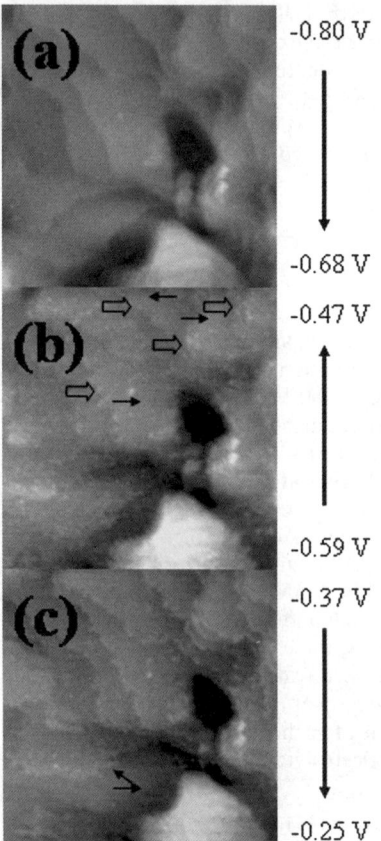

Fig. 2 *In situ* EC-STM images of the Cu(111) surface were obtained in 1 mM HNO$_3$ + 0.1 M HF during anodic potential sweeping between −0.8 V−−0.25 V (a), −0.59 V−−0.47 V (b), and −0.37 V−−0.25 V (c). Image size is 150 nm × 150 nm. The tip bias and tunneling current are 10 mV and 10 nA, respectively.

which are 0.22 ± 0.03 nm high. The smooth step edges seen in Fig. 2a turn into irregular or sea-saw-like shapes. Comparison of the step positions with that of Fig. 2a shows that the first Cu(111) layer expands laterally. The presence of the particles makes step edges harder to discern. However, no defects could be seen on the terraces, which suggest that Cu has not dissolved into solution at these potentials.

The above observations are consistent with those found in the literature.[20,21,27,39] Friebel *et al.*[20] observed a roughened surface and lateral expansion of the top layer of Cu(111) in neutral sulfate solution resulting from growth of the first oxide layer on the Cu(111) surface. Marcus *et al.*[21,27,39] also found evidence for Cu(I) oxide formation on Cu(111) in NaOH solution. These authors studied oxidation of Cu single crystal surfaces and observed formation of both dark and brighter islands on the Cu(111) terrace. The darker islands were associated with an adsorbed O-containing species, while the lighter ones were associated with Cu(I) oxide. The height (0.22 ± 0.03 nm) of the small white particles we find here is close to the value (0.25 nm) of Cu(I) oxide grown on Cu(111) in NaOH.[21] Thus, we suggest that the small white particles and the roughened surface of terrace are attributed to Cu(I) oxide and the O species adsorption, respectively. The roughened structure of the terrace can be attributed to an electronic effect observed by STM for O species adsorbed on metals.[40] Interestingly, the Cu(111) terrace is quite heterogeneous, with substantial areas found free of particles (denoted by the closed arrows in Fig. 2b) and with the apparent nucleation of bigger particles at step edges (open arrows).

In Fig. 2c, obtained as the potential was swept between −0.37 V and −0.25 V, the number of the small white particles on the terraces has decreased relative to images obtained at more negative potentials. However, larger particles are still observed on the step edges. Some particles are still found on the terraces. In Fig. 2c, the steps are well resolved in contrast to the image in Fig. 2b. At these more positive potentials, the step edges begin to erode quickly, as demarcated by the arrows in Fig. 2c. At this point, the potential cycle was reversed, and the potential returned to more negative values.

Fig. 3 shows *in situ* EC-STM images obtained from the Cu(111) surface during the cathodic potential sweep. The bottom portion of the EC-STM image of Fig. 3a reveals that the surface consists of irregular terraces and some bigger white particles. The step edges are very irregular as indicated by arrow A. Around −0.39 V, new bigger white particles start to form (arrow B). At a slightly more negative potential (−0.41 V), the bigger particles begin to nucleate and form disordered terraces (arrow C). The step edges expand and form the saw-tooth structure again. Arrow D indicates a terrace which is denser relative to the area marked by arrow C and contains a few dark holes.

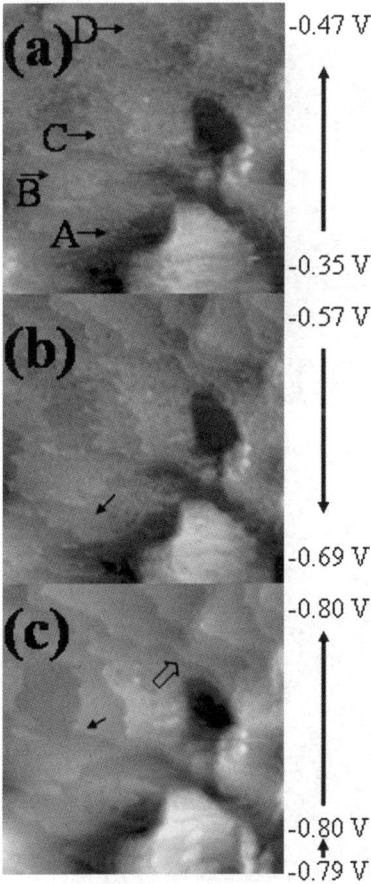

Fig. 3 *In situ* EC-STM images of the Cu(111) surface were obtained in 1 mM HNO$_3$ + 0.1 M HF during cathodic potential sweeping between −0.35 V−−0.47 V (a), −0.57 V−−0.69 V (b), and −0.79 V−−0.80 V (c). Image size is 150 nm × 150 nm. The tip bias and tunneling current are 10 mV and 10 nA, respectively.

After reaching a potential of −0.57 V on the cathodic sweep, the terraces become flat and clean. The image in Fig. 3b additionally shows that the terraces are fully covered with round holes and the step edges retain the saw-tooth structure. Moving the potential to −0.69 V led to a decrease in the number of holes on the terraces, one of which is marked with an arrow in Fig. 3b.

Fig. 3c shows *in situ* EC-STM image obtained at −0.8 V. The surface is atomically flat and steps of a Cu(111) monoatomic height are well resolved, though there are still a few round holes of a monoatomic height. As time went by at −0.8 V the saw-like structures at the step edges gradually became smooth. Additionally, atomic resolution images of the Cu(111) lattice were obtained easily from the terrace.

Following a complete potential cycle, we can compare the EC-STM image of Fig. 3c with that of Fig. 2a. The steps are located approximately at same positions but the steps in the image in Fig. 3c are expanded in some areas (closed arrow) and shrunk in the other areas (open arrow). Additionally, the image of Fig. 3c shows a number of holes on terraces in contrast to that of Fig. 2a.

3.2 *In situ* STM—Cu(100)

Fig. 4 and 5 show potential dependent *in situ* EC-STM images of Cu(100) obtained in the same solution as the Cu(111) case described above. The white compass in

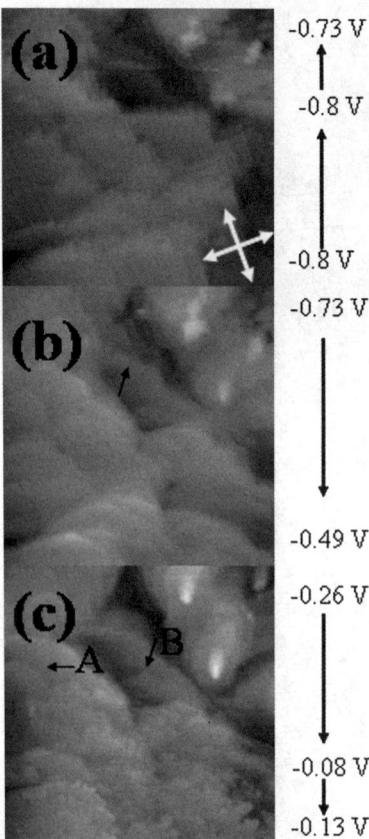

Fig. 4 *In situ* EC-STM images of the Cu(100) surface were obtained in 1 mM HNO$_3$ + 0.1 M HF during potential sweeping between −0.80 V−−0.73 V (a), −0.73 V−−0.49 V (b), and −0.26 V−−0.08 V−−0.13 V (c). Image size is 300 nm × 300 nm. The tip bias and tunneling current are 10 mV and 10 nA, respectively.

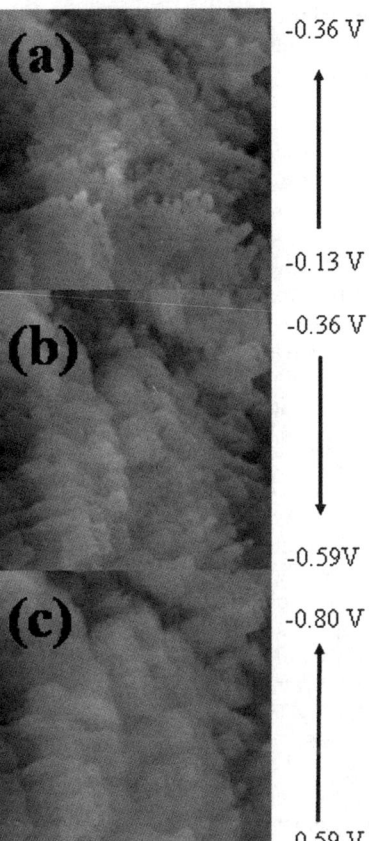

Fig. 5 *In situ* EC-STM images of the Cu(100) surface were obtained in 1 mM HNO$_3$ + 0.1 M HF during cathodic potential sweeping between −0.13 V−−0.36 V (a), −0.36 V−−0.59 V (b), and −0.59 V−−0.80 V (c). Image size is 300 nm × 300 nm. The tip bias and tunneling current are 10 mV and 10 nA, respectively.

Fig. 4a indicates the {110} directions of the Cu(100) crystal. The image in Fig. 4a exhibits wide meandering terraces and steps similar to those seen on Cu(111) except for rectangular step edges. The rectangular step edges have also been found on Cu(100) in chloride-containing solution.[41,42] The behavior of the lattice found here, however, is substantially different from that found in Cl$^-$ containing solution. First, the rectangular steps are found only at potentials more negative than *ca.* −0.67 V in 1 mM HNO$_3$. By way of contrast, the Cl$^-$ adlayer is desorbed at potentials more negative than −0.4 V *vs.* Ag/AgCl.[43] Second, the rectangular steps are oriented in the {110} direction here, but in the case of a chloride adlayer the straight steps of Cu(100) are aligned toward the {100} direction.[31] Thus, the rectangular steps edges are likely not associated with Cl$^-$ in this case.

Fig. 4b shows the result of moving the applied potential to values between −0.49 V and −0.73 V. Here, the step edges begin to change. An arrow in Fig. 4b (obtained when the surface was at a potential of *ca.* −0.67 V) indicates a step in which the rectangular step has disappeared and become round. Interestingly, the straight steps of the Cu(100) lattice become round at the same potential as that in which the terraces of Cu(111) were covered with the small white particles as shown in Fig. 2a. The steps of the Cu(100) surface in Fig. 4a are oriented toward the {110} direction, which means that a (111) face is exposed on the step. This result suggests that the Cu(111) face is oxidized at the same potential in both systems. At more positive potentials, the rectangular structures of terrace step edges cannot be discerned. Fig. 4c shows *in situ* EC-STM images obtained at a potential close to the OCP. The steps are still round and evince 'frizzy' behavior seen in other contexts.[44] Comparison of a terrace (arrow A) at the equivalent location in Fig. 4b confirms that the terrace steps retracted slightly. The step edges start to erode quickly around −0.16 V (arrow B). As the applied potential reaches −0.08 V, the steps retract quickly and isotropically, and some peninsulas and free-standing islands are formed. Equivalent step flow features for Cu(100) in sulfuric acid have been reported previously.[29] However, the rate of reaction in sulfuric acid is apparently much slower than that found in nitrate, since these authors found that the step flow features developed over the course of *ca.* 40 min. In contrast, the step flow seen here occurred as soon as the potential was swept to this region. This observation suggests that nitrate ion plays an important role in the step flow reaction.

Fig. 5 shows *in situ* EC-STM images of Cu(100) surface during the cathodic potential sweep. The peninsulas found in Fig. 4c start to expand isotropically and then meet with others, forming junctions where a few defects occur. However, by the time a potential of −0.36 V is achieved, most of the defects have disappeared. This disappearance is in contrast to the Cu(111) case (Fig. 3b) in which many holes remain. Moving the potential to more negative values, as shown in Fig. 5b and c, leads to further development of the surface. In Fig. 5b, the round peninsulas turn into smooth rectangular terraces but the points of the terraces still exhibit a rounded aspect. Through the images of Fig. 5b and c, the terraces became wide and the holes disappear. Around −0.70 V, the points of the rectangles became sharp and perfect rectangular structures reappear. There is little difference between the image in Fig. 5c obtained after the potential cycle and that in Fig. 4a, obtained at the beginning.

4. Discussion

The results presented here show that the Cu(100) and Cu(111) surfaces react somewhat differently relative to each other in nitric acid solution. Differences include: (a) delayed reduction current on Cu(100) relative to Cu(111), (b) formation of pits in terraces on Cu(111) following oxidation, and (c) the degree of surface oxidation observed on the surfaces. The results speak to different evolution of Cu(111) and Cu(100) surfaces in potential regions associated with surface oxidation and nitrate reduction.

The CVs obtained from Cu(111) or Cu(100) in nitric acid exhibit different features relative to each other. The CV of the Cu(100) surface in nitrate solution shows a featureless double layer region until −0.3 V and reduction current from −0.3 V to −0.72 V with a peak at −0.58 V. In the Cu(111) case, a small reduction current was observed, even close to the OCP, and the main reduction current was shifted to positive potentials. The delayed onset of nitrate reduction activity means that the electron transfer event associated with nitrate reduction is delayed or inhibited on the Cu(100) surface relative to Cu(111). One reason for this delay may be associated with the amount of free Cu^+ available near the electrode surface at these potentials. Cu^+ is known to be a catalyst for nitrate reduction.[45] The obvious oxidation-like event on the Cu(111) surface at potentials relevant to nitrate reactivity is contrasted with the paucity of these on Cu(100).

EC-STM images obtained from Cu(100) do not show the terrace surface evolution during the cathodic potential excursions except at the step edges. In contrast, the Cu(111) surface evolves over the entire potential range. The Cu(111) surface evinces the oxide-like structures seen in other electrolytes and associated with the oxidation of the surface. In the absence of nitric acid, but in acid electrolyte, these structures are not seen except at substantially more positive potentials. During cathodic potential sweeping, the surface is covered with coarse particles and round holes, and reorganizes to clean terraces.

Differential activity of Cu(111) and Cu(100) was also observed in the anodic oxidation of Cu single crystals in NaOH,[39] as well as in the oxygen induced oxidation of the Cu surfaces.[24] In the case of oxygen induced oxidation, the difference arises as a result of the structures of the oxygen chemisorbed layer, oxygen surface diffusion, surface energy, and interfacial strain energy,[24] while the anodic oxidation case is attributed to the higher stability of the $Cu_2O(111)$ precursor than that for Cu(100).[39]

With this background we consider the origin of the differential oxidation activity on Cu(111) and Cu(100) surfaces in nitric acid. Among the reasons, oxygen surface diffusion and interfacial strain energy can not explain the differences we report here because nitrate ion does not interact with bulk layers of the Cu crystals.

The difference in surface energies plays an important role in the initial oxidation of the surface.[24] The surface energy of Cu(111) is 1170 ergs cm^{-2}, which is lower than that of Cu(100) (1280 ergs cm^{-2}).[46] It is expected that the Cu(111) surface is more stable and thus oxidation of the surface will not be easier than the Cu(100) case. Therefore, we can also exclude the surface energy as an origin for the difference we found here.

Having rejected these two explanations for the origin of the differential reactivity between the two Cu faces, we next focus on the chemisorbed oxygen layer. The presence of the Cu oxide-like and roughened structure shown in Fig. 2 and 3 is unique to the (111) face of Cu. The unique feature here is the presence of this oxide-like layer at relatively negative potentials in the acid electrochemical environment, arising as a consequence of nitrate reduction. Chu et al.[19] observed that oxide forms a single monolayer thick structure on Cu(111) in pH 4.5 aqueous solution at positive potentials. Friebel et al.[20] also reported that oxygen containing species adsorb on Cu(111) in acidic solution again at relatively positive potentials. The origin of the more stable oxide on Cu(111) may again relate to the relative stability of the $Cu_2O(111)$ structure relative to that formed on Cu(100).

The fact that the oxide adlayer appears on Cu(111) in the potential range where nitrate is reduced, but is absent without the nitrate suggests that the oxide layer plays a role in the electroreduction process. We observed adlattices of nitrate, nitrite, and other intermediates during the nitrate reduction process on Cu(100).[17] However, no adlattice structures were observed on Cu(111), in part because of the oxide features reported above. We therefore suggest that the surface evolution of Cu(111) results from the formation of Cu oxide during nitrate reduction.

On many surfaces, it is thought that the nitrate to nitrite conversion is the rate-determining step:[7]

$$NO_3^- + 2e^- + 2H^+ \rightarrow H_2O + NO_2^- \qquad (2)$$

in which the nitrate ion first adsorbs on the Cu surface before it is reduced. The observation of the nitrate adlattice conversion to nitrite on Cu(100) supports this mechanism. But in the case of Cu(111), exposure of the surface to nitrate leads to apparent oxide or oxygen species adsorption on the surface. This suggests that a oxygen atom of the nitrate ion is dissociated and adsorbs on the Cu(111) surface:

$$NO_3^- + Cu(111) \rightarrow NO_2^- + Cu(111)O \qquad (3)$$

The genesis of this mechanism might be the relatively high stability of the $Cu_2O(111)$ precursor oxide.

At the most positive potentials the surface is fully oxidized. As the potential is then swept negatively, the terrace starts to divide into coarse particles and then reorganizes. During the formation of the coarse particles, the steps expand laterally again as saw-like structures. We attribute this surface evolution to the reduction of Cu oxide monolayer to Cu metal. The terrace must expand to make room for the intermediates between Cu oxide and Cu metal, such as Cu–OH, which may induce the terrace to divide to the coarse particles. The coarse particles move to reorganize and form again the compact terraces of Cu(111) surface. The presence of holes in the Cu(111) terraces and to a lesser extent on Cu(100) undoubtedly arises from incomplete electrochemical annealing as the reduced oxide diffuses along the terrace.

Conclusion

We examined the interaction between nitrate ion and Cu single crystal surfaces by using CV and *in situ* EC-STM. The CV of Cu(100) surface in nitrate solution shows a featureless double layer region until -0.3 V and reduction current from -0.3 V to -0.72 V with a peak at -0.58 V. In case of the Cu(111) surface, a small reduction current was observed, even close to the OCP, and the main reduction current was shifted to positive potentials relative to the Cu(100) case. EC STM images demonstrate that the terrace of both Cu(111) and Cu(100) are atomically flat at potentials more negative than -0.7 V. In the presence of nitrate, Cu(111) evinces oxide-like features at the potentials at which nitrate is reduced on this surface. These features are not found on Cu(100), except at the (111)-oriented step edges. These observations suggest that reduction of nitrate and oxidation of the Cu surface is enhanced on Cu(111), due to the formation of the surface oxide which is more facile on the (111) face. During cathodic sweeping on Cu(111), the oxide monolayer is reduced, in the course of which terraces evince coarse particles, reorganize, and then form flat surfaces exhibiting holes with monoatomic depth. At a more negative potential than -0.7 V, the holes on terraces gradually disappear.

Acknowledgements

SEB acknowledges the Korea Research Foundation for a fellowship (No. KRF-2005-214-C00068). This work was funded by the NSF (CHE-06-03675), which is gratefully acknowledged.

References

1. M. J. Moorcroft, J. Davis and R. G. Compton, *Talanta*, 2001, **54**, 785–803.
2. W. F. Plieth, *Encyclopedia of Electrochemistry of the Elements*, Marcel Dekker, New York, 1978.
3. D. Reyter, G. Chamoulaud, D. Belanger and L. Roue, *J. Electroanal. Chem.*, 2006, **596**, 13–24.
4. G. Shugar and J. Ballinger, *Chemical Technicians Ready Reference Handbook*, McGraw-Hill, New York, 1996.
5. W. Kern, *J. Electrochem. Soc.*, 1990, **137**, 1887–1892.
6. Y. Ein-Eli and D. Starosvetsky, *Electrochim. Acta*, 2007, **52**, 1825–1838.
7. G. E. Dima, A. C. A. de Vooys and M. T. M. Koper, *J. Electroanal. Chem.*, 2003, **554**, 15–23.
8. D. Pletcher and Z. Poorabedi, *Electrochim. Acta*, 1979, **24**, 1253–1256.
9. N. G. Carpenter and D. Pletcher, *Anal. Chim. Acta*, 1995, **317**, 287–293.
10. J. Davis, M. J. Moorcroft, S. J. Wilkins, R. G. Compton and M. F. Cardosi, *Analyst*, 2000, **125**, 737–741.
11. M. da Cunha, J. P. I. De Souza and F. C. Nart, *Langmuir*, 2000, **16**, 771–777.
12. I. R. Moraes, M. daCunha and F. C. Nart, *J. Braz. Chem. Soc.*, 1996, **7**, 453–460.
13. M. daCunha, M. Weber and F. C. Nart, *J. Electroanal. Chem.*, 1996, **414**, 163–170.
14. P. M. Castro and P. W. Jagodzinski, *J. Phys. Chem.*, 1992, **96**, 5296–5302.
15. S. Cattarin, *J. Appl. Electrochem.*, 1992, **22**, 1077–1081.
16. K. Bouzek, M. Paidar, A. Sadilkova and H. Bergmann, *J. Appl. Electrochem.*, 2001, **31**, 1185–1193.
17. S.-E. Bae, K. L. Stewart and A. A. Gewirth, *J. Am. Chem. Soc.*, 2007, **129**, 10171–10180.
18. S.-J. Hsieh and A. A. Gewirth, *Langmuir*, 2000, **16**, 9501–9512.
19. Y. S. Chu, I. K. Robinson and A. A. Gewirth, *J. Chem. Phys.*, 1999, **110**, 5952–5959.
20. D. Friebel, P. Broekmann and K. Wandelt, *Phys. Status Solidi A*, 2004, **201**, 861–869.
21. J. Kunze, V. Maurice, L. H. Klein, H. H. Strehblow and P. Marcus, *J. Phys. Chem. B*, 2001, **105**, 4263–4269.

22 H. Y. H. Chan, C. G. Takoudis and M. J. Weaver, *Electrochem. Solid-State Lett.*, 1999, **2**, 189–191.
23 H. Y. H. Chan, C. G. Takoudis and M. J. Weaver, *J. Phys. Chem. B*, 1999, **103**, 357–365.
24 G. W. Zhou and J. C. Yang, *J. Mater. Res.*, 2005, **20**, 1684–1694.
25 Z. Q. Tian, B. Ren and D. Y. Wu, *J. Phys. Chem. B*, 2002, **106**, 9463–9483.
26 M. Pourbaix, *Atlas of Electrochemical Equilibria in Aqueous Solutions*, National Association of Corrosion Engineers, Houston, 1974.
27 V. Maurice, H. H. Strehblow and P. Marcus, *Surf. Sci.*, 2000, **458**, 185–194.
28 M. C. Kang and A. A. Gewirth, *J. Phys. Chem. B*, 2002, **106**, 12211–12220.
29 M. R. Vogt, A. Lachenwitzer, O. M. Magnussen and R. J. Behm, *Surf. Sci.*, 1998, **399**, 49–69.
30 W. Polewska, M. R. Vogt, O. M. Magnussen and R. J. Behm, *J. Phys. Chem. B*, 1999, **103**, 10440–10451.
31 D. W. Suggs and A. J. Bard, *J. Phys. Chem.*, 1995, **99**, 8349–8355.
32 W. Suggs and A. J. Bard, *J. Am. Chem. Soc.*, 1994, **116**, 10725.
33 J. B. Cotton and I. R. Scholes, *Br. Corros. J.*, 1967, **2**, 1–5.
34 M. Edwards, T. Meyer and J. Rehring, *J. Am. Water Works Assoc.*, 1994, **86**, 73–81.
35 S. S. El-Egamy, K. M. Ismail and W. A. Badawy, *Corros. Prev. Control*, 2004, **51**, 89–97.
36 L. I. Antropov, M. I. Donchenko and T. I. Motronyuk, *Prot. Met.*, 1984, **20**, 27–32.
37 B. J. Cruickshank, D. D. Sneddon and A. A. Gewirth, *Surf. Sci. Lett.*, 1993, **281**, L308–314.
38 Z. D. Schultz, S. K. Shaw and A. A. Gewirth, *J. Am. Chem. Soc.*, 2005, **127**, 15916–15922.
39 J. Kunze, V. Maurice, L. H. Klein, H. H. Strehblow and P. Marcus, *Corros. Sci.*, 2004, **46**, 245–264.
40 R. Wiesendanger, *Scanning Probe Microscopy and Spectroscopy*, Cambridge University Press, Melbourne, 1994.
41 M. R. Vogt, F. A. Moller, C. M. Schilz, O. M. Magnussen and R. J. Behm, *Surf. Sci.*, 1996, **367**, L33–L41.
42 O. M. Magnussen, *Chem. Rev.*, 2002, **102**, 679–725.
43 M. R. Vogt, F. A. Moller, C. M. Schilz, O. M. Magnussen and R. J. Behm, *Surf. Sci.*, 1996, **367**, L33–L41.
44 M. Giessen, M. Dietterle, D. Stapel, H. Ibach and D. M. Kolb, *Surf. Sci.*, 1997, **384**, 168.
45 E. V. Filimonov and A. I. Shcherbakov, *Prot. Met.*, 2004, **40**, 280–284.
46 S. M. Foiles, M. I. Baskes and M. S. Daw, *Phys. Rev. B: Condens. Matter Mater. Phys.*, 1986, **33**, 7983–7991.

Molecular structure at electrode/electrolyte solution interfaces related to electrocatalysis

Hidenori Noguchi, Tsubasa Okada and Kohei Uosaki*

Received 3rd March 2008, Accepted 1st May 2008
First published as an Advance Article on the web 11th August 2008
DOI: 10.1039/b803640c

The potential dependence of the interfacial water structure at Pt and Au thin film electrodes was investigated by sum frequency generation (SFG) spectroscopy in internal reflection mode. In the case of the Pt electrode, two broad peaks were observed in the OH stretching region at *ca.* 3200 cm^{-1} and *ca.* 3400 cm^{-1}, which are known to be due to the symmetric OH stretching (v_1) of tetrahedrally coordinated, *i.e.*, strongly hydrogen bonded "ice-like" water, and the asymmetric OH stretching (v_3) of water molecules in a more random arrangement, *i.e.*, weakly hydrogen bonded "liquid-like" water, respectively. In the case of the Au electrode, however, a 3400 cm^{-1} band was dominant in the SFG spectrum, suggesting that the interaction between water molecules and Au and Pt are different, *i.e.*, water molecules are more disordered at the Au surface. The potential dependence of interfacial water during the methanol oxidation reaction on a Pt electrode was also investigated. SFG intensity strongly depended on electrode potential. Several possibilities are suggested for the potential dependence of the SFG intensity.

1 Introduction

To fully understand the mechanism of electrochemical reactions, information about structures of molecules at electrode/electrolyte interfaces, including short-lived intermediates and the solvent, which is water in most electrochemical reactions, is essential. Actually, water is not just a solvent but also a reactant or product in many important electrocatalytic reactions. Thus, molecular-level understanding of the structural arrangement of water molecules at an electrode/electrolyte solution interface is one of the most important issues in electrochemistry. The presence of oriented water molecules, caused by an interaction between water dipoles and the strong electric field in the double layer has been proposed.[1-3] It has been also proposed that water molecules are present at the electrode surface in the form of clusters.[4,5] While scanning probe microscopy is very useful to determine the two-dimensional structure of the electrode surface, three-dimensional structural information of the interface and molecular structure cannot be obtained by this technique. Despite the numerous studies on the structure of water at metal electrode surfaces using various techniques such as surface enhanced Raman spectroscopy (SERS),[6,7] surface infrared spectroscopy,[8,9] surface enhanced infrared spectroscopy (SEIRAS)[10] and X-ray diffraction,[11,12] the exact nature of the structure of water at electrode/solution interface is still not fully understood.

Sum frequency generation (SFG) is a second order nonlinear optical process, in which two photons of frequencies ω_1 and ω_2 generate one photon of sum frequency

Physical Chemistry Laboratory, Division of Chemistry, Graduate School of Science, Hokkaido University, Sapporo, 060-0810, Japan. E-mail: uosaki@pcl.sci.hokudai.ac.jp

($\omega_3 = \omega_1 + \omega_2$). The second order nonlinear processes, including SFG, are inhibited in media with inversion symmetry under the electric dipole approximation and take place only at the interface between these media where the inversion symmetry is necessarily broken. By using visible light of fixed wavelength and tunable IR light as the two input light sources, SFG spectroscopy can be surface sensitive vibrational spectroscopy, as proved by Shen and co-workers.[13,14] SFG spectroscopy is particularly useful to study the structure of water molecules at various interfaces where the presence of a much larger amount of bulk water than interfacial water makes the measurement of interfacial water by other vibrational techniques very difficult.[15-20] Furthermore, SFG is free from the ambiguity associated with the choice of reference spectrum and can avoid the necessity of working with a rough electrode as required for SEIRAS and SERS. Thus, SFG spectroscopy is an ideal technique to investigate the structure of the water at metal electrode/electrolyte solution interfaces.[21-26]

Tadjeddine and his coworkers applied SFG spectroscopy to investigate Pt single- and poly-crystalline electrode/electrolyte solution interfaces in external reflection mode with a thin electrolyte solution layer. They reported that SFG peaks corresponding to water dimmers bonded to three hydrogen terminals on Pt (Pt–H) by hydrogen bonding at Pt/H_2SO_4 solution interfaces.[21] They also mentioned that the SFG signal in OH stretching was quite weak in perchlorate solution and proposed that water in the hydration shell predominates the interfacial water.[21,24]

We examined the structure of water at a Au/10 mM H_2SO_4 solution interface under potential control in internal reflection mode using a thin gold film electrode.[25] A broad band centered around 3500 cm^{-1} with a shoulder around 3250 cm^{-1} was observed, indicating that water molecules at Au electrode are rather weakly hydrogen bonded, *i.e.*, disordered. It was also found that the SFG intensity of the OH stretching becomes minimum at the potential of zero charge (pzc) and increases as the potential becomes more negative or positive than the pzc but the orientation of the water molecules at Au electrode does not flip when the charge of the electrode changes its sign because sulfate anion is adsorbed on the Au electrode when the potential of the electrode is more positive than the pzc, resulting in the net surface charge being negative. We also recently reported the potential dependent structure at a Pt/0.1 M $HClO_4$ solution interface determined by SFG.[26]

Gewirth and his colleagues observed four different OH stretching modes in SFG spectra at Ag(100)/0.1 M NaF solution interface under potential control measured in external reflection mode and assigned them to: (1) weakly hydrogen bonded water, (2) strongly hydrogen bonded water, (3) water specifically adsorbed on the Ag surface, and (4) hydronium cation, H_3O^+, which was observed only at potentials near the pzc.[27]

Here, we compare the potential dependent interfacial water structure at Pt and Au thin film electrodes in 0.1 M $HClO_4$ solution determined by SFG measurements in the OH stretching region in internal reflection mode. Furthermore, the potential dependent interfacial water structure at a Pt electrode during the methanol oxidation reaction was also determined by SFG. While two broad peaks of almost equal intensity were observed at *ca.* 3200 cm^{-1} and *ca.* 3400 cm^{-1}, which are known to be due to the strongly hydrogen bonded, *i.e.*, "ice-like", water and the weakly hydrogen bonded, *i.e.*, "liquid-like" water, respectively, at the Pt electrode, the 3400 cm^{-1} band was dominant in the SFG spectrum at the Au electrode, suggesting that water molecules are more highly oriented at the Pt electrode than at the Au electrode. Parabolic behavior of SFG intensity *vs.* potential was observed with a minimum close to the pzc of the electrode in $HClO_4$ solution. The SFG intensity of OH stretching decreased as anodic oxide was formed. At the Au electrode, an anomalously sharp peak was observed at *ca.* 3550 cm^{-1} just before the oxide was formed, in addition to the broad band at 3400 cm^{-1}. In methanol solution, SFG intensity was relatively strong as long as CO was on the electrode surface but decreased as adsorbed CO was oxidatively removed and anodic oxide was formed.

2. Experimental

Electrochemical and SFG measurements were carried out using a spectroelectrochemical cell made of Kel-F, as shown in Fig. 1. Before use, the cell was cleaned in conc. H_2SO_4 followed by thorough rinsing with Milli-Q water. 10 nm thick Pt or Au films evaporated on an IR-grade fused quartz hemi cylindrical shaped prism (Daico MFG Co., Ltd.) with a 5 nm titanium buffer layer was used as a working electrode. XPS measurements showed that there was no Ti or Ti oxide at the surface. An Ag/AgCl (saturated NaCl) and a Pt wire were employed as a reference electrode and a counter electrode, respectively. The electrolyte solutions were prepared using reagent grade $HClO_4$ and CH_3OH (Wako Pure Chemicals) and purified water provided by a Milli-Q system (Millipore Inc.), and deaerated by bubbling high-purity Ar gas (99.999%) through them for at least 30 min prior to the spectroelectrochemical measurements. The electrode potential was controlled with a potentiostat/functiongenerator (Toho Technical Research, PS-07).

The SFG system used in the present study was described in detail elsewhere.[17] Briefly, a picosecond Nd:YAG laser (EKSPLA, PL2143B) with a 25 ps pulse width and repetition rate of 10 Hz was employed to pump an OPG/OPA/DFG (EKSPLA, PG401 VIR/DFG) system, which generates tunable infrared pulses. The loosely focused "visible" (100 μJ per pulse@532 and 1064 nm) and IR (100 μJ per pulse@3000 nm) beams were overlapped at the sample surface. While 532 nm light was used as the visible beam for the Pt electrode, 1064 nm light was used for the Au electrode to avoid a strong non-resonance signal at 532 nm from the Au substrate.[25] The incident angles of the visible and IR beams were about 70 and 50 degrees, respectively. The SF signal was separated from the reflected visible and IR pulses by passing through irises and a monochromator (Oriel instruments, MS257) and was detected by a photomultiplier tube (Hamamatsu, R3896) and normalized to the intensities of the IR and visible pulses. Temporal and spatial overlaps were adjusted by monitoring the SFG signal from a quartz plate. The polarization combination was (ppp), where the first, second and third letter in the parenthesis denote the polarization of SF, visible and IR beams, respectively. SFG measurements were carried out at room temperature (*ca.* 22 °C).

The electrode potential, current, and SFG signal were recorded using a personal computer (DELL, Dimension 3000) through a 13 bit AD converter (Stanford Research System, SR245).

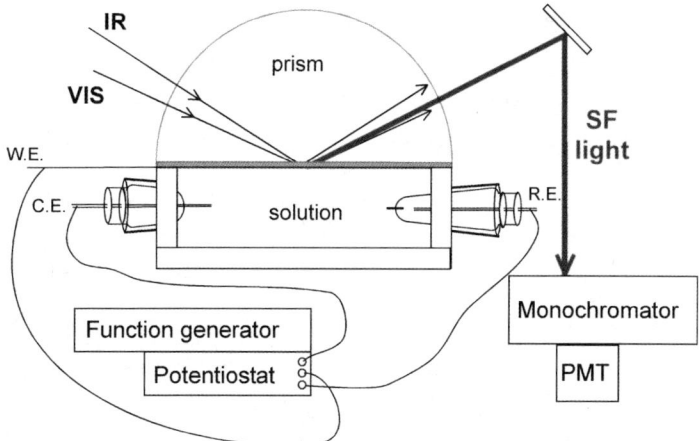

Fig. 1 Schematic diagram of a spectroelectrochemical cell and measuring arrangement used in the present study to detect SFG signal at metal/electrolyte solution interface.

3. Results and discussion

3.1. Potential dependent SFG spectra at Au/0.1 M HClO$_4$ and Pt/0.1 M HClO$_4$ solution interfaces

The solid lines in Figs. 2 and 3 show typical cyclic voltammograms of thin Au and Pt film electrodes in a 0.1 M HClO$_4$ solution, respectively. They are in good agreement with CVs of poly-crystalline Au and Pt electrode, respectively, with waves due to oxide formation and reduction at both electrodes and hydrogen adsorption and

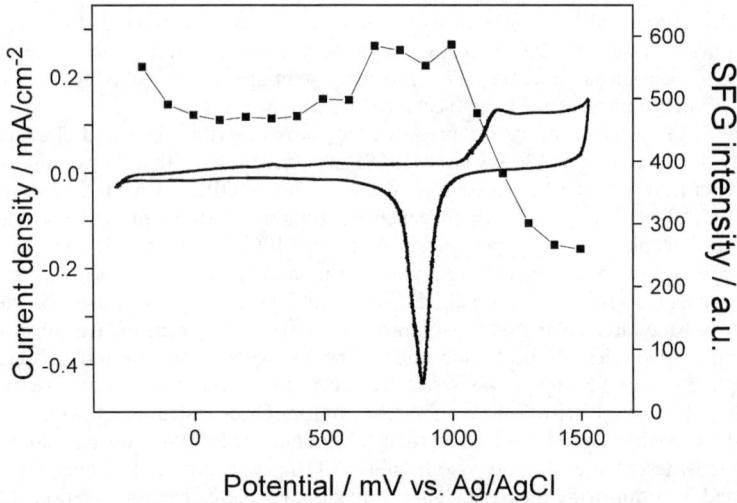

Fig. 2 Cyclic voltammogram obtained with a sweep rate of 50 mV s^{-1} (solid line) and potential dependence of integrated SFG intensity in the OH stretching region (●) of an Au thin film electrode in 0.1 M HClO$_4$ solution.

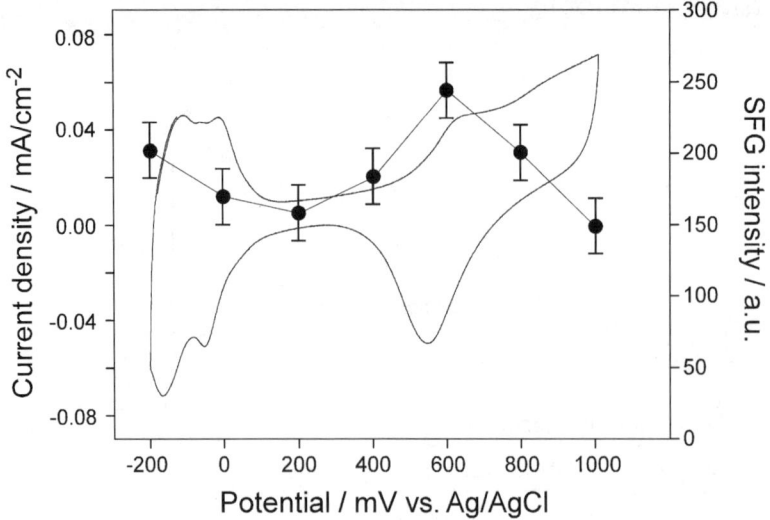

Fig. 3 Cyclic voltammogram obtained with a sweep rate of 50 mV s^{-1} (solid line) and potential dependence of integrated SFG intensity in OH stretching region (●) of a Pt thin film electrode in 0.1 M HClO$_4$ solution.

desorption at Pt electrode. These results confirmed that the conductivities of the Au and Pt thin films were good enough to be used as electrodes.

Generally, SFG spectra in the OH stretching vibration region are featured by two broad peaks at ca. 3200 cm^{-1} and at 3400 cm^{-1}.[17] The former and the latter have been assigned to the vibration of OH oscillators of three coordinated hydrogen bonded water i.e., less ordered "liquid-like" water, molecules and that of the four coordinated hydrogen bonded water, i.e., highly ordered "ice-like" water molecules, respectively, based on the IR study of water clusters.[28] Thus, the intensity ratio between these two peaks can be considered as an index of the order of the interfacial water.[29]

Fig. 4 shows an SFG spectra in the OH stretching region (2800 cm^{-1}–3800 cm^{-1}) obtained at the Au electrode in a 0.1 M HClO$_4$ solution at various potentials. The SFG spectra are dominated by a broad peak centered around 3500 cm^{-1}, as was the case in the SFG spectra obtained at an Au/H$_2$SO$_4$ solution interface.[25] On the other hand, the SFG spectra in the OH stretching region obtained at the Pt electrode in a 0.1 M HClO$_4$ solution at various potentials showed two broad peaks at ca. 3200 cm^{-1} and ca. 3450 cm^{-1} as shown in Fig. 5.[26,30] SFG spectra were fitted by two broad OH bands centered at ca. 3200 and 3450 cm^{-1} using eqn (1)

$$I_{SFG} \propto \left| \chi_{NR}^{(2)} + \frac{A_0 e^{i\varphi}}{\omega - \omega_0 + i\Gamma_0} \right|^2 \quad (1)$$

as shown by solid lines in Fig. 4 and 5. The present results and that at the Au/H$_2$SO$_4$ solution interface suggest that water molecules at the Au electrode/electrolyte solution interface are less ordered than those at the Pt electrode/electrolyte solution interface.

Although there is no direct spectroscopic evidence to show that the interactions between water molecules and Au and Pt are different, we think that structure of water molecules in the first layer or in the vicinity of metal surface is very important to the whole interfacial water structure. STM studies showed a well ordered H$_2$O film on a clean Pt surface[31] but an amorphous H$_2$O film on a clean Au surface,[32] suggesting that the interaction of water molecules with the Pt surface is strong and that with the Au surface is weak. The stronger interaction between water molecules and Pt surface might be the origin of the higher order of interfacial water at the Pt surface. A theoretical study on the interactions between water molecules and metal substrates is underway in our group.

The shape of the SFG spctra did not change significantly with potential, but the intensity seemed to be affected by potential. To clarify the potential dependencies of the SFG intensity, the integral intensity of SFG spectra between 2800 cm^{-1} to 3800 cm^{-1} of the Au electrode taken from Fig. 3 and that of Pt electrode taken from Fig. 4 were plotted against electrode potential, as shown also in Fig. 2 and 3, respectively. Since the off-resonance SFG intensity did not change with electrode potential, the contribution from the non-resonant component can be ignored in the results. As the potential was changed from negative to positive, the SFG intensity decreased initially, reached a minimum, increased again, reached a maximum, and then decreased again. The position of the SFG intensity minimum was around 300 mV at the Au electrode, which is close to the pzc of an Au electrode in an HClO$_4$ solution,[33–35] and around 200 mV at the Pt electrodes, which is close to the pzc[36,37] of a Pt electrode in an HClO$_4$ solution.[38–40] A similar result was observed at an Au/H$_2$SO$_4$ solution interface,[25] although the measurements were limited to a relatively narrow potential region.

Previously, we proposed that SFG intensity due to interfacial water at quartz/water interfaces reflects the number of oriented water molecules within the electric double layer and in turn the double layer thickness, based on the pH dependence of the SFG intensity[17] and a linear relation between the SFG intensity and (ionic strength)$^{-1/2}$.[18] In the case of the Pt/electrolyte solution interface, the drop of the

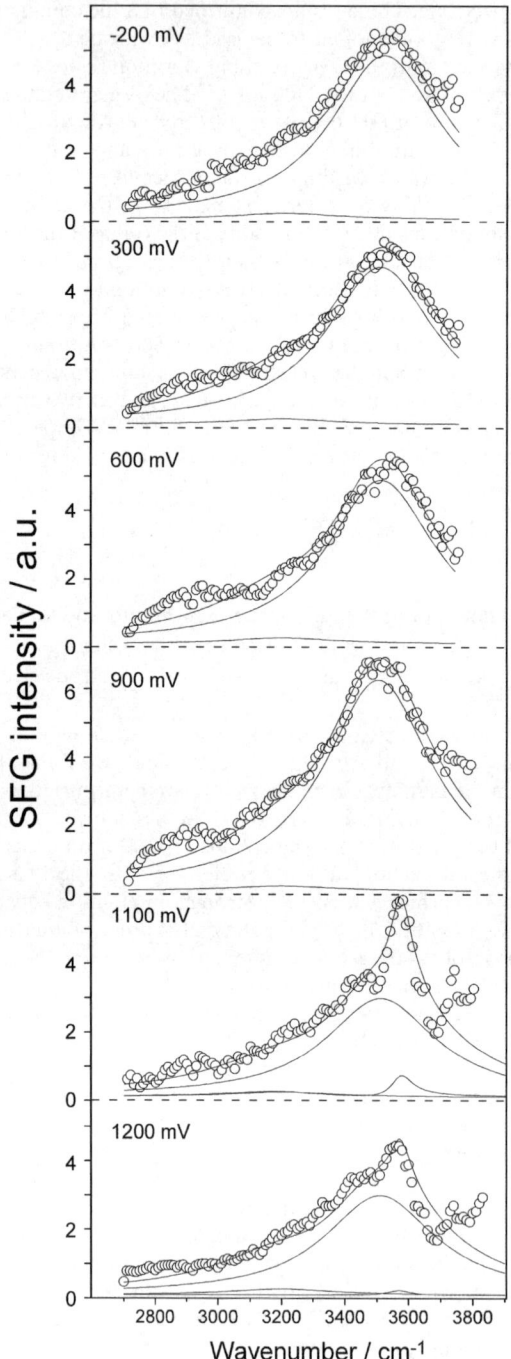

Fig. 4 SFG spectra in the OH stretching region at an Au electrode at various potentials in 0.1 M HClO$_4$ solution.

potential profile at the vicinity of the electrode becomes precipitous as the electrode becomes more highly charged. Thus, the ordered water layer at the vicinity of the electrode surface becomes thinner as the electrode becomes more highly charged.

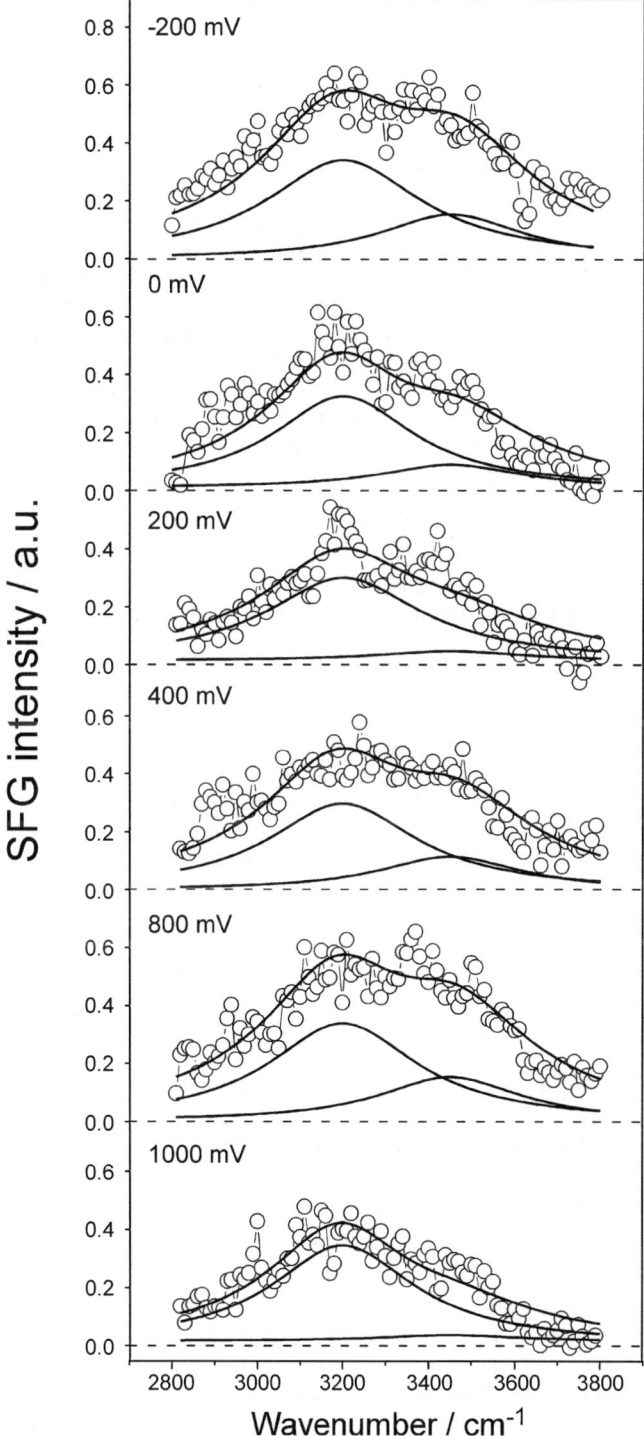

Fig. 5 SFG spectra in the OH stretching region at a Pt electrode at various potentials in 0.1 M HClO$_4$ solution.[26]

Since the number of ordered water molecules becomes smaller, the SFG intensity should become weaker at the potentials away from the pzc. This is contrary to the experimental result.

When the electrolyte concentration is relatively high, potential dependence of the double layer thickness is low and the potential dependence of the fraction of oriented water predominantly determines that of the SFG intensity. Since the polarization of IR in the present experiment is "*p*", water orienting normal to the surface is effectively detected by SFG. Water molecules are lying parallel to the surface around the pzc and they will reorient from "oxygen up" to "oxygen down" as the surface charge of the electrode surface changes from negative to positive, as long as no specific adsorption of ions takes place. An IR study[10] as well as computer simulations[41,42] also suggest that water molecules at the metal electrode surface have an oxygen up and oxygen down orientation on negatively and positively charged surfaces, respectively, on average. Ataka *et al.* also investigated the potential dependent structure change of water at an Au/HClO$_4$ electrolyte solution interface and concluded that water molecules change their orientation from oxygen-up to oxygen-down as the potential is varied from more negative to more positive than the pzc, *i.e.*, the surface charge changes from negative to positive, based on the result of the potential dependence of integrated intensities of δHOH bands.[43]

Of course, the orientation of water molecules is determined not only by the electric field in the double layer, *i.e.*, the charge at the electrode, but also by chemical interactions such as water–metal and water–water interactions. If we consider the adsorption of water molecules with their oxygen end toward the surface (oxygen up) under equilibrium conditions, the fraction of oxygen up water molecules can be expressed by:[44]

$$N_\uparrow/N_T = \theta_\uparrow = \exp[-\Delta G_\uparrow^0/RT] \quad (2)$$

where $N_\uparrow$ is the number of oxygen up water, N_T is the total number of sites for water adsorption, $\theta_\uparrow$ is the fraction of oxygen up water, and $\Delta G_\uparrow^0$ is the standard free-energy change associated with the adsorption of water in the oxygen up state. $\Delta G_\uparrow^0$ can be resolved into three contributions. (1) The chemical work of adsorption, ΔG_c^0, (2) the electrical work, $\mu X \cos\alpha$, where μ is the dipole moment of adsorbed water, X is the double layer field, and α is the angle between the electric field and the dipole moment of water molecules, and (3) the work of lateral interaction between water molecules, U_c.

Thus, eqn (2) for the oxygen up water where $\alpha = \pi$ can be written as:

$$\theta_\uparrow = \exp\{-[(\Delta G_c^0)_\uparrow/kT - \mu X/kT + U_c(\theta_\uparrow - \theta_\downarrow)/kT]\} \quad (3)$$

and similarly, the fraction of the oxygen down water where $\alpha = 0$ can be given by:

$$\theta_\downarrow = \exp\{-[(\Delta G_c^0)_\downarrow/kT + \mu X/kT - U_c(\theta_\uparrow - \theta_\downarrow)/kT]\} \quad (4)$$

The double layer field, X, is given by

$$X = q_M/\varepsilon\varepsilon_0 \quad (5)$$

where q_M is the charge density on the electrode and ε is the dielectric constant of the adsorbed water layer and ε_0 is the permittivity of the free space ($\varepsilon_0 = 8.854 \times 10^{-12}$ C^2 J^{-1} m^{-1}).

Excess fractional of water dipole pointing one way (up or down), θ, can be expressed as

$$\theta = \theta_\uparrow - \theta_\downarrow = \exp\{-[(\Delta G_c^0)_\uparrow/kT - \mu X/kT + U_c(\theta_\uparrow - \theta_\downarrow)/kT]\} \\ - \exp\{-[(\Delta G_c^0)_\downarrow/kT + \mu X/kT - U_c(\theta_\uparrow - \theta_\downarrow)/kT]\} \quad (6)$$

At the potential of minimum SFG, *i.e.*, minimum orientation, θ must be 0. Eqn (6) shows that the potential of minimum orientation does not necessary equal the potential where $X = 0$, *i.e.*, "pzfc". Habib and Bockris reported that they differ by 0.1 V.[45]

SFG intensity in OH stretching region at both Au and Pt electrodes decreased as oxide was formed, as shown in Fig. 2 and 3. There are several possibilities for this decrease. One is the disruption of the well-ordered hydrogen bonded network structure of water molecules at a roughened electrode surface compared to an atomically flat surface, since it is well known that the atomically flat surfaces of Au[46] and Pt[47] are roughened by surface oxide formation. The other possibility is the electric effect. Since metal oxides are semiconducting/insulating,[48] an additional potential drop occurred within the metal oxides, resulting in a smaller electric field within the double layer. Furthermore, surface charge should be also affected by the oxide formation.

At the Au electrode, when the potential became more positive than 1000 mV, a rather sharp band centered around 3550 cm^{-1} appeared on top of the broad OH band. Thus, the SFG spectra of Au electrode observed at 1100 and 1200 mV were fitted with two broad OH bands and one narrow band centered at 3550 cm^{-1}. This kind of narrow band is usually observed in SFG spectra when non-hydrogen bonded OH group, so-called "free OH" exists on the surface. Free OH was detected by SEIRAS in Au/0.1 M $HClO_4$ solution at 3612 cm^{-1} by Ataka *et al.*[43] They observed a sharp peak at 3612 cm^{-1} at more negative potential. Since the ClO_4^- anion is a poor hydrogen bond acceptor, the co-adsorbed anion is believed to disrupt the hydrogen-bonding between interfacial water molecules. In this case, adsorption of ClO_4^- is an important factor for free OH to be observed. However, the narrow peak was not observed in the anion adsorption potential region, and also 3612 cm^{-1} is a much higher wavenumber than that of the present peak position. Since this narrow peak was only observed in the Au oxide formation region and it disappeared when Au oxide was reduced, it is reasonable to think that this peak is related to Au oxide formation. It is known that the OH stretching band of $Au(OH)_2$ and AuOH will give an peak at 3565 cm^{-1} and 3569 cm^{-1}, respectively.[49] The peak positions of these bands are close to that of the experimentally observed peak. Actually, potential dependent X-ray reflectivity measurements at Au(111) and Au(100) electrodes in H_2SO_4 solution at various potentials showed formation of a monolayer of oxygen species at potentials less negative than the oxide formation potential, in agreement with the surface X-ray scattering results.[46]

3.2. Potential dependent SFG spectra at a Pt/0.1 M $HClO_4$ solution interface during methanol oxidation reaction

Fig. 6 shows a typical cyclic voltammogram of a thin Pt film electrode in a 0.1 M $HClO_4$ solution containing 0.1 M methanol. The oxidation of methanol starts at around 300 mV with a current peak at 700 mV in the positive direction scan. In the negative direction scan, the anodic current started to increase from 600 mV with a peak at 500 mV and decreased to almost zero at 200 mV. Since the hydrogen waves were observed in the potential range between −200 and *ca.* 100 mV, this suggested that the Pt surface was not completely covered by CO derived from methanol.

Fig. 7 shows SFG spectra in OH stretching region (2800 cm^{-1}–3800 cm^{-1}) obtained at the Pt electrode in a 0.1 M $HClO_4$ solution containing 0.1 M methanol at various potentials. The shape of the spectra is essentially same as that of the Pt electrode in a 0.1 M $HClO_4$ solution without methanol. Two broad peaks were observed at around at 3200 cm^{-1} and at 3450 cm^{-1}. Although the *S*/*N* ratio of spetra was relatively low, the potential dependent intensity changes of the SFG spectrum are clearly seen in Fig. 7. The SFG spectrum observed at 200 mV was of the interfacial water co-adsorbed with CO. Although Shiroishi *et al.* reported the presence of non-hydrogen bonded OH at 3658 cm^{-1} due to the weak interaction between CO and water molecules when water molecules are co-adsorbed on Pt surface with

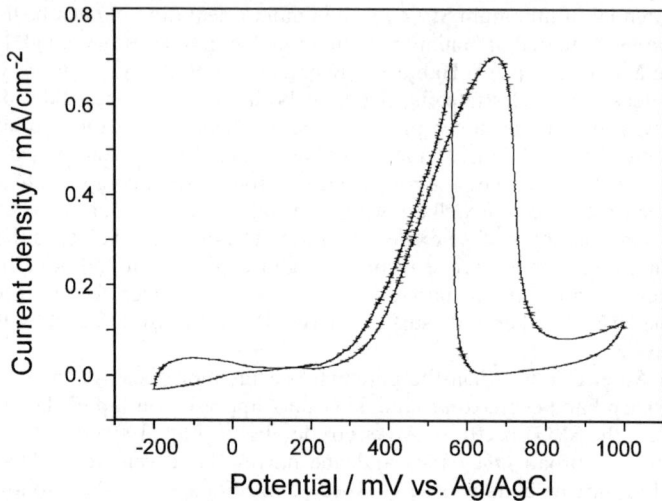

Fig. 6 Cyclic voltammogram of a Pt thin film electrode in 0.1 M HClO$_4$ solution containing 0.1 M methanol. Sweep rate: 50 mV s^{-1}.

CO by SEIRAS,[50] no such peak corresponding to non-hydrogen bonded OH was observed in the present study.

Fig. 8 shows the potential dependence of the integrated SFG intensity in OH stretching region taken from Fig. 7. The steady state currents observed at various potentials during SFG measurements are also plotted in the same figure. SFG spectra were fitted by two broad OH bands centered at *ca.* 3200 and 3450 cm^{-1}, as shown in Fig. 7, but due to the relatively low S/N ratio of the spectra, it was difficult to discuss the potential dependencies of individual peaks. Instead, the integral SFG intensities between 2800 cm^{-1} to 3800 cm^{-1} were plotted against electrode potential. Since the off-resonance SFG intensity did not change with electrode potential, the contribution from the non-resonant component can be ignored in the results. SFG intensity was almost constant at potentials between −200 and 400 mV, where the surface was covered by CO. It started to decrease significantly as the potential became more positive than 600 mV and finally reached a constant SFG intensity at potentials more positive than 700 mV. The former potential is related to the oxidation of surface adsorbed CO and the latter to the Pt oxide formation. These results imply that the structure of the water layer is greatly affected by the decrease of surface CO. Unfortunately, we were not able to probe CO stretching region simultaneously, since a fused quartz prism was used in this experiments. We are now planning to use a CaF$_2$ prism so that CO and interfacial water can be probed at the same time. As the potential becomes more positive than 600 mV, oxidative desorption of adsorbed CO from Pt surface takes place by:[51]

$$\text{Pt–CO(ad)} + \text{OH(ad)} \rightarrow \text{CO}_2 + \text{H}^+ + \text{e}^- \qquad (7)$$

Adsorbed CO reacts with an oxygen donor, as adsorbed OH species on a Pt surface derived from dissociation of water. A random oxidation mechanism was proposed based on IRAS measurements, in which the shift of peak wavenumber of the CO stretching band was observed during CO oxidation, which results from the decrease of dipole–dipole coupling in adsorbed layer.[52] Since the SFG signal is sensitive to the ordering structure of water at the interface, a significant decrease of the SFG signal during the CO oxidation reaction was assigned to the disordering of the CO–water network.

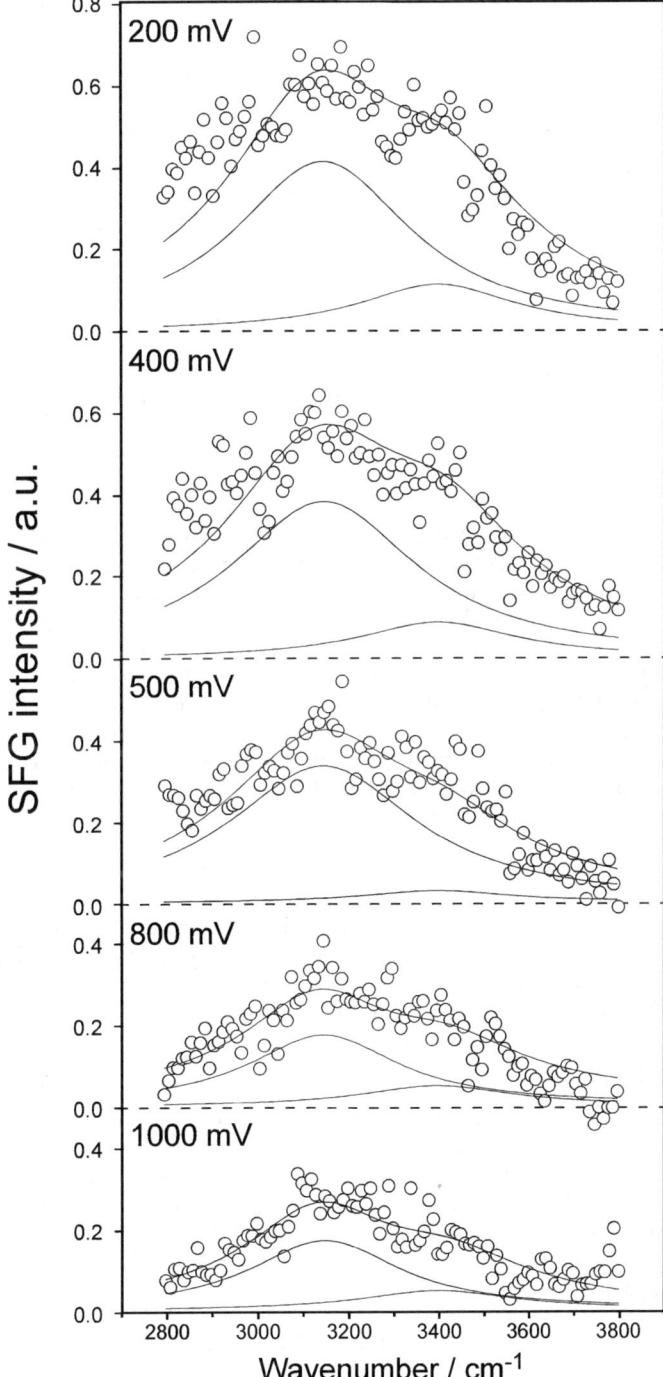

Fig. 7 SFG spectra in OH stretching region at a Pt electrode at each potential in 0.1 M HClO$_4$ solution containing 0.1 M methanol.

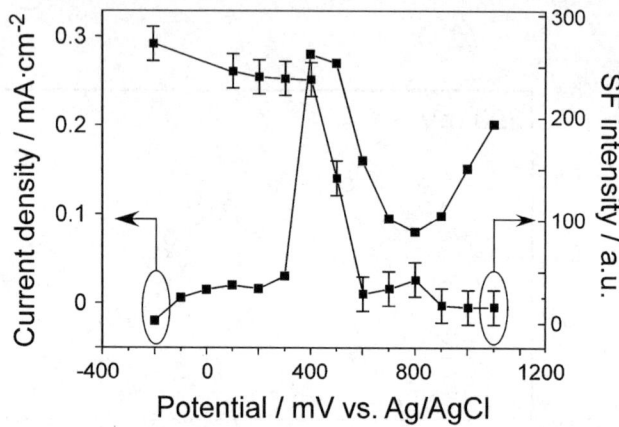

Fig. 8 Potential dependence of integrated SFG intensity in OH stretching region and current flow during SFG measurements of a Pt thin film electrode in 0.1 M HClO$_4$ solution containing 0.1 M methanol.

4. Conclusions

In conclusions, we investigated the structure of interfacial water molecules at Pt and Au thin film electrode/electrolyte solution interfaces by electrochemical SFG. While two broad peaks were observed in the OH stretching region at *ca.* 3200 cm^{-1} and *ca.* 3400 cm^{-1}, which are known to be due to the strongly hydrogen bonded, *i.e.*, "ice-like", water and the weakly hydrogen bonded, *i.e.*, "liquid-like" water, respectively, at the Pt electrode, the 3200 cm^{-1} band was dominant in the SFG spectra at the Au electrode, showing that water molecules are more highly oriented at the Pt electrode than at the Au electrode.

Parabolic behavior of SFG intensity *vs.* potential was observed with a minimum close to the pzc of the electrode in the HClO$_4$ solution. The SFG intensity of OH stretching decreases as the potential becomes more positive in an oxide formation region.

In the case of the Au electrode, an anomalous peak at *ca.* 3550 cm^{-1} was observed in the Au oxide formation region in addition to the broad peak.

During the methanol oxidation reaction on the Pt electrode, a significant decrease of SFG intensity of water was observed. Since the SFG signal is sensitive to the ordering structure of water at the interface, a decrease of the SFG signal during CO oxidation reaction was assigned to the disordering of the CO–water network.

Acknowledgements

This work was partially supported by the Grants-in-Aid for Scientific Research (A) (No. 18205016), for Young Scientists (B) (No. 19750054), and by the Global COE Program (Project No. B01: Catalysis as the Basis for Innovation in Material Science), and for Priority Aria "Molecular Nano Dynamics" from Ministry Education, Culture Sports, Science and Technology (MEXT), Japan.

References

1 J. O'M. Bockris, M. A. V. Devanathan and K. Müller, *Proc. R. Soc. London, Ser. A*, 1963, **A274**, 55.
2 N. F. Mott and R. J. Watts-Tobin, *Electrochim. Acta*, 1961, **4**, 79.
3 B. B. Damaskin and A. N. Frumkin, *Electrochim. Acta*, 1974, **19**, 173.
4 J. O'M. Bockris and M. A. Habib, *Electrochim. Acta*, 1977, **22**, 41.
5 K. Kunimatsu and A. Bewick, *Ind. J. Technol.*, 1986, **24**, 131.

6 B. Pettinger, M. R. Philpott and J. G. Gordon, II, *J. Chem. Phys.*, 1981, **74**, 934.
7 S. Z. Zou, Y. X. Chen, B. W. Mao, B. Ren and Z. Q. Tian, *J. Electroanal. Chem.*, 1997, **424**, 19.
8 A. Bewick and J. W. Russel, *J. Electroanal. Chem.*, 1982, **132**, 329.
9 T. Iwasita and X. Xia, *J. Electroanal. Chem.*, 1996, **411**, 95.
10 K. Ataka and M. Osawa, *Langmuir*, 1998, **14**, 951.
11 M. F. Tony, J. N. Howard, J. Richer, G. L. Borges, J. G. Gordon, O. R. Melroy, D. G. Wiesler, D. Yee and L. B. Sorensen, *Nature*, 1994, **368**, 444.
12 M. Ito and M. Yamazaki, *Phys. Chem. Chem. Phys.*, 2006, **8**, 3623.
13 X. D. Zhu, H. Suhr and Y. R. Shen, *Phys. Rev. B: Condens. Matter Mater. Phys.*, 1987, **35**, 3047.
14 Y. R. Shen, *Nature*, 1989, **337**, 519.
15 G. L. Richmond, *Chem. Rev.*, 2002, **102**, 2693.
16 Y. R. Shen and V. Ostroverkhov, *Chem. Rev.*, 2006, **106**, 1140.
17 S. Ye, S. Nihonyanagi and K. Uosaki, *Phys. Chem. Chem. Phys.*, 2001, **3**, 3463.
18 S. Nihonyanagi, S. Ye and K. Uosaki, *Electrochim. Acta*, 2001, **146**, 3057.
19 K. Uosaki, T. Yano and S. Nihonyanagi, *J. Phys. Chem. B*, 2004, **108**, 19086.
20 V. Ostroverkhov, G. A. Waychunas and Y. R. Shen, *Phys. Rev. Lett.*, 2005, **94**, 046102.
21 A. Peremans and A. Tadjeddine, *J. Chem. Phys.*, 1995, **103**, 7197.
22 S. Baldelli, G. Mailhot, P. N. Ross and G. A. Somorjai, *J. Am. Chem. Soc.*, 2001, **123**, 7697.
23 W. Q. Zheng, O. Pluchery and A. Tadjeddine, *Surf. Sci.*, 2002, **502–503**, 490.
24 W. Zheng and A. Tadjeddine, *J. Chem. Phys.*, 2003, **119**, 13096.
25 S. Nihonyanagi, S. Ye, K. Uosaki, L. Dreesen, C. Humbert, P. Thirty and A. Peremans, *Surf. Sci.*, 2004, **573**, 11.
26 H. Noguchi, T. Okada and K. Uosaki, *Electrochim. Acta*, 2008, **53**, 6841.
27 Z. D. Schultz, S. K. Shaw and A. A. Gewirth, *J. Am. Chem. Soc.*, 2005, **127**, 15916.
28 U. Buch and F. Huisken, *Chem. Rev.*, 2000, **100**, 3863.
29 Q. Du, E. Freysz and Y. R. Shen, *Phys. Rev. Lett.*, 1994, **72**, 238.
30 The SFG spectra shown in Fig. 5 are different from those report by Tadjeddine and his coworkers.[21,24] This discrepancy maybe due to the difference in the experimental configuration. Strong IR absorption by the water layer, although it was very thin, makes the SFG measurements in the OH stretching region in external reflection mode employed by Tadjeddine's group.
31 M. Morgenstern, T. Michely and G. Comsa, *Phys. Rev. Lett.*, 1996, **77**, 703.
32 N. Ikemiya and A. A. Gewirth, *J. Am. Chem. Soc.*, 1997, **119**, 9919.
33 H. A-Kozlowska, B. E. Conway, A. Hamelin and L. Stoicoviciu, *Electrochim. Acta*, 1986, **31**, 1051.
34 H. A-Kozlowska, B. E. Conway, A. Hamelin and L. Stoicoviciu, *J. Electroanal. Chem.*, 1987, **228**, 429.
35 D. M. Kolb and J. Schneider, *Electrochim. Acta*, 1986, **31**, 929.
36 In the case of the Pt electrode, where the adsorption processes involving charge transfer is expected, two types of pzc should be considered.[38,39] One is the potential of zero free charge (pzfc), at which the truly free, electronic excess charge density on the metal surface equals zero, and the other is the potential of zero total charge (pztc), at which the sum of the free, electronic excess charge density and the charge density transferred in adsorption processes equals zero. Only pztc is experimentally accessible.
37 A. N. Frumkin and O. A. Petrii, *Electrochim. Acta*, 1975, **20**, 347.
38 V. Climent *et al.*, in *"Interfacial Electrochemistry"* ed. A. Wieckowski, Marcel Dekker, New York, 1999, p. 463.
39 A. Cuesta, *Surf. Sci.*, 2004, **572**, 11.
40 G. A. Attard and A. Ahmadi, *J. Electroanal. Chem.*, 1995, **389**, 175.
41 G. Nagy and K. Heinzinger, *J. Electroanal. Chem.*, 1992, **327**, 25.
42 R. Akiyama and F. Hirata, *J. Chem. Phys.*, 1998, **108**, 4904.
43 K. Ataka, T. Yotsuyanagi and M. Osawa, *J. Phys. Chem.*, 1996, **100**, 10664.
44 J. O'M. Bockris and A. K. N. Reddy, *Modern Electrochemistry 2*, Plenum, New York, 1970.
45 M. A. Habib and J. O'M. Bockris, *Langmuir*, 1986, **2**, 388.
46 T. Kondo, J. Morita, K. Hanaoka, S. Takakusagi, K. Tamura, M. Takahasi, J. Mizuki and K. Uosaki, *J. Phys. Chem. B*, 2007, **111**, 13197.
47 K. Sashikata, N. Furuya and K. Itaya, *J. Vac. Sci. Technol., B*, 1991, **9**, 457.
48 A. Damjanovic, V. I. Birss and D. S. Boudreaux, *J. Electrochem. Soc.*, 1991, **138**, 2549.
49 X. Wang and L. Andrews, *Inorg. Chem.*, 2005, **44**, 9076.
50 H. Shiroishi, Y. Ayato, K. Kunimatsu and T. Okada, *J. Electroanal. Chem.*, 2005, **581**, 132.
51 T. Iwashita, *Electrochim. Acta*, 2002, **47**, 3663.
52 K. Kunimatsu, H. Seki, W. G. Golden, J. G. Gordon II and M. R. Philpott, *Surf. Sci.*, 1985, **158**, 596.

PAPER

A comparative *in situ* ^{195}Pt electrochemical-NMR investigation of PtRu nanoparticles supported on diverse carbon nanomaterials

Fatang Tan,[†a] Bingchen Du,[a] Aaron L. Danberry,[a] In-Su Park,[b] Yung-Eun Sung[b] and YuYe Tong[*a]

Received 21st February 2008, Accepted 2nd May 2008
First published as an Advance Article on the web 15th September 2008
DOI: 10.1039/b803073a

This paper reports a detailed *in situ* ^{195}Pt electrochemical-nuclear magnetic resonance (EC-NMR) study of PtRu nanoparticles (NPs) that had a nominal atomic ratio of Pt : Ru = 1 : 1 and were supported on carbon nanocoils and carbon black (Vulcan XC-72) respectively. The particle sizes of the two samples were determined by X-ray diffraction using the Sherrer equation: 3.6 nm for the former and 3.2 nm for the latter, which were further corroborated by transmission electron microscope measurements. By taking advantage of a unique correlation between the spectral frequency of the ^{195}Pt NMR resonance and the radial atomic position in a particle, qualitatively- and spatially-resolved local Pt atomic fractions in the particles were deduced by using a Ruderman–Kittel–Kasuya–Yosida (RKKY) *J*-coupling-based method as a function of different electrode potentials. The results indicated that both samples had Pt-enriched cores and Pt-deprived surfaces and, most importantly, the local Pt concentration varied as the electrochemical environment changed. The spatially-resolved Fermi level local densities of states (E_f-LDOS), which are a measure of the electronic frontier orbitals in metals, were deduced across the NMR spectrum and correlated with the EC activity in methanol electro-oxidation. The results were also compared to those obtained previously from Pt/Ru NPs supported respectively on carbon and graphite nanofibers.

1 Introduction

In heterogeneous catalysis and electrocatalysis, bimetallic catalysts, prevailingly in the form of nanoparticles (NPs), can in principle offer more tunability, better selectivity, and higher activity than their respective monometallic counterparts.[1–3] For instance, Pt/Ni[4] or Pt/Pd[5] is better in catalyzing oxygen reduction reaction than Pt, Ni, or Pd does individually; and Pt/Ru is more tolerant to CO-poisoning than Pt and, therefore, does a better job in catalyzing methanol (MeOH) electro-oxidation.[6] Understanding the fundamentals that govern such generally observed superior catalytic performance is the necessary premise for the rational design of better nanoscale bimetallic catalysts. In this regard, local elemental composition and the

---

[a]*Department of Chemistry, Georgetown University, 37th & O Streets, NW, Washington DC, 20057, USA. E-mail: yyt@georgetown.edu*
[b]*School of Chemical & Biological Engineering, Seoul National University, Seoul 151-744, S. Korea*

† Permanent address: School of Materials Science & Engineering, Huazhong University of Science and technology, Wuhan, Hubei 430074, P. R. China.

associated electronic properties are amongst the most important physical attributes that need to be known in order to establish a useful working knowledge on the physical properties–catalytic performance correlation for a given catalyst.

Although catalytic actions take place on the NP surface, it is still highly desirable to access not only the surface but also the *spatially-resolved* compositional and associated physical properties across an entire NP. Since the latter define the electronic properties of the NP, they may also play an important role in shaping the surface catalytic performance. However, it is still very much a technical challenge to do so. Geometrically, a spherical metal NP can be mentally considered as being made of onion-like atomic layers starting from a central atom. But no available three-dimensional imaging techniques can deliver a spatial resolution at the atomic layer scale, not even at the nanometer scale. Extended X-ray absorption fine structure spectroscopy (EXAFS), which is probably the current method of choice for accessing nanoscale compositional information, can only measure overall average elemental specific coordination numbers in NPs.[7]

Very recently, we have proposed a ^{195}Pt nuclear magnetic resonance (NMR)-based method that can in principle access spatially-resolved compositional and electronic information on Pt-based bimetallic NP systems[8,9] by utilizing a unique *monotonic* relationship between a given ^{195}Pt resonant frequency (f) and Pt atoms of a given atomic layer (r) within the NPs.[10] Such information is essential for identifying what are the key physical parameters that control the catalytic performance of Pt-based bimetallic NPs and is the central focus of the present study. To better illustrate the f–r relationship and its use, we discuss first, briefly, a known case for a commercial carbon black (Vulcan XC-72)-supported Pt NP sample whose average particle size was 2.5 nm.[11]

1.1 The ^{195}Pt NMR layer-model: the f–r relationship

In Fig. 1A, we show the point-by-point ^{195}Pt NMR spectrum of the Pt NP sample, which is very broad (approx. 3 MHz broad in a 400 MHz magnet). A spherical 2.5 nm Pt NP can be considered as made of four atomic layers plus a central

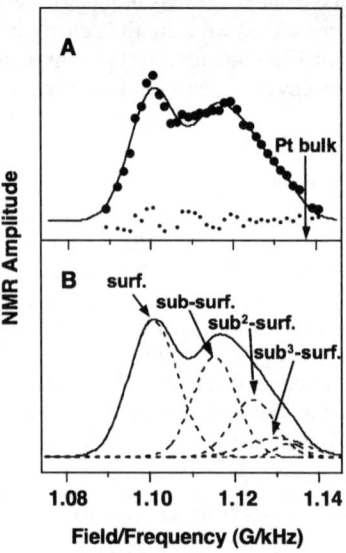

Fig. 1 The point-by-point ^{195}Pt NMR spectrum (A) and its layer-model simulation (B) of a 2.5 nm commercial Pt/Vul XC-72 sample. The small dots in (A) represent the difference between the experimental data (solid circles) and the simulation (solid curve). Adapted with permission from ref. 11.

atom. Since, generally speaking, atoms in a given layer experience a different chemical environment as compared to atoms in the other layers, their nuclear spins are therefore expected to resonate at different frequencies. It happens to Pt (which is unique among known metals) that the nuclear spins of the surface Pt atoms resonate at the lowest field (or the highest frequency) and as the Pt atoms move away from the surface towards the inside of the NP, the resonance frequency of their nuclear spins moves up-field monotonically towards that of bulk Pt as indicated by the arrow in Fig. 1A.

The results of a multi-layer simulation[10,11] are presented in Fig. 1B, assuming that the monotonic change in the peak position of each atomic layer follows an exponential rule $K_n(r) = K_\infty - (K_\infty - K_0)\exp(-n/m)$ where K_∞ and K_0 are respectively the Knight shifts of bulk and surface Pt atoms, n is the number of the layer starting from zero for the surface layer and increasing as the atomic position moves inwards, and m is the so-called 'healing length' characterizing the rate of recovery towards the Knight shift of bulk Pt atoms. As can be seen, different resonant frequencies can be associated with a given atomic layer: surface, sub-surface, sub^2-surface, *etc.* Or, in more qualitative terms, the information for geometric position is encoded in the resonance frequency: the lower the field (or the higher the frequency) of the resonating Pt spins, the closer these spins are to the surface.

For Pt-based bimetallic NPs, such detailed layer-model analysis is still lacking. But available data on Pt/Rh,[12] Pt/Pd,[13] as well as Pt/Ru[9] (also see discussions in this paper) strongly suggest that a similar monotonic f–r relationship exists. The broadness of the overall spectrum ensures a sufficient qualitative spatial resolution that makes our following discussions on spatially-resolved information more tangible than other existing techniques.

1.2 The RKKY *J*-coupling measurements: the local Pt atomic fraction

At a given spectral position, which can in principle be associated with a given geometric position (*i.e.* atomic layer), the local Pt atomic fraction as seen by the resonating nuclear spins can be determined by the so-called Ruderman–Kittel–Kasuya–Yosida (RKKY)[14–16] *J*-coupling (which is a metal analogous to the ubiquitous heteronuclear *J*-coupling in solution NMR) measurements. The broad ^{195}Pt resonance in Pt-based NPs ensures that the difference between the frequencies of neighboring Pt nuclear spins is much larger than the *J*-coupling constant (approx. 3–4 kHz for Pt) so they can be considered as being magnetically inequivalent. In this case, the *J*-coupling manifests itself in a form called 'slow-beat' that causes the normal spin–spin relaxation (T_2) curve to oscillate. In the simplest description, the relaxation curves can be expressed as[17]

$$S(\tau)/S_0 = \exp(-2\tau/T_2)\left\{P_0 + \exp\left[-(\tau/T_{2J})^2\right]\sum_{n=1}^{12} P_n\cos^n(J\tau)\right\} \quad (1)$$

where $S(\tau)/S_0$ is the normalized spin echo amplitude, τ is the time interval between the two pulses in the conventional Hahn spin-echo sequence, T_2 is the nuclear spin–spin relaxation time, T_{2J} accounts for a Gaussian-type spread in J due to the inevitable environmental heterogeneity, and P_n is the probability of having n nearest neighboring nuclear spins and therefore a function of their local concentration. For all practical purposes the effect caused by the spins beyond the nearest neighbors can be neglected. Thus, only P_0, P_1, and P_2 are retained for the T_2 relaxation curve fittings, but the constraint $P_2 = 1 - P_0 - P_1$ applies.

By determining the P_n, one can in principle access concentration of the nuclear spin under observation. According to Slichter and co-workers, P_n can be expressed as[17]

$$P_n = \sum_{i=1}^{12} A_{ni} Q_n \quad (2)$$

provided that the Knight shift gradient ∇K is large enough. In eqn (2), Q_n is the probability of having n nearest neighbors whose spins can be flipped by the radiating radio-frequency field B_1, A_{ni} is the probability of i out of the n nearest neighbors being the ^{195}Pt isotope (whose natural abundance is 0.337), and the maximum value of n for the model is 6.[17] More specifically, A_{ni} is a function of the Pt atomic fraction C_{Pt} ($= 1$ for bulk Pt) and Q_n is a function of δ, a parameter related to the Knight shift gradient: $\delta = \omega_1/(a\omega_0|\nabla K|)$ where a is the distance between nearest neighbors, ω_1 and ω_0 are the Larmor frequencies under B_1 and the external static field B_0. Eqn (2) applies to the condition $\delta \ll 1$, which holds for the Pt-based NPs where the ∇K is sufficiently large. It is further assumed that δ is a constant across the spectrum and takes an *ad-hoc* value of 0.1. The C_{Pt} was then deduced from the P_0 by solving eqn (2) under the constraint of $\sum P_n = 1$. The consistency of the method has been discussed in a previous publication.[8]

In the following, we are going to use the above-described method to analyse *in situ* ^{195}Pt EC-NMR data obtained on Pt/Ru NPs supported on carbon nanocoils (CNCs) and carbon black Vulcan XC-72 (Vul) as a function of the electrode potential to determine the corresponding compositional and electronic properties and to correlate these properties with the observed electrocatalytic activities. We will also compare them with the previously published results obtained on Pt/Ru NPs supported on graphite nanofibers (GNFs) and carbon nanofibers (CNF).[8,9] We found that the variations of the spatially-resolved local Pt atomic fraction over the dimension of the NPs was sample dependent, despite the fact that all of them had the same nominal atomic ratio of Pt : Ru = 1 : 1, although supported on different carbon supports. More importantly, the local metal composition distributions also depended on the EC treatment, *i.e.* the electrode potential. In addition, based on the across-the-spectrum spin–lattice relaxation (T_1) measurements, the spatially-resolved Fermi level local densities of states (E_f-LDOSs), which are a measure of the electronic frontier orbital in metals, were deduced using the phenomenological two-band model.[18,19] Among the four samples studied, higher MeOH activity was found to be correlated with the higher values and variations in the surface E_f-LDOS, which highlights the important role that the electronic frontier orbitals may play in determining the surface catalytic activity.

2 Experimental

2.1 PtRu NPs synthesis

Two PtRu NPs with a nominal atomic ratio Pt : Ru = 1 : 1 and supported on carbon nanocoils (CNCs) and carbon black Vulcan XC-72 (Vul) respectively were synthesized by the conventional borohydride reduction method. Briefly, laboratory-synthesized[20] CNCs and commercial Vul were used as support materials. First, the carbon materials were dispersed in ultrapure water (18.2 M$\Omega\cdot$cm), and then a given amount (calculated to give a nominal total metal loading of 80 wt%) of Pt salt (H$_2$PtCl$_6\cdot x$H$_2$O, Aldrich Chem. Co.) and Ru salt (RuCl$_3\cdot x$H$_2$O, Aldrich Chem. Co.) were added to the solution. After being mixed for approx. 1 h at room temperature, the metal salts were reduced by NaBH$_4$ while the solution was stirred vigorously. The resulting precipitates were washed with deionized water and freeze-dried without any heat treatment. The particle sizes as determined by applying the Sherrer equation to the X-ray diffraction (XRD) peaks (Fig. 2A) were 3.6 and 3.2 nm for the PtRu/CNC and PtRu/Vul NPs respectively. XRD measurements were carried out with a Rigaku D/MAX 2500 operated with a Cu Kα source (λ = 1.541 Å) operating at 40 kV and 200 mA. The samples were scanned from 20 to 80° (2θ) with a step width of 0.02° and a scan speed of 4 ° min^{-1}. The respective transmission electron microscope (TEM) images are shown in Fig. 2B (PtRu/CNC) and 2C (PtRu/Vul) which gave average particle sizes of 3.9 and 2.9 nm respectively

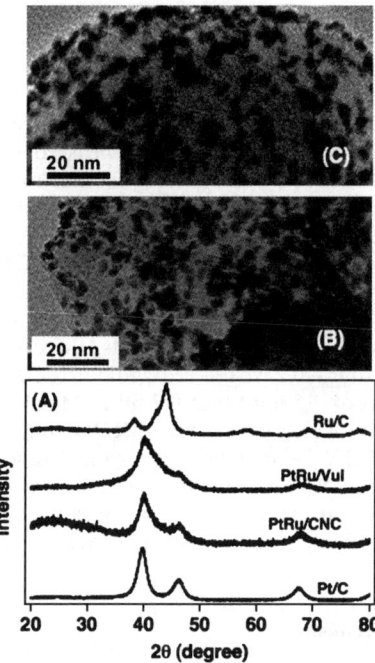

Fig. 2 (A) The XRD data of the PtRu/CNC and PtRu/Vul NPs. The data on the monometallic Pt/carbon (60 wt%, E-Tek) and Ru/carbon (60 wt%, E-Tek) are also presented for comparison. (B) and (C) are the respective TEM images of the PtRu/CNC and PtRu/Vul NPs.

(200 counts), corroborating well with the NP sizes determined by the XRD measurements. The atomic ratios as determined by energy dispersive X-ray spectroscopy (EDS) were Pt : Ru = 57 : 43 for the former and 50 : 50 for the latter. The metal weight specific areas as determined by CO stripping (*vide infra*, using 0.42 mC cm^{-2} as the specific charge for CO electro-oxidation) were respectively 33 and 43 m^2 g^{-1} which were in good agreement with the NP sizes determined by XRD and TEM if one takes into consideration that the surface in direct contact with the support is not accessible by the CO. For a given type of support, the samples were from the same batch.

2.2 Electrochemical characterizations

For electrochemical (EC) characterizations of the PtRu NPs, the working electrode (WE) was prepared as follows. An aqueous suspension of the PtRu NPs was first prepared by dispensing approx. 5 mg PtRu sample into 5 ml of 2-propanol containing about 50 μl 5% Nafion® solution (Aldrich), followed by approx. 10 min sonication. A 20 μl of suspension was then drop-cast using a micropipette onto a 5 mm glassy carbon (GC) electrode which had been pre-cleaned by successively polishing it using alumina oxide powders of decreasing size of 1, 0.3, and 0.05 μm respectively. The loaded amounts were 23 μg for PtRu/CNC and 15 μg for PtRu/Vul.

All EC measurements were carried out using either a CHI 760C or a CHI 660C potentiostat (CHI Instruments Inc.) in a conventional three-electrode setup with continuous Ar blanking. Ag/AgCl (3 M) and Pt gauze were used as reference and counter electrodes (RE and CE) respectively. If not specified, all potential values reported were against the Ag/AgCl (3 M) reference. All the current densities were normalized by the exposed Pt surface areas as determined by CO stripping experiments. The supporting electrolyte was 0.1 M HClO$_4$ prepared with Milli-Q water

(18.2 MΩ·cm). In order to avoid any possible Ru dissolution, the electrode potential was always kept below 0.7 V.

For CO stripping experiments, multiple cyclic voltammetries (CVs; about 20) were first run (50 mV s^{-1} and between −0.25 and 0.7 V) under CO bubbling until a stable and reproducible CO stripping peak was obtained (CO annealing). The cell was then purged with ultrapure Ar until no CO oxidation peak could be observed. The final CO stripping CV was obtained by first holding the electrode potential at 0 V while bubbling CO for 2 min followed by Ar purging for at least 5 min and subsequently recording three consecutive CVs.

For MeOH electro-oxidation measurements, 0.5 M MeOH in 0.1 M HClO$_4$ solution was used. For CVs, multiple cycles (approx. 10) between −0.25 and 0.7 V were needed to generate stable and reproducible peak shapes and currents as reported here. For chronoamperometry (CA) experiments, the electrode was first cleaned by holding the potential at −0.2 V in a cell without MeOH until the current died down to minimum (approx. 15 min) then 0.2 ml of MeOH was added into the cell under the same potential which gave a final MeOH concentration of 0.5 M. The potential was held at −0.2 V for another 15 min with ultrapure Ar bubbling through that mixed the added MeOH well with the solution. After that, measurements of a 60 min CA experiment at 0.2 V were taken. Background current was recorded in the same fashion but without the presence of MeOH and subtracted from the CA current.

2.3 ^{195}Pt NMR measurements

For NMR sample preparation, about 100 mg of the PtRu NPs was loaded into a 5 mm (d) × 25 mm (l) NMR glass sample cell which was then attached to a three-electrode EC setup as a working electrode compartment (see the diagram shown in Fig. 3). Electrical contact was achieved by burying a gold wire into the sedimented PtRu NPs in the sample cell. The supporting electrolyte was 0.5 M HClO$_4$ and the RE and CE were a commercial Ag/AgCl (3 M) electrode and a Pt gauze respectively. The open circuit potential (OCP) for as-prepared samples was about 0.7 V. Once the measurements on the as-prepared samples were finished, the sample was connected back to the EC setup and electrochemically treated by holding at different potentials until the current decayed down to and stabilized at about 60 μA. Since the setup produced a large IR drop (usually about 200 mV overpotential was applied), the actual potential of the sample was measured with the stabilized OCP. In addition to the as-prepared samples, NMR measurements of the PtRu/CNC NPs whose OCPs were 200 and −5 mV respectively and of the PtRu/Vul

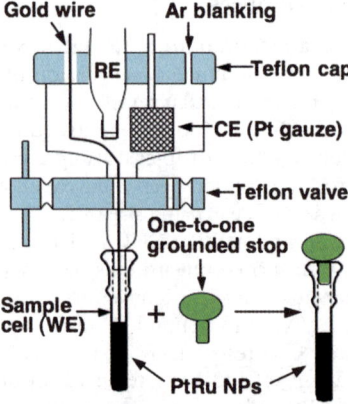

Fig. 3 Diagram of EC-NMR sample preparation setup. Adapted with permission from ref. 9.

NPs whose OCP was 190 mV only were carried out and compared. For simplicity, the above, different EC conditions will be abbreviated in the following as EC200, EC−5, EC190, and as-prepared respectively.

During the EC treatments, the cell was periodically (including the last 30 min) blanked by ultrapure Ar. After the EC treatment, the NMR sample cell filled with the supporting electrolyte was detached from the EC setup under the blanking of Ar, sealed immediately with a one-to-one grounded glass stop, inserted into the NMR probe, and then loaded down to the cryostat which was pre-cooled at 80 K. After the NMR measurements, the sample was *immediately* re-attached to the EC cell and the OCP was measured. No change of the stabilized OCP was observed, ensuring that the EC environment did not change during the NMR measurements.

All ^{195}Pt NMR measurements reported here were carried out at 80 K on a 'home-assembled' spectrometer equipped with an active-shielded 9.395 T widebore superconducting magnet, an Oxford SpectrostatCF cryostat (Oxford Instruments, UK), an AMT (Lancaster, PA) 1 kW power amplifier, a Tecmag (Houston, TX) Appollo data acquisition system, and a home-built single-channel solenoid probe. The conventional '$\pi/2-\tau-\pi-\tau$–echo' Hahn spin-echo sequence was used to acquire data at a specific spectral position. The values of $\pi/2$ pulse length and τ were 3 and 25 μs respectively. For spin–spin relaxation measurements, the τ in the Hahn spin-echo sequence was varied and typically 23 different τ values were used for measuring a given relaxation curve. For nuclear spin–lattice relaxation measurements, a saturation comb composed of four $\pi/2$ pulses with varying intervals between two consecutive pulses was used to saturate the NMR signal. After a waiting period t, the recovered nuclear spin magnetic moment was monitored with the Hahn echo acquisition sequence. Fourteen data points (*i.e.* t values) were acquired for each relaxation curve and the relaxation time T_1 was obtained *via* the standard three-parameter exponential saturation–recovery fit.

3 Results and discussions

3.1 The ^{195}Pt NMR spectra and relaxation measurements

The ^{195}Pt NMR spectra. Fig. 4 presents the area-normalized point-by-point ^{195}Pt NMR spectra of the two samples with different stabilized OCPs. Fig. 4A compares the spectra of the PtRu/Vul NPs at the as-prepared and the EC190 OCPs, Fig. 4B the spectra of the PtRu/CNC NPs at the EC200, EC−5, and as-prepared OCPs, and Fig. 4C the spectra of the PtRu/Vul NPs at the EC190 OCP and the PtRu/CNC NPs at the EC200 OCP. The red solid curves in (A) and (B) are the double-Lorentzian fits of the as-prepared samples whose surfaces were expected to be covered by surface oxides (OCP stabilized at about 0.7 V). This is corroborated by the smaller peaks at about 1.089 G kHz^{-1}. The areas under the smaller peaks are 10 and 17% of the total spectral areas of the PtRu/CNC and the PtRu/Vul NPs respectively. These values should be compared to the expected value of about 35% had the Pt atoms distributed homogeneously in the NPs. Thus, we can conclude that the as-prepared samples had severely Pt-deprived surfaces and it was more marked for the PtRu/CNC than for the PtRu/Vul NPs.

One would expect that the EC treatments that led to the OCPs of the PtRu NPs stabilized at and below 200 mV should clean the surfaces of Pt oxides for both samples. It was indeed the case as clearly indicated by the disappearance of the peaks at 1.089 G kHz^{-1} in Fig. 4A and 4B. At the same time, the corresponding ^{195}Pt NMR spectra moved up-field, more so for the PtRu/Vul than for the PtRu/CNC. This may reflect the fact that there were more exposed Pt surface atoms for the former than for the latter so the reduction of the Pt surface oxide could generate effects going deeper in the NPs, as would be expected for the pure Pt NPs.[21]

For the PtRu/CNC NPs, the more negative stabilized OCP pushed the NMR spectrum further up-field. At the EC200 OCP, there is also a small shoulder above

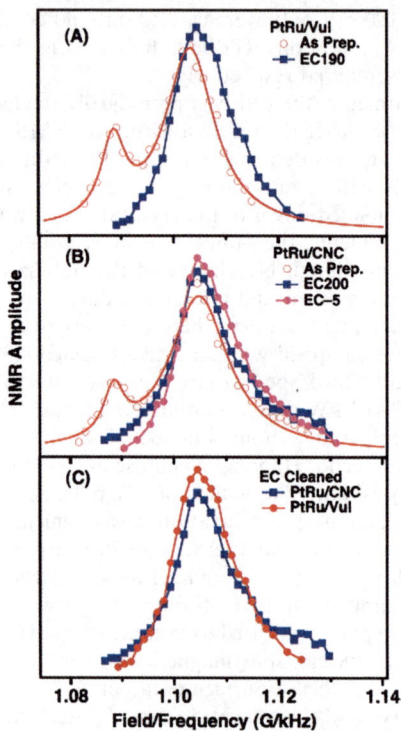

Fig. 4 The area-normalized point-by-point ^{195}Pt NMR spectra of (A) the PtRu/Vul NPs at the as-prepared (red open circles) and the EC190 (blue squares) OCPs and (B) the PtRu/CNC NPs at the as-prepared (red open circles), EC200 (blue squares), and EC−5 (solid magenta circles) OCPs respectively. In (C) the spectra of the PtRu/CNC (blue squares) at the EC200 OCP and the PtRu/Vul (red circles) at the EC190 OCP. The red curves in (A) and (B) are the double-Lorentzian fits of the as-prepared samples and the smaller peaks at about 1.089 G kHz^{-1} are from the surface Pt oxides.

1.12 G kHz^{-1}, indicating a segregation of Pt deep inside the NPs, although the involved number of Pt atoms was small. However, this small shoulder disappeared at the EC−5 OCP, suggesting a re-distribution of the Pt atoms (*vide infra*).

While the stabilized OCPs were very similar for the PtRu/CNC at the EC200 and the PtRu/Vul at the EC190 OCP (*i.e.* 200 *vs.* 190 mV), their NMR spectra differ in some details (Fig. 4C). This is most likely a reflection of the difference in the local distribution of the Pt atoms (*vide infra*). But the overall lineshapes (the peak position and the width) are very similar to each other and to previously published spectra.[8,9,22] It is still unclear why the ^{195}Pt NMR spectra of PtRu NPs are much less variant as a function of the NP size than the pure Pt NPs.

The relaxation measurements. Fig. 5 presents collections of the across-the-spectrum spin–lattice relaxation time T_1 values (A) and of the RKKY *J*-coupling constants (B) measured on different samples that include the previously published data on the PtRu/GNF and PtRu/CNF samples.[8,9] The *J*-coupling constants were obtained by fitting the eqn (2) to the oscillatory spin–spin relaxation T_2 curves (or slow beats), some of which are represented in Fig. 6. The dashed straight lines are only for eye-guiding purpose and the dashed oval highlights the shorter T_1 measured on the PtRu/CNC and PtRu/GNF at the low-field end of the spectrum at which the surface Pt nuclear spins most likely resonate.

Several important observations can be made from these measurements. First, despite being obtained from a collection of seven different data sets, the T_1 values

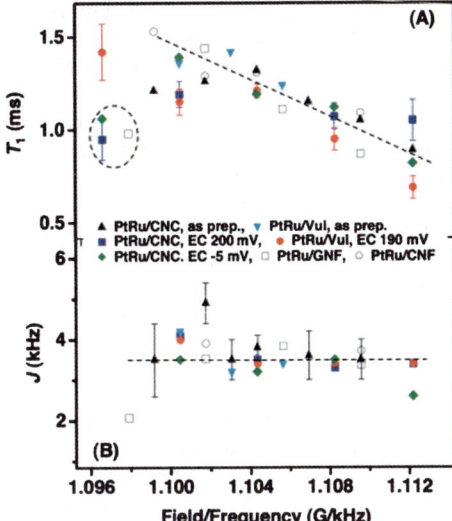

Fig. 5 (A) The spin–lattice relaxation time T_1 values and (B) the RKKY J-coupling constants obtained from fitting the spin–spin relaxation T_2 curves measured across the spectrum for a diverse set of samples. The dashed straight lines are for eye-guiding and the dashed oval is for highlighting some shorter T_1 values measured at the low-field end of the spectrum which are highly likely from the surface Pt atoms.

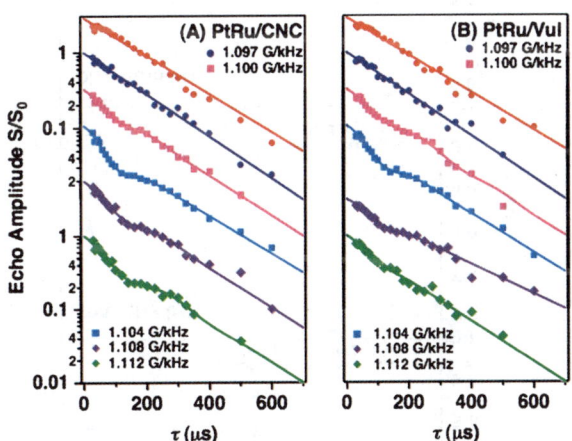

Fig. 6 The representative oscillatory spin–spin relaxation T_2 curves (slow beats) measured at spectral positions as indicated and their fitting results (solid curves) to eqn (2). (A) The PtRu/CNC NPs at the EC200 OCP and (B) the PtRu/Vul at the EC190 OCP. The red solid circles and the straight fitting lines are the results measured at the Pt oxide positions of the respective samples.

in Fig. 5A show a remarkable general pattern. That is, the higher the field (or lower frequency) the Pt nuclear spins resonate at, the shorter their T_1 is, as highlighted by the accompanying eye-guiding straight dashed line. This is exactly the same pattern as observed in pure Pt and PtPd NPs[13,18] where the monotonic f–r relationship has been well established. This observed pattern, as we have argued previously,[9] strongly suggests that a similar monotonic f–r relationship holds for the PtRu NPs, although more differentiating experiments are highly desirable and are planned in the near future. Second, except for a couple of points, the J-coupling constants deduced from the slow beat fittings (Fig. 6) are basically a constant across the spectrum. This invariance is largely expected and had been frequently observed previously in the ^{195}Pt NMR of Pt-based bulk alloys as the fraction of the alloying elements varies.[23] This is because the J-coupling is mainly determined by the Pt–Pt pair-wise interaction and should be less variant once a Pt nuclear spin has a value like that of its neighbor. This observation enhances substantially the confidence in the fitting of eqn (2) and the P_n data so obtained. Third, provided that the monotonic f–r relationship holds, the much shorter T_1 values observed at the low-field end of the spectrum on the PtRu/CNC and PtRu/GNF samples, as highlighted by the dashed oval, are quite unique and should reflect interesting surface electronic properties that may have important ramifications in differentiating the catalytic activity difference observed with these samples.

3.2 Spatially-resolved local Pt atomic fraction

With the f–r relationship rather convincingly established and the detailed spin–spin relaxation measurements recorded (Fig. 6), let us now look into how the local Pt atomic fraction as deduced using eqn (1) and eqn (2) varies. Fig. 7 shows the spatially-resolved (encoded in the spectral frequencies via the monotonic f–r relationship) local Pt atomic fractions deduced by the method discussed in Section 1.2.[8]

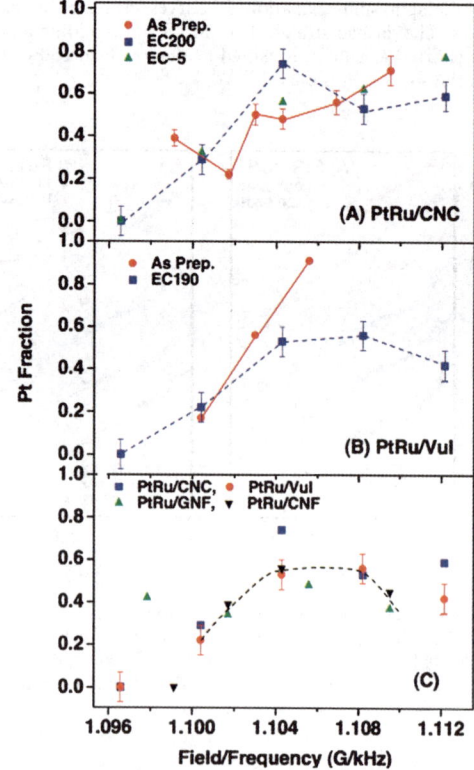

Fig. 7 The deduced local Pt atomic fractions of the same sample but at different stabilized OCPs: (A) as-prepared (approx. 700 mV, red circles), 200 mV (blue squares), and −5 mV (green triangles) for the PtRu/CNC NPs and (B) as-prepared (red circles) and 190 mV (blue squares) for the PtRu/Vul NPs. (C) The results for different samples but at similar potentials are compared: 190 mV for the PtRu/Vul (red circles) and 200 mV for the PtRu/CNC (blue squares), PtRu/GNF (green triangles), and PtRu/CNF (black inverted triangles). The data on PtRu/GNF and PtRu/CNF were published previously[8,9] but are presented here for comparison. All lines in (A)–(C) are for eye-guiding purposes.

While the RKKY *J*-coupling constants deduced from the same sets of data were almost invariant over the dimension of the NPs, it was certainly not the case for the local Pt atomic fractions, which varied from the sample to sample and from one stabilized OCP to the other.

As discussed above, the analysis of the ^{195}Pt NMR spectra of the as-prepared samples indicated that there was severe Pt deprivation at the NP surface for both the PtRu/CNC and PtRu/Vul samples. This is further supported by the T_2 measurements at the surface oxide positions where no slow beats were observed and the relaxation curves can be well represented by a single exponential decay (red solid circles and the associated straight lines in Fig. 6). After the surface Pt oxide was reduced, evidence of the surface Pt deprivation remained: no slow beats were observed at the low-field end of the spectral positions (1.097 G kHz^{-1}, see the blue solid circles in Fig. 6), *i.e.* the Pt nuclear spins resonating at that spectral position did not see the presence of neighboring Pt atoms.

The most striking observation is that *the variation in local Pt fraction across the dimension of the NPs for a given sample depends on the electrochemical environments*, *i.e.* the stabilized OCP (Fig. 7A and 7B). The PtRu/CNC NPs (Fig. 7A) showed different oscillatory behaviours at the as-prepared (red circles) and the EC200 (blue squares) OCPs, which was smoothed out at the EC−5 (the green triangles) OCP at which an increased segregation of Pt at the core of the NPs (high-field end) was observed. The variation was even larger for the PtRu/Vul NPs (Fig. 7B). At the as-prepared OCP, there was a strong segregation of Pt at the middle of the spectrum that was smoothed out when the NPs were at the EC190 OCP.

However, when the variations in the local Pt fraction obtained from the four different samples but at very similar stabilized OCPs are compared in Fig. 7C (190 mV for PtRu/Vul and 200 mV for PtRu/CNC, PtRu/GNF, and PtRu/CNF),

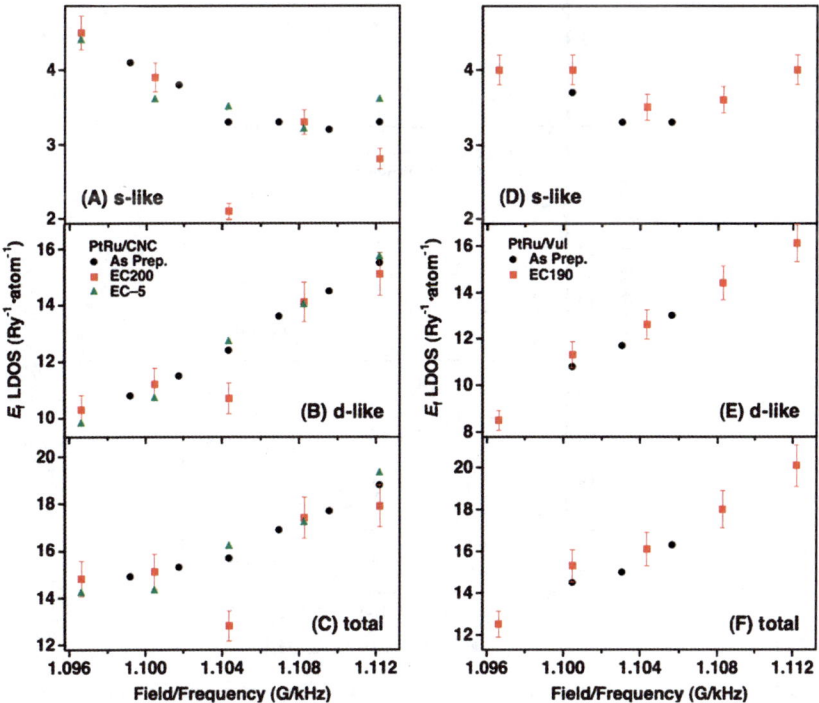

Fig. 8 The E_f-LDOS values of the PtRu/CNC (A)–(C) and of the PtRu/Vul (D)–(F) NPs under different stabilized OCPs: (A) and (D) the s-like, (B) and (E) the d-like, and (C) and (F) the total E_f-LDOSs.

it becomes obvious that there is a remarkable overlap of the local Pt fraction distributions measured at the spectral positions between 1.100 and 1.110 G kHz^{-1}, as highlighted by the dashed curve. Since this spectral region is right in the middle of the spectrum where the peak position locates and covers about 50% of the resonating Pt atoms (Fig. 4), the observed overlap may suggest that the Pt atomic distribution in the middle of the NP is much less variant for a given electrode potential. On the other hand, such overlap disappears at the surface spectral region (low-field end) and the NP core spectral region (high-field end).

3.3 The E_f-LDOS and EC activity

A unique feature of metal (^{195}Pt) NMR is that it enables the determination of the E_f-LDOS, which is an important physical attribute for gauging the electronic frontier orbitals in metals. It is therefore interesting to see if there is any correlation between the surface E_f-LDOS and the corresponding catalytic activity. Fig. 8 collects the E_f-LDOS data deduced from the T_1 data shown in Fig. 5A by using the phenomenological two-band model[18] for both the PtRu/CNC and PtRu/Vul samples under different stabilized OCPs. For a given sample, very similar E_f-LDOS values were obtained. This is actually a surprising observation considering the variation in the local Pt fraction observed in Fig. 7 and warrants further theoretical studies. On the other hand, the monotonic increasing trend of the d-like E_f-LDOS as the spectral position moves towards the high-field end is

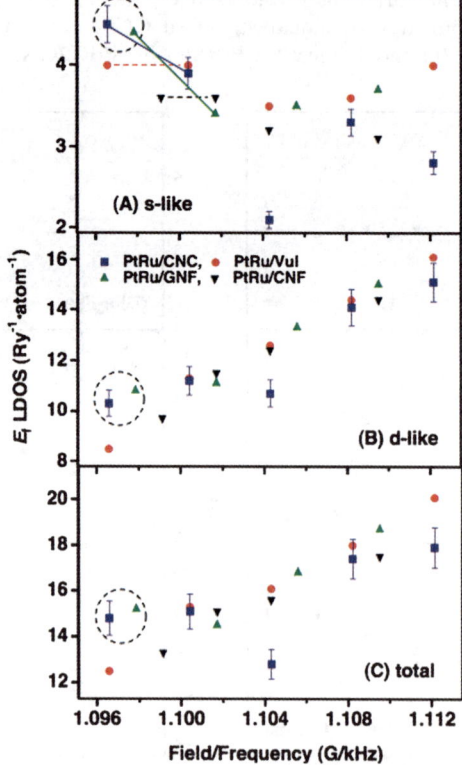

Fig. 9 A comparison of the E_f-LDOSs for the PtRu/CNC (blue squares), PtRu/Vul (red circles), PtRu/GNF (green triangles), and PtRu/CNF (black inverted triangles) NPs: (A) the s-like, (B) the d-like, and (C) the total E_f-LDOSs. The dashed circles in (A)–(C) and the straight solid and dashed lines in (A) highlight respectively the higher E_f-LDOS values and the steeper change of the s-like surface E_f-LDOS for the PtRu/CNC and PtRu/GNF samples.

expected for the monotonic *f–r* relationship as convincingly established for the pure Pt NPs.[24]

The difference in the E_f-LDOSs between the two samples is quite obvious. First, the pattern of variation for the s-like E_f-LDOSs of the PtRu/CNC (Fig. 8A) is evidently different from that of the PtRu/Vul (Fig. 8D). Second, both the s- and d-like E_f-LDOSs measured at the surface spectral region (low-field end) are larger for the former than for the latter. These differences are further compared and contrasted in Fig. 9 together with the previously published data obtained on the PtRu/GNF and PtRu/CNF samples.[8,9]

Several interesting observations can be made here. First, at the low-field-end spectral position where the surface Pt nuclear spins are supposed to resonate, both the PtRu/CNC and Pt/GNF samples have higher E_f-LDOS values than the other two samples. Second, the variations in the surface s-like E_f-LDOSs as estimated by the difference between the two leftmost data points (Fig. 9A) are also higher for the same two samples (the slopes of the green and blue straight lines). The estimated variations for the other two samples are almost zero as highlighted by the two horizontal dashed lines in Fig. 9A. Third, at the high-field end, the PtRu/Vul showed highest E_f-LDOS values, which may indicate a higher Pt segregation at the core of the NPs.

The surface E_f-LDOS is a measure of the electronic frontier orbitals at the NP surface. It would be to argue that a higher surface E_f-LDOS offers more electrons

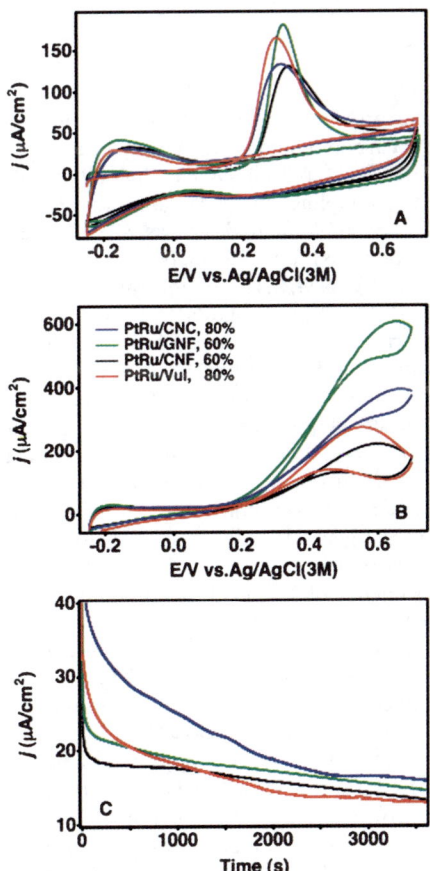

Fig. 10 The CO stripping CVs (A), MeOH electro-oxidation CVs (B), and the CAs of the MeOH electro-oxidation observed on PtRu/CNC (blue curves), PtRu/GNF (green curves), PtRu/CNF (black curves), and PtRu/Vul (red curves) NPs.

available for participating catalytic reactions at the *least* cost of energy, and a larger variation in the s-like surface E_f-LDOS implies the availability of more diversified surface sites for accommodating different energetic requirements for different steps of reactions. Thus, one would expect that the PtRu/CNC and PtRu/GNF could be more reactive than the other two samples.

Interestingly, this is indeed the case observed for the MeOH electro-oxidation, as shown in Fig. 10. While the CO stripping CVs (Fig. 10A) are less conclusive in differentiating the most active samples, both MeOH electro-oxidation CVs and CAs have shown that the PtRu/CNC and PtRu/GNF NPs have higher intrinsic (Fig. 10B) and better long-term activities (*i.e.* better CO tolerance) measured at 60 min (Fig. 10C) in terms of MeOH electro-oxidation. The former had an order of activity as: PtRu/GNF > PtRu/CNC > PtRu/CNF > PtRu/Vul and the latter as: PtRu/CNC (*ca.* 16 μA cm^{-2}) > PtRu/GNF (*ca.* 15 μA cm^{-2}) > PtRu/CNF (*ca.* 13 μA cm^{-2}) ≈ PtRu/Vul (*ca.* 13 μA cm^{-2}). The higher intrinsic MeOH activity for the PtRu/GNF sample is consistent with its higher surface Pt content (Fig. 7C) while the higher CO tolerance for the PtRu/CNC sample is likely to do with it having a superior bimetallic surface composition that compromises better the two conflicting demands: higher intrinsic activity offered by Pt and higher CO eliminating capability offered by Ru. Although how much predictive power of such a correlation can provide is still an open question for debate, it is nonetheless very suggestive in pointing out that the surface E_f-LDOS is, among others, an important physical parameter to look at when discussing the catalytic activity of metal surfaces.

Conclusions

In summary, we have presented results obtained from a detailed *in situ* [195]Pt EC-NMR study of the carbon nanocoil (CNC)- and carbon black (Vul)-supported PtRu NPs under different stabilized OCPs. Spatially-resolved local Pt atomic fractions, the RKKY *J*-coupling constants, and the E_f-LDOS values were all deduced and the results were compared/contrasted with the previously published ones obtained on the PtRu/GNF and PtRu/CNF NPs,[8,9] which has arguably showcased that the *in situ* [195]Pt EC-NMR is a very information-rich and powerfully investigative tool to study Pt-based bimetallic NPs. Based on the above analyses and discussions, we summarize our important conclusions as follows:

• The local distributions of the Pt atomic fraction within the NPs were shown for the first time to vary as the electrode potential changed (Fig. 7A and 7B). This may account for the often observed large variations in the catalytic performance of given bimetallic NPs with the same nominal metal composition and carbon support, because the possible variations in the local distributions of metal composition, as demonstrated here, have rarely been taken into consideration and controlled. This observation also exposes a general challenge to our current mechanistic studies of the catalytic performance of nanoscale bimetallic systems because such electrochemical environment dependence of the local metal composition distributions would make the results of mechanistic studies less tractable.

• A strong Ru surface segregation was observed for three (PtRu/CNC, PtRu/CNF, and PtRu/Vul) of the four PtRu samples studied (Fig. 7C). This is in contrast to a previous study that concluded instead that a Pt surface segregation was more likely.[22]

• A large portion (approx. 50%) of the local Pt atomic fractions, most likely in a region between the core and the surface, appeared to be much less variant from sample to sample at a constant given electrode potential. But no such invariance was observed for the surface and core regions (Fig. 7C).

• The higher surface E_f-LDOS and larger variation in it observed for the PtRu/CNC and the PtRu/GNF NPs (Fig. 9) could be correlated with their higher MeOH activity (Fig. 10B and 10C). This is a potentially important observation because it highlights the important role that the metal surface frontier orbitals may play in determining the surface catalytic activity.

- For a given sample, however, the local E_f-LDOS was found to be almost independent of the electrode potentials (Fig. 8) despite the dependency of the local Pt atomic fraction (Fig. 7A and 7B). This is indeed a surprising observation and warrants further experimental and theoretical investigations.
- However, the observed sample-to-sample variations in terms of the local metal composition distributions unfortunately masked the possible effects that might be exerted by the different carbon supports.

Acknowledgements

This research is supported by DOE (DE-FG02-07ER15895). F. T. is a fellow of the Georgetown-CSC (China Scholarship Council) post-doctoral fellowship program. A. L. D. was an NSF REU summer student from Minnesota State University at Mankato. I.-S. P. and Y.-E. S. thank financial support by the Ministry of Science and Technology, the KOSEF through the Research Center for Energy Conversion and Storage, and KRF (KRF-2004-005-D00064). The authors also gratefully acknowledge Professor Kyung-Won Park from Soongsil University, Korea for providing the carbon nanocoils used in synthesizing the PtRu/CNC NPs.

References

1 J. H. Sinfelt, *Bimetallic Catalysts: Discoveries, Concepts, and Applications*, John Wiley & Sons, New York, 1983.
2 C. T. Campbell, *Annu. Rev. Phys. Chem.*, 1990, **41**, 775–837.
3 V. Ponec and G. C. Bond, *Catalysis by Metal and Alloys*, Elsevier, Amsterdam, 1995.
4 V. R. Stamenkovic, B. Fowler, B. S. Mun, G. Wang, P. N. Ross, C. A. Lucas and N. M. Markovic, *Science*, 2007, **315**, 493–497.
5 J. Zhang, M. B. Vukmirovic, Y. Xu, M. Mavrikakis and R. R. Adzic, *Angew. Chem., Int. Ed.*, 2005, **44**, 2132–2135.
6 J. S. Spendelow and A. Wieckowski, *Phys. Chem. Chem. Phys.*, 2004, **6**, 5094–5118.
7 Y. Zhang, M. L. Toebes, A. van der Eerden, W. E. O'Grady, K. P. de Jong and D. C. Koningsberger, *J. Phys. Chem. B*, 2004, **108**, 18509–18519.
8 A. L. Danberry, B. Du, I. S. Park, Y. E. Sung and Y. Y. Tong, *J. Am. Chem. Soc.*, 2007, **129**, 13806–13807.
9 B. Du, A. L. Danberry, I.-S. Park, Y.-E. Sung and Y. Tong, *J. Chem. Phys.*, 2008, **128**, 052311–052317.
10 J. P. Bucher, J. Buttet, J. J. van der Klink and M. Graetzel, *Surf. Sci.*, 1989, **214**, 347–357.
11 Y. Y. Tong, C. Rice, N. Godbout, A. Wieckowski and E. Oldfield, *J. Am. Chem. Soc.*, 1999, **121**, 2996–3003.
12 Z. Wang, J.-P. Ansermet, C. P. Slichter and J. H. Sinfelt, *J. Chem. Soc., Faraday Trans.*, 1988, **84**, 3785–3802.
13 Y. Y. Tong, T. Yonezawa, N. Toshima and J. J. v. d. Klink, *J. Phys. Chem.*, 1996, **100**, 730–733.
14 M. A. Ruderman and C. Kittel, *Phys. Rev.*, 1954, **96**, 99–102.
15 C. Froidevaux and M. Weger, *Phys. Rev. Lett.*, 1964, **12**, 123–125.
16 C. P. Slichter, *Principles of Magnetic Resonance*, Springer-Verlag, Heidelberg, 1990.
17 H. T. Stokes, H. E. Rhodes, P.-K. Wang, C. P. Slichter and J. H. Sinfelt, *Phys. Rev. B: Condens. Matter Mater. Phys.*, 1982, **26**, 3575–3581.
18 J.-P. Bucher and J. J. van der Klink, *Phys. Rev. B: Condens. Matter Mater. Phys.*, 1988, **38**, 11038–11047.
19 J. J. van der Klink and H. B. Brom, *Prog. Nucl. Magn. Reson. Spectrosc.*, 2000, **36**, 89–201.
20 T. Hyeon, S. Han, Y. E. Sung, K. W. Park and Y. W. Kim, *Angew. Chem., Int. Ed.*, 2003, **42**, 4352–4356.
21 Y. Y. Tong, C. Rice, A. Wieckowski and E. Oldfield, *J. Am. Chem. Soc.*, 2000, **122**, 11921–11924.
22 P. K. Babu, H. S. Kim, E. Oldfield and A. Wieckowski, *J. Phys. Chem. B*, 2003, **107**, 7595–7600.
23 G. C. Carter, L. H. Bennett and D. J. Kahan, *Metallic Shifts in NMR*, Pergamon Press, Oxford, 1977.
24 J. J. van der Klink, *Adv. Catal.*, 2000, **44**, 1–117.

PAPER | www.rsc.org/faraday_d | Faraday Discussions

Spectroelectrochemical flow cell with temperature control for investigation of electrocatalytic systems with surface-enhanced Raman spectroscopy

Bin Ren,* Xiao-Bing Lian, Jian-Feng Li, Ping-Ping Fang, Qun-Ping Lai and Zhong-Qun Tian

Received 27th February 2008, Accepted 23rd April 2008
First published as an Advance Article on the web 21st August 2008
DOI: 10.1039/b803366h

We describe a method for investigating the reaction mechanism of fuel cell systems by designing a spectroelectrochemical cell with functions of temperature and flow control to mimic the reaction condition of fuel cell systems and utilizing Au core Pt shell (Au@Pt) nanoparticles to enhance the Raman signal of the surface species on the surface of electrocatalysts. The cell consists of three parts: a thin-layer spectroelectrochemical reaction chamber with an optical window for Raman measurement, the heating chamber right beneath the reaction chamber, and a long spiral flow channel to preheat the solution to the desired temperature and effectively exchange the solution. The temperature of the solution can be easily controlled from room temperature to 80 °C, and the flow rate can be as high as 945 μl s^{-1}. The temperature and flow control is demonstrated by monitoring the changes in the cyclic voltammograms and the Raman signals. By synthesizing Au@Pt nanoparticles and assembling them on a Pt substrate, we can significantly enhance the Raman signal of surface species on the Pt shell surface, which allows us to detect strong signal of CO as the dissociative product of formic acid as well as the intermediate species of the oxidation process. The further development and perspectives of using SERS to study the electrocatalytic systems are discussed.

Introduction

C1 molecules have attracted extensive attention in the research and development of fuel cells due to their special advantages such as abundant resources, easy storage and transportation, and high energy density.[1-5] In aid of optimizing the reaction condition and designing the electrocatalysts, various types of electrochemical methods and *in situ* and *ex situ* surface techniques, including NMR, DEMS, and IR spectroscopy, have been used to study the adsorption and reaction of C1 molecules on platinum group metals or their alloys.[6-12] Different mechanisms have been proposed to understand the phenomenon observed under different experimental conditions.

Among these methods, vibrational spectroscopic techniques manifest themselves with their special advantages in understanding the electrocatalytic mechanisms with the molecular level information. IR has been most widely applied to investigate

State Key Laboratory for Physical Chemistry of Solid Surfaces, College of Chemistry and Chemical Engineering, Department of Chemistry, Xiamen University, Xiamen, 36100, China. E-mail: bren@xmu.edu.cn; Fax: +86-592-2085349; Tel: +86-592-2186532

the electroxidation of methanol and formic acid on bi- and multi- component catalysts.[7,8] It can conveniently obtain the vibrational information in the high frequency region (higher than 1000 cm^{-1}) of the adsorbed species on single crystal surfaces, smooth electrode surfaces and the surfaces of low roughness. It has provided abundant valuable data for the identification of the adsorbed species on surfaces.[8-12] However, IR has its own limitations. For example, it is very difficult to obtain the information in the low frequency region reflecting the interaction between the substrate and the adsorbates unless a very strong light source, such as synchrotron, is used and it can normally only be applied to the surfaces of a very high reflectivity.

SERS is complementary to the IR in these aspects. It can be applied not only to the electrode surfaces of high roughness and low reflectivity that is very close to the practically used materials, but also to detect the vibrations reflecting the interaction between the substrate and adsorbates in the low frequency region.[13-17] These two unique features make SERS advantageous in the study of electrocatalytic systems. Previous SERS studies were mainly limited to Ag, Cu and Au surfaces. In the past 10 years, we have developed special surface preparation methods for generating SERS-active surfaces of transition metals and their alloys, including Pt, Pd, Rh and Pt-Ru, and performed electrochemical Raman study of the oxidation of carbon monoxide, methanol and formic acid on the surfaces.[13,14,18-20] We have demonstrated using SERS that the dual-path reaction mechanism and the bi-functional mechanism operates on Pt and Pt-Ru surfaces, respectively, during the oxidation of methanol.[21]

In parallel, Weaver's group developed a strategy to coat a transition-metal thin layer over SERS-active Au substrates to improve the SERS detection sensitivity of the surface species on transition-metal surfaces. This strategy allowed them to study various electrocatalytic systems, including the adsorption or dissociative adsorption and electrooxidation of CO, methanol, and formic acid.[15,22,23]

It should be pointed out that most of these previous studies, especially in the case of SERS, were performed at room temperature and in a static solution. However, in real fuel cell systems, the reaction normally takes place at a high temperature (ca. 80 °C) and under a flow condition.[24] Therefore, in order to understand the reaction mechanisms in the fuel cell systems, it is important to perform the SERS studies under a similar condition to the real reaction.

It is not difficult for an electrochemical cell to have a function of temperature and flow control. But in a spectroelectrochemical system, the detection sensitivity becomes the major concern as there are only a monolayer species at the surface.[25] It becomes a challenge to include the temperature and flow controls while maintaining a good potential control over the electrochemical system and allowing an efficient detection of the spectral signal.

In 2001, Weaver's group reported a thin layer flow cell to allow a rapid replacement of solution in 1 s, allowing study of electrochemical processes down to this time scale by means of temporal SERS sequences.[26] From 2000, our group have also concentrated on developing new types of cell for *in situ* SERS studies. We reported a three-phase Raman cell that can work under the conditions of both the solid/liquid interface with a potential control and the solid/gas interface with a convenient gas flow control.[19] This cell was further applied to solid/liquid/gas interfaces with a convenient exchange of the solution and gas flow. We found that the dissociative adsorbed CO generated at the solid/liquid interface can migrate across the three-phase region to the solid/gas interface. In order to realize temperature control on these two types of Raman cells, the source solution has to be heated up in a separate chamber and pumped into the Raman cell for measurement. This design will on one hand complicate the experiment, and on the other hand lead to a significant drop of the solution temperature in the flow path to the cell.

In this work, we design a spectroelectrochemical flow cell with a convenient temperature control so that the room temperature electrolyte can be heated up after being purged inside the cell body. The temperature of the solution can be controlled from room temperature up to 80 °C by a thermocouple put right underneath the working electrode. The flow rate of the electrolyte is controlled by changing the relative height of the source solution and the exhaust solution or the gas pressure of the source solution chamber. Electrochemical and SERS measurements are made to demonstrate the effectiveness of the temperature and flow control. In order to further enhance the surface Raman signal of the surface species on transition-metal surfaces, we ultilize a "borrowing SERS" strategy, which is to coat a very thin layer of Pt over a highly SERS-active Au nanoparticle core (Au@Pt) to produce 2 orders of magnitude enhancement over the normal electrochemically roughened pure Pt eletrode.

Experimental

Electrochemical cyclic voltammograms (CV) were recorded on a CHI631a electrochemical workstation. Raman spectra were obtained using a confocal microprobe Raman system (LabRam I, Jobin-Yvon). The excitation line was 632.8 nm from an external He–Ne laser. A 50× long working-length objective (8 mm) was used in the present experiment. The width of the slit was 200 μm and the diameter of the pinhole was 800 μm.

A polycrystalline Pt electrode with an exposed diameter of about 2 mm was used as the working electrode. The Pt electrode was prepared by melting a commercially available Pt rod with a purity of 99.99% and subsequently removing impurities concentrating on the electrode surface in a *regia aqua* solution After several rounds of melting and removing, the formed Pt sphere was inserted into a Teflon shroud. Then, the top part of the sphere was polished to obtain a mirror finish and leakage free Pt disk electrode. The counter and reference electrodes were a platinum ring and a saturated calomel electrode (SCE), respectively.

HCOOH, $K_4Fe(CN)_6$, $K_3Fe(CN)_6$, KCl used in this study were of analytical grade reagents. H_2SO_4 is guaranteed reagent grade. All solutions were prepared with Mill-Q water.

Results and discussion

Cell design

As mentioned above that most of previous spectroelectrochemical cells had a separate heating chamber to the cell to simplify the cell design. However, during the transportation of the high temperature solution in the very thin pipe to the cell, the temperature will drop significantly. So a temperature monitoring device, such as a thermocouple, has to be put in the cell. On the other hand, during the experiment, several setups have to be constructed together, which will lead to a difficulty in handling. To overcome these problems, we designed and fabricated a spectroelectrochemical flow cell with potential control.

Fig. 1 shows a scheme of the design of the cell. According to its function, the cell can be described in three parts: the spectroelectrochemical cell, the flow control, and the temperature control. The spectroelectrochemical cell part consists of, from top to bottom, the optical window, the cell body, the working electrode, the counter electrode, the reference electrode, and the reaction chamber to accommodate all the electrodes and the electrochemical reaction. Due to the special requirement of the Raman measurement and the flow efficiency, the reaction chamber has to be as small as possible. For example, to improve the detection sensitivity, the solution layer between the surface of the working electrode and the optical window has to be as thin as possible, for example, 0.5–0.7 mm, which produces a volume of the reaction chamber less than 30 μL. The flow rate of the cell was simply controlled by elevating the bottle containing the fresh source solution or by changing the gas pressure to produce a continuous solution flow rather than a pulse flow when a pump is used. In a flow cell, the pressure inside the cell will be higher than the atmospheric pressure. To avoid solution leakage, the connection between the quartz window,

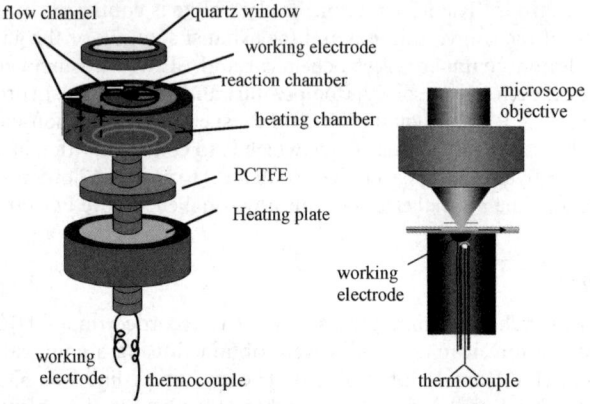

Fig. 1 The scheme of a spectroelectrochemical flow cell with temperature control, consisting of three major parts: the reaction chamber for the electrochemical control and Raman measurement, the flow channel to control the flow rate and heat exchange, and the temperature control unit with a heating plate and a thermocouple.

the working electrode and the reference electrode with the cell body were made with rubber O-rings. As the solution layer is very thin, it will be very difficult to purge the generated gas bubbles out of the cell body. To solve this problem, we designed the reaction chamber in an elliptic shape, and the working electrode is placed asymmetrically at one narrow end right facing the spiral outlet (or inlet of reaction chamber). Therefore, the outlet forms almost a V shape so that the bubbles can be squeezed out of it. The counter electrode was placed in the lower stream of the cell, so that the reaction products generated on the counter electrode will not contaminate the working electrode.

In the cell design, the most challenging part is the temperature control unit. It consists of a spiral channel, a PCTFE thin plate, a heater, and a supporting base. The spiral channel forms a close space by the PCTFE plate realized by using an O-ring. The solution is travelling from the outer spiral inlet to the central sprial outlet and further pushed up to the reaction chamber. The purpose of the spiral channel is to increase the length of the flow path to ensure a sufficient exchange of heat to reach the controlled temperature. The heating plate is also a self-designed unit, with a hole in the center for the working electrode to go through. The power of the heating plate can be adjusted by changing the applied voltage. The temperature is monitored with a thermcouple placed right underneath the Pt sphere working electrode. As Pt is a good heat conductor, the temperature difference between the surface of the working electrode and the sensing point of the thermocouple should be very small.

Demonstration of the temperature control

In a real electrocatalytic system, it is important to have an accurate control over the reaction temperature. For this purpose, we calibrated the temperature by simply using the intensity ratio of the anti-Stokes (I_{AS}) to Stokes (I_S) Raman signal intensities of the 520.6 cm^{-1} band of a silicon single crystal wafer. It is well known in Raman spectroscopy that the ratio depends on the temperature:[27]

$$\frac{I_{AS}}{I_S} = \left(\frac{\nu_0 + \nu_k}{\nu_0 - \nu_k}\right)^4 \exp\left(\frac{-1.4 \times \nu_k}{T}\right) \quad (1)$$

where ν_0, ν_k and T are the incident laser frequency, the Raman shift frequency, and the temperature, respectively. Therefore, if we measure the ratio, we can easily obtain the real temperature at the sample surface. The result is shown in Fig. 2.

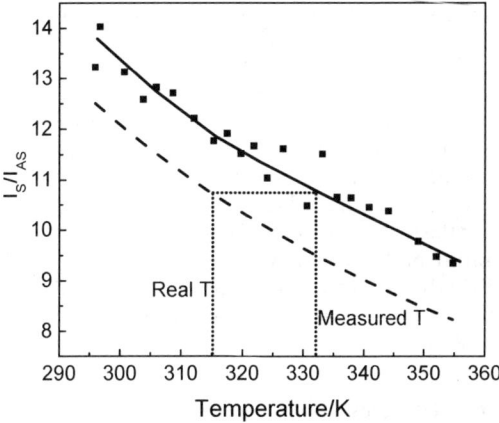

Fig. 2 The calibration curve of the cell temperature. The filled squares are the measured intensity ratio of the anti-Stokes to the Stokes bands of a silicon wafer using 1st phonon band at 520 cm^{-1}. The upper line is the fitting result of the measured experimental ratio and the dash line are the calculated ratio at the temperature of interest.

The dash line is the calculated value based on the eqn (1), the filled squares are the measured ratio. As the experimental temperature was read from the thermocouple, and the thermocouple was at a closer distance to the heating plate than the sample surface, the temperature obtained from the thermocouple will be slightly higher than the real value estimated from anti-Stokes to Stokes ratio, *i.e.*, the real temperature on the sample surface. The calculated curve and the experimental curve can serve as a working curve in the future experiment. For example, from the Raman measurement, we found a deviation of about 17 °C in the real temperature to the read temperature. So, in the future experiment, we can simply read the temperature from the thermocouple and, subtracting the difference read from the two curves, we can obtain the accurate temperature of the electrode surface.

To demonstrate the effect of temperature on the reaction of the electrocatalytic system, we used the oxidation of formic acid as a model system by monitoring the CV, and the result is shown in Fig. 3. It can be seen from the Figure that the reaction

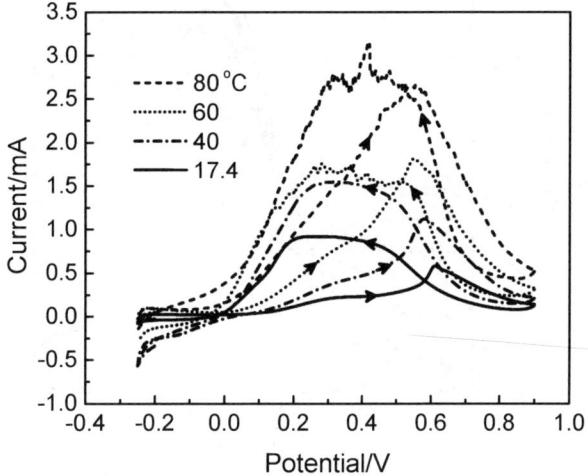

Fig. 3 Cyclic voltammograms of a Pt electrode at elevated temperatures as indicated in the Figure in the solution of 0.1 M HCOOH + 0.5 M H$_2$SO$_4$. The scan rate is 50 mV s^{-1}.

current increases progressively with the increasing temperature both in the positive and negative potential scan. The oxidation potential was also obviously negatively moved, indicating the oxidation reaction can take place more easily on the Pt surface. Furthermore, at temperatures higher than 60 °C, due to the vigorous oxidation of formic acid on the Pt surface, some noise on the CV curve can be found, which may be due to the formation of bubbles on the surface during the oxidation process. Another major difference is that the shape of the curve also changes quite significantly with the temperature, which may indicate a difference in the reaction process, especially at the lower potentials. The result demonstrates that we can easily control the temperature of the system by integrating the heating unit in the spectroelectrochemical cell body. It also demonstrates that it is necessary to control the reaction temperature in order to study the oxidation mechanism that may change with the temperature.

Demonstration of flow efficiency

In an electrocatalytic system, such as the oxidation of formic acid, the formic acid is rapidly oxidized on the Pt surface and produces CO_2 and other intermediate species or by-products. The existence of products on the surface may alter the condition on the Pt electrode surface and interfere the analysis of the reaction process or even lead to the complication of the reaction mechanism. Therefore, it is necessary to use a flow cell, especially to simplify the analysis of reaction mechanism.

A very important parameter for characterizing a flow cell is the flow efficiency, which can be reflected by monitoring the reaction current from the CV curves.[23] To simplify the analysis, we performed a flow rate dependent cyclic voltammetric study of $K_4Fe(CN)_6$/$K_3Fe(CN)_6$ redox system, and the result is shown in Fig. 4. The currents appearing at the positive potentials and negative potentials correspond to the oxidation and reduction of $K_4Fe(CN)_6$ and $K_3Fe(CN)_6$, respectively. It can be seen from the figure that the current increases progressively with the increasing flow rate. When the flow rate is slower than 45 μL s^{-1}, we can still observe the current

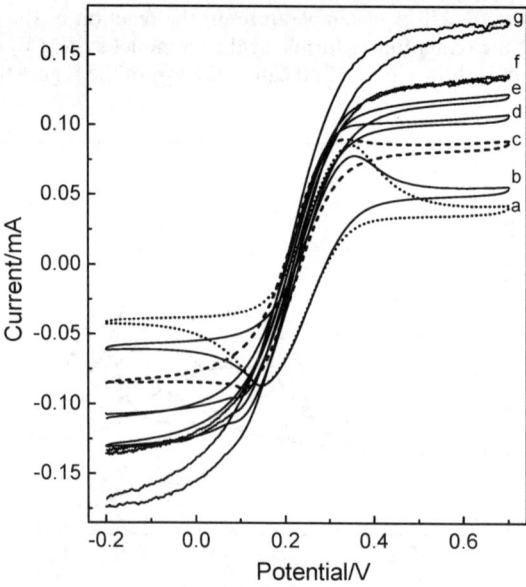

Fig. 4 Cyclic voltammograms of a Pt electrode obtained in a flow cell at different flow rates: (a) 17 μL s^{-1}; (b) 45 μL s^{-1}; (c) 127 μL s^{-1}; (d) 183 μL s^{-1}; (e) 274 μL s^{-1}; (f) 650 μL s^{-1}; (g) 945 μL s^{-1}. The solution is 1 mM $K_4Fe(CN)_6$ + 1 mM $K_3Fe(CN)_6$ + 0.1 M KCl. The scan rate is 100 mV s^{-1}.

peak indicating the diffusion controlled process. However, when the flow rate is higher than 127 µL s^{-1}, no peak was observed. Instead, we only see plateaus on both potential extremities. Further increase of the flow rate does not change the curve form too much, but a progressive increase in the plateau current indicated a forced convection process. The phenomenon is similar to the case of a rotating-disk electrode configuration. As has been stated above, the volume of the reaction chamber of the cell is only about 30 µL, we would expect the solution can be totally replaced at the flow rate of 45 µL s^{-1}. However, from Fig. 4 we find that a complete exchange of the solution might occur at a flow rate between 45 µL s^{-1} and 127 µL s^{-1}. This value is in accordance with the 4.4 ml s^{-1}, *i.e.* 73 µL s^{-1} reported by Weaver's group.[23] It demonstrates that we can easily achieve the effective solution exchange at the temperature controlled cell by simply changing the relative height of the source solution to the exhaust solution or the pressure in the source solution chamber. When the flow rate is controlled at a condition similar to the working flow of fuel cell systems at the desired temperature, we can mimic the reaction process of an electrocatalytic system.

Borrowing SERS strategy to enhance the detection sensitivity

From above, we have demonstrated that we can conveniently control the temperature and flow rate of the desired system by designing a proper spectroelectrochemical flow cell with integrated heating unit. Then to use the cell for providing molecular information using SERS, it is necessary to have a substrate that can provide sufficiently high detection sensitivity.

About 12 years ago, our group have demonstrated that we were able to obtain SERS from an electrochemically roughened Pt surface,[18] which make it possible to obtain SERS signal of some molecules with very weak SERS activities. We have been able to obtain SERS signals of oxidation and adsorption of the dissociative product (carbon monoxide) of methanol on Pt. However, we have not been able to obtain SERS signal of intermediate species due to the lack of sufficient sensitivity.

Recently, a borrowing SERS strategy has been proposed, which is to coat a thin layer of transition metals (such as Pt, Pd, Ni, Co) on a highly SERS-active Au nanoparticle core surface, forming a core–shell structure.[14,28–33] This kind of nanoparticles have the chemical properties of the shell materials (such as Pt) and can still benefit from the SERS enhancement of the Au core due to the long-range effect of the electromagnetic enhancement. By using this method, we have been able to obtain the very strong signals of pyridine, thiocyanide and carbon monoxide from such nanoparticle surfaces. It provides an enhancement about 2 orders of magnitude higher than that of pure transition-metal surfaces. The core–shell structure shows particular advantages in investigating the system with very weak signals.

The Au core Pt shell nanoparticles (denoted as Au@Pt) were prepared by the following method: first, Au nanoparticles with a diameter of 55 nm were synthesized following the Frens method, and the obtained nanoparticles were used as the core or seed. Then, to 30 ml solution containing the Au core, different amounts (to obtain different shell thickness) of 1 mM H_2PtCl_6 was added, and the mixture was heated up to 80 °C. Then, half the volume of 10 mM ascorbic acid to that of H_2PtCl_6 was slowly dropped into above mixtures through a syringe controlled by a step motor while stirring. The mixtures were then stirred for another 30 min to ensure a complete reduction of H_2PtCl_6. The color of the mixtures turned from red brown to dark brown, indicating formation of products. The core–shell structure and the control of the thickness has been well-documented in our previous paper.[14]

The Au@Pt nanoparticles sol was then centrifuged for three times to remove excess reactants and to obtain a clean surface. Then, 25 µL aliquot of the remaining sol was cast on a mechanically polished and electrochemically cleaned smooth Pt electrode, which was dried in a desiccator for 30 min. Such an electrode is ready for electrochemical or SERS measurements. Fig. 5(a) shows the SEM image of

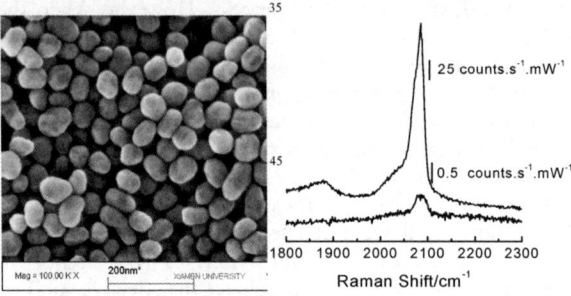

Fig. 5 (Left) SEM image of Au@Pt nanoparticles assembled on a Pt electrode(Au@Pt/Pt). (Right) SERS spectra of CO adsorbed on a roughened pure Pt electrode (bottom) and a Au@Pt/Pt electrode surface (top).

55 nm Au@0.7 nm Pt nanoparticles dispersed as a thin film on Pt (denoted as 55 nm Au@0.7 nm Pt/Pt), which shows a very uniform surface assembled with nanoparticles of very narrow size distribution. The surface shows a very uniform light golden color under daylight visible to naked eyes.

It should be noted that the Au@Pt nanoparticles shows very good SERS activity over a pure Pt electrode. To examine the SERS activity of the Au@Pt/Pt electrode, we carried out a comparative experiment on the electrochemically roughened Pt electrode and 55 nm Au@0.7 nm Pt/Pt using CO as the probe molecule. The result is shown in Fig. 5(b). The band at about 2086 cm^{-1} is due to the on-top adsorbed CO and that at about 1880 cm^{-1} is from the bridge-bound CO, as well recognized according to the results obtained in a previous SERS study[34] and IR spectroscopy.[35] By comparing the spectral intensity, we further found that the SERS signal from the 55 nm Au@0.7 nm Pt/Pt is much more intense, about 50 fold, than the pure massive Pt electrode and the bridge-bound CO can be clearly identified. The absolute height of the band is meaningless as the data acquisition time and power is different. In the former, a shorter time and a lower laser power was used, but the intensity is much higher than the latter. This result convincingly demonstrates that by using a core–shell strategy, we could dramatically enhance the detection sensitivity of the electrocatalytic system.

Investigation of the electrocatalytic system—CO and formic acid oxidation on Au@Pt/Pt electrode surfaces

Since formic acid is attracting increasing interest in using as a fuel, we use the electrooxidation of formic acid on the Au@Pt/Pt surface to demonstrate the function of our cell. Fig. 6 shows the SERS spectra obtained on the Au@Pt nanoparticles surface when the electrode potential was changed from negative to positive values at 40 °C. Similar to case of the CO system, we observed two bands at 2057 and 495 cm^{-1}, which can be routinely assigned to CO and Pt–C stretch of the on-top CO.[13,34] Compared with the case of CO, we find the frequency of the CO band is lower than the pure CO case, and the Pt–C band is higher. This difference can be explained by a lower CO coverage on the surface, as there may exist steric resistance during the dissociative adsorption, *i.e.*, the dissociation of formic acid to CO needs more than one surface Pt atom.[21,35,36] The intensity increases with the positive movement of the electrode potential up to 0.1 V, which can be understood by the restructuring of the surface CO. Further positive movement of the electrode potential leads to the oxidation of CO, whereas a significant increase of the SERS signal in the frequency region of 1100 to 1650 cm^{-1}.

Our control experiment in the 0.5 M H_2SO_4 solution without formic acid only shows a very weak SERS signal in this region, most possibly from the very trace

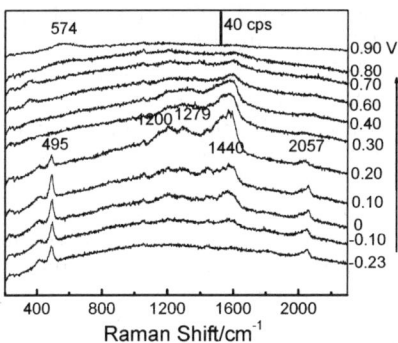

Fig. 6 SERS spectra obtained on a Au@Pt/Pt electrode in a solution containing 0.1 M HCOOH and 0.5 M H_2SO_4 at 40 °C. The acquisition time is 5 s.

amount of citrate left from the synthesizing process. It should be especially noted that the Au@Pt nanoparticles have been cleaned by centrifugation three times and water rinsing before assembly. Furthermore, after the nanoparticles were assembled on the surface, the electrode was further kept at a negative potential of −1.0 V to repel the contaminants from the surface and then the solution was changed to a fresh one. These efforts are to eliminate as much as possible the interference of contaminants. The comparative experiment may lead to a conclusion that the signal observed in frequency region of 1100 to 1650 cm^{-1} may have a combined contribution of intermediate species during the oxidation of formic acid and the trace amount of citrate. At present, it is still difficult to have a good assignment of the observed SERS signal in this region. The spectra were obtained with an integration time of 5 s and at a shorter intergration time, the spectra show a fluctuation character. A combination of these fluctuation feature leads to several broad bands in this region. Apparently, we need more experimental data to reach a statistic conclusion of the intermediate species.

It should also be pointed out that we have no evidence to exclude the possibility that the minor amount of impurities from formic acid may also be adsorbed on the surface to contribute to the signal, which also needs further investigation.

Conclusions and future development

A spectroelectrochemical cell with integrated functions of flow and temperature control was designed and fabricated. The electrochemical cyclic voltammetric results indicates that the flow rate can be effectively controlled from 0 to 945 μl s^{-1}, and the temperature can be controlled from room temperature up to 80 °C. The temperature of the electrode was calibrated by using the intensity ratio of Stokes to the anti-Stokes Raman bands, which is slightly lower than the temperature read from the thermal couple placed underneath the working electrode. By synthesizing Au core Pt shell nanoparticles, the SERS activity of Au can be borrowed to enhance the SERS activity of the Pt shell, so that the SERS signal of surface species can be significantly enhanced. It allows us to obtain an intense SERS signal of CO at the Au@Pt/Pt electrode surface in comparison with the pure Pt electrode surface. The cell was further used to study the electrooxidation of formic acid and the dissociative adsorption product, CO, of formic acid was detected. The control experiment demonstrates that the signal obtained in the region from 1100 to 1650 cm^{-1} may have a contribution from the intermediate species.

To further develop the Raman flow cell with variable temperature into a general technique for electrochemistry, there is still space to further improve the design and the substrate considering the following points.

First, the reaction chamber can be further minimized, so that the solution exchange can be more efficient, and the detection sensitivity of Raman measurement can be further improved. This can be done by thinning the inlet and outlet of the cell and by designing a special type of counter electrode.

Second, by using a special type of pump, so that the flow control can be integrated into the cell body. Thereafter, it is not necessary to have a flow pipe, which may increase the convenience of the experiment.

Third, since the nanoparticles were synthesized by chemical reduction method using citrate, it is inevitable that the surface will be covered with some organic contaminants. Although we have been using centrifugation or a hydrogen bubbling method to clean or desorb the contaminants, we still can not absolutely get rid of the contaminants at the present stage. It will be important to find a method to obtain contaminant free highly SERS-active substrates.

Four, in the present study, the Au core was completely covered by a Pt shell. It has been reported that small Au clusters may be good catalysts for the oxidation of CO. Therefore, if we reduce the amount of Pt so that submonolayer of Pt was formed on the Au surface, then the partial exposed Au substrate with the Pt surrounding may be good electrocatalysts. In addition, when the intermediate species are produced on the Pt surface, the intermediate species can diffuse to the Au surface with the highest enhancement, hopefully allowing detection the intermediate species that are very important for understanding the electrooxidation mechanisms.

Acknowledgements

The authors acknowledge the financial support by Natural Science Foundation of China and the Ministry of Science and Technology of China (20433040 and 20673086), the Ministry of Education of China(NCET-05-0564) and Fok Ying Tung Foundation (101015).

References

1 R. Parsons and T. Van der Noot, *J. Electroanal. Chem.*, 1988, **257**, 7.
2 S. Wasmus and A. Kuver, *J. Electroanal. Chem.*, 1999, **461**, 14.
3 B. Beden, C. Lamy and J.-M. Leger, in *Modern Aspects of Electrochemistry*, ed. J. O'M. Bockris, B. E. Conway and R. E. White, Plenum, New York, 1992, 22, p. 97.
4 K. A. Friedrich, K.-P. Geyzers, U. Linke, U. Stimming and J. Stumper, *J. Electroanal. Chem.*, 1996, **402**, 123.
5 A. Hamnett, in *Interfacial Electrochemistry*, ed. A. Wieckowski, Marcel Dekker, New York, 1999, 843.
6 Y. Y. Tong, A. Wieckowski and E. Oldfield, *J. Phys. Chem. B*, 2002, **106**, 2434.
7 H. Wang and H. Baltruschat, *J. Phys. Chem. C*, 2007, **111**, 7038.
8 J. M. Leger, S. Rousseau, C. Coutanceau, F. Hahn and C. Lamy, *Electrochim. Acta*, 2005, **50**, 5118.
9 Y. X. Chen, S. Ye, M. Heinen, Z. Jusys, M. Osawa and R. J. Behm, *J. Phys. Chem. B*, 2006, **110**, 9534.
10 D. Kardash and C. Korzeniewski, *Langmuir*, 2000, **16**, 8419.
11 N. Tian, Z. Y. Zhou, S. G. Sun, Y. Ding and Z. L. Wang, *Science*, 2007, **316**, 732.
12 S. Park, A. Wieckowski and M. J. Weaver, *J. Am. Chem. Soc.*, 2003, **125**, 2282–2290.
13 Z. Q. Tian, B. Ren and D. Y. Wu, *J. Phys. Chem. B*, 2002, **106**, 9463.
14 Z. Q. Tian, B. Ren, J. F. Li and Z. L. Yang, *Chem. Commun.*, 2007, 3514.
15 M. J. Weaver, S. Z. Zou and H. Y. H. Chan, *Anal. Chem.*, 2000, **72**, 38A.
16 R. L. Birke, T. Lu and J. R. Lombardi, in *Techniques for Characterization of Electrodes and Electrochemical Processes*, ed. R. Varma and J. R. Selman, John Wiley, New York, 1991, 211.
17 B. Pettinger, in *Adsorption at Electrode Surface*, ed. J. Lipkowski and P. N. Ross, VCH, New York, 1992, 285.
18 Z. Q. Tian, B. Ren and B. W. Mao, *J. Phys. Chem. B*, 1997, **101**, 1338.
19 B. Ren, X. F. Lin, J. W. Yan, B. W. Mao and Z. Q. Tian, *J. Phys. Chem. B*, 2003, **107**, 899.
20 Z. Liu, Z. L. Yang, L. Cui, B. Ren and Z. Q. Tian, *J. Phys. Chem. C*, 2007, **111**, 1770–1775.

21 C. X. She, J. Xiang, B. Ren, Q. L. Zhong, X. C. Wang and Z. Q. Tian, *J. Korean Electrochem. Soc.*, 2002, **5**, 221.
22 S. Park, Y. Xie and M. J. Weaver, *Langmuir*, 2002, **18**, 5792.
23 H. Luo and M. J. Weaver, *J. Electroanal. Chem.*, 2001, **501**, 141.
24 S. K. Kamarudin, W. R. W. Daud, S. L. Ho and U. A. Hasran, *J. Power Sources*, 2007, **163**, 743.
25 B. Ren, X. F. Lin, Y. X. Jiang, P. G. Cao, Y. Xie, Q. J. Huang and Z. Q. Tian, *Appl. Spectrosc.*, 2003, **57**, 419.
26 B. Ren, L. Cui, X. F. Lin and Z. Q. Tian, *Chem. Phys. Lett.*, 2003, **376**, 130.
27 G. I. Pangilinan and Y. M. Gupta, *Appl. Phys. Lett.*, 1997, **70**, 967.
28 Y. X. Jiang, J. F. Li, D. Y. Wu, Z. L. Yang, B. Ren, J. W. Hu, Y. L. Chow and Z. Q. Tian, *Chem. Commun.*, 2007, 4608.
29 B. Zhang, J. F. Li, Q. L. Zhong, B. Ren, Z. Q. Tian and S. Z. Zou, *Langmuir*, 2005, **21**, 7449.
30 J. F. Li, Z. L. Yang, B. Ren, G. K. Liu, P. P. Fang, Y. X. Jiang, D. Y. Wu and Z. Q. Tian, *Langmuir*, 2006, **22**, 10372.
31 J. W. Hu, Y. Zhang, J. F. Li, Z. Liu, B. Ren, S. G. Sun, Z. Q. Tian and T. Lian, *Chem. Phys. Lett.*, 2005, **408**, 354.
32 J. W. Hu, J. F. Li, B. Ren, D. Y. Wu, S. G. Sun and Z. Q. Tian, *J. Phys. Chem. C*, 2007, **111**, 1105.
33 S. Kumar and S. Z. Zou, *Langmuir*, 2007, **23**, 7365.
34 B. Ren, X. Q. Li, C. X. She, D. Y. Wu and Z. Q. Tian, *Electrochim. Acta*, 2000, **46**, 193.
35 Y. Y. Yang, S. G. Sun, Y. J. Gu, Z. Y. Zhou and C. H. Zhen, *Electrochim. Acta*, 2001, **46**, 4339.
36 Q. L. Zhong, P. Huang, B. Zhang, X. Y. Yang, Y. M. Ding, H. H. Zhou, B. Ren and Z. Q. Tian, *Acta Phys. Chim. Sin.*, 2006, **22**, 291.

PAPER

Mesoscopic mass transport effects in electrocatalytic processes†

Y. E. Seidel,[a] A. Schneider,[a] Z. Jusys,[a] B. Wickman,[b] B. Kasemo[b] and R. J. Behm[*a]

Received 15th April 2008, Accepted 14th May 2008
First published as an Advance Article on the web 8th August 2008
DOI: 10.1039/b806437g

The role of mesoscopic mass transport and re-adsorption effects in electrocatalytic reactions was investigated using the oxygen reduction reaction (ORR) as an example. The electrochemical measurements were performed on structurally well-defined nanostructured model electrodes under controlled transport conditions in a thin-layer flow cell. The electrodes consist of arrays of Pt ultra-microelectrodes (nanodisks) of defined size (diameter $\sim$100 nm) separated on a planar glassy carbon (GC) substrate, which were fabricated employing hole-mask colloidal lithography (HCL). The measurements reveal a distinct variation in the ORR selectivity with Pt nanodisk density and with increasing electrolyte flow, showing a pronounced increase of the H_2O_2 yield, by up to 65%, when increasing the flow rate from 1 to 30 $\mu L\ s^{-1}$. These results are compared with previous findings and discussed in terms of a reaction model proposed recently (A. Schneider *et al.*, *Phys. Chem. Chem. Phys.*, 2008, **10**, 1931), which includes (i) direct reduction to H_2O on the Pt surface and (ii) additional H_2O_2 formation and desorption on both Pt and carbon surfaces and subsequent partial re-adsorption and further reduction of the H_2O_2 molecules on the Pt surface. The potential of model studies on structurally defined catalyst surfaces and under well-defined mass transport conditions in combination with simulations for the description of electrocatalytic reactions is discussed.

1 Introduction

Electrocatalytic reactions have been investigated in much detail over many decades.[1,2] In recent years, the interest in various electrocatalytic reactions has suddenly increased because of their potential applications in fuel cell technology.[3–14] The kinetics and mechanism of these reactions have been investigated by numerous techniques, including purely electrochemical methods as well as combined electrochemical and *in-situ* spectroscopic techniques ('hyphenated techniques') such as various forms of *in-situ* IR spectroscopy,[11,15–23] mass spectrometric techniques such as on-line differential electrochemical mass spectrometry (DEMS)[11,14,18,23–27] or, most recently, X-ray absorption spectroscopy[28–33] and X-ray diffraction.[34–37] These measurements, together with increasingly sophisticated theoretical studies,[38–44] have resulted in a wealth of mechanistic information on the individual reaction steps and

---

[a] *Institute of Surface Chemistry and Catalysis, Ulm University, Ulm, D-89069, Germany. E-mail: juergen.behm@uni-ulm.de*
[b] *Department of Applied Physics, Chalmers University of Technology, Gothenburg, SE-41296, Sweden*

† Electronic supplementary information (ESI) available: Table S1, collection efficiencies. See DOI: 10.1039/b806437g

in an increasingly better understanding of the molecular processes contributing to them.[19]

Despite this insight and the increasingly better agreement between theory and experiment, these model studies mostly miss one important point essential for a realistic description of the ongoing reaction, namely the transport steps required for transport of reactants to the electrodes, or of products away from the electrodes. In experimental model studies, electrolyte transport is enforced, *e.g.*, in different types of flow cells, such as channel flow cells or other types of flow cells,[45–47] which are employed for electrochemical measurements[48–56] or for *in-situ* spectroscopic studies (DEMS,[57] IR[58]), or in measurements using other hydrodynamic methods. The latter include, *e.g.*, measurements using the rotating (ring) disk electrode[5,10,59,60] or wall-jet measurements.[61–65]

It is well accepted that transport of reactants to the electrode may affect the measured reaction rate by putting an upper limit on the rate, which results in the formation of a transport limited current (or 'limiting current').[45–48,66] It is less realized, however, that transport effects may also affect the reaction pathways and their contributions to the total reaction, *i.e.*, the selectivity of the reaction. First of all, a limited removal of reaction products may result in an accumulation of product species in front of the electrode surface and subsequently in back-reactions of the products or in blocking of the surface by adsorbed product species. Second, in reactions producing soluble, reactive reaction intermediates or reaction side products, the removal of these species from surface near regions may affect the reaction and, in particular, the product distribution by reducing the probability for re-adsorption and further reaction of these species (re-adsorption effects). Simple examples for such cases are the oxidation of methanol or formaldehyde, which may result in the incomplete oxidation products formaldehyde and formic acid or in the complete oxidation product CO_2 for methanol oxidation, or in formic acid and CO_2 for formaldehyde oxidation (see ref. 27 and references therein). Recently, we have shown that for methanol oxidation on a thin-film carbon supported Pt catalyst electrode and under enforced electrolyte flow, the product distribution changes significantly upon varying the catalyst loading, at otherwise constant reaction conditions, from predominantly formaldehyde production at low catalyst loadings to increasingly higher yields of CO_2 (and hence lower formaldehyde yields) at higher catalyst loadings.[67] The highest formaldehyde yield was obtained for a massive Pt electrode.[68] This was explained by a concept involving desorption of the incomplete oxidation products formaldehyde and formic acid and their subsequent re-adsorption and further oxidation to formic acid (formaldehyde) and CO_2 (formaldehyde and formic acid) ('desorption–re-adsorption–reaction' concept). With increasing surface roughness or with increasing thickness of the catalyst layer, the probability for re-adsorption and hence for further oxidation of these reactive intermediates increases, and therefore we expect also the fraction of complete oxidation products to increase, as observed experimentally. Similar trends were also observed later for formaldehyde oxidation[27,69] for acetaldehyde oxidation,[70] and for ethanol oxidation,[71] where the yield of incomplete oxidation products formic acid or acetaldehyde/acetic acid was higher on massive Pt electrodes than on thin-film Pt/C catalyst electrodes under otherwise similar reaction conditions.

Following the above 'desorption–re-adsorption–reaction' concept, one would expect a decrease of the complete oxidation product yield also for an increasing electrolyte flow rate, because of the increasingly thinner diffusion layer, which makes it increasingly more probable that a desorbing reaction intermediate will leave the diffusion layer (or 'boundary layer') and is transported away rather than being re-adsorbed again. Such a trend was indeed also observed experimentally for ethanol oxidation on a thin-film Pt/C electrode.[71]

Furthermore, these studies also revealed that the reactant concentration may affect the product distribution.[68,71] In this case, the explanation is less straightforward, since the probability for re-adsorption and further reaction of reaction intermediates is not altered. An increasing reactant concentration, however, will increase the reactant arrival rate at the electrode surface. If the reaction rate of adsorbed reactants or intermediates is not too high, an increasing adsorption rate will most simply result in an increasing steady-state coverage of these (re-)adsorbed reaction intermediates, which in turn may have consequences on the reaction characteristics and therefore also on the product distribution. For instance, the increasing CO_2 yield in potentiodynamic measurements of the ethanol oxidation reaction on a thin-film Pt/C electrode[71] was explained by an almost constant contribution of CO_2 formation *via* formation (at low potentials) and oxidation (at higher potentials) of CO_{ad}, while the formation of incomplete oxidation products acetaldehyde and acetic acid decreased significantly with decreasing ethanol concentration.[71] Effects induced by an adsorption induced change in steady-state adlayer composition and coverage have to be considered of course also when varying the electrolyte flow rate, which in addition to the modified re-adsorption probabilities will also change the reactant adsorption rate.

While we can understand these trends on a qualitative scale, a more quantitative understanding is still missing. It requires new experimental and theoretical approaches, which include not only the transport from the (flowing) bulk electrolyte to the electrode and *vice versa*, but also the desorption of incomplete oxidation/reduction products, their lateral diffusion in the diffusion layer, and the re-adsorption processes discussed above. This is illustrated schematically in Fig. 1a. Recently, we have introduced nanostructured, glassy carbon (GC) supported Pt model electrodes (Pt/GC)[69,72–74] for studies of transport processes in electrocatalytic reactions. These nanostructured electrodes, which consist of Pt nanodisks of fixed, defined diameter between 70 and 150 nm (height $\sim$20 nm) assembled in adlayers with a rather narrow distribution of separations between the Pt nanodisks, are produced *via* colloidal lithography (CL)[72,73,75–78] or, more recently, *via* hole-mask colloidal lithography (HCL).[79,80] Because of their well-defined and rather regular structures,

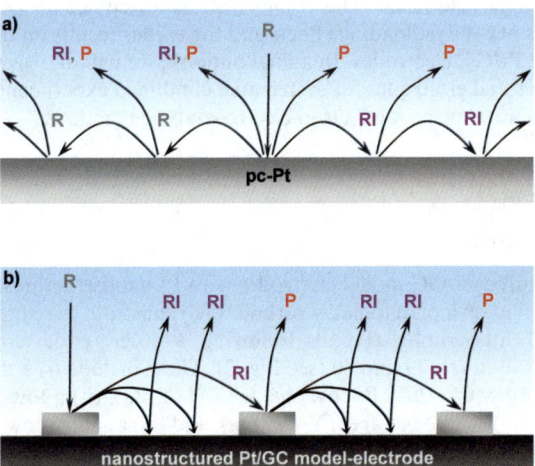

Fig. 1 Schematic description of the desorption, diffusion, re-adsorption and reaction processes contributing to an electrocatalytic reaction which includes reaction pathways where volatile, reactive reaction intermediates are formed ('desorption–re-adsorption–reaction' process) (a) on a pc-Pt electrode and (b) on a nanostructured Pt/GC model electrode. R: reactant molecule arriving at the electrode surface, right sequence: multiple desorption, re-adsorption and further reaction of the reaction intermediate RI to product P before its final off-transport through the diffusion layer into the flowing electrolyte; left sequence: desorption, multiple re-adsorption and further reaction of reactant R to reaction intermediate RI before its final off-transport through the diffusion layer into the flowing electrolyte.

these electrodes are ideally suited for model studies on lateral transport effects, e.g., by investigating the change in reaction characteristics upon varying the density and/or size of the nanodisks (compare Fig. 1b) or the electrolyte flow rate. From the same structural reasons, they are particularly attractive as (experimental) model systems for quantitative comparison with simulations of the contributing transport and reaction processes. The latter would be very close to the simulations of diffusion processes to arrays of ultra-microelectrodes in a flow cell situation reported previously (see e.g., ref. 81–84), complemented by lateral transport processes between the microelectrodes and by the surface reactions.

In the present contribution, we illustrate and discuss mass transport effects in model studies of the oxygen reduction reaction (ORR) on nanostructured Pt/GC model electrodes, focusing on changes in the reaction characteristics (activity, selectivity) due to variations in the Pt nanodisk density and/or in the electrolyte flow rate. The nanostructured electrodes were prepared by hole-mask colloidal lithography (HCL) rather than by colloidal lithography (CL) in order to avoid the formation and presence of additional undesired Pt nanoparticles of 3–5 nm diameter on the GC substrate in the areas between the Pt nanodisks, which were shown recently to be formed during CL processing and which may have significant impact on the reaction characteristics.[80] First, measurements of the ORR performed on CL-prepared electrodes have shown that on these model electrodes significant amounts of H_2O_2 are formed at potentials where the carbon substrate is non-reactive, up to 0.8 V_{RHE}, and that the H_2O_2 yield is affected by the density of the nanodisks.[74] These findings were examined by the formation of H_2O_2 on the Pt nanodisks, its desorption, re-adsorption and further reduction of H_2O_2 on the other Pt nanodisks, which leads to a lower H_2O_2 yield with increasing density of Pt nanodisks.[74] Clear trends, however, were hard to evaluate, most likely due to effects introduced by the (undesired) Pt nanoparticles. In the present paper, we report on the systematic expansion and continuation of these preliminary measurements, employing improved, HCL-prepared nanostructured electrodes, together with a polycrystalline Pt and a Pt-free GC electrode as reference, to investigate effects of varying the Pt nanodisk density and the electrolyte flow on the ORR reaction characteristics. The results will be discussed in comparison with previous results of catalyst loading effects and the earlier results on the CL-prepared nanostructured Pt/GC electrodes. In a final outlook, we will discuss the potential of these nanostructured electrodes for systematic, combined experimental and theoretical studies of mass transport effects in electrocatalytic reactions.

2 Experimental

Electrode preparation

The nanostructured Pt/GC model electrodes with Pt nanostructures of ~100 nm in diameter deposited on a planar glassy carbon (GC) substrate were prepared via hole-mask colloidal lithography (HCL), following a recently developed procedure described in detail in ref. 79 and 80 (see Fig. 2). These include (i) a medium-loading sample (HCL-20) with ~20% Pt coverage (∅ ~ 118 nm), (ii) a low-loading sample (HCL-10) with ~10% Pt coverage (∅ ~ 95 nm), and (iii) an ultra-low loading sample (HCL-01) with ~1% Pt coverage (∅ ~ 95 nm). All experiments were performed on three different samples of each type to ensure reproducibility. As references, we also included a polycrystalline Pt sample (pc-Pt) and a polished GC substrate.

Prior to the fabrication of the Pt nanodisks, the GC substrates (∅ 9 mm, Sigradur Hochtemperatur Werkstoffe GmbH) were pre-treated as described previously,[73] including polishing with alumina slurry down to 0.3 μm grid, cleaning by immersion in concentrated H_2SO_4 and 5 M KOH, multiple rinsing in Millipore Milli-Q water, and sonicating in Millipore Milli-Q water. Subsequently, they were dried in a N_2 stream, and finally treated in an oxygen plasma (50 W, 250 mtorr, 2 min). The Pt nanostructures were fabricated along the hole-mask colloidal lithography procedure

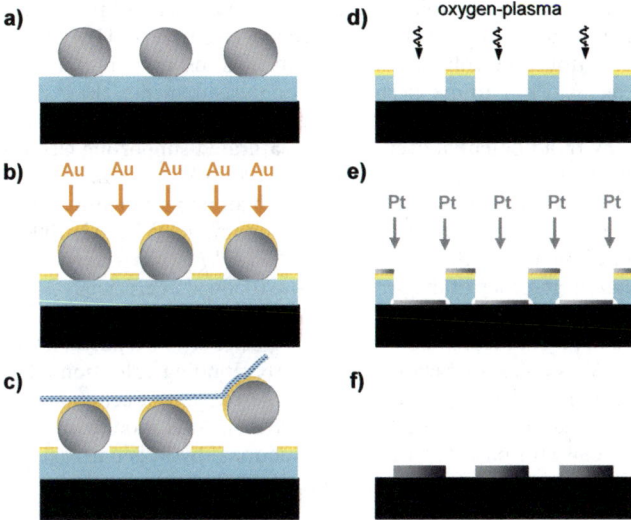

Fig. 2 Schematic presentation of the hole-mask colloidal lithography (HCL) fabrication procedure for the preparation of the nanostructured model electrodes. It starts with a polished GC substrate on which a sacrificial resist PMMA layer is deposited by spin-coating and masked with PS-particles. In the next step, an Au hole-mask is evaporated on top. The PS-particles are removed by tape-stripping resulting in the hole-mask. Afterwards, oxygen plasma treatment is used for assigning the holes into the PMMA layer, followed by the Pt material evaporation. In the last step the hole-mask is lifted-off in acetone.

schematically illustrated in Fig. 2. A sacrificial resist PMMA (polymethylmethacrylate) layer and, on top of this, masking PS beads were deposited on the pre-treated GC substrate (Fig. 2a). This was done by spin coating with PMMA and subsequent treatment in an oxygen plasma (5 s) to increase the hydrophilicity of the surface, followed by pipetting a thin polyelectrolyte layer on the PMMA surface, and finally the charged PS beads were deposited. Afterwards, a plasma resistant, 20 nm thick Au film was evaporated on top (Fig. 2b), and the PS beads were removed by tape-stripping (Fig. 2c). This results in a masking Au film with well-defined holes (Au hole-mask) on the PMMA layer. During a subsequent oxygen plasma etching step, the Au covered areas are protected, while the unprotected, bare PMMA film in the holes of the Au mask is removed (Fig. 2d). In the next step, Pt was deposited through the mask onto the GC surface (Fig. 2e). After removal of the hole-mask by a lift-off in acetone, the Pt islands located in the former holes of the hole-mask stick to the surface, while the remaining part of the Pt film is removed together with the PMMA/Au film. This results in an array of Pt nanostructures with well-defined size and lateral distribution (Fig. 2f). The morphology of the nanostructured model electrodes was examined by SEM using a LEO 1550 (Zeiss) instrument (10 kV operation energy).

Electrochemical measurements

The experiments were performed in a dual thin-layer flow cell,[54,85] which is based on the flow cell design described in ref. 57. It allows for further reaction of the electroactive reaction products generated at the working electrode (generator) on a second electrode (collector), and thus determination of the amount of oxidizable (reducible) product species, similar to a rotating ring disk electrode (RRDE) set-up, but at a higher collection efficiency.[54]

Two Pt counter electrodes were connected to the cell *via* separate ports at the inlet and outlet of the cell. The reference electrode (saturated calomel electrode, SCE) was connected through a Teflon capillary at the outlet of the cell. All potentials, however, are quoted *vs.* that of the reversible hydrogen electrode (RHE). The potentials of both generator and collector were controlled using a bi-potentiostat from Pine Instruments. Prior to the electrocatalytic measurements, the generator (polycrystalline Pt or nanostructured model electrodes) and the collector electrodes were pre-cleaned by rapid potential cycling in Ar-saturated supporting electrolyte in the potential range 0.06 to 1.36 V (scan rate 100 mV s^{-1}). In potentiodynamic measurements, the scan rate was 10 mV s^{-1}, with an upper potential limit of 1.16 V. The electrolyte flow was driven by a syringe pump (model WPL AL-1000) connected to the outlet of the flow cell.

The collection efficiency of the present set-up (defined as efficiency $N = |I_{coll}|/|I_{gen}|$) was determined by hydrogen evolution/oxidation experiments in Ar-saturated electrolyte, employing the sample as generator electrode and a polycrystalline Pt disk biased at 0.3 V as collector electrode. The corresponding collection efficiencies are listed in Table S1 (in the ESI†). The hydrogen peroxide production at the generator electrode in O_2-saturated 0.5 M H_2SO_4 electrolyte was monitored using a polycrystalline Pt collector biased at 1.2 V. The fractional hydrogen peroxide yield $x_{H_2O_2}$ in the ORR, which describes the fraction of H_2O_2 formed relative to the total amount of O_2 consumed, was calculated as

$$x_{H_2O_2} = (2\ I_{coll}/N)/(I_{gen} + I_{coll}/N) \qquad (1)$$

The mass transport-normalized kinetic ORR currents at the working (generator) electrode were determined as

$$I_k = I_{lim}\ I_{gen}/(I_{lim} - I_{gen}) \qquad (2)$$

with I_{lim} denoting the ORR mass transport limited current at 0.2 V. They are presented in Tafel plots as (i) geometric electrode area normalized and (ii) active Pt surface area normalized kinetic ORR current densities, where the latter were calculated *via* the active Pt surface area determined from the underpotential hydrogen deposition (H_{upd}) charge.

The supporting electrolyte (0.5 M sulfuric acid solution) was prepared using ultrapure sulfuric acid (Merck suprapur) and Millipore Milli-Q water. It was dearated by high-purity argon gas (Westfalen Gase, N 6.0) before and during the electrochemical experiments. For O_2 reduction experiments, the base solution was saturated with O_2 (MTI Gase, N 5.7) in the electrolyte supply bottle. All experiments were performed at room temperature.

3 Results and discussion

The morphology of the resulting model electrodes is illustrated by the SEM images in Fig. 3. The large scale images (Fig. 3a, b) demonstrate the homogeneous distribution of the Pt nanodisks over the GC substrate surface for the model electrodes with higher surface coverage. Only for the ultra-low loaded sample depicted in Fig. 3c, the Pt nanodisks are essentially randomly distributed. High resolution images (Figs. 3d–f) show the morphology of the Pt nanodisks and the glassy carbon substrate. They clearly demonstrate the absence of Pt nanoparticles in between the Pt nanostructures.

The electrochemical properties of these electrodes were characterized by base voltammetry and by pre-adsorbed CO_{ad} oxidation ('CO_{ad} stripping'). The cyclic voltammograms obtained in 0.5 M H_2SO_4 solution (Fig. 4) are characteristic for polycrystalline Pt, with distinct signals for hydrogen adsorption/desorption in the H_{upd} range between 0 and 0.35 V, a double-layer region up to ~0.8 V, and the signals for Pt oxidation/Pt oxide reduction at potentials anodic of 0.8/0.6 V. The H_{upd} and the Pt oxidation/reduction signals decrease with decreasing Pt coverage. These

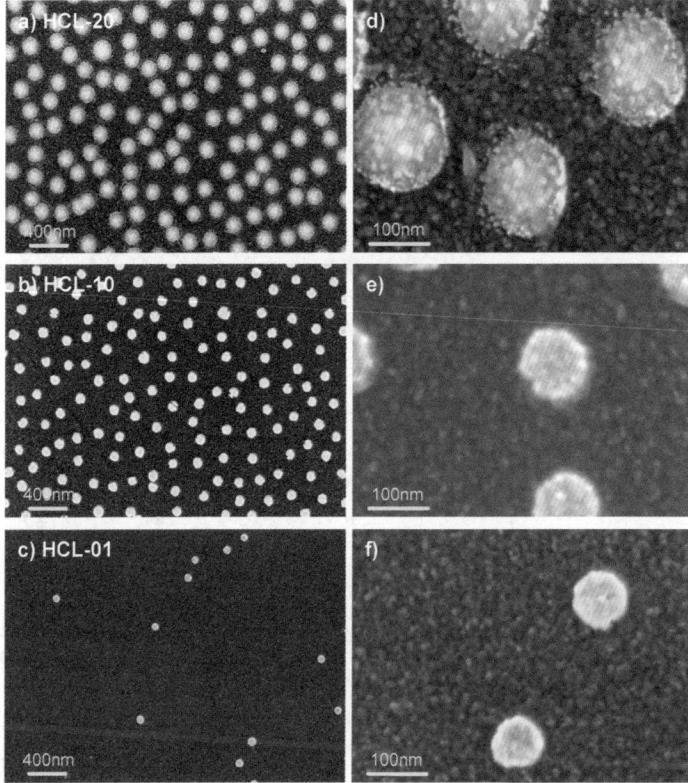

Fig. 3 Representative large-scale (a–c) and high-resolution (d–f) SEM images of the HLC-prepared Pt/GC model-catalysts. (a, d) Medium Pt coverage (~20% Pt, HCL-20), (b, e) low Pt coverage (~10% Pt, HCL-10) and (c, f) ultra low Pt coverage (~1%, HCL-01). Large scale images 2.6 μm × 3.6 μm; high resolution images 420 nm × 580 nm.

results closely resemble previous findings for similar type nanostructured electrodes.[69,80] From the H_{upd} charges, the active Pt surface areas were determined to 0.45 cm^2 (pc-Pt), 0.23 cm^2 (HCL-20), 0.11 cm^2 (HCL-10) and 0.006 cm^2 (HCL-01), respectively, assuming a monolayer hydrogen charge of 0.21 mC cm^{-2} (ref. 86) and a H_{upd} coverage of 0.77 ML at the onset of H_2 evolution.[87] The active surface areas follow exactly the SEM determined geometric surface area of the Pt nanodisks, underlining the identical morphology of the Pt nanodisks and the absence of additional, undesired Pt nanoparticles on the GC areas in between the Pt nanodisks. The latter would lead to a pronounced increase of the active Pt surface area compared to the Pt coverage at low Pt coverages, as was observed in an earlier study on CL-prepared nanostructured Pt/GC model electrodes.[80]

These findings are qualitatively and quantitatively supported by previous CO_{ad} stripping measurements, which equally showed signals characteristic for polycrystalline Pt and yielded active Pt surface areas comparable to those determined by the H_{upd} uptake (see also ref. 80). Furthermore, these data and also energy dispersive X-ray emission (EDX) measurements confirmed that (undesired) Pt nanoparticles between the (desired) Pt nanodisks are essentially absent.[80]

The ORR activity (a, d, g, k) and selectivity (b, e, h, l and c, f, i, m) of the polycrystalline Pt electrode (a–c), the nanostructured Pt/GC samples (d–m), and the Pt-free glassy carbon substrate (k–m) in the positive-going potential scan, measured as the Faradaic current for O_2 reduction (generator current) and as the H_2O_2

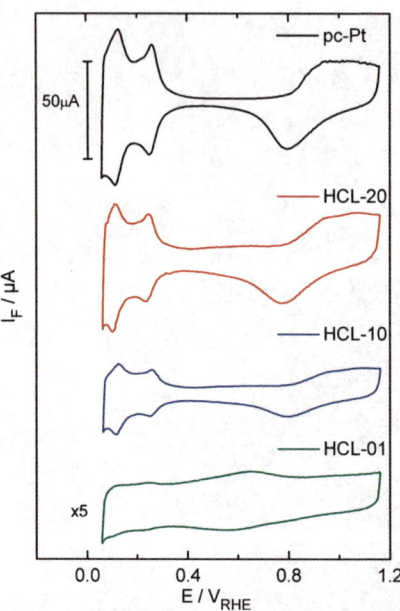

Fig. 4 Base voltammograms (scan rate 100 mV s^{-1}) recorded on a pc-Pt electrode (top line), on the nanostructured HCL-20 (~22% Pt surface coverage), HCL-10 (~10% Pt surface coverage) and the HCL-01 (~1% coverage, bottom line; note, that the currents for that sample were multiplied by a factor of 5) sample.

oxidation current on the collector (collector current), respectively, are plotted in Fig. 5. To more specifically address mass transport effects, a wide range of electrolyte flow rates was employed, ranging to very low values (1 μL s^{-1}). The ORR Faradaic currents on the pc-Pt and on the nanostructured Pt/GC electrodes (Fig. 5a, d, g, k) display the general features characteristic for the ORR on Pt under enforced mass transport, namely, a mass transport limited region at lower potential and a mixed mass transport/kinetically limited region at more anodic potentials, where the Faradaic current decays to zero on the PtO blocked surface at potentials positive of 0.9 V. These findings agree well with earlier data for the ORR on massive Pt electrodes,[88] on carbon supported Pt/C catalyst electrodes,[89] and with the data obtained on the CL-prepared nanostructured Pt/GC electrodes.[74] Comparing the Faradaic currents on the different samples, two important trends can be distinguished. First, on the nanostructured Pt/GC electrodes the mass transport limited current gradually decreases with the decrease in Pt coverage, while those of the pc-Pt sample and of the HCL-20 sample are approximately identical. Furthermore, the potential for transition from the purely mass transport limited regime to the mixed kinetically/mass transport limited regime depends sensitively on the Pt loading and on the electrolyte flow rate, shifting to lower potentials with decreasing Pt coverage and with increasing electrolyte flow.

The clear decrease of the mass transport limited ORR current with decreasing Pt coverage differs distinctly from the observation of a constant mass transport limited current on CL-prepared nanostructured Pt/GC electrodes.[74] The decay of the limiting ORR current can be understood by comparison with the results of previous simulations of the diffusion controlled current on arrays of ultra-microelectrodes on an inert substrate.[81] These simulations clearly identified two limiting situations, one where the individual microelectrodes are far apart and where the transport to each electrode can be described by hemispherical diffusion spheres, and a second one where the electrodes are close enough for an effective overlap of the diffusion

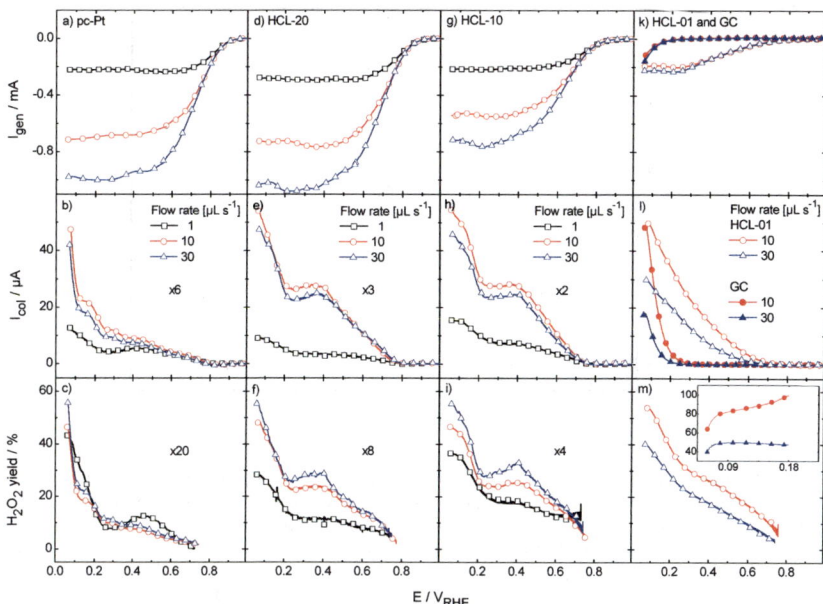

Fig. 5 Oxygen reduction reaction activity (a,d,g,k), hydrogen peroxide oxidation at the smooth polycrystalline Pt collector electrode (geometric area 1.33 cm^2, biased at 1.2 V) (b,e,h,l), and the corresponding hydrogen peroxide yields (c,f,i,m) simultaneously monitored in the double-disk thin-layer flow-cell over a smooth polycrystalline Pt electrode (a–c), on nano-structured Pt/GC model electrodes: (HCL-20 (d–f); HCL-10 (g–i) and HCL-01), and on a blank GC substrate (k–m) at different electrolyte flow rates (1, 10, and 30 µL s^{-1}, for assignments see figure). Solution: 0.5 M H$_2$SO$_4$ (O$_2$-saturated), exposed electrode area 0.3 cm^2, potential scan rate 10 mV s^{-1} (positive-going scan), room temperature.

spheres. In the first case, the total diffusion limited current is identical to the sum of the currents to separated individual microelectrodes, in the latter case it can be described by the diffusion limited current to a planar electrode.[81] Hence, the mass transport limited current starts to decrease only below a critical density of the microelectrodes. On the CL-prepared nanostructured Pt/GC electrodes, the additional Pt nanoparticles present on the GC surface between the Pt nanodisks were apparently sufficient to keep the Pt coverage/density above that critical value, and the mass transport limited current remained constant despite a significant decrease in Pt nanodisk density.[74] On the present, HCL-fabricated nanostructured electrodes, where these Pt nanoparticles are absent, the critical density of Pt nanodisks is passed already for the HCL-10 sample at higher electrolyte flow rates. The decrease of the transport limited current at potentials cathodic of 0.2 V, which is particularly obvious at the higher flow rates on the nanostructured electrodes, is attributed to H$_2$O$_2$ under these conditions, which reduces the Faradaic current accordingly (2-electron pathway for H$_2$O$_2$ instead of the 4-electron pathway for H$_2$O).[74] ORR measurements on a Pt-free glassy carbon substrate, which are included for comparison, show a small ORR activity at potentials negative of 0.2 V (Fig. 5 k, filled symbols).

For flow cell measurements on a non-structured electrode, it is well known that the mass transport to the electrode, and thus the mass transport limited current, decay with the cube root of the electrolyte flow rate v ($I \propto v^{1/3}$).[90–92] Therefore, we plotted the mass transport limited current at 0.2 V, *i.e.*, above the onset of the ORR current decrease due to H$_2$O$_2$ formation, as a function of the cube root of the electrolyte flow rate (Fig. 6, note that each data point is averaged over several measurements). The data clearly lie on straight lines, underlining that this relation is

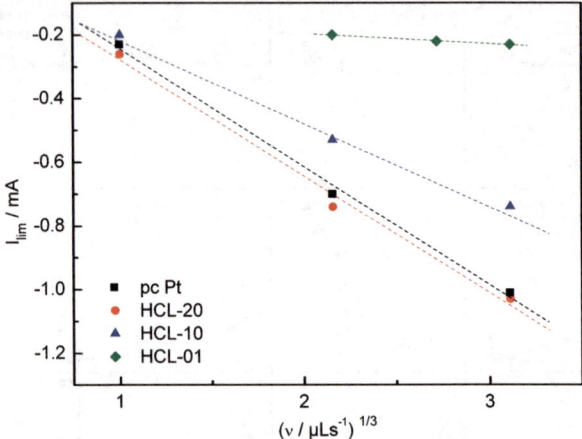

Fig. 6 Dependence of the mass transport limited ORR current (at 0.2 V) vs. the cube root of the electrolyte flow rate in a dual thin-layer flow cell configuration over pc-Pt electrode and over a nanostructured Pt catalysts supported on planar glassy carbon substrate (see figure). Solution: O_2-saturated 0.5 M H_2SO_4, room temperature.

valid also for the nanostructured electrodes, and also for those with Pt densities below the critical density discussed above. The data for the pc-Pt sample and the HCL-20 sample, where the Pt nanodisk density is above that limit under present transport conditions, are essentially identical. The slopes of the lines for the HCL-10 and the HCL-01 lines, however, are significantly lower than those of the pc-Pt and the HCL-20 sample. This can be explained by inspecting the equation describing the correlation between mass transport limited current I_{lim} and electrolyte flow rate v:[90–92]

$$I_{lim} = knFcv^{1/3} (DA/b)^{2/3} \quad (3)$$

where k is the cell constant ($k = 1.467$ for a channel-flow thin-layer cell), n is the number of electrons, F is the Faraday constant, c is the concentration of the species involved in the reaction, v is the electrolyte flow rate, D is the diffusion coefficient of the reacting species, A is the electrode surface area, and b is the thickness of the thin-layer gap, respectively. Since the other factors are not changed in these measurements, it is obviously a change in the (effective) surface area A below the critical Pt nanodisk density which is responsible for the decreasing slope with lower Pt coverage.

The shift of the transition potential between the two reaction regimes and the related broadening of the mixed mass transport/kinetically limited region to lower potentials with decreasing Pt coverage can be explained by increasing kinetic limitations related to the lower Pt surface area available for the reaction. This can be simply demonstrated by removing mass transport effects and normalizing the resulting kinetic currents to the active Pt surface areas. Tafel plots of the kinetic current densities normalized to the geometric surface area, $i_{k,geom.}$, which is identical for all samples (geometric surface area 0.283 cm^2), and the active Pt surface area normalized kinetic ORR currents densities, $i_{k,act}$, which decrease with Pt coverage, are presented in Fig. 7a and b, respectively, for an electrolyte flow rate of 30 µL s^{-1}. As expected, the slope of the lines is around 60 mV dec^{-1} in the high potential region (low overpotential, 0.9–0.86 V), while it is around 120 mV dec^{-1} in the lower potential region (0.8–0.7 V), in agreement with a one- or two-electron transfer in the rate limiting step, respectively.[89,93] The almost identical Tafel curves obtained for the active Pt surface area normalized kinetic ORR currents densities on the different samples clearly confirm that the wider range of the mixed transport

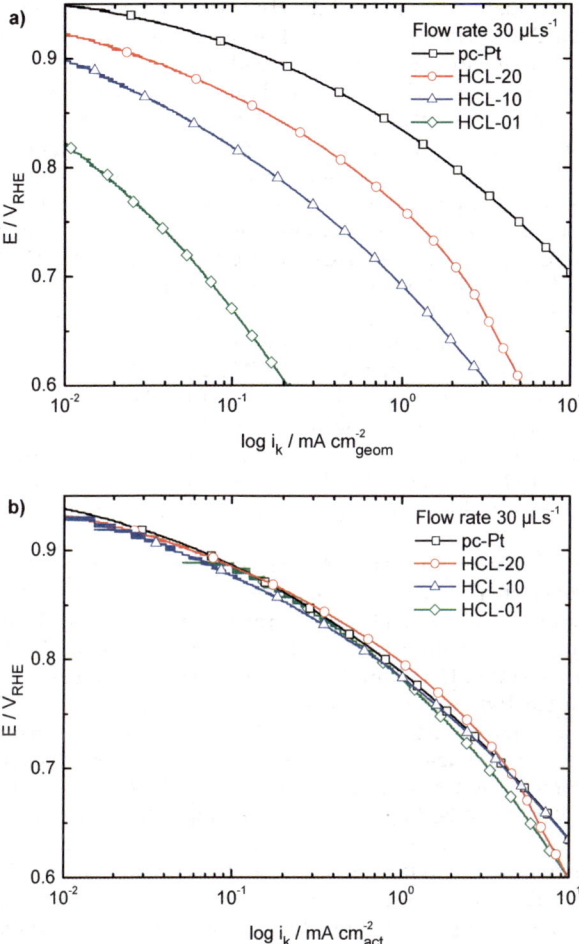

Fig. 7 Tafel plots for geometric (a) and active (b) surface area normalized kinetic oxygen reduction current over smooth polycrystalline Pt (squares), and nanostructured Pt/GC model electrodes: HCL-20 (circles), HCL-10 (triangles) and HCL-01 (diamonds). The data are extracted from Fig. 5 a, d, g, k. The limiting currents were taken at 0.2 V.

limited/kinetically limited region is simply a consequence of the competition being kinetic control and mass transport control, where the former depends on the active Pt surface are, while the latter depends mainly on the geometric surface area and, at low Pt nanodisk densities, also on the active surface area A. Similar trends were observed for the lower flow rates.

The H_2O_2 peroxide production during the ORR over the pc-Pt electrode decreases considerably with decreasing flow rate down to 1 µL s^{-1} (Fig. 5b). Correcting, however, for the simultaneous decrease in Faradaic current, the resulting H_2O_2 yields do not change significantly with flow rate. The H_2O_2 production on pc-Pt electrode was attributed to surface blocking effects, reducing the probability for finding neighboring empty Pt sites for O_2 dissociation. This favors the pathway for H_2O_2 formation, where only one adsorption site is assumed to be required.[93] Surface blocking starts by (bi)sulfate adsorption in the double-layer potential region, and increases rapidly at potentials <0.3 V by H_{upd} adsorption. It is important to note that at potentials relevant for fuel cell cathode operation (0.7–0.8 V), H_2O_2 formation on the pc-Pt electrode is negligible (Fig. 5c), and also in the double-layer range the H_2O_2 oxidation current and the H_2O_2 yields are very low on this electrode (see the expanded scales in Fig. 5b and c). For the nanostructured Pt/GC electrodes, the H_2O_2 production increases significantly with decreasing Pt coverage (Fig. 5e, h, i), resulting in a corresponding increase in H_2O_2 yields even at potentials relevant to cathode operation. A similar trend, though much less pronounced, was observed recently on CL-prepared nanostructured Pt/GC electrodes and explained by a reaction mechanism where (i) O_2 can be reduced to both H_2O and to H_2O_2 on the Pt surface, and where (ii) the resulting H_2O_2 intermediate can re-adsorb on other Pt sites and be further reduced to H_2O ('desorption–re-adsorption–reaction' concept).[74] This is illustrated in Fig. 1, with H_2O_2 as reaction intermediate RI. At potentials below 0.2 V, H_2O_2 can be formed also on the carbon areas of the nanostructured Pt/GC electrodes, as evidenced by the onset of H_2O_2 formation on the Pt-free GC electrode (see Fig. 5l, m). Because of the low absolute currents, the contribution to the total H_2O_2 yield on the nanostructured Pt/GC electrodes, however, will be small. In the above reaction mechanism, the increase of the H_2O_2 yield with decreasing Pt coverage is simply due to the decreasing probability for desorbed (dissolved) H_2O_2 molecules to re-adsorb and further react on Pt sites. Correspondingly, the probability for H_2O_2 to escape into the bulk electrolyte and be removed from the reaction cell increases with decreasing probability for re-adsorption and hence lower Pt coverage.

Although the probability for H_2O_2 formation on Pt in a single reaction event can not be extracted from these data without extensive modeling (see outlook in chapter 4), it must be quite high, as evidenced by the rather high H_2O_2 formation on the nanostructured Pt/GC samples with low Pt coverage (Fig. 5i, m). Correspondingly, the negligible amount of H_2O_2 formation on massive Pt electrodes[93] and on realistic carbon supported Pt/C catalysts,[89] in particular at potentials > 0.3 V, points to a high probability for re-adsorption and further reduction of desorbable (dissolved) H_2O_2 intermediates on these electrodes/catalyst layers. Our results agree fully with previous findings of catalyst loading effects on thin-layer Pt/C catalyst electrodes by Inaba et al.[94] Very recently, Bonakdarpour et al. reported a pronounced increase of H_2O_2 formation at low catalyst loadings for the ORR on $RuSe_x/C$ catalyst electrodes and attributed this to the formation of H_2O_2 reaction intermediates.[95] While for the latter catalyst, significant H_2O_2 formation may appear as 'expected', considering the amount of H_2O_2 formation already on 'normal loading' thin-layer catalyst electrodes,[65,96] the observation of a high H_2O_2 yield in the ORR on Pt, even at potentials around 0.5–0.6 V, seems to be in complete contrast to the previous mechanistic understanding of the ORR on Pt.[97,98] This clearly demonstrates the important role of re-adsorption effects (re-adsorption and further reaction of desorbable, reaction intermediates as described by the 'desorption–re-adsorption–further reaction' concept and illustrated in Fig. 1) for the mechanistic understanding of electrocatalytic reactions.

The much more pronounced effects on the HCL-prepared nanostructured Pt/GC electrodes compared to the CL-prepared ones can be simply explained by the presence of significant amounts of additional Pt nanoparticles on the GC surface, which are distributed in the surface areas between the Pt nanodisks on the latter samples. These nanoparticles, which are likely to be produced during sputtering step in the nanostructuring process,[80] limit the decrease in re-adsorption probability when reducing the density of the Pt nanodisks.

This mechanism is fully supported by the trend for H_2O_2 formation upon varying the electrolyte flow rate. In contrast to the ORR on CL-prepared Pt/GC electrodes, not only the H_2O_2 formation but also the H_2O_2 yield increases clearly for the nanostructured HCL-20 and HCL-10 Pt/GC electrodes with increasing electrolyte flow (Fig. 5f, i). For the HCL-01 sample (Fig. 5m) we expect a similar general trend, but in this case the situation is more complicated because of the significant relative contributions from the (partly kinetically limited) ORR on the carbon substrate areas at $E < 0.25$ V and the fact that, due to the very low Pt loading, the mixed transport and kinetically limited region ranges down to a potential of ≈ 0.25 V. Within the mechanism discussed above, that can be easily explained by the increasing probability for re-adsorption at lower electrolyte flow. Due to the increasing concentration gradient/thinner diffusion layer at higher electrolyte flow the probability for H_2O_2 formed at the electrode has a higher chance for escaping into the bulk electrolyte and be transported out of the reaction cell with increasing electrolyte flow. The probability for re-adsorption on Pt areas and further reduction to H_2O decreases accordingly. This effect is clearly visible on all nanostructured samples. At high Pt coverages, either due to high Pt loading (pc-Pt, on realistic Pt/C catalysts or on nanostructured Pt/GC electrodes with additional Pt nanoparticles on the surface), where also the H_2O_2 yield is rather small, these electrolyte flow effects are essentially not detectable. On the present HCL-prepared Pt/GC electrodes, in contrast, they are clearly visible. Considering also that there is no H_2O_2 formation on the GC area at potentials >0.2 V, these findings convincingly support the 'desorption–re-adsorption–reaction' concept introduced for the ORR on Pt recently.[74] On massive Pt electrodes and at commonly used electrolyte flow rates, the probability for re-adsorption of H_2O_2 and further reduction to H_2O is so high that H_2O_2 formation is hardly observed, despite the obviously rather high probability for H_2O_2 formation in a single adsorption/reaction step. Based on the data for the HCL-01 electrode, this probability must be higher than 30% at 0.5 V!

Finally, it should be noted that although the measurements described in the present communication were performed on model surfaces to elucidate the effects of mass transport on the ORR activity and selectivity under defined reaction conditions, the results also have consequences for practical applications, e.g., in polymer electrolyte fuel cell (PEFC) cathodes. They clearly point out that with decreasing catalyst loading, which is strived for from economical reasons, there is an increasing risk of H_2O_2 production on the cathode catalyst even under typical cathode reaction conditions, which may have disastrous consequences on the lifetime of electrode or membrane due to corrosion effects.[99,100] It is important to note that for a correct assessment of the H_2O_2 related corrosion effects, the macroscopic H_2O_2 yield measured at the exhaust of the catalyst is not decisive, but rather the local steady-state concentration in the electrode, which due to the re-adsorption effects may be much higher than the former.

4 Outlook

As mentioned in the Introduction, measurements on well-defined nanostructured samples under controlled transport conditions similar to those presented here provide an ideal basis for further simulations. Simulations of vertical transport to a non-structured sample in a similar flow cell geometry[101] or in a rectangular cell geometry[56] were published recently, further simulations coupling these transport/diffusion processes with a subsequent catalytic surface reaction (CO oxidation) are currently in progress.[102] On the other hand, (vertical) diffusion to arrays of regularly arranged ultra-microelectrodes of different shapes was calculated in a number of studies, starting from a fixed concentration at a defined plane above the electrode surface[81–84] (see Fig. 8a). This is identical to the situation in a flow cell, where above the diffusion layer the reactant concentration can be assumed to be constant and identical to the bulk concentration in the flowing electrolyte. A logical next step will be to combine these approaches and to include (i) lateral transport and re-adsorption processes according to Fig. 1 and 8b, and (ii) the specific geometry of the nanostructured model electrode surface in the simulation. Obviously, this requires a highly complex fully three-dimensional description of the ongoing transport processes. In order to avoid further complications, such calculations could first be applied to simulate the processes in a transport limited single-step reaction such as H_2 oxidation with negligible steady-state coverages of reactants and reaction intermediates, similar to the approach used in ref. 56. These simulations should map

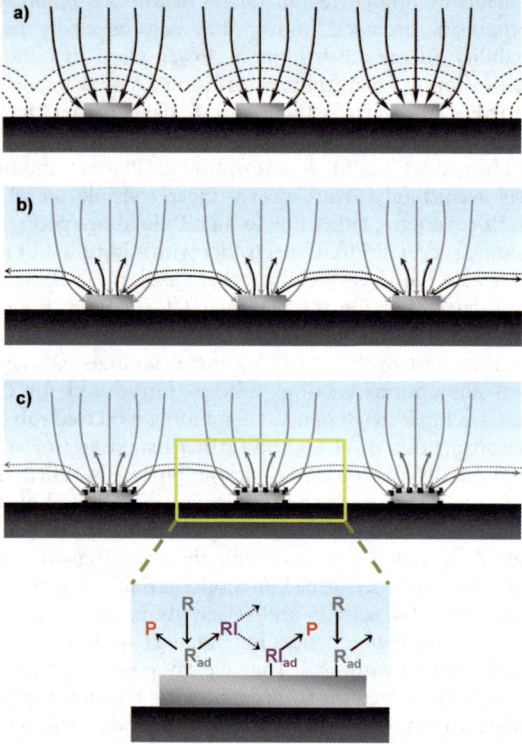

Fig. 8 Hierarchy of model studies and simulations on transport effects in electrocatalytic reactions on structurally well-defined nanostructured model electrodes (arrays of ultramicroelectrodes) under defined transport conditions in a realistic cell geometry. (a) Vertical transport of reactants (R) to the model electrode without reaction intermediate formation or product re-adsorption and the surface (experimental example: H_2 oxidation); (b) as (a), but in addition also formation, desorption, lateral transport and re-adsorption/reaction of volatile reaction intermediates on the non-modified active surface of the model electrode (experimental example: O_2 reduction oxidation in the mass transport limited potential regime); (c) as (b) but in addition consideration of the surface processes, *e.g.*, by density functional calculations, and the build-up of an adlayer formed by adsorbed reactants, reaction intermediates, and reaction side products (experimental examples with increasing complexity: CO oxidation (no RI), O_2 reduction in the mass transport limited potential regime (RI: H_2O_2), oxidation of methanol or formaldehyde (RI: formaldehyde (methanol oxidation only), formic acid).

out the macroscopic and mesoscopic (vertical) diffusion to a nanostructured surface in a realistic cell geometry and identify the influence of lateral transport processes and surface reaction of the reactants under conditions where effects related to changes in the composition and coverage of the adlayer can be neglected (see Fig. 8a). In a next step, one would include the formation, desorption, lateral transport and re-adsorption/further reaction of reaction intermediates (RI), as illustrated in Fig. 1 and 8b, but would still neglect possible changes in the composition/coverage of the adlayer under different steady-state conditions. For this level of simulation, where details of the surface processes can still be neglected, the ORR on nanostructured Pt/GC electrodes (in the transport-limited potential range), as described in the present contribution, would be a suitable experimental system. Finally, in the last step, also details of the ongoing surface processes need to be included, at least to an extent that variations in the adlayer composition/coverage and the related changes in the catalytic properties of the active electrode areas can be accessed (see Fig. 8c). Suitable experimental examples would be the ORR in the kinetically limited potential region on nanostructured Pt/GC samples or, even more complex, the oxidation of methanol or formaldehyde on the same electrodes. This requires a complex multi-scale approach for the combination of molecular scale surface processes and mesoscopic transport processes. The final goal of these kinds of combined experimental/theoretical approach would be a quantitative description of the experimental reaction characteristics (overall activity and product distribution under different reaction conditions) on these structurally simple and well-defined nanostructured model electrodes under controlled transport conditions, including a three dimensional (3D) description of the macroscopic and mesoscopic transport processes in the electrolyte above the electrode and of the molecular scale surface processes.

In total, this would be an enormous step towards a molecular understanding of electrocatalytic reactions, similar to the recent progress in heterogeneous catalysis reached by efforts to bridge the materials and pressure gap between surface science model studies under ultrahigh vacuum conditions on single-crystal model catalysts and catalytic studies on realistic supported catalysts and under realistic reaction conditions (see ref. 103 and subsequent articles in that special issue). One important difference between the two areas of heterogeneous catalysis and electrocatalysis highly relevant for the present considerations is that in the latter case even the proper description of electrocatalytic reactions under model conditions and on model electrodes requires inclusion of transport processes, which is not the case in surface science model studies performed under UHV conditions.

Finally, it should not be forgotten that despite the experimental and theoretical complexity of the tasks described above, we have so far obtained a proper description of an electrocatalytic reaction on a two dimensional (2D) electrode with regularly distributed active centers. The extension of this approach to the description of a 3D electrode with active centers, either in a regular structure or even in a disordered arrangement, as experienced experimentally in a thin film catalyst layer or in a porous fuel cell electrode, would again represent a major task.

5 Conclusions

We have investigated the role of mesoscopic transport effects in the oxygen reduction reaction on Pt by measuring the activity and selectivity (H_2O_2 formation) of the O_2 reduction reaction on nanostructured Pt/GC model electrodes in a dual thin-layer flow cell under controlled and varied electrolyte flow conditions and at different Pt coverages. The Pt/GC electrodes were fabricated by hole-mask colloidal lithography and consist of Pt nanodisks of defined diameter supported on a glassy carbon substrate, which are arranged in a rather regular array with a narrow distribution of separations.

In addition to the expected increase of the transport limited current with increasing Pt coverage in the low to medium Pt coverage regime and, for all Pt coverages, with increasing electrolyte flow rate v, the data show a quantitative agreement with the $I \propto v^{1/3}$ predicted for non-structured electrodes in a channel flow, with the slope depending on the Pt coverage on the nanostructured electrodes. This is explained by the geometric factor in the dependence of the limiting current on the electrolyte flow v.[90] Furthermore, they show a clear trend to lower H_2O_2 yields with decreasing electrolyte flow. The same is observed for decreasing Pt coverages, in agreement with previous findings. The results are explained in terms of a reaction model proposed recently, which includes (i) direct reduction to H_2O on the Pt surface, and (ii) significant formation and desorption of H_2O_2 on the Pt surface, at potentials ≤ 0.3 V also on the carbon surface, its subsequent partial re-adsorption and further reduction to H_2O on the Pt surface ('desorption–re-adsorption–reaction' concept). Because of the high probability for re-adsorption and further reduction of the volatile H_2O_2 reaction intermediates, the experimentally observed H_2O_2 formation is negligible on extended massive Pt electrodes or on carbon supported Pt catalysts, although it is clearly formed on Pt at potentials < 0.8 V Pt. These findings are compared to recent observations of an increased formation of complete oxidation products compared to incomplete oxidation products in the oxidation of different small organic molecules with increasing catalyst loading or decreasing electrolyte flow, which can be explained in similar terms, *i.e.*, by an increasing probability for re-adsorption and further oxidation of volatile reaction intermediates to the complete oxidation product CO_2 with increasing catalyst loading or decreasing electrolyte flow.

In an outlook, the potential and further development of combined experimental and theoretical model studies on these nanostructured model electrodes for the understanding and description of electrocatalytic processes on model systems was mapped out. In the end, the description should, in a multi-scale approach, include (i) macroscopic and mesoscopic transport of the reactant in a realistic reaction cell and under defined electrolyte conditions, (ii) desorption, lateral transport, re-adsorption and further reaction of volatile reaction intermediates according to the 'desorption–re-adsorption–reaction' concept, and (iii) a molecular scale description of the ongoing surface processes and the adlayer resulting under reaction conditions.

Note added in proof

We would like to add that, though in a very different experiment, an increasing amount of H2O2 formation in the ORR on Pt with increasing electrolyte transport was reported and discussed also by S. Chen and A. Kucernak.[104]

Acknowledgements

This work was supported by the Landesstiftung Baden-Württemberg *via* the Kompetenznetz Funktionelle Nanostrukturen (project B9), MISTRA (Contract No. 95014) and the Swedish Energy Agency (Grant No. P12554–1). We gratefully acknowledge A. Minkow (Institute of Micro- and Nanomaterials, Ulm University) for the SEM images.

References

1 J. O. Bockris and H. Wroblowa, *J. Electroanal. Chem.*, 1964, **7**, 428.
2 W. Vielstich, *Fuel Cells*, Wiley-Interscience, London, 1965.
3 S. Gottesfeld and T. A. Zawodzinski, in *Advances in Electrochemical Science and Engineering*, ed. R. C. Alkire, H. Gerischer, D. M. Kolb and C. W. Tobias, Wiley-VCH, Weinheim, 1997, 5, 195.
4 W. Vielstich, T. Iwasita, in *Handbook of Heterogenous Catalysis*, Eds.: G. Ertl, H. Knözinger, J. Weitkamp, Wiley-VCH, Weinheim, 1997, 4, 2090.
5 R. R. Adzic, in *Electrocatalysis*, Eds.: J. Lipkowski, P. N. Ross, Wiley-VCH, Inc., New York, 1998, 197.
6 C. Lamy and J.-M. Léger, *J. Chem. Phys.*, 1991, **88**, 1649.
7 M. J. Weaver, *J. Phys. Chem.*, 1996, **100**, 13079.
8 T. D. Jarvi and E. M. Stuve, in *Electrocatalysis*, ed. J. Lipkowski and P. N. Ross, Wiley-VCH, Heidelberg, 1998, 75.
9 C. Lamy, J.-M. Léger, S. Srinivasan, in *Modern Aspects of Electrochemistry*, ed. J. O. Bockris, B. Conway and R. E. White, Kluwer Adcademic/Plenum Publishers, New York, 2001, 34, 53.
10 N. M. Markovic and P. N. Ross Jr., *Surf. Sci. Rep.*, 2002, **45**, 117.
11 T. Iwasita-Vielstich, in *Advances in Electrochemical Science and Engineering*, Ed. H. Gerischer and C. W. Tobias, VCH Verlagsgesellschaft, Weinheim, 1990, 127.
12 *Handbook of Fuel Cells. Fundamentals, Technology and Applications - Vol. 1: Fundamentals and Survey of Systems*, Eds.: W. Vielstich, H. A. Gasteiger, A. Lamm, Wiley & Sons, Chichester, 2003, p. 1.
13 *Handbook of Fuel Cells. Fundamentals, Technology and Applications – Vol. 2: Electrocataysis*, ed. W. Vielstich, H. A. Gasteiger and A. Lamm, Wiley & Sons, Chichester, 2003, p. 1.
14 R. J. Behm and Z. Jusys, *J. Power Sources*, 2006, **154**, 327.

15 B. Beden, C. Lamy, A. Bewick and K. Kunimatsu, *J. Electroanal. Chem.*, 1981, **121**, 343.
16 I. Villegas, N. Kizhakevariam and M. J. Weaver, *Surf. Sci.*, 1995, **335**, 300.
17 A. Hamnett, in *Interfacial Electrochemistry: Accomplishments and Challenges*, ed. A. Wieckowski, Marcel Dekker Inc., New York, 1999, p. 843.
18 T. Iwasita and F. C. Nart, in *Advances in Electrochemical Science and Engineering*, ed. H. Gerischer and C. W. Tobias, VCH, Weinheim, 1990, p. 123.
19 S.-G. Sun, in *Electrocatalysis*, ed. J. Lipkowski and P. N. Ross, Wiley-VCH, New York, 1998, p. 243.
20 C. Korzeniewski, in *Interfacial Electrochemistry*, ed. A. Wieckowski, Marcel Dekker Inc., New York, USA, 1999, 345.
21 F. C. Nart and T. Iwasita, in *Encyclopedia Electrochemistry – Interfacial Kinetics and Mass Transport*, ed. A. J. Bard, M. Stratmann and E. J. Calvo, VCH, Weinheim, 2003, vol. 2, p. 243.
22 K. Kunimatsu, H. Uchida, M. Osawa and M. Watanabe, *J. Electroanal. Chem.*, 2006, **587**, 299.
23 M. Heinen, Y. X. Chen, Z. Jusys and R. J. Behm, *Electrochim. Acta*, 2007, **52**, 5634.
24 O. Wolter and J. Heitbaum, *Ber. Bunsen-Ges. Phys. Chem.*, 1984, **88**, 2.
25 H. Baltruschat, *J. Am. Soc. Mass Spectrom.*, 2004, **15**, 1693.
26 T. H. M. Housmans, A. H. Wonders and M. T. M. Koper, *J. Phys. Chem. B*, 2006, **110**, 10021.
27 Z. Jusys and R. J. Behm, in *Fuel Cell Catalysis: A Surface Science Approach*, ed. M. T. M. Koper, 2008.
28 P. G. Allen, S. D. Conradson, M. S. Wilson, S. Gottesfeld, I. D. Raistrick, J. Valerio and M. Lovato, *Electrochim. Acta*, 1994, **39**, 2415.
29 J. W. Couves and P. Meehan, *Physica B*, 1995, **208 & 209**, 665.
30 N. Alonso-Vante, M. Fieber-Erdmann, H. Rossner, E. Holub-Krappe, C. Giorgetti, A. Tadjeddine, E. Dartyge, A. Fontaine and r. Frahm, *J. Phys. IV*, 1997, **7**, 887.
31 W. E. O'Grady, P. L. Hagans, K. I. Pandya and D. L. Mariche, *Langmuir*, 2001, **17**, 3047.
32 N. Alonso-Vante, I. V. Malakhov, S. G. Nikitenko, E. R. Savinova and D. I. Kochubey, *Electrochim. Acta*, 2002, **47**, 3807.
33 C. Roth, N. Martz, A. Morlang, R. Theissmann and H. Fuess, *Phys. Chem. Chem. Phys.*, 2004, **6**, 3557.
34 C. A. Lucas, N. M. Markovic and P. N. Ross, *Phys. Rev. Lett.*, 1996, **77**, 4922.
35 N. M. Markovic, B. N. Grgur, C. A. Lucas and P. N. Ross, *J. Phys. Chem. B*, 1999, **103**, 487.
36 C. A. Lucas, N. M. Markovic and P. N. Ross, *Surf. Sci.*, 2000, **448**, 77.
37 J. X. Wang, I. K. Robinson, B. M. Ocko and R. R. Adzic, *J. Phys. Chem. B*, 2005, **109**, 24.
38 M. T. M. Koper and W. Schmickler, in *Electrocatalysis*, ed. J. Lipkowski, P. N. Ross, Wiley-VCH, New York, 1998, p. 243.
39 S. A. Wasileski, M. T. M. Koper and M. J. Weaver, *J. Phys. Chem. B*, 2001, **105**, 3518.
40 J. K. Nørskov, T. Bligaard, A. Logadottir, S. Bahn, L. B. Hansen, M. Bollinger, H. Bengaard, B. Hammer, Z. Sljivancanin, M. Mavrikakis, Y. Xu, S. Dahl and C. J. H. Jacobsen, *J. Catal.*, 2002, **209**, 275.
41 J. Greeley and M. Mavrikakis, *Nat. Mater.*, 2004, **3**, 810.
42 C. D. Taylor and M. Neurock, *Curr. Opin. Solid State Mater. Sci.*, 2005, **9**, 49.
43 V. Stamenkovic, B. S. Mun, K. J. J. Mayrhofer, P. N. Ross, N. M. Markovic, J. Rossmeisl, J. Greeley and J. K. Nørskov, *Angew. Chem.*, 2006, **45**, 2897.
44 V. R. Stamenkovic, B. S. Mun, M. Arenz, K. J. J. Mayrhofer, C. A. Lucas, G. Wang, P. N. Ross and N. M. Markovic, *Nat. Mater.*, 2007, **6**, 241.
45 W. J. Albery, C. C. Jones and A. R. Mount, in *Chemical Kinetics*, ed. R. G. Compton, and A. Hamnett, Elsevier Publ., Amsterdam-Oxford-New York-Tokyo, 1986, vol. 29, p. 129.
46 C. M. A. Brett and A. M. F. C. Oliveira Brett, in *Chemical Kinetics*, ed. C. H. Bamford and R. G. Compton, Elsevier Publ., Amsterdam-Oxford-New York-Tokyo, 1986, vol. 26, p. 355.
47 J. A. Cooper and R. G. Compton, *Electroanalysis*, 1998, **10**, 141.
48 H. Gerischer, I. Mattes and R. Braun, *J. Electroanal. Chem.*, 1965, **10**, 553.
49 K. Aoki, K. Tokuda and H. Matsuda, *J. Electroanal. Chem.*, 1977, **79**, 49.
50 P. R. Unwin and R. G. Compton, in *Chemical Kinetics*, ed. C. H. Bamford and R. G. Compton, Elsevier Publ., Amsterdam-Oxford-New York-Tokyo, 1986, vol. 29, p. 173.
51 N. Wakabayashi, H. Uchida and M. Watanabe, *Electrochem. Solid-Sate Lett.*, 2002, **5**, E62.
52 T. H. Madden, N. Arvindan and E. M. Stuve, *J. Electrochem. Soc.*, 2003, **150**, E1.
53 T. H. Madden and E. M. Stuve, *J. Electrochem. Soc.*, 2003, **150**, E571.
54 Z. Jusys, J. Kaiser and R. J. Behm, *Electrochim. Acta*, 2004, **49**, 1297.
55 N. Wakabayashi, M. Takeichi, H. Uchida and M. Watanabe, *J. Phys. Chem. B*, 2005, **109**, 5836.
56 J. Fuhrmann, H. Zhao, E. Holzbecher, H. Langmach, M. Chojak, R. Halseid, Z. Jusys and R. J. Behm, submitted.

57 Z. Jusys, H. Massong and H. Baltruschat, *J. Electrochem. Soc.*, 1999, **146**, 1093.
58 Y.-X. Chen, M. Heinen, Z. Jusys and R. J. Behm, *Angew. Chem., Int. Ed.*, 2006, **45**, 981.
59 A. N. Frumkin, L. Nekrasov, V. G. Levich and J. Ivanov, *J. Electroanal. Chem.*, 1959, **1**, 84.
60 T. J. Schmidt, H. A. Gasteiger and R. J. Behm, *J. Electrochem. Soc.*, 1999, **146**, 1296.
61 J. Yamada and H. Matsuda, *J. Electroanal. Chem.*, 1973, **44**, 189.
62 M. Bergelin, J. M. Feliu and M. Wasberg, *Electrochim. Acta*, 1998, **44**, 1069.
63 U. Koponen, T. Peltonen, M. Bergelin, T. Mennola, M. Valkiainen, J. Kaskimies and M. Wasberg, *J. Power Sources*, 2000, **86**, 261.
64 C. L. Green and A. Kucernak, *J. Phys. Chem. B*, 2002, 11446.
65 L. Colmenares, Z. Jusys and R. J. Behm, *J. Phys. Chem. C*, 2007, **111**, 1273.
66 V. G. Levich, *Physicochemical Hydrodynamics*, Prentice Hall, Eaglewood Cliffs, NJ, 1962.
67 Z. Jusys, J. Kaiser and R. J. Behm, *Langmuir*, 2003, **19**, 6759.
68 H. Wang, T. Löffler and H. Baltruschat, *J. Appl. Electrochem.*, 2001, **31**, 759.
69 R. W. Lindström, Y. E. Seidel, Z. Jusys, M. Gustavsson, B. Kasemo and R. J. Behm, to be published.
70 H. Wang, Z. Jusys and R. J. Behm, *J. Appl. Electrochem.*, 2006, **36**, 1187.
71 H. Wang, Z. Jusys and R. J. Behm, *J. Phys. Chem. B*, 2004, **108**, 19413.
72 M. Gustavsson, H. Fredriksson, B. Kasemo, Z. Jusys, C. Jun and R. J. Behm, *J. Electroanal. Chem.*, 2004, **568**, 371.
73 Y. E. Seidel, R. Lindström, Z. Jusys, M. Gustavsson, P. Hanarp, B. Kasemo, A. Minkow, H. J. Fecht and R. J. Behm, *J. Electrochem. Soc.*, 2008, 155–K50.
74 A. Schneider, L. Colmenares, Y. E. Seidel, Z. Jusys, B. Wickman, B. Kasemo and R. J. Behm, *Phys. Chem. Chem. Phys.*, 2008, **10**, 1931.
75 C. Werdinius, L. Österlund and B. Kasemo, *Langmuir*, 2003, **19**, 458.
76 P. Hanarp, M. Käll and D. S. Sutherland, *J. Phys. Chem. B*, 2003, **107**, 5768.
77 L. Österlund, S. Kielbassa, C. Werdinius and B. Kasemo, *J. Catal.*, 2003, **215**, 94.
78 L. Österlund, A. Grant and B. Kasemo, in *Nanocatalysis*, ed. U. Heiz and U. Landman, Springer Verlag, Berlin, 2007, p. 4.
79 H. Fredriksson, Y. Alaverdyan, A. Dmitiev, C. Langhammer, D. S. Sutherland, M. Zäch and B. Kasemo, *Adv. Mater.*, 2007, **19**, 4297.
80 Y. E. Seidel, M. Müller, Z. Jusys, B. Wickman, P. Hanarp, B. Kasemo, U. Hörmann, U. Kaiser and R. J. Behm, *J. Electrochem. Soc.*, 2008, DOI: 10.1149/1.2956326.
81 W. E. Morf, *Anal. Chim. Acta*, 1996, **330**, 139.
82 E. J. F. Dickinson, I. Streeter and R. G. Compton, *J. Phys. Chem. B*, 2008, **112**, 4059.
83 W. E. Morf, *Anal. Chim. Acta*, 1997, **341**, 121.
84 W. E. Morf, M. Koudelka-Hep and N. F. de Rooij, *J. Electroanal. Chem.*, 2006, **590**, 47.
85 Z. Jusys, J. Kaiser and R. J. Behm, *J. Phys. Chem. B*, 2004, **108**, 7893.
86 V. S. Bagotzky, Y. B. Vassiliev and O. A. Khazova, *J. Electroanal. Chem.*, 1977, **81**, 229.
87 T. Biegler, D. A. J. Rand and R. Woods, *J. Electroanal. Chem.*, 1971, **29**, 269.
88 H. Angerstein-Kozlowska, B. E. Conway and W. B. A. Sharp, *J. Electroanal. Chem.*, 1973, **43**, 9.
89 U. A. Paulus, T. J. Schmidt, H. A. Gasteiger and R. J. Behm, *J. Electroanal. Chem.*, 2001, **495**, 134.
90 S. G. Weber and J. T. Long, *Anal. Chem.*, 1988, **60**, 903A.
91 A. J. Bard, L. R. Faulkner, *Electrochemical Methods – Fundamentals and Applications*, John Wiley & Sons, New York, 1980.
92 C. H. Hamann, A. Hamnett, W. Vielstich, *Electrochemistry*, Wiley-VCH, Weinheim, 1998.
93 N. M. Markovic, H. A. Gasteiger and P. N. Ross, *J. Phys. Chem.*, 1995, **99**, 3411.
94 M. Inaba, H. Yamada, J. Tokunaga and A. Tasaka, *Electrochem. Solid-State Lett.*, 2004, **7**, A474.
95 A. Bonakdarpour, C. Delacote, R. Yang, A. Wieckowski and R. J. Dahn, *Electrochem. Commun.*, 2008, **10**, 611.
96 L. Colmenares, Z. Jusys and R. J. Behm, *Langmuir*, 2006, **22**, 10437.
97 A. Damjanovic, in Modern Aspects of Electrochemistry, ed. J. O. M. Bockris and B. E. Conway, Plenum Press, New York, vol. 5, p. 369.
98 M. Gattrell and B. MacDougall, in *Electrocatalysis*, ed. W. Vielstich, H. A. Gasteiger and A. Lamm, Wiley & Sons, Chichester, 2003, vol. 2, (p. 30), 443.
99 A. B. La Conti, M. Hamdan and R. C. McDonald, in *Handbook of Fuel Cells – Fundamentals Technology and Applications*, ed. W. Vielstich, A. Lamm and H. A. Gasteiger, Wiley, Chichester, 2003, vol. 3, p. 49, 647.
100 L. M. Roen, C. H. Paik and T. D. Jarvi, *Electrochem. Solid-State Lett.*, 2004, **7**, 19.
101 J. Fuhrmann, H. Zhao, E. Holzbecher and H. Langmach, *J. Fuel Cell Sci. Technol.*, 2008, **5**, 021008.
102 D. Zhang, O. Deutschmann, Y. E. Seidel and R. J. Behm, submitted.
103 R. Imbihl, R. J. Behm and R. Schloegl, *Phys. Chem. Chem. Phys.*, 2007, **9**, 3459.
104 S. Chen and A. Kucernak, *J. Phys. Chem. B*, 2004, **108**, 3262.

General Discussion

Professor Wieckowski opened the discussion of the paper by Sang-Eun Bae and Andrew A Gewirth*:
1. Why is surface oxidation formation made more facile on Cu(111) and not on Cu(100) (an open surface) face?
2. Is the voltammetry for NO_3^- reduction similar to that for CO_2 reduction on Cu(111) and Cu(100)?

Professor Gewirth replied:
1. Certainly, the stability of the fully formed oxide on Cu(100) is greater than that on Cu(111) as UHV results show.[1] However, the thermodynamic stability of the final product doesn't necessarily have implications for that of the precursor or intermediate species. Both the Wandelt and Maurice groups addressed the facile formation of a Cu(I) oxide on Cu(111) relative to Cu(100), and suggested that the Cu(I) precursor was more stable on this surface relative to Cu(100). At this stage, we can only repeat their argument, but it certainly is an open question.
2. CO_2 electroreduction on Cu surfaces, emphasizing the work of Hori, has recently been reviewed.[2] The potentials for CO_2 reduction are substantially more negative than those for nitrate. This means that nearly every voltammetric response found for CO_2 reduction on Cu is convolved with hydrogen evolution which in turn makes establishing a Tafel slope or performing the most rudimentary electrochemical characterization problematic. Thus, there is no direct comparison yet of the nitrate and carbon dioxide voltammetry on any Cu face.

1. F. Besenbacher and J. K. Nørskøv, *Prog. Surf. Sci.*, 1993, **44**, 5.
2. M. Gattrell, N. Gupta, and A. Co, *J. Electroanal. Chem.*, 2006, **594**, 1.

(87:86) **Professor Koper** asked:
Would the idea of stabilising a certain oxide in order to accelerate the reduction of nitrite to nitrate transfer to other good catalysts for nitrate reduction, like rhodium?
Would such an oxide also play a role in a similar reaction as CO_2 reduction, which is also catalysed by copper?

Professor Gewirth answered:
Would the idea of stabilising a certain oxide in order to accelerate the reduction of nitrate to nitrate transfer to other good catalyst for nitrate reduction, like rhodium?
This is a good question, and I don't know the answer now. Certainly, we should check.
Would such an oxide also play a role in a similar reaction as CO_2 reduction, which is also catalysed by copper?
The fact that Cu is a good catalyst for both nitrate and carbon dioxide electroreduction has been noted in passing by many people. This effect may have to do with similarities in the interaction of the anions being reduced (NO_3^- and CO_3) with the Cu surface or with the oxide. As noted above, the potentials at which CO_2 is reduced on Cu are substantially more negative than those for nitrate. This makes the presence of a putative oxide in this case somewhat more problematic.

Professor Schiffrin commented:
(1) The results that you show in Fig. 1 in your paper refer only to cyclic voltammetry transients. Electrode poisoning for this reaction on polycrystalline copper has been extensively documented. Have you observed a similar effect? It would be interesting to compare the steady state current for nitrate reduction in the two

surface orientations studied to see if the differences observed relate to transient effects of the two surfaces.

(2) How safe do you consider is the assignment of the bright features observed with STM to a Cu_2O layer? These only indicate a lowering of the tunnelling barrier and could be caused by other effects, for example, the adsorption of nitrate or of its reduction products on Cu(111).

Professor Gewirth answered:

In our first paper on the subject,[1] we did compare rotating disk voltammetry for Cu(100), and Cu(poly). We found that the exchange current density substantially lower on Cu(100) (56 μA cm^{-2}) relative to Cu(poly) (149 μA cm^{-2}). The corresponding Tafel slope for Cu(100) was 226 mV per decade while for Cu(poly) the slope was 203 mV per decade for nitrate reduction, identical with that reported previously.[2] Finally, the onset potential for Cu(poly) was 200 mV more positive than that for Cu(100). There is certainly an open question about association of the features we observe to a Cu_2O layer. This association is made only by comparison with the results of the Wandelt and Maurice groups—referenced in the paper—who looked at oxidation of Cu(111) in different environments. Our features look like those reported in these other cases. We are certain, of course, that nitric acid oxidizes Cu, as is well known. What intermediates do we get going from Cu(0) to the Cu(II) final product?

1. S.-E. Bae, K. L. Stewart and A. A. Gewirth, *J. Am. Chem. Soc.,* 2007, **129**, 10171.
2. G. E. Dima, A. C. A. de Vooys and M. T. M. Koper, *J. Electroanal. Chem.*, 2003, **15**, 554–555.

Professor Schiffrin said:

Nitrate reduction on copper presents unusual characteristics. Pletcher and others have demonstrated that nitrate is quantitatively reduced to ammonia in acid media on this surface.[1,2] However, nitrate reduction is inhibited on aged surfaces but if the reduction takes place simultaneously with copper electrodeposition from low concentration solutions of Cu(II), high electrocatalytic activity is observed.[3–5] The origin of this effect is probably the formation of a reactive Cu(I) intermediate or the presence of very reactive centres on the freshly deposited Cu surface, for example, at steps. We have investigated a similar reaction but using silver instead of copper. Similarly to Cu, the reduction on freshly deposited silver is much faster than on the bulk metal (see Fig. 1). As shown in Fig. 2, the number of electrons exchanged per mol of nitrate is $n = 8$ and hence the reduction product, ammonia, is the same as that observed by Pletcher.[1,5] Thus, there must be a similarity in the mechanism of nitrate reduction on these two metals. It is proposed that in both

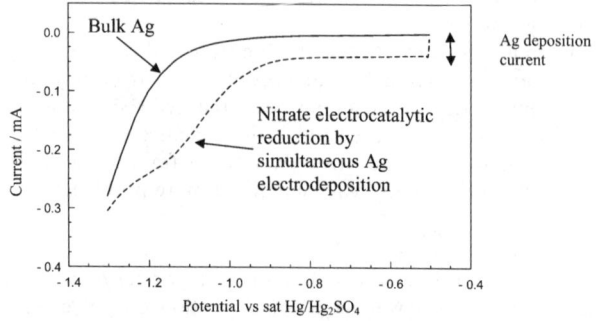

Fig. 1 Nitrate reduction at a silver rotating disk electrode in 0.1 M $NaClO_4$ + $HClO_4$ + 0.15 mM $NaNO_3$ (solid line); the same but containing also 0.1 mM $AgClO_4$ (dashed)

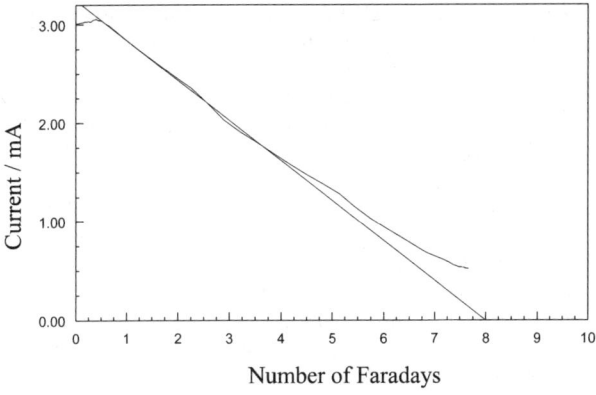

Fig. 2 Determination of n from a long-term electrolysis experiment. Conditions the same as in Fig. 1.

cases, the freshly deposited metal ions greatly enhance the electrocatalytic reduction of nitrate. This can have an electronic origin caused by the morphology of the evolving freshly electrodeposited surface and these metal surface structures are readily intercepted by the nitrate ion. In the case of silver, examples of this evolution of electronic properties can be seen in the intermediates formed during the reduction of Ag$^+$ to yield silver nanoparticles e.g., Ag$_2^+$, Ag$_4^+$, or Ag$_8^+$.[6,7]

1. D. Pletcher and Z. Poorabedi, *Electrochim. Acta*, 1979, **24**, 1253–1256
2. W. J. Albery, B. G. D. Hagget, C. P. Jones, M. J. Pritchard and L. R. Svanberg, *J. Electroanal. Chem.*, 1985, **188**, 257–263
3. M. E. Bodini and D. T. Sawyer, *Anal. Chem.*, 1977, **49**, 485–489
4. A. G. Fogg, S. P. Scullion, T E. Edmonds and B. J. Birch, *Analyst*, 1991, **116**, 573–579.
5. N. G. Carpenter and D. Pletcher, *Anal. Chim. Acta*, 1995, **317**, 287–293.
6. B. G. Ershov, E. Janata, A. Henglein, *J. Phys. Chem.*, 1993, **97**, 339–343.
7. A. Henglein, *Chem. Mater.*, 1998, **10**, 444–450.

Professor Pletcher remarked:

The reduction of nitrate to nitrite should not be treated as a 'simple reaction'. Firstly, the reaction involves two electrons and the cleavage of a N–O bond as well as protonation steps and little is known about the order of these events. Secondly, with a copper electrode in acid solution, the major product is the ammonium ion formed in a 8e$^-$/10H$^+$ reduction and the evidence shows that the reduction of nitrite is rapid compared to that of nitrate. Hence, the copper surface may have a coverage by many intermediates from all steps of the conversion of nitrate to ammonia when the 'reduction of nitrate to nitrite' is being studied. The literature on the reduction of nitrate on polycrystalline copper shows conclusively that the rate of reduction is enhanced by the presence of freshly deposited copper. This can result from the presence of Cu^{2+} in the electrolyte or by pulsing a copper electrode between appropriate potentials. It is therefore tempting to suggest that metastable or readily poisoned 'active sites' are important for nitrate catalysis. I do not believe that there is any evidence that Cu(I) is an important intermediate in the formation of these active sites although the formation of a Cu(I) layer on the surface during anodic polarisation would provide a ready source of material for the formation of freshly deposited copper.

Professor Gewirth answered:

There are two points raised here: first, that nitrate reduction should not be considered a 'simple' reaction and second, that multiple intermediates might be present on the surface, given the relative rates of nitrate and nitrite reduction. With regards to the first statement, one can only agree that the nitrate reduction reaction is substantially more complicated than, say, the ORR, which itself is (as this conference has shown) not easy. However, the presence of a complicated reaction should be viewed as a challenge rather than an impediment to present and future research. If our community rejected every not 'simple' reaction, we would not be addressing reactivity important to societal needs.

Of course the commenter is correct that at different points in the reactivity, different intermediates will undoubtedly be present on the Cu electrode surface. It is reasonable that the number and type of intermediates will depend on the electrode potential. As we showed in our previous work[1], the number of electrons passed depends very strongly on the applied potential. At the potentials used here, we found 2 electrons were passed. If we used more negative potentials, then more electrons were passed. Of relevance is the SERS spectra, which showed only nitrate and nitrite on the surface at the relatively modest potentials used here. Of course, this situation changes at more negative potentials. The part of the comment starting at "The literature…" is hard to address. On the one hand, the commenter says that freshly deposited Cu enhances the rate of nitrate reduction. On the other, he says that he 'do[es] not believe' that Cu(I) is responsible, even though Cu(I) can be formed during the deposition of Cu metal. There is no justification given in this comment for the latter statement of belief. What we are doing here is examining this reaction in detail, utilizing tools not previously directed at this problem. What we show is that there is in fact a signature for a Cu(I) oxo species present on the surface in the potential region that nitrate reduction occurs. This signature is not present in the absence of nitrate. Does this prove that Cu(I) is the responsible species? Of course not, and we don't suggest this level of proof. All we can do at the moment is make the association.

1. S.-E. Bae, K. L. Stewart and A. A. Gewirth, *J. Am. Chem. Soc.*, 2007, **129**, 10171.

Professor Schiffrin remarked:

It would appear from your results that Cu(I) could indeed be a mediator for nitrate reduction. It is proposed that the difference between the (111) and the (100) surfaces is due to the higher stability of an oxide layer. How stable do you think this layer would be in the acid media that you have used in your work? I am not sure that such a layer would be stable in the conditions of your experiments. For polycrystalline copper in slightly alkaline solutions, there is good ellipsometric evidence indicating that monolayer oxidation takes place at potentials negative to Cu_2O formation.[1] It is not necessary to assume, however, the presence of an oxide layer in acid media since anion adsorption can lead to the stabilisation of surface species that formally can be regarded to correspond to an oxidation state higher than Cu(0). I wonder if the electrocatalytic properties of copper for nitrate reduction are related to a surface redox catalytic mechanism involving the coordination of these surface species, formally Cu(I), to the nitrate ion. In this respect, it is interesting to notice that $Cu(I)NO_3$ does not appear to be a stable species and I could only found a couple of references to this compound in very old literature.[2] It should be noticed that anions such as Cl^-, Br^- or I^- strongly inhibit the rate of reduction. During this Discussion, Markovic[3] has argued on the key role played by adsorbed anions on electrocatalytic reaction and Koper[4] has proposed that inhibition is due to competitive adsorption. It could be that, as proposed above, that coordination by anions is responsible for the stabilisation of the reactive Cu(I) species in solution.

1. Silvia Ceré, Susana R. de Sanchez, David J. Schiffrin, *J. Electroanal. Chem.*, 1995, **386**, 165–171
2. Leopold Gmelin, Gmelins Handbuch der anorganischen Chemie, Deutschen Chemischen Gesellschaft, Verlag Chemie, Weinheim, c1926-1982, p. 3.
3. N. M. Markovic, *Faraday Discuss.*, 2008, **140**, DOI: 10.1039/b803714k.

4. G. E. Dima, A. C. A. de Vooys and M. T. M. Koper, *J. Electroanal. Chem.*, 2003, **554–555**, 15–23

Professor Gewirth replied:
This is a good comment and raises a number of important questions. First, there is the issue of stability of the putative Cu(I) species in the acid environment we are working with here. Of course, the Pourbaix diagram shows clearly that there is no bulk stability of a Cu(I) oxide at the low pH values used here. And we agree that monolayer oxidation occurs at more negative potentials that bulk oxide formation, at least at higher pH values.

So this comment asks about the nature of the Cu(I) oxide species. What do we really know about the intermediates on the Cu surface? We've performed SERS measurements and these may show evidence for a Cu–N bond, but this evidence isn't definitive. The SERS measurements also show the presence of Cu(I)–O features is dependent on the addition of nitrate to the solution. But what is stabilizing the Cu(I) species? Does this species exhibit general stability, or is it some sort of transient? At this point, we don't know. Is the origin of the difference between Cu(100) and Cu(111) due to anion (particularly Cl$^-$) inhibition? The only things we can say are the following: (a) *in situ* STM images from Cu(100) in nitrate-containing solution don't show the directionality of the (2 × 2) adlattice expected for a Cl$^-$ decorated surface and (b) the polycrystalline voltammetry is in complete agreement with that in the literature. Of course, we can't prove that a halide anion isn't present here, and Markovic in other contexts has proposed that halides dominate nearly all reactivity.

The final point to make here is one I've heard attributed to Jack Halpern. He proposed that when an observed species is proposed as an intermediate in a catalytic reaction, the only thing one can say about that species is that it exhibits the kind of stability that makes it certain it is not an intermediate. We are still very much in the dark regarding many aspects of electrocatalysis.

Professor Buess-Herman asked:
1. Can you make a correlation between the presence of the hysteresis in the CV curve of Fig 1 in the paper and your STM data? We have experienced that the occurrence of the hysteresis is a function of the initial potential and could be due to traces of oxidized copper.

2. Our CV curves recorded on Cu(poly), Cu (111), Cu (100) in HClO$_4$ 0.1 M + 5 mM NaNO$_3$ indicate that larger overpotentials are required at the Cu(111) compared to Cu(100) and Cu(poly). The electroreduction of NO$_3^-$ is of course a multi step reaction and even the first reduction step is complex. In neutral solution the response is split in two and crystallography of the surface has a different effect on the peaks with still a larger over voltage on Cu(111) for the first peak. Extensions of EC-STM measurements to higher pH would give more insight on the first step.

Professor Gewirth replied:
1. We have experienced that the occurrence of the hysteresis is a function of the initial potential and could be due to traces of oxidized copper. We didn't find that the hysteresis on Cu(100) was dependent on the initial potential after cycling. The STM images showing the nitrate to nitrite interconversion on Cu(100) were obtained at −0.220 and −250 mV *vs.* Ag/AgCl. These potentials are just where the voltammetry deviates from zero current on the cathodic sweep. The images showing the putative Cu(I) oxide feature on Cu(111) were obtained at somewhat more negative potentials.

2. We don't have any insight into the nitrate reduction mechanism at higher pH values. It may well be that the mechanism is different at different pHs. There is precedent here in the ORR, which appears to work differently at high pH relative to the acid situation.

Professor Pletcher commented:

I would caution strongly against the change in pH as a parameter for investigating the mechanism of nitrate reduction. There is ample evidence that (a) the product of nitrate reduction at a copper cathode depends strongly on the ratio of nitrate/proton in the solution under study (b) if a large excess of protons is not present, complications arise from proton starvation in the reaction layer at the cathode surface. Neither observation is surprising since the reduction of nitrate always consumes protons while the formation of the ammonium ion requires $10H^+/NO_3^-$.

Professor Gewirth answered:

While I agree with this comment in that the already complex nitrate reduction situation gets even more complex at higher pH values, I have one caveat: a real nitrate reduction catalyst will likely be required for natural waters at or near neutral pH. If we restrict our study to the less practically relevant acid situation, we will not address this important societal need.

Professor Sun asked:

STM is generally sensitive to atoms (space resolution) but is not sensitive to molecules, *i.e.* it does not have the ability to determine molecules on a surface. How could it be possible to identify from STM image the nitrate and nitrite ions on the surface?

Professor Gewirth replied:

The commenter is absolutely correct in that it is very difficult with STM to make a definitive identification of any species on the surface. However, the STM has been utilized to image molecules—even in the electrochemical environment—for many years.[1] These molecules have different shapes and electron density profiles relative to the bare surface or surfaces modified in other ways. In the present case, we showed that addition of nitrate to a Cu(100) surface gave rise to an adlattice different from (a) the bare surface and (b) the halide modified surface. Likewise, the addition of nitrite to Cu(100) gave yet another adlattice structure. It is reasonable to associate these adlattices to the two different anions on the surface.

1. A. A. Gewirth and B. K. Niece, *Chem. Rev.*, 1997, **97**, 1129.

Professor Wieckowski opened the discussion of the paper by Kohei Uosaki*, Hidneori Noguchi** and Tsubasa Okada:

Former Russian literature (Frumkin *et al.*) should be reviewed in terms of PZC in the context of the new SFG data which correlate the SFG intensity and PZC.

Professor Noguchi replied:

We added a new reference (A. N. Frumkin and O. A. Petrii, *Electochim. Acta*, 1975, **20**, 347.) to explain the PZC of Pt electrode in the text.

Professor Tsirlina remarked:

Let me present a more general comment concerning platinum metal pzc, in addition to the previous comment concerning the paper of N. Markovic (*Faraday Discuss.*, 2008, **140**, DOI: 10.1039/b803714k). We should not mix **two types of pzc** (pzfc and pztc for free and total charge, respectively) and a possible existence of **two pzfc** in solution of certain composition. The latter situation (mentioned today by Andrzej Wieckowski) had been first predicted in 1936.[1] Direct experimental observations **of two pzfc** were reported for Ir in alkaline iodide-containing solutions,[2] Pt in sulphate solution, pH 6,[3] and Rh in chloride-containing solutions, pH 5–9.[4] There are many examples of free charge *vs.* potential curves with the pronounced decrease of free charge in the oxygen adsorption region, when more positive pzfc was approached, but not attained. The reason for the second pzfc is oxygen adsorption (or hydroxide adsorption with charge transfer). The residual negative charge of oxygen atoms decreases the adsorption of anions and supports the adsorption of cations. Finally the charge of adsorbed ions changes its sign, which means that free electrode charge passes a new ("inverted") zero point. When superposition of hydrogen and oxygen adsorption is very pronounced (*e.g.* in alkaline media), the anomalous adsorption of cations starts at lower potentials, and these systems usually demonstrate no pzfc in the overall potential range.

General consideration of electrochemical surface thermodynamics can be found in ref. 5–7. Interrelations of various thermodynamic quantities form the basis for experimental techniques of pzfc and pztc quantitative determination. Unfortunately, for pzfc these techniques are still limited to high surface area electrodes and solutions with excess of supporting salt electrolyte. This is the reason why only pztc values are available for single crystalline surfaces in acids, when pzfc for these systems remain uncertain. The most complete sets of data on pH-dependent pzfc and pztc are available for Pt in sulphate and chloride media.[3,8] For pH below 5–6 pzfc is always lower than pztc. At low pH, the difference amounts to several dozens of mV and decreases with pH. For sulfate media, extrapolation of these dependences to pH 0–1 results in pztc values of 0.32–0.33 V RHE and pzfc values of 0.22–0.23 V RHE. For perchloric solutions, application of the abovementioned techniques is strongly complicated by chloride formation taking part in parallel with H UPD. The potential dependencies of platinum free charge for 0.005 M H_2SO_4 + 0.5 M Na_2SO_4 and 0.01 M $HClO_4$ + 1 M $NaClO_4$ are compared in ref. 9. The values of pzfc are very close (~0.3 V RHE), despite much lower perchlorate adsorption.

1. A. Shlygin, A. Frumkin and V. Medvedovskii, *Acta Physicochim. URSS*, 1936, **4**, 911.
2. O. A. Petrii and Nguyen Van Thieu, *Elektrokhmiya*, 1970, **6**, 408.
3. T. Ya. Kolotyrkina, O. A. Petrii and V. E. Kazarinov, *Elektrokhimiya*, 1974, **10**, 1352
4. R. Notoya and O. A. Petrii, *Dokl. Akad. Nauk SSSR*, 1976, **226**, 1117.
5. A. N. Frumkin, O. A. Petrii and B. B. Damaskin, *J. Electroanal. Chem.*, 1970, **27**, 81.
6. A. N. Frumkin and O. A. Petrii, *Electrochim. Acta*, 1970, **15**, 391.
7. A. N. Frumkin and O. A. Petrii, *Electrochim. Acta*, 1975, **20**, 347.
8. A. N. Frumkin, O. A. Petrii and T. Ya. Kolotyrkina-Safonova, *Dokl. Akad. Nauk SSSR*, 1975, **222**, 1159.
9. S.Ya. Vasina and O.A. Petrii, *Elektrokhimiya*, 1970, **6**, 242.

Dr Biedermann asked:

How thick is the water layer that is probed by the SFG experiment? Is it possible to estimate the percentage of orientated water molecules corresponding to the SFG intensity observed (*e.g.* 10% of a monolayer)?

Professor Noguchi responded:

According to the theoretical calculation of interfacial water by A. Morita *et al.*, SFG signal comes from the one or two monolayer of interfacial water. (A. Morita *et al.*, *Chem. Phys.*, 2000, **258**, 371) However, in our previous SFG experiments on water, we found linear relation between the intensity of OH band and (ionic strength)$^{1/2}$, which is proportional to the double layer thickness (S. Nihonyanagi *et al.*, *Electrochim. Acta*, 2001, **46**, 3057). This result suggest that all the molecules in the double layer region will correspond to the SFG signal. Thus, we may say that all the oriented water molecules are corresponding to the SFG signal. But we also need to consider the orientation of water dipole, which becomes SFG active and further quantitative analysis to estimate the number of oriented water molecules depending on electrode potential is still now under consideration.

Dr Ikeshoji asked:

We have simulated Pt-liquid water interface by the first principles molecular dynamics at 80 °C under several electric potentials. The results are shown in the cover article of PCCP (M. Otani *et al.*, *Phys. Chem. Chem. Phys.* 2008, **25**). Water molecules at no bias (*i.e.* at PZC) are mainly parallel to the Pt surface and oriented randomly in another directions. When Pt surface is negatively charged with H_3O^+ in water, surface water molecules are in the H-down configuration, and more structured at the higher bias. But, they are not in the ice, though they are connected by the hydrogen bond. In our calculation, we have only 12 surface Pt atoms in the unit cell (36 Pt atoms and 32 water molecules in total) under the periodic boundary conditions and the simulation time is short (4 ps for each bias). So, on the larger surface, we cannot avoid a possibility of coexistance of two phases, ice and randomly oriented structure. My question is whether the two SFG peaks observed by Noguchi *et al.* can be assigned as peaks from the structured water layer obtained in our-simulation; it is neither completely iced nor in random.

Professor Noguchi replied:

The SFG spectra of water at Pt surface showed two broad peaks centered *ca.* 3200 cm^{-1} and *ca.* 3450 cm^{-1}. These results suggest that two phase "ice-like" ordered water (not "ice") and "liquid-like" disordered water co-exist at the Pt surface. We also think that the *ca.* 3200 cm^{-1} peak is originated from an ordered water layer which are connected by the hydrogen bond. Thus, we think your conclusions obtained from the theoretical calculations support our SFG results.

Professor Wieckowski commented:

What is the origin of the relation between the SFG minimum and the PZC? Water desorbs to UHV at higher temperature from Pt than from Au. Therefore, differences between water spectra from Pt and Au are expected.

Professor Noguchi responded:

We think that the SFG intensity is related to the orientation of the water dipole. Since we used "p" polarized IR light for SFG measurements, the water dipole which is normal to the electrode surface becomes SFG active. Thus, minimum SFG intensity at pzc suggests that at pzc, orientation of the water dipole is almost parallel to the surface. Water desorption measurements carried out in UHV conditions suggest that the interaction between Pt and water molecules is stronger than Au. We also think that difference between Pt and Au originates from the interaction between the metal surface and water molecules. Recent IRAS measurement reported by M. Nakamura *et al.* (*J. Phys. Chem. C*, 2008, **112**, 9458) measuring the potential dependence of the peak wavenumber shift of water δHOH bending mode at Pt and Au electrode also support our view.

Professor Korzeniewski remarked:

The free O–H stretching mode for water is frequently observed in SFG spectra of water at liquid–liquid interfaces. Why is a band for the mode absent in your spectra of water–Pt and water–Au interfaces?

Professor Noguchi replied:

The free OH stretching peak can also be observed in SFG spectra of water at solid–liquid interfaces. However, the appearance of free OH band depends on the wettability of the solid surface. When the surface is hydrophobic in nature, *e.g.* quartz–water interface, no free OH band was observed. We think this is the case in Pt–water and Au–water interfaces. Clean Pt and Au surfaces are both hydrophobic. However, when the surface was hydrophilic in nature, *e.g.* octadecyltrichlorosilane (OTS) modified quartz–water interface, a free OH band appears in the SFG spectra (*Phys. Chem. Chem. Phys.*, 2001, **3**, 3463). This situation is similar to the liquid–liquid interface.

Professor Tryk commented:

Is it possible, and if so, do you plan to carry out measurements with the single crystal or single crystal-like surfaces? This is important because we need to try to bridge the gap between theory and experiment and the use of single crystal surfaces makes this process easier. Is this possible in the ATR mode, for example, with a very thin epitaxial layer?

Professor Noguchi replied:

I also agree with you that the single crystal experiments are very important. At the present moment, however, it is impossible to use a single crystal surface in the ATR configuration. When thin metal layers were deposited on flat substrates, all surfaces become polycrystalline. All the single crystal measurements are carried out in external reflection configuration. Prof. Gewirth's group are studying the interfacial water using single crystal surfaces in external reflection configuration (*J. Am. Chem. Soc.*, 2005, **127**, 15916). We are also planning to use single crystal in external reflection configuration.

Professor Neurock commented:

As you are scanning the potential and examining the structure of the water and the water–metal interface. I would imagine that at higher potentials you would see anion adsorption from the electrolyte. This would change the structure of water at the water/metal interface. You would also see differences of where this occurs for Pt *vs.* Au. Can you please comment on the effects of anions?

Professor Noguchi responded:

We also think that the effect of anions on the structure of interfacial water is important. Previously, we measured the potential dependence of SFG spectra of water at the Au–H_2SO_4 interface. (S. Nihonyanagi *et al.*, *Surf. Sci.*, 2004, **573**, 11) From careful analysis of SFG spectra, it was found that water molecules do not flip-flop as the electrode potential changes from more negative to more positive than pzc. We concluded that the water molecules are incorporated with sulfate anions with their hydrogen atoms pointing to the surface in the negative potential region. Thus, it is true that the structure of interfacial water will be affected by anions. At the present moment, we think that the different water structure between Pt and Au originates from the difference of the interaction between metal surface and water molecules. We couldn't say anything about the effect of anions on the difference between Au and Pt. We must improve the quality of SFG spectra to discuss more about the structure of water including the effect of anion.

Dr Fernandes Gomes commented:
Is there any evidence of C–H peaks in methanol?

Professor Noguchi answered:

We have no evidence of CH bands in the methanol oxidation experiments. CH stretching peaks generally appears in the 2800–3000 cm^{-1} region. This region is in a low wavenumber edge of the broad OH stretching peak coming from the interfacial water. Thus, we think the reason why we couldn't observe CH bands are (1) CH bands are buried in the intense OH band. (2) Adsorbates are in an SFG inactive configuration. May be we should try the methanol oxidation reaction in D_2O, which makes the CH bands region more clear.

Professor Chen commented:

1. In page 8, you argued that the SFG intensity should become weaker at potentials away from the pzc since the number of ordered water molecules at the Pt–electrolyte solution interface would decrease as the electrode becomes more highly charged. This is contrary to your experimental result. Actually, the water molecules adjacent to electrode surface should become more oriented with increased surface charge due to the increased strength of the interfacial electric field. Therefore, the number of ordered water molecules increases as the potential departs

from pzc. If the SFG intensity increases with the number of ordered water molecules, you would observe an increase of SFG intensity at potentials away from pzc. This matches your experimental results?

2. When you correlate the SFG signal to the structures of the Pt–H_2O interface, you may have to bear in mind that the water molecules at the Pt electrode surface have a bi-layer structure rather than a homogeneous monolayer. In the water bi-layer structure, one half of the water molecules orient parallel to the electrode, while another half are normal to electrode surface. Moreover, the parallel ones and the normal ones are located at different distances away from electrode surface and these distances vary with electrode potential. Do you think there are possibilities that the observed SFG peaks and intensities are related to the water bi-layer structure and the potential dependent distance of water molecules to the electrode surface.

Professor Noguchi answered:

1. From the linear relation between the SFG intensity and the (ionic strength)$^{-1/2}$, we know that SFG intensity reflects the number of oriented water molecules in double layer region (*Electrochim. Acta*, 2001, **146**, 3057). At the electrode–electrolyte solution interface, however, the drop of the potential profile at the vicinity of the electrode surface become precipitous as the surface becomes more charged. This means that the thickness of the double layer becomes thinner as the electrode becomes more highly charged and we should observe the weakening of SFG intensity away from the pzc. However, this was contrary to our results. Thus, as we mentioned in our paper, the potential dependence of SFG intensity is not only related to the thickness of the double layer but also related to the orientation change of the water dipole. Since the polarization of the IR in the present experiment is "p", water oriented normal to the surface is effectively detected by SFG.

2. Since we are using a polycrystalline Pt surface in our experiments, we don't think a well ordered bi-layer structure of water molecules is formed on the Pt surface. However, if we are able to use single crystal surface, maybe we need to consider about the bi-layer structure. It would be very interesting if we can probe the potential dependent distance change of bi-layer water molecules on Pt surface. However, we also observed the same potential dependence of SFG intensity even at an Au electrode. So at this moment, we don't think a bi-layer structure is important to explain the potential dependence of SFG intensity.

Professor Gross opened the discussion of the paper by Fatang Tan, Bingchen Du, Aaron L Danberry, In-Su Park, Yung-Eun Sung and YuYe Tong*:

Your results of a Pt deprivation at the surface of the PtRu particles are rather surprising. First, Pt atoms are bigger than Ru atoms (the nearest-neigbour distance in Pt bulk is 2.5% larger than in Ru bulk). Hence there is a much better strain relief if the Pt atoms are at the surface and not inside the PtRu particles. Second, the cohesive energy of Ru is larger than the one of Pt and, furthermore, the Pt–Ru interaction is stronger than the Pt–Pt interaction.[1] Therefore the energy gain at the higher coordinated sites in the interior of the PtRu particles should be larger for Ru than for Pt, so that the Pt atoms should be preferentially located at the lower coordinated surface sites. Do you have an explanation for your surprising results?

1. M. Lischka, C. Mosch and A. Gross, *Electrochim. Acta*, 2007, **52**, 2219.

Professor Tong responded:

This is a good question but unfortunately I do not have a good answer at this moment. Thermodynamically, your reasoning is very sound. But for bimetallic nanoparticles, surprising experimental observations in terms of local composition and segregation have not been uncommon in the literature. For instance, for bulk

Pt–Rh alloy systems, it is expected that Pt would be enriched in the surface by following the same reasoning you have put forward and has indeed been observed experimentally on bulk Pt–Rh alloy surface in ultra-high vacuum. However, Slichter, Sinfelt and co-workers observed two decades ago that there was instead a significant enrichment of Pt in the core of PtRh nanoparticles of about 8 nm in diameter and a nominal atomic Pt : Rh ratio of 4 : 1.[1] A more recent example of such a counter-intuitive result was obtained on AuPd nanoparticles where Pd-rich shell/Au-rich core morphologies were observed,[2] although thermodynamic considerations would predict the contrary. For PtRu nanoparticles, the situation is further complicated by the easily formed multiple phases of Ru oxides during particle synthesis, thermal treatments, and electrochemical processes.

1. Z. Wang, J.-Ph. Ansermet, C. P. Slichter and J. H. Sinfelt, *J. Chem. Soc., Faraday Trans. 1*, 1988, **84**, 3785.
2. A. A. Herzing, M. Watanabe, J. K. Edwards, M. Conte, Z.-R. Tang, G. H. Hutchings and C. J. Kiely, *Faraday Discuss.*, 2008, **138**, 337.

Professor Russell remarked:

Do you have any additional characterisation of your catalysts? Do you know from XRD *etc.* that the particles are really 1 : 1 Pt : Ru? It could be that the reason you see less Pt at the surface then expected is that your particles are poorly mixed to start off with, with perhaps a surface layer enriched with Ru or Ru oxides.

Professor Tong responded:

Yes, we do. The as synthesized samples were characterized by both TEM/EDS and XRD. The atomic Pt : Ru ratios as determined by EDS were about 0.53 : 0.47, which was very close to the nominal 1 : 1 ratio. The alloyed fraction of Ru as determined by XRD was about 0.32. This suggests that part of the Ru was unalloyed. However, Pt-195 NMR cannot tell whether or not any form of Ru oxide has existed on the surface.

Professor Strasser asked:

Question about Fig 1B in your paper: Resonance frequency is correlated to position within the nanoparticles. But in a particle ensemble with a size distribution, there could be Pt atoms at different locations within the particle, yet similar frequencies. Is this possible and how would that effect the labelling in your Fig 1B? Would large particles not contain large number of Pt atoms with identical frequencies, so there should be a larger peak at higher frequencies?

Professor Tong answered:

For clean-surface pure Pt nanoparticles, it has been observed that the resonance positions of surface, sub-surface, sub^2-surface, *etc.* are largely independent of the particle size. The NMR layer-model analysis, as shown in Fig. 1B, is rather solidly based on sound experimental observations[1] and theoretical calculations.[2] But for Pt-based bimetallic nanoparticles, such a frequency-geometric location correlation is still largely an assumption.

1. J. P. Bucher, J. Buttet, J. J. van der Klink and M. Graetzel, *Surf. Sci.*, 1989, **214**, 347.
2. M. Weinert and A. J. Freeman, *Phys. Rev. B*, 1983, **28**, 6262.

Professor Wieckowski remarked:

1. For Pt/Ru nanoparticles the Urbana EC-NMR data show that platinum is separated to the top and that Ru is in the core of the nanoparticles.
2. The Tong NMR data and Urbana NMR data are identical as the same samples were looked at (JM samples).

Professor Tong replied:

The PtRu samples that we have studied are actually not the exactly the same samples as studied by the Urbana group, who was the first to use ^{195}Pt EC-NMR to investigate the PtRu nanoparticles. Our samples were synthesized in house while the Urbana group's were commercial samples. Since the exact local composition may well depend on the history of the samples, it would not be surprising that different conclusions arose.

Professor Behm commented:

A comment to the question raised by A Gross. If at the last moment when segregation in the nanoparticles was possible, oxygen was present on the surface (or oxygen containing adsorbates), they would, because of the strong bonding to Ru, stabilise Ru at the surface, in contrast to the situation of adsorbate free nanoparticles where Pt would be more stable.

Professor Tong responded:

I agree. I have no doubt that this would happen at a high temperature. However, how fast Pt can exchange with the initially surface-enriched Ru and segregate in the surface at room temperature in the situation of free adsorbates depends critically on how big is the kinetic barrier for such an exchange, which is actually unknown.

Dr Thompsett remarked:

How well alloyed are the samples (PtRu) reported in the paper? Fig 2 in the paper shows the XRD patterns for the various PtRu samples studied, however no lattice parameter have been reported to indicate the degree of Pt and Ru mixing. If the samples do not show a lattice parameter close to 3.85 Å this would suggest that the sample may contain unalloyed Ru (hydrous Ru oxide) which may obscure surface Pt atoms. If the samples are not completely alloyed it would be interesting to also look at well alloyed samples, perhaps treated to different temperatures to probe surface segregation changes.

Professor Tong replied:

As I indicated when addressing Professor Russell's question, all as synthesized samples showed a 2θ value of 68.2° for Pt(220) XRD peak which corresponds to a lattice constant of 3.88 Å. This suggests that the fraction of alloyed Ru is about 0.32, which is smaller than the Ru content of about 0.47, as determined by TEM/EDS, and indicates the existence of unalloyed Ru, as you pointed out. It would be highly interesting to see what the NMR results would be for a completely alloyed system, which will be on our to-do list in the near future.

Professor Savinova said:

First of all, it is necessary to specify whether equilibrium or non-equilibrium structures are discussed. In fuel cell related research, reactions are often carried out at ambient or slightly elevated temperatures. Thus, metal particles may preserve their non-equilibrium component distribution and shapes (set by the preparation conditions) for a long time. This can be explained by the high activation energies required for the transition into the equilibrium structures.

Second, it is now well established that segregation strongly depends on the adsorption. For example, for PtRu particles it is expected that hydrogen adsorption favours Pt segregation to the surface, while oxygen adsorption favours Ru segregation. This might explain the differences between the conclusions of this work and those of the group of A. Wieckowski.

Third, recently it became apparent that the segregation phenomena depend also on the particle size. It is a challenging task to determine the phase diagrams of bimetallic nanomaterials, and find out how they depend on the particle size. This matter is now being explored by the physicists and can only be understood through concerted efforts of theoreticians and experimentalists.

Professor Tong replied:
I agree completely with your comments

Dr Buder opened the discussion of the paper by Fatang Tan, Bingchen Du, Aaron L Danberry, In-Su Park, Yung-Eun Sung and YuYe Tong*:
The Pt/Ru fraction changes with potential. Since PtRu is used as a catalyst in fuel cells this result is important for the development of these catalysts. Do you have a strategy for improving PtRu catalysts for fuel cells? The change of the Pt/Ru fraction indicates a migration of Pt particles or Ru particles. Do you have an idea where they are migrating?

Professor Tong responded:
It is indeed a critical challenge that we face if we want to go beyond the trial-and-error approach in designing and developing more active and stable electrocatalysts. I hope that as we understand better the PtRu systems, which still has a long way to go, we may be able to devise a viable strategy to address adequately this challenge. As to your second question, the change of Pt fraction does not imply the migration of Pt particles or Ru particles, but indicates a diffusion of Pt atoms within the PtRu nanoparticles.

Professor Scherson opened the discussion of the paper by Bin Ren*, Xiao-Bing Lian, Jian-Feng Li, Ping-Ping Fang, Qun-Ping Lai and Zhong-Qun Tian:
Why not use a micro-electrode instead of flowing liquid through the cell? The net rates or mass transport can greatly exceed that associated with flow through the cell.

Professor Ren replied:
We have not tried that, but it is a good idea to use a micro-electrode to achieve the similar function of a flow cell. In order to ensure a high collection efficiency in electrochemical *in-situ* SERS study, the solution layer is very thin, usually in the order of several hundred microns. The existence of the optical window will alter, to some extent, the mass transport in the thin solution layer. Therefore, at present we still prefer a flow cell that can be controlled to a condition comparable to the real working conditions of fuel cells. The second problem is that it is still a challenge to disperse metal nanoparticles uniformly over the micro-electrode to ensure the reproducibility in different measurements.

Professor Korzeniewski commented:
Where is the reference electrode placed in your flow cell? Do you take precautions to maintain the electrode at a constant temperature and the stable potential?

Professor Ren responded:
The reference electrode is placed in the down stream of the flow cell to eliminate the contamination and the reference electrode chamber is on the peripheral of the flow cell. Due to this kind of configuration, the temperature of the reference electrode is only slightly above the room temperature even when the cell is at the elevated temperature. Therefore, we have not paid attention to this problem. Anyway, it is really worthwhile to calibrate the temperature in the reference electrode chamber so that the electrode potential can be comparable at different temperature.

Professor Wieckowski asked:
Avoid SERS! Further progress in infrared measurements is most welcome. In the next generation of infrared measurements, new windows need to be used/constructed to open up the spectral range for higher and lower frequencies than those currently provided by conventional windows.

Professor Ren answered:

I do not agree with the comment "avoid SERS!". Every technique has its advantage and disadvantage and up to now there is still no perfect technique that can meet all our demand for obtaining the information from the electrochemical interfaces. Among the vibrational spectroscopies that can be applied for *in-situ* study, SFG and IR are suitable for single crystal surface of various materials with very strict surface selection rules. But both of them are extremely difficult to use to obtain the low frequency information regarding the metal and adsorbate bonds. Although the use of a synchrotron IR source may provide an alternative way to solve the problem, there are still limited successful reports indicating the great difficulty of such kind of study.[1,2] Furthermore, the synchrotron facility is not accessible by most scientists. Therefore, IR using a synchrotron source has an intrinsic barrier to become a general technique in electrochemistry.

To further increase the detection sensitivity of IR spectroscopy for the electrochemical interfaces, there is a growing interest of using surface-enhanced infrared spectroscopies (SEIRA and AIREs)[3,4] that require the surface to be rough or contain nanostructures, similar to SERS. Therefore, the problems associated with SERS will also be associated with surface-enhanced infrared spectroscopies. Undoubtedly, it will be highly desirable to further improve the detection sensitivity of infrared by designing new types of windows that will allow a high throughput in the low frequency region. But, up to now, there is still no immediate solution to the problem. On the other hand, infrared, Raman and SFG are complementary vibrational techniques. If one is able to employ more than one technique to investigate a complex system, the electrochemical system, it may be very beneficial for obtaining a complete view of the studied system. As SERS is an effect originated from surface with nanoscale structures, it is especially advantageous in studying the rough surfaces commonly used in electrocatalysis. The newly emerging techniques, such as tip-enhanced Raman spectroscopy, have provided a unprecedented chance for investigating the process on the single crystal surfaces.

1 C. A. Melendres and F. Hahn, *J. Electroanal. Chem.*, 1999, **463**, 258.
2 F. Hahn F, Y. L. Mathis, A. Bonnefont, F. Maillard and C. A. Melendres, *Infrared Phys. Technol.*, 2008, **51**, 446.
3 Y. X. Chen, A. Miki, S. Ye, H. Sakai and M. Osawa, *J. Am. Chem. Soc.*, 2003, **125**, 3680.
4 G. Q. Lu, S. G. Sun, L. R. Cai, S. P. Chen, Z. W. Tian and K. K. Shiu, *Langmuir*, 2000, **16**, 778.
5 B. Ren, G. Picardi, B. Pettinger, R. Schuster and G. Ertl, *Angew. Chem., Int. Ed.*, 2005, **44**, 139.

Professor Schiffrin asked:

The main origin of the Raman enhancement observed is due to the local increase in electromagnetic field caused by the Au core. This effect should be clearly reflected in the UV-Vis absorbance of the nanoparticles, and the main 523 nm plasmon band of gold should be a good measure of the ability of the core–shell structure to induce local electromagnetic field enhancement. Have you observed such a relationship?

Professor Ren answered:

Yes, it is true that there are various attempts to make a correlation of the plasmon band of metal nanoparticles with the surface enhancement Raman effect. For example, Van Duyne's group employed a highly ordered AgFON method to obtain a SERS substrate with the defined surface structure, which allows them to correlate the surface enhancement with the plasmon band absorption.[1] It was found that when the excitation line is located slightly to the blue of the maximum of plasmon band, the maximum surface enhancement can be achieved. However, a good correlation is still limited to Ag. In the case of Au, there are still controversies as to whether the SPR absorption wavelength can be related to the maximum SERS enhancement. The common practice for measuring the absorption spectra of Au nanoparticles is made with Au sol. The obtained 523 nm absorption band is characteristic of isolated Au nanoparticles in sol. Whereas, SERS measurement is generally made over Au nanoparticles dispersed over a solid substrate. Therefore, Au nanoparticles are generally in an aggregated state. The aggregation of Au nanoparticles will lead to a red shift of the absorption band. Therefore, the absorption measurement and SERS measurement is usually not comparable due to the different state of Au nanoparticles in the measurement. We have made correlated SERS and absorption measurements on highly dispersed Au nanoparticles immobilized over an ITO surface, i.e., Au nanoparticles are in separated state, and we can only detect an extremely weak SERS signal, although the absorption spectrum clearly indicates a peak absorption at ~530 nm. The main reason may be the overlapping of the interband transition with the SPR, which lead to the degradation of SPR quality and the SERS enhancement.

1 A. D. McFarland, M. A. Young, J. A. Dieringer and R. P. Van Duyne, *J. Phys. Chem. B*, 2005, **109**, 11279.

Professor Sun commented:
In the case of the SERS study of catalysts of Pt group metal you put Au and Ag as core and Pt metals as shell. This kind of configuration means that there is a need for Au metal to enhance the signal and detection sensitivity. Can it give some information about the SERS nature? In comparison with IR spectroscopy the enhancement of IR absorption (both in SEIRA and AIREs) the Pt metals nanostructure are more significant than that in SERS. How do you interpret this effect?

Professor Ren replied:
Differently from the strong and effective surface plamon resonance and therefore the strong SERS of Au, Ag and Cu in the visible region, the SERS from transition metal surfaces is much weaker. The SERS enhancement of transition metals is mainly from the combined contribution from the weak surface plasmon resonance, lightning rod effect, and electromagnetic coupling among nanoparticles, and the chemical enhancement. Among all the Pt-group metals, only Pd shows a distinct surface plasmon resonance with absorption in the visible light region on either rough Pd surface or Pd nanoparticles, but the overall enhancement is still weak. The second part of the question is not clear enough to give a definite answer. The SPR as well as the coupling between nanoparticles depend on the wavelength. Furthermore, IR intensity is proportional to E^2 and SERS intensity is roughly proportional to E^4. The combination of these two effects leads to a difference in the enhancement in the SERS and AIREs. Actually, nanostructured Pt shows a larger surface enhancement in SERS (*ca.* 1000) than in AIREs (<50).

Professor Wittstock opened the discussion of the paper by Y E Seidel, A Schneider, Z Jusys, B Wickmann, B Kasemo and R. J. Behm*:
Does the fractional hydrogen peroxide yield $x(H_2O_2)$ also depend on the macroscopic dimension of the microelectrode array? If yes, what consequence has that for the significance of $x(H_2O_2)$ values determined (and often used)? Have such values any quantitative value if compared between set ups (when macroscopic geometries are not identical)?

Professor Behm answered:
We definitely expect the fractional yield of H_2O_2 to depend on the macroscopic dimension of the microelectrode array, and of the electrode in general. The quantitative values are significant for the present sample dimension, coverage/density of nanostructures (100% coverage = massive Pt electrode), O_2 concentration and electrolyte flow. Experiments performed on different set-ups and under different conditions (reaction/transport parameters) will therefore lead to different fractional yields.

This does not mean, however, that the quantitative values are of no significance. First, they provide the experimental basis of any successful modeling of the underlying transport processes. Furthermore, it was also the purpose of this work to alert the community to the existence of such transport effects and to the fact that they may considerably change/affect the mechanistic interpretation of previous data of electrocatalytic reactions (see also my response to question from Professor Wieckowski later in this discussion).

Dr Wang commented:
Have you carried out the H_2O_2 measurements in $HClO_4$, besides the H_2SO_4 solution? It should be interesting to see whether (bi)sulfate adsorption promotes the desorption of H_2O_2 by blocking nearby Pt sites for the O–O bond breaking, and by lateral repulsion between bisulfate and adsorbed intermediates.

Professor Behm answered:
So far, all measurements were performed in sulfuric acid. But we fully agree that this, and anion effects in general, would be an interesting point for future work. It is on the agenda.

Professor Gewirth commented:
Is it possible that there are Pt sites with different reactivity in your constructs, especially at the Pt–C interface? Could this account for the peroxide production?

Professor Behm answered:
Based on all evidence, including microscopic and electrochemical characterization, the nanostructures consist of polycrystalline platinum. The size of the nanostructures (diameter 100–150 nm) is far above that of the nanoparticle in supported catalysts (typical diameter 3–4 nm). Therefore, in a local structural picture, there should be no difference in the reaction behavior of the nanostructures used here and that of polycrystalline Pt (no particle size effects). Experimentally, however, there are significant differences, which increase with decreasing density/coverage of the nanostructures on the support (increasing H_2O_2 formation with lower density). These differences, which obviously are not based on differences in the local reaction behavior, are the basis of the transport model for the reaction presented here. So far, this model was qualitatively confirmed also in other reactions we had investigated (formaldehyde oxidation, methanol oxidation).

Professor Savinova said:
First of all, congratulations on a very good and important work. However, I would like to mention that the issue of the interplay of reaction/diffusion and re-adsorption events during the ORR was earlier brought up by Anthony Kucernak from Imperial College London (S. L. Chen and A. Kucernak, *J. Phys. Chem. B*, 2004, **108**, 3262–3276), and the authors must cite this reference.

Professor Behm answered:
We apologize for this mistake; the paper was added at the proof stage. For more, see response to Dr Kucernak below.

Dr Kucernak remarked:
Professor Savinova was kind enough to mention my previous work discussed in reference.[1] This work supports the work presented in the paper presented by Behm in showing that at high mass transport (as expected in his system when his electrodes are small and well separated) hydrogen peroxide production is enhanced. In our work, we deposited single platinum particles of different sizes on ultra-small insulated carbon electrodes and studied the oxygen reduction reaction as a function

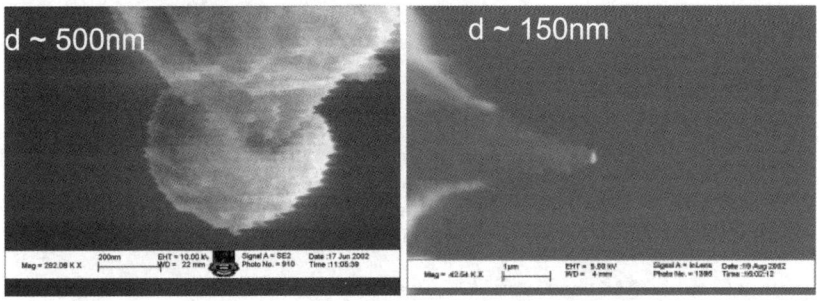

Fig. 3 Platinum particles of diameter 500 nm and 150 nm deposited at the tip of carbon insulated carbon microelectrodes.

of particle size. Examples of two such electrodes are shown in Fig. 3 below, and their manufacture is described in ref. 2.

Experiments were performed on platinum particles from about 50 nm–20 μm in radius, *i.e.* much larger than the size at which the oft quoted "particle size effects" are seen. In measuring the diffusion limiting currents for oxygen reduction at 0.4 V (RHE) we noticed that the effective number of electrons transferred per oxygen molecule decreased from 4 (for large particles, $r > 5$ μm) down to about 3.4 for the smallest particles, Fig. 4 below. We interpret this as the effect of mass transport away from the electrode enhancing the 'loss' of hydrogen peroxide from the electrode—*i.e.* this is a mass transport effect. Smaller platinum particles have higher rates of mass transport (both towards them and away from them), and thus the concentration of hydrogen peroxide immediately adjacent to the surface of the platinum particles is less at smaller particles. This shifts the equilibrium to enhance the production of hydrogen peroxide. In Fig. 4 we show the calculated

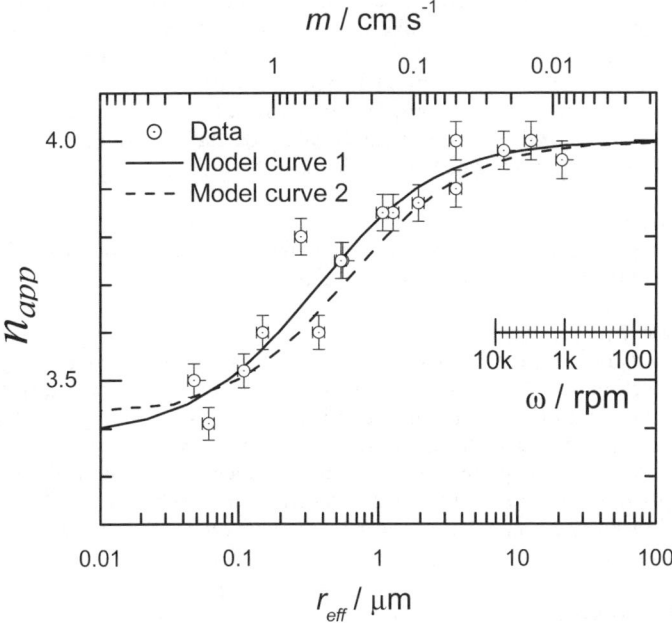

Fig. 4 Variation of the value of n_{eff} with the effective radius of single carbonsupported Pt particles for the ORR at 0.4 V (RHE) in O_2-saturated 0.1 mol dm^{-3} H_2SO_4.

mass transport coefficients and also a bar representing the equivalent rotation rates for a rotating disk electrode. We see that even at the fastest rotation rate, the mass transport rate is still too low to produce significant amounts of hydrogen peroxide and the oxygen reduction reaction almost exclusively produces water. Thus, the geometry of these systems is somewhat different, but the effect is the same—as one increases mass transport of hydrogen peroxide away from the electrode, the rate of production of hydrogen peroxide is enhanced. It would be interesting to see if one could calculate the mass transport coefficients in the paper presented by Behm, and see if the hydrogen peroxide generation rates are similar to those seen by us.

1. Shengli Chen and Anthony Kucernak, *J. Phys. Chem. B*, 2004, **108**, 3262-3276.
2. Shengli Chen and Anthony Kucernak, *J. Phys. Chem. B*, 2003, **107**, 8392-8402.

Professor Behm replied:

First of all we would like to apologize for not having cited this very nice paper initially, was had been corrected in the meantime.

We would like to note, however, that the general concept of transport effects affecting the reaction characteristics, *via* desorption, re-adsorption and further reaction of desorbable reaction intermediates, is a very general one. To the best of our knowledge, we had first formulated that mechanism for methanol oxidation in 2003 (Z. Jusys *et al.*, *Langmuir*, 2003, **19**, 6759), where we reported results of DEMS model studies on the MOR on thin-film Pt/C catalyst electrodes, which showed that increasing the catalyst loading resulted in increasing CO_2 and decreasing formaldehyde yields. In the meantime, similar effects were observed also for other comparable reactions (formaldehyde oxidation, ethanol oxidation *etc.*).

The nanostructured electrodes used in the present study were developed to study such effects on electrodes which are structurally simpler and better defined than a thin catalyst layer, and where the structural parameters can be modified in a very controlled way. In that respect, the work by Chen *et al.* represents a different approach to tackle these kinds of transport effects using a structurally well defined model system, and we are more than happy that this group arrived at comparable conclusions for the oxygen reduction reaction. Finally, a last little comment. The important parameter one would like to obtain from such kind of experiments is the probability for H_2O formation *vs.* H_2O_2 desorption in a single O_2 adsorption event (*e.g.*, for comparison with DFT type calculations), or the corresponding branching ratios in other reactions. We are convinced that this number can only be obtained by introducing novel kinds of modeling, including also stochastic motion of the reaction intermediate in the diffusion layer.

Dr Herrero asked:

According to the results presented in the paper, is it possible to discard the direct pathway for the ORR on platinum electrodes?

Professor Behm responded:

The results presented in this paper clearly demonstrate that the indirect pathway for O_2 reduction, *via* H_2O_2 formation, its subsequent desorption, re-adsorption, and further reduction to H_2O, is important and contributes significantly over the entire potential range of the reaction. They do not allow us, however, to exclude the direct pathway. Based on the present data, on might conclude on a contribution of the indirect pathway of >25% at 0.5 V RHE. A definite answer on the question whether the direct pathway can be ruled out or not will require measurements at significantly higher space velocities.

Professor Wieckowski commented:

Will the modelling of the Behm type lead to providing prediction on the optimised cathode catalyst loading (in terms of the peroxide escape from the cathode)?

Professor Behm replied:

The final objective of this combined experimental and theoretic approach is twofold. First, we would like to alert the community to the importance of transport effects in electrocatalytic reactions and to establish the use of this model ('desorption–readsorption–further reaction model') and/or the underlying ideas in the mechanistic discussion of electrocatalytic reactions wherever adequate. Second, we would like to reach a state where the influence of the dimension, morphology and active material loading in an electrode can be described at least semi-quantitatively, which would allow us to predict optimized electrode structures in a systematic way. It needs to be stated, however, that this will be much more complicated than modeling the transport conditions above the well defined 2D models used in the present study, and hence this will only be the result of very extensive work. Nevertheless, the objective was properly stated.

Applications would involve not only minimizing H_2O_2 escape at the cathode outlet, but also minimizing H_2O_2 formation in the electrode to reduce electrode corrosion, and a number of other reactions, *e.g.*, methanol oxidation, to exclude formaldehyde escape at the outlet of a DMFC at low dimensions and catalyst loading.

Dr Kucernak asked:

An important aspect to realise is that the entire oxygen reduction mechanism can be described without the need to invoke the "Direct Pathway" *i.e.* the results are consistent with all of the oxygen progressing through a peroxide intermediate, labelled "H_2O_2*" in the kinetic scheme below,[1] Fig. 5. This discussion is more fully described in ref 2.

For oxygen reduction occurring under the diffusion limited regime, we can define n_{app} as a function of the reaction kinetics. n_{app} is expected to vary between 2 (all oxygen reduced to H_2O_2) and 4 (all oxygen molecules reduced to water). Intermediate values represent a situation where both water and hydrogen peroxide are formed.

$$n_{app} = \frac{i}{i_{4e}} \times 4 = 4 - \frac{2}{\left(1 + \frac{k_1}{k_2}\right)(1+x)} \tag{1}$$

Where i is the measured current under conditions of defined mass transport and i_{4e} is the current when all of the oxygen reaching the electrode is reduced to water. x is defined in the equation below

$$x = \frac{k_3}{a_{2-}}\left(1 + \frac{a_2}{m_{H_2O_2}}\right) = \frac{k_3}{a_{2-}}\left(1 + \frac{a_2 r_{eff}}{D_{H_2O_2}}\right) \tag{2}$$

Where the critical diffusion step (labelled "diffusion") in Fig. 5 is the mass transport coefficient for hydrogen peroxide in the system under study. For a microelectrode,

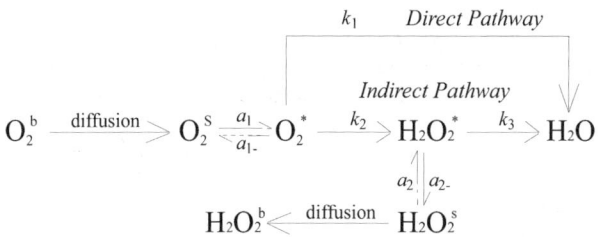

Fig. 5 Wroblowa scheme for the pathways of the oxygen reduction reaction

the mass transport coefficient is $D_{H_2O_2}/r_{eff}$, where r_{eff} is the effective radius of the microelectrode and $D_{H_2O_2}$ is the diffusion coefficient of hydrogen peroxide in the medium used.

Phenomenologically, the two-electron reduction to H_2O_2 would appear to become more pronounced when H_2O_2 mass transport in the vicinity of the electrode surface is increased or adsorption of H_2O_2 is modified (*e.g.* by other adsorbed species). Our data for n_{app} as a function of particle size (and thus mass transport coefficient) is shown in Fig. 4 above. Eqn (1) has five parameters, and thus is over parameterised for fitting the curve in this diagram, but it is important to note that a pure series mechanism for the four-electron reduction of oxygen requires $k_1/k_2 \sim 0$. Model fits to the data require that $k_1/k_2 < 2.5$ to obtain $n_{app} \sim 3.5$. Hence the experimental data can be fit assuming a series mechanism (*i.e.* ignoring any direct 4 electron reduction). This is not proof that the direct 4-electron reduction process does not occur, merely that we do not need to invoke it to explain the results.

1. H. S. Wroblowa, Y. C. Pan and J. J. Razumney, *J. Electroanal. Chem.*, 1976, **69**, 195.
2. Shengli Chen and Anthony Kucernak, *J. Phys. Chem. B*, 2004, **108**, 3262–3276

Professor Morgan addressed Professor Behm:
What is the status of attempted modelling? What are you missing for a successful model?

Professor Behm responded:
At present, we have a complete description of the electrolyte flow in the circular flow cell used in these experiments and of the diffusion transport to a homogeneous electrode under conditions of complete reactant conversion (mass transport limited reaction rate) (J. Fuhrmann *et al.*, *J. Fuel Cell Sci. Technol.*, 2008, **5**, 020301, see also J. Fuhrmann *et al.*, *Phys. Chem. Chem. Phys.*, 2008, **10**, 3784, whose results fully agree with our experimental findings. Results on reactions with incomplete reactant consumption (CO oxidation), also on a homogenous electrode, were obtained and are currently being written up. At present, we are missing the description of the lateral transport of reaction intermediates between the nanostructures (and re-adsorption on the same nanostructure), which would not occur in models (and codes) based on classic diffusion theory (see also my response to Dr Fermin later in this discussion). This part has to be developed, *e.g.*, on the basis of Monte Carlo techniques, and to be implemented into the description of the diffusion processes (reactant diffusion to the electrode surface *etc.*).

Professor Strasser commented:
Would your results not suggest the use of an electrochemical space velocity just like in heterogeneous catalysis? Your results suggest the electrode dimension may affect mechanistic conclusions. Large electrodes show production of H_2O_2 later than small electrodes with otherwise the same conditions. Is that accurate and what does that mean for mechanistic conclusions?

Professor Behm replied:
First question: I fully agree that the use of an electrochemical space velocity would be highly desirable, since it would help to decide on comparable experimental conditions. Even more important, however, is the fact that in many electrochemical model studies the space velocity is rather ill defined, since they are performed in stagnant electrolyte, and the diffusion to the electrode surface in a cyclic voltammogram depends significantly on the experimental parameters and set-up.

Second question: I fully agree with the first conclusion, that the dimension and catalyst loading will affect the selectivity of the reaction (not only the activity, as expected from common considerations). Hence, large electrodes should show less H_2O_2 formation than small ones. It should be noted that the critical size depends very much on the reaction and catalyst. In the present case, under 'normal' transport conditions (electrolyte flow) already a massive Pt electrode is sufficient to essentially suppress H_2O_2 formation, *i.e.*, under these conditions, the probability for readsorption and further reduction is sufficiently high to reach very low H_2O_2 contents. For practical applications, this means that in electrodes of macroscopic dimensions, H_2O_2 formation at the outlet of the fuel cell will be negligible. It may become a problem, however, for micro-structured designs of very low catalyst loading. More important for practical applications is the fact that, although not detected at the outlet, H_2O_2 will be continuously formed in the electrode and may this way result in chemical catalyst/carbon corrosion before being further reduced to H_2O_2.

Regarding mechanistic conclusions, this means that adsorbing a single O_2 molecule and its subsequent reaction will lead with a significant probability (>25% at 0.5 VRHE) to the formation and desorption of H_2O_2, which may, depending on the reaction and transport conditions, re-adsorb and react further to H_2O or not. This probability for formation/desorption of H_2O_2 should be reflected, *e.g.*, in DFT-type calculations of the reaction pathways and the selectivity, as these calculation describe the 'single hit' situation and do not include readsorption of volatile intermediates. The remaining fraction may consist of either direct oxidation to H_2O or indirect oxidation, but without desorption of the adsorbed H_2O_2 species, as described in the reaction scheme depicted below (Fig. 6).

In that scheme, the amount of desorbing H_2O_2 depends on (i) ratio between $H_2O_{2,ad}$ formation and O–O bond splitting (step 3) and (ii) the ratio between $H_2O_{2,ad}$ desorption and further reaction of $H_2O_{2,ad}$ to H_2O_{ad} (+desorption) and OH_{ad} (step 4, upper route).

Dr Fermin said:

Referring to Fig. 8 in the paper, it is difficult to envisage the lateral diffusion of intermediates considering the effective convolution of the diffusion fields around individual particles. The results in Fig. 5(a) and (d) in the paper demonstrate that the nanostructured electrode behaves as a continuous Pt film, indicating that the diffusion profiles around the particles are convoluted in the timescale of the experiment (microelectrode array-type behaviour). In terms of conventional diffusion arguments, the intermediates will not diffuse towards a neighbour particle where there is high local concentration of the same intermediate. Can the author clarify this point?

Professor Behm responded:

We fully agree with the author of this question that applying standard diffusion theory there would be no lateral diffusion of dissolved reaction intermediates between the different Pt nanostructures, because of the absence of concentration gradients. The lateral migration of intermediates is better described by Brownian

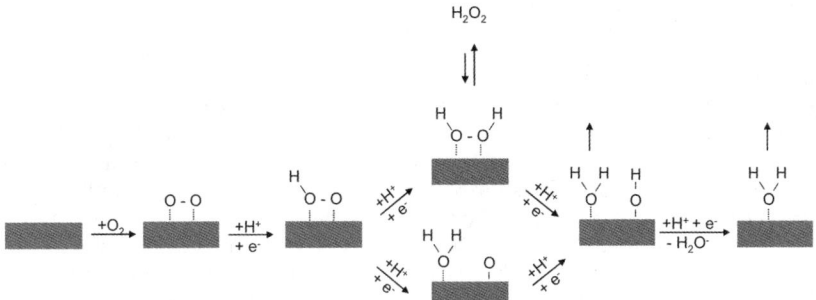

Fig. 6 Tentative reaction scheme for the O_2 reduction reaction on Pt including transport effects (see also paper by Rossmeisl *et al.* in this issue).

motion, which results in zero net transport, but nevertheless allows the respective molecules to reach the neighboring nanostructures. From the same reason, re-adsorption on the same Pt nanostructure it has originated from is possible as well. Once re-adsorption took place, there will be a given probability for further reaction to the final product, in this case H_2O. From these reasons, it is not possible to describe the underlying transport processes by conventional diffusion models. Instead, techniques allowing a random walk description should be used, which could be based on Monte Carlo techniques. In cooperation with other groups we are currently working on implementing such model descriptions (see also my response to the comment of Professor Morgan earlier in the discussion).

Dr Kucernak said:
Further to my comment previously, I should point out that Pletcher et al.[1] were the first to show that hydrogen peroxide yield increased with mass transport during oxygen reduction at microelectrodes and rotating disk experiments in neutral and alkaline solutions.

1. D. Pletcher and S. Sotiropolous, *J. Electroanal. Chem.*, 1993, **356**, 109

Professor Schiffrin said:
I would question the correctness of employing a macroscopic diffusional model derived, for example, from Fick's law, to describe mass transport in a structure consisting of reactive nanoparticles with distances between them in the nanometre regime. This is particularly important when considering the important case discussed in this paper, where "re-adsorption", or reaction in adjacent nanoparticles, is very important for describing the properties of the ensemble. The transport problem becomes highly non-linear and modelling is required.

Professor Behm replied:
We fully agree that the transport phenomena in these experiments can not be modeled simply by employing Fick's diffusion laws, due to the problems arising from lateral transport (and possible re-adsorption) of reaction intermediates between the individual nanostructures. This is actually the central problem which requires a new approach for modeling (see also response to Professor Morgan earlier). On the other hand, it should be possible to describe the diffusion of reactants to the nanostructured electrode surface (through the diffusion layer) using models based on the diffusion laws. The distances between the nanostructures are still considerably larger than molecular dimensions. Diffusion to an array of microelectrodes from a flowing electrolyte was modeled, *e.g.*, by W. E. Morf in a series of papers (*J. Electroanal. Chem.*, 2006, **590**, 47 and references therein). Going to even smaller dimensions than in the present paper, however, other approaches have to be used (see response to Professor Morgan).

Professor Tryk commented:
Model studies of this type are very important in helping to understand what is happening inside a polymer electrolyte fuel cell cathode. The differences in peroxide generation behavior exhibited by a low loading of Pt on glassy carbon *vs.* bulk Pt appear to stem from the fact that any peroxide generated at the Pt nanoparticle has a high probability of escaping, as you mention. Also, the hope that the Pt nanoparticle behaves like bulk Pt is supported by Fig. 7(b) in your paper, in which the ORR Tafel plots, after being normalized to the electrochemically active surface area, are superimposed fairly closely. However, I would like to play the Devil's Advocate for a moment and ask the question: isn't it possible for trace impurities, even at the ppm level, either in solution or emerging from the carbon as corrosion products, to be efficiently transported to the Pt nanoparticle surface, adsorb, and without completely blocking the surface, to allow more peroxide to be produced?

This could occur simply *via* a slight decrease in the peroxide adsorption energy. Watanabe and coworkers found, in channel-flow measurements in the generation–collection mode, with extremely thin layers of Pt/C electrocatalyst material immobilized with Nafion on gold substrates, that the peroxide yields were close to 1% over a wide range of temperatures (0–110 C) at 0.7 V *vs.* RHE for both the Pt/C electrocatalysts and bulk Pt (Yano *et al.*, *J. Phys. Chem. B*, 2006, **110**, 16544). In fact, they considered that Nafion itself was responsible for a kind of impurity effect, because, without it, there was no detectable peroxide. Of course, this was for a 19 wt% loading of Pt on C, so the particles were not separated by long distances and may not have acted as isolated microelectrodes. In any case, these types of possible impurity effects should be examined further. For example, one way to avoid the effects of possible carbon corrosion could be to use a boron-doped diamond support, as first proposed by Swain and coworkers and later by several other groups (see Spataru *et al.*, *J. Electrochem. Soc.*, 2008, **155**, B264, and Gonzalez-Gonzalez *et al.*, *Diamond Relat. Mater.*, 2006, **15**, 275).

Professor Behm responded:

First part of the comment: We disagree that the experimental findings on clean bulk electrodes have been ignored. In fact, for such electrodes our data also indicate that O_2 is fully reduced to H_2O (4 electrons per O_2 molecule). The measured number of electrons per educt molecule does not differ between a direct reaction process and a sequential process, as long as the overall reaction process results in the same product. Our results clearly show that only for model catalysts with a much lower Pt loading (lower fraction of the electrode surface consists of Pt, rest carbon) H_2O_2 formation becomes increasingly important.

Second part of the comment: Contaminations are of course always a possible source of experimental discrepancies, and it is very hard to fully rule them out. In the present case, the model electrodes were potential cycled in base electrolyte until no CO/CO_2 formation could be detected (checked by DEMS and the Faradaic current), and the CV showed the characteristics of polycrystalline Pt. Furthermore, in the series of measurements, the differences occur only for samples with lower Pt loading, while higher loading samples including pc Pt are in full agreement with previous data. Therefore we have, even applying very critical standards, no evidence that the general phenomenon of transport effects in electrocatalytic reactions and of increasing H_2O_2 formation in the ORR under conditions of increasing space velocities, are due to contaminations. This is supported also by comparable findings of the group around A. Kucernak (*J. Chem. Phys. B*, 2004, **108**, 3262.

Third part of the comment: We fully agree that other support materials could be used and would be interesting as well, and this is also planned for the future. We would like to emphasize, however, that also for the glassy carbon supports, which were used from experimental reasons, we have no indications of measurable contamination effects (see above).

On the catalysis of the hydrogen oxidation†

E. Santos,[ab] Kay Pötting[b] and W. Schmickler[b]

Received 11th February 2008, Accepted 25th March 2008
First published as an Advance Article on the web 14th August 2008
DOI: 10.1039/b802253d

A recently developed model for electrocatalysis is combined with results of quantum chemical calculations to investigate the effect of the electrode's electronic structure on the rate of the hydrogen oxidation reaction. Model calculations have been performed for three metals with widely differing properties: Cd(0001), Au(111) and Pt(111). In line with experimental findings, the energy of activation decreases in this order. These results are explained in terms of the interaction of the bonding orbital of the hydrogen molecule with the d band of the electrode as it passes the Fermi level.

1 Introduction

The oxidation of hydrogen and its reverse, the generation, are amongst the most important electrochemical reactions, and not only on account of their role in technological applications. Because of their relative simplicity, they are often considered to be the prototype of an electrocatalytic reaction, whose rate depends strongly on nature of the electrode material. Indeed, the rates on non-catalytic metals such as mercury and cadmium on the one hand, and good catalysts like platinum and palladium on the other hand, differ by more than six orders of magnitude. Much research effort has been directed at understanding the catalysis of the hydrogen reaction but, until very recently, with little success. Various attempts have been made to correlate the reaction rate with other quantities. The best known correlation is probably the volcano plot of the reaction rate *versus* the energy of adsorption of hydrogen as proposed by Trasatti[1,2] and others before him,[3-6] but even this is flawed. All the metals on the descending branch in Trasatti's plot (Ti, Ta, Nb) are covered by an oxide film, which greatly reduces the rate, a fact that was not known when this relation was established. The justification for the volcano curve is usually given in terms of Sabatier's principle, which states that for a reaction to proceed rapidly the intermediates should have an intermediate energy of adsorption; a weak adsorption proceeds too slowly, a strong adsorption blocks the surface. However, this principle cannot be applied to the electrochemical hydrogen reaction in the same manner as for gas-phase surface reactions, since the free energy of adsorption of hydrogen from the solution varies with the electrode potential, and a metal that adsorbs hydrogen weakly at the equilibrium potential will adsorb strongly at more negative potentials. Taken to its logical conclusion, this would result in volcano-shaped current potential curves,[7] which is absurd. There are more elaborate and justifiable arguments for volcano plots in electrochemistry, but in the absence of any empirical evidence we need not invoke them.

[a]*Faculdad de Matemática, Astronomía y Física, Universidad Nacional de Cordoba, Cordoba, Argentina*
[b]*Institute of Theoretical Chemistry, University of Ulm, Ulm, D-89069, Germany*
† The HTML version of this article has been enhanced with colour images.

Recently, two of us have presented a model for the catalysis of electrochemical electron transfer reactions, and proposed a mechanism, by which a d band situated near the Fermi level may reduce the activation energy.[8,9] The application of our model to hydrogen oxidation at ten different metals,[10] with the metal d bands and their coupling to hydrogen as the main ingredients, showed a good correlation between the calculated energies of activation and the exchange current densities, the first such correlation based on a theory. Here, we want to follow up on this work and improve it in the following way: the previous calculations were based on a model Hamiltonian that did not include exchange and correlation interactions between hydrogen and the substrate. While there is good evidence from theoretical surface science that these terms do not vary greatly between metals,[11] the neglect of these terms makes it impossible to establish an absolute energy and electrode potential scale. Here we shall show how these interactions can be included into the theory by combining the model with results obtained from standard density functional theory, so that calculations can now be performed for definite overpotentials. At the same time, we shall also abandon the tight-binding approximation, by which we had previously described the hydrogen molecule.

2 The mechanism of electrocatalysis

While the emphasis of this work is on the catalytic properties of the metal, we would like to start by pointing out the important role that the solvent plays. The oxidation of hydrogen: $H_2 \rightarrow 2H^+ + 2e^-$ requires almost 32 eV; about 22 eV are provided by the hydration of the proton, 9–10 eV, twice the work function, by the metal, and the rest by the potential drop between the electrode and the bulk of the solution, which is the only part that an experimentalist can control. Thus, solvation plays the dominant part in the energetics, and any model for the hydrogen reaction that neglects the solvent leaves out a most important part. In order to illustrate the catalysis mechanism proposed by Santos and Schmickler, and the role played by the solvent, we consider a simple electron transfer of the type: $A \rightarrow A^+ + e^-$, where A is an adsorbed species, and the final state is a solvated ion; the electrochemical desorption of hydrogen is an obvious example. The valence orbital of the initial state is filled and lies below the Fermi level of the electrode (see Fig. 1). Due to the interaction with the metal, the orbital is broadened and can be characterized by its density of states (DOS) $\rho_a(\varepsilon)$, where ε is the electronic energy. In order for the reaction to occur, a thermal fluctuation must lift the valence orbital to the Fermi level, so that an electron can be transferred from the reactant to the metal. The resulting ion interacts strongly with the solvent, the solvation shell relaxes towards equilibrium, and in the

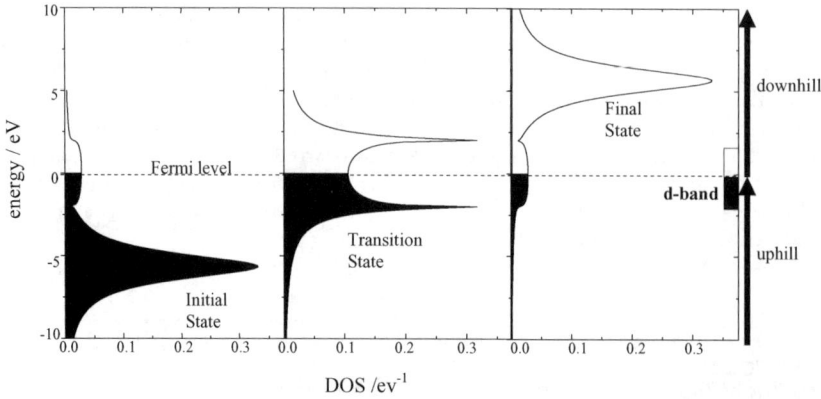

Fig. 1 Mechanism of electrocatalysis by a d band near the Fermi level.

final state the valence orbital lies well above the Fermi level. Thus, the lifting of the valence level to the Fermi level is energetically uphill, the solvent relaxation resulting in a further rise of the orbital is downhill. The energy of activation required to lift the valence orbital to the Fermi level can be greatly reduced by a metal d band that lies close to the Fermi level and interacts strongly with the reactant's valence orbital. Such an interaction leads to a considerable broadening of the valence level, and if it is sufficiently strong, even to a splitting into a bonding and antibonding orbital. When the overall reaction is in equilibrium, the orbital is about half-filled at the transition state. The part of the density of states that lies below the Fermi level reduces the energy of activation. The same mechanism also operates in other electron transfer reactions like a two-electron step $A_2 \rightarrow 2A^+ + 2e^-$. We shall return to this mechanism below when we discuss the results of our model calculations.

3 Formalism

We use a slightly simplified version of the model proposed by Santos et al.,[8,9] and consider just one reactant orbital interacting with the metal surface; in the model calculations reported below this will either be the 1s orbital of the hydrogen atom or the bonding orbital of the H_2 molecule. Also, we do not include the Coulomb repulsion U between two electrons on the same orbital, since this will be handled by DFT calculations. The density of states of the reactant, labeled a, is then:

$$\rho_a(\varepsilon) = -\frac{1}{\pi}\Im\frac{1}{\varepsilon - \varepsilon_a - 2q\lambda - \Lambda(\varepsilon) + i\Delta(\varepsilon)} \tag{1}$$

where ε_a is the electronic level of the reactant, λ is the energy of reorganization of the solvent, q is the solvent coordinate, and $\Im$ denotes the imaginary part. The interaction of the level with the metal is described by the level broadening:

$$\Delta(\varepsilon) = \pi\sum_k |V_{ak}|^2 \delta(\varepsilon - \varepsilon_k) \tag{2}$$

where k labels the states on the metal, and V_{ak} couples the reactants level to the metal. The level shift $\Lambda(\varepsilon)$ is obtained from:

$$\Lambda(\varepsilon) = \frac{1}{\pi}\mathscr{P}\int_{-\infty}^{\infty}\frac{\Delta\varepsilon')}{\varepsilon - \varepsilon'} \tag{3}$$

where $\mathscr{P}$ denotes the principle value. When the electronic level ε_a, the function $\Delta(\varepsilon)$ and the solvent reorganization energy λ are given, the energy of the system can be calculated as a function of the solvent coordinate q:

$$E_{\text{model}}(q) = \int_{-\infty}^{0}\varepsilon\rho_a(\varepsilon)\,d\varepsilon + \lambda q^2 + 2\lambda zq \tag{4}$$

The first term is the electronic energy of the reactant, and the second is the potential energy of the solvent; z is the charge number when the orbital is empty. The Fermi level of the metal has been set to zero, so that the integral is over all occupied states. Since this is essentially a one-electron model, it will contain a certain error ΔE caused by exchange and correlation effects. For $q = 0$, solvation effects are absent, and the energy of the system can be obtained from DFT calculations, which account for these effects. We write

$$E_{\text{DFT}} = E_{\text{model}}(q=0) + \Delta E(q=0) \tag{5}$$

to get the best possible estimate of the error for this case. Fluctuations of the solvent coordinate q shift the energy level, and large fluctuations will change the occupancy

$\langle n_a(q) \rangle$ of the orbital. The error must depend on this occupancy: as long as the occupancy is unchanged, the bonding is unchanged, and the electronic energy is not affected. On the other hand, when the orbital is empty, all the electronic energy, including ΔE, must vanish. It is therefore natural to assume:

$$\Delta E(q) = \Delta E(q = 0) \times \langle n_a(q) \rangle / z_a \qquad (6)$$

where z_a is the occupancy of the orbital for $q = 0$. Naturally this is an approximation, but there is no reason to expect it to be worse than the approximations made for the exchange and correlation energies in DFT itself.

Eqn (1) through (6) define a procedure to calculate the system energy for various values of the solvent coordinate q. In order to obtain the reactant orbital energy ε_a and the broadening $\Delta(\varepsilon)$ we have proceeded in the following way: the density of states of the metal has been separated into a broad sp band and a d band. The sp band has been assumed to be the same on all metals: a semielliptic band of width 20 eV centered at the Fermi level, and inducing a broadening $\Delta_{sp} = 0.5$ V. To obtain the interaction with the d band, we performed DFT calculations, from which we obtained the adsorption energy of the reactant, the density of state of the metal d band $\rho_d(\varepsilon)$, and the density of states $\rho_a(\varepsilon)$ projected onto the valence orbital of the reactant. This should correspond to the density of states of eqn (1) for $q = 0$. We fitted $\rho_a(\varepsilon)$ to this form by assuming that the part of $\Delta(\varepsilon)$ pertaining to the interaction with the d band is given by:

$$\Delta_d(\varepsilon) = \pi |V_{eff}|^2 \rho_d(\varepsilon) \qquad (7)$$

where V_{eff} is the effective coupling constant between the d band and the orbital a, which is taken to be independent of the electronic energy. This gives two parameters to be obtained by the fit: ε_a and V_{eff}. This fitting procedure works requires a fairly high number of k-points in the unit cell (see below).

4 Model calculations

4.1 Reaction of molecular hydrogen

For explicit calculations, we chose three different metal surfaces: Cd(0001), Au(111), and Pt(111), corresponding to a bad, a mediocre, and an excellent catalyst for the hydrogen reaction. All DFT calculations were performed with the dacapo code,[12] which uses a plane-wave expansion of the Kohm–Sham wave functions and the PBE functional. The calculations were performed for a 2×2 unit cell of a five layer slab, whose lattice constant was kept fixed at the calculated equilibrium distance (Pt: 2.83 Å, Cd: 3.18 Å, Au: 2.96 Å). The ionic cores were represented by ultrasoft pseudopotentials; $8 \times 8 \times 1$ k points corresponding to a 1×1 unit cell were employed, and the plane wave cut-off was set at 350 eV. The d band surface densities of states of these metals are shown in Fig. 2: The width of the d band increases from Cd to Au to Pt, and so does the energy of the d band center, the d band of Pt extending right up to the Fermi level.

For each metal we performed a series of calculations for the adsorption and dissociation of the H_2 molecule. Following Hammer and Nørskov,[13] we placed the molecule at a fixed distance d over a surface atom, let the hydrogen atoms relax in the plane parallel to the surface, and calculated the minimum energy with these restraints, as well as the density of states projected onto the bonding orbital of H_2. Fig. 3 shows the adsorption energy as a function of the distance d for the three metals. As the distance to the surface decreases, the separation between the two hydrogen atoms increases, until they are finally adsorbed in the threefold fcc hollow sites. On platinum, hydrogen dissociates practically without a barrier, on gold and cadmium dissociation is unfavorable and further requires the passing of an energy

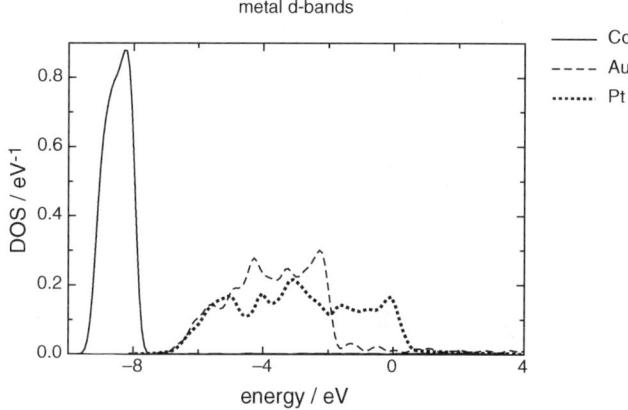

Fig. 2 The d band surface densities of states of Cd(0001), Au(111) and Pt(111).

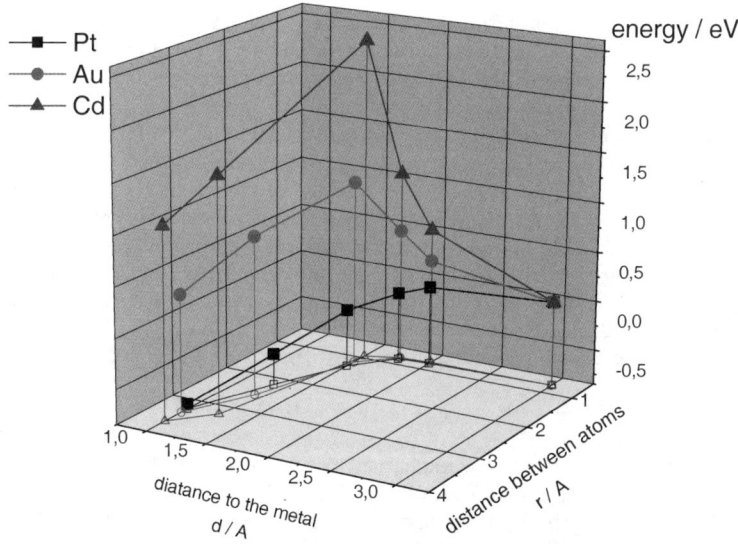

Fig. 3 Energy of an adsorbed H_2 as a function of the bond length r and of the distance d to the surface; at the shortest distance d, the molecule is dissociated, with the two H atoms in the fcc threefold hollow sites.

barrier; both the barrier and the energy of the adsorbed atom are much higher on cadmium than on gold. The curves for platinum and gold in Fig. 3 are exactly the same as those published by Hammer and Nørskov,[13] which is not surprising since we used the same software. For a full discussion of hydrogen on gold and platinum we refer to their work. To understand the behavior of cadmium, we show in Fig. 4 the density of states of the H_2 binding orbital when this molecule is at a distance of 1.5 Å from the surface. The interaction with the low-lying d band splits this orbital into a bonding and antibonding part; both are filled, and so no energy is gained. Instead, the overlap between the hydrogen and cadmium orbitals produces a repulsion.

For each distance d we fitted the hydrogen densities of states to our model from eqn (1) using eqn (7), and thus obtained all required parameters as a function of the distance d of the molecule from the surface or, equivalently, of the separation r between the two hydrogens. In order to calculate potential energy surfaces we

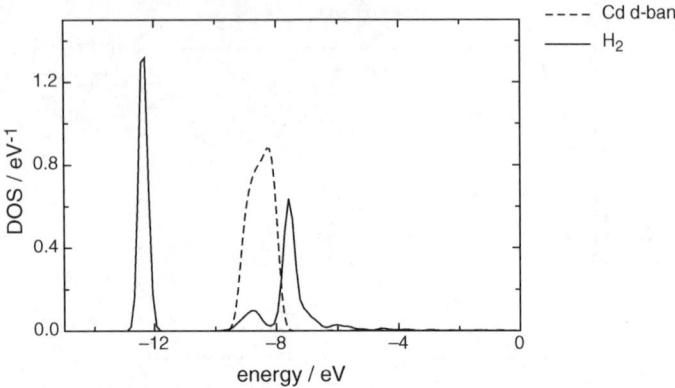

Fig. 4 Density of states of cadmium d band and of the binding orbital of H$_2$ situated at a distance of 1.5 Å over an atop site. The interaction with the d band splits the latter into a bonding and an antibonding part.

have to consider one more term: the interaction of the final state, the two protons, with its surroundings. In electron transfer theory, this is divided into two parts: the slow part, which is represented by the energy of reorganization, and the fast part, which comprises the electronic polarizability of the solvent, the image interaction with the metal, and the effect of the electrostatic potential.[14,15] In order to obtain an expression for the fast part, we consider equilibrium between initial and final state. As the energy of the initial state we take the energy of the H$_2$ molecule, which is about $E_i = -31.73$ eV; its interaction with the solvent is negligibly small. This must be equal to the energy of the final state. The contribution of the energy of reorganisation to the energy of two protons is -4λ and has already been accounted for. In the Marcus theory of electron transfer and its extensions,[15] the interactions with the environment are linear, therefore the contribution of the fast interaction is linear in the occupation $\langle n_a \rangle$ and takes the form:

$$V_f = (2 - \langle n_a \rangle)(E_i + 4\lambda)/2 \qquad (8)$$

It vanishes when the reactant is uncharged ($\langle n_a \rangle = 2$), and results in the correct free energy for the final state, two protons with $\langle n_a \rangle = 0$.

Finally, we have to specify the energy of reorganisation λ of the solvent. The total energy of solvation of the proton is about 11 eV; generally, one estimates from the Marcus formula that about one half of this energy comes from the slow modes and pertains to the reorganisation, but this rule does not necessarily hold for the proton. At the surface, the solvation is considerably reduced, and partially replaced by the image force as the stabilising interaction. We have performed calculation for λ in the range of 3–4 eV. The trends and the order of magnitude for our results do not depend on the exact value, and here we show results for $\lambda = 3.5$ eV.

Using the parameters obtained from DFT, we have calculated free energy surfaces for the overall reaction H$_2$ → 2H$^+$ + 2e$^-$ as a function of the solvent coordinate q and the interatomic separation r. As an example, we show the surface for the reaction on Au(111) in Fig. 5. At $q = 0$ and $r = 0.76$ Å, we see a minimum corresponding to the stable molecule; the valley centered at $q = -2$ corresponds to 2H$^+$. The direct oxidation of the molecule according to H$_2$ → 2H$^+$ + 2e$^-$ would require an activation energy of about 3.5 eV, the dissociation requires about 1.1 eV and is therefore favoured, even though it leads to an intermediate state with a higher energy. For the subsequent oxidation, we have to keep in mind that this diagram is for the simultaneous oxidation of two hydrogen atoms, which is less favourable than a consecutive oxidation of two atoms. It is well-known from Marcus theory[14] that the energy

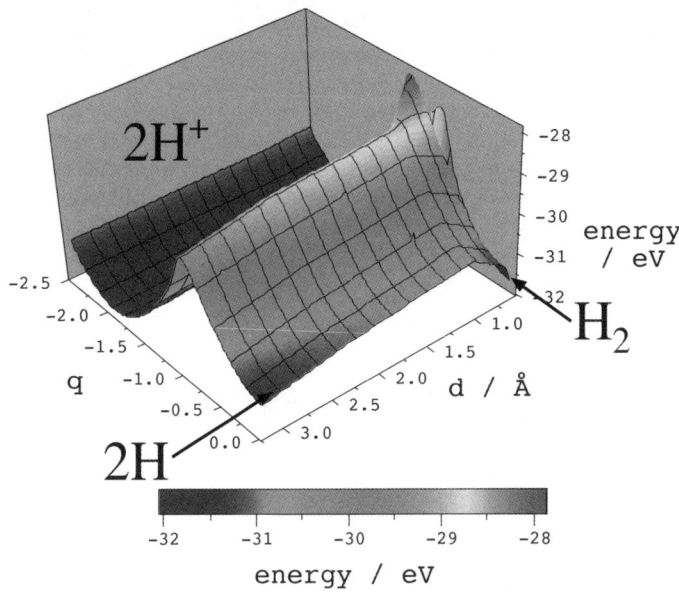

Fig. 5 Free energy surface for the hydrogen oxidation on Au(111).

required for the reorganisation of the solvent scales with the square of the transferred charge. Therefore we can infer from this diagram that the preferred mechanism is:

$$H_2 \rightarrow 2H \quad 2H \rightarrow 2H^+ + 2e^- \tag{9}$$

but the reaction for the second step has to be calculated separately. We found the same mechanism on all three metals; and this seems to be in line with experimental results. An alternative would be the Heyrowsky route with: $H_2 \rightarrow H + H^+ + e^-$. This would require more extensive calculations than we have performed so far, and we therefore leave this to future work.

4.2 Reaction of a single hydrogen atom

In order to investigate the oxidation of a single hydrogen atom, we have performed DFT calculations for a single atom adsorbed on a fcc hollow site, and varied its separation from the surface. At each position we calculated the adsorption energy, the DOS projected onto the 1s orbital of hydrogen, and from the latter we obtained the model DOS of eqn (1) by fitting in the same way as for the molecule. As expected, the interaction between the atom and the metal decreases with the distance.

We want to calculate free energy curves for the reaction as a function of the solvent coordinate. However, as the degree of solvation increases the separation from the surface increases as well, but the relation is not known. As long as we do not know the details of proton solvation at the surface, we cannot hope to calculate absolute values for activation energies. We can, however, calculate orders of magnitude, trends, and in particular the effect of the electronic structure of the metal. In a separate study,[16] we have developed a model for proton transfer to a metal based on a semi-empirical valence bond theory. This study indicates that the decisive step occurs when the proton is about 1.5 Å from the surface. We have performed calculations with parameters derived from metal–atom separations in the range 1.4–1.6 Å; as expected, they give very similar free energy curves; here we show the results for 1.4 Å.

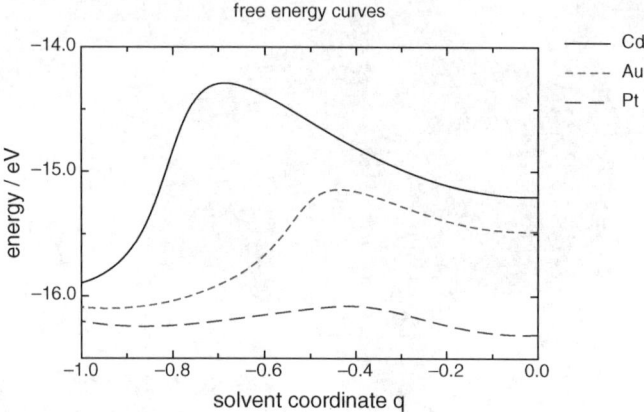

Fig. 6 Free energy curves for the oxidation of a H atom.

The resulting free energy curves are shown in Fig. 6. The dissociation of the hydrogen molecule on Cd requires an energy of about 0.72 eV per atom, and the energy of activation for the oxidation of the atom is about 0.8 eV higher. The dissociation on gold is also endothermic by about 0.38 eV, and the additional energy of activation for the oxidation about 0.3 eV. In contrast, the dissociation on platinum is exothermic, and the further oxidation requires very little activation energy. These results agree very well with the known experimental behaviour of these metals: hydrogen does not adsorb on Cd or Au except as a short-lived intermediate during hydrogen evolution, while on Pt it adsorbs before the onset of hydrogen evolution. On Cd, both hydrogen oxidation and evolution are strongly inhibited, on platinum they proceed rapidly, and gold is a rather mediocre catalyst. At the oxidized state $q = -1$, the energies at the three metals are not quite the same because there is still some interaction with the metal.

The reason for the large differences in the energy of activation can be seen from Fig. 7, which shows the densities of states of the hydrogen atom on these three metals as the occupation of the valence orbital passes the value of 1/2, *i.e.* when it is centered at the Fermi level. At Cd, the DOS is narrow, and has a high peak right at the Fermi level. At this state, the interaction with the d band, which lies more than 7 eV below E_F, is negligible. In contrast, the strong interaction with the d band of Pt induces

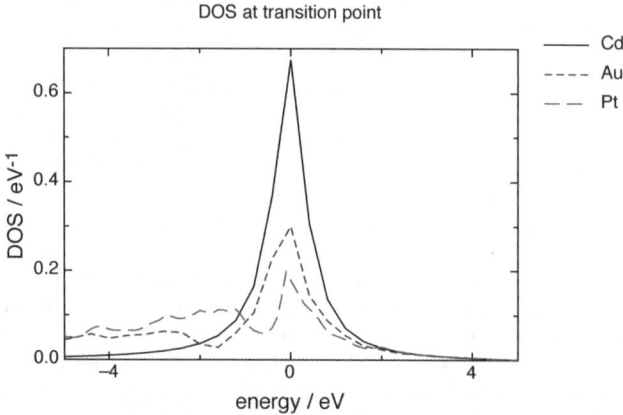

Fig. 7 Density of states of the hydrogen atom as its valence orbital passes the Fermi level.

a long tail in the DOS that extends several eV below E_F, thus significantly reducing the energy at the transition state. This behaviour is completely in line with the qualitative discussion in section 2.

5 Conclusions

In this work, we have suggested a method to combine our model for d band electrocatalysis with the results of DFT calculations. In this way we have obtained accurate values for the adsorbed molecule and atom, and thus established an absolute energy scale, which makes it possible to perform model calculations for a particular electrode potential.

Thus, within our model we can calculate the effect that the electronic structure of the electrode has on the dissociation of hydrogen (Tafel reaction) and on the discharge of the proton (Volmer reaction). In principle, our method is not limited to single crystal surfaces, but the effects of steps, overlayers and alloying can be incorporated as well, and we plan to do so in future works. Since the calculations presented here were limited to a small coverage with adsorbed hydrogen, we have not modelled the Heyrowsky reaction $H_2 \rightarrow H_{ad} + H^+ + e^-$, but again, this is no principle limitation.

We believe that it is the unique advantage of our method that it makes it possible to calculate the effect of the electronic structure on the energy of activation. This is of utmost importance, since practically all important electrode reactions are activated processes. DFT alone can give very useful information about the thermodynamics of species adsorbed on an electrode surface (see *e.g.* ref. 17 and 18) but it cannot handle the activation processes, which involve substantial solvent reorganisation.

During the last few years many theoretical investigations have been aimed at the electronic and structural properties of electrode surfaces, and much progress has been made. What is lacking is an equally detailed understanding of the role of the solvent, its reorganisation during charge transfer, how it delivers a proton to the electrode, or how it accepts one from it. This lack of knowledge also limits our present work. In order to make quantitative calculations for the activation energies, we would require exact, and not just plausible, values for the reorganisation energy, and know further details about the configuration of the solvent during the activation. Therefore, we are limited to calculating the variation of the reaction rate with the electronic structure, and using approximate values for the reorganisation energy to get estimates of the absolute values. Nevertheless, we believe that this constitutes important progress in our understanding of electrocatalytic reactions.

Acknowledgements

Financial support by the Deutsche Forschungsgemeinschaft (Schm 344/34-1 and Sa 1770/1-1), and of the European Union under COST is gratefully acknowledged. E. S. thanks CONICET for continued support.

References

1 S. Trasatti, *J. Electroanal. Chem.*, 1972, **39**, 163.
2 S. Trasatti, *Adv. Electrochem. Electrochem. Eng*, 1977, **10**, 213.
3 J. Horiuti and M. Polanyi, *Acta Physicochim. URSS*, 1935, **2**, 505.
4 H. Gerischer, *Z. Phys. Chem., Neue Folge*, 1956, **8**, 137.
5 R. Parsons, *Trans. Faraday Soc.*, 1958, **94**, 1059.
6 L. Krishtalik, *Elektrokhimiya*, 1966, **2**, 616.
7 W. Schmickler and S. Trasatti, *J. Electrochem. Soc.*, 2006, **153**, L31.
8 E. Santos and W. Schmickler, *ChemPhysChem*, 2006, **7**, 2282.
9 E. Santos and W. Schmickler, *Chem. Phys.*, 2007, **332**, 39.
10 E. Santos and W. Schmickler, *Ang. Chem., Int. Ed.*, 2007, **46**, 8262.

11 B. Hammer and J. K. Nørskov, *Adv. Catal.*, 2000, **45**, 71.
12 B. Hammer, L. B. Hansen and J. K. Nørskov, *Phys. Rev. B: Condens. Matter Mater. Phys.*, 1999, **59**, 7413. The code is described at: www.fysik.dtu.dk/campos.
13 B. Hammer and J. K. Nørskov, *Surf. Sci.*, 1995, **343**, 211.
14 R. A. Marcus, *J. Chem. Phys.*, 1956, **24**, 966.
15 W. Schmickler, *Chem. Phys. Lett.*, 1995, **237**, 152.
16 F. Wilhelm, W. Schmickler, R. R. Nazmutdinov and E. Spohr, *J. Phys. Chem.*, in press.
17 J. K. Nørskov, J. Rossmeisl, A. Logadottir, L. Lindqvist, J. Kitchin, T. Bligaard and H. Jonsson, *J. Phys. Chem. B*, 2004, **108**, 17886.
18 P. Vassilev, R. A. van Santen and M. T. M. Koper, *J. Chem. Phys.*, 2005, **122**, 054701.

PAPER

Hydrogen evolution on nano-particulate transition metal sulfides†

Jacob Bonde,[a] Poul G. Moses,[b] Thomas F. Jaramillo,[c] Jens K. Nørskov[b] and Ib Chorkendorff[*a]

Received 5th March 2008, Accepted 28th March 2008
First published as an Advance Article on the web 21st August 2008
DOI: 10.1039/b803857k

The hydrogen evolution reaction (HER) on carbon supported MoS_2 nanoparticles is investigated and compared to findings with previously published work on Au(111) supported MoS_2. An investigation into MoS_2 oxidation is presented and used to quantify the surface concentration of MoS_2. Other metal sulfides with morphologies similar to MoS_2 such as WS_2, cobalt-promoted WS_2, and cobalt-promoted MoS_2 were also investigated in the search for improved HER activity. Experimental findings are compared to density functional theory (DFT) calculated values for the hydrogen binding energies (ΔG_H) on each system.

Introduction

Research efforts to develop electrocatalysts for energy conversion reactions have increased substantially in recent years. Platinum, the ubiquitous electrocatalyst used in PEM fuel cells, is both expensive and scarce, prompting widespread efforts to discover cost-effective materials to replace Pt. In this work we focus on non-noble metal sulfide catalysts for the hydrogen evolution reaction (HER) under acidic conditions, a reaction catalyzed most effectively by Pt-based materials.[1]

Previously, MoS_2 has been studied as a catalyst in hydrodesulfurisation[2] and in the photo-oxidation of organics.[3,4] In electrocatalysis, it has recently been shown that the edge structure of nanoparticulate MoS_2 is active for the HER, mimicking the active sites/co-factor of the hydrogen evolving enzymes nitrogenase and hydrogenase.[5,6] This work aims to extend the investigation on carbon-supported nanoparticulate MoS_2 for the HER. Unlike the case of Au(111) supported MoS_2 studied by STM in previous work,[6] the catalysts probed herein are more commercially relevant, which also implies that they are less homogeneous and more difficult to image on the atomic scale. As knowledge of the concentration of active sites on a catalyst surface is paramount to elucidating structure–composition–activity relationships, the first aim of this work is to utilize electrochemical oxidation to probe MoS_2 surface area, distinguishing between basal plane and edge sites. In developing this methodology to quantify active sites on a macroscopic scale, we then direct our attention to related catalyst systems, namely WS_2, cobalt-promoted WS_2, and cobalt-promoted MoS_2. We end by comparing experimentally determined activity data to predictions

[a]*Center for Individual Nanoparticle Functionality (CINF), Department of Physics, Technical University of Denmark, Lyngby, DK-2800, Denmark. E-mail: ibchork@fysik.dtu.dk*
[b]*Center for Atomic-scale Materials Design (CAMd), Department of Physics, Technical University of Denmark, Lyngby, DK-2800, Denmark*
[c]*Department of Chemical Engineering, Stanford University, 381 North-South Mall, Stauffer III, Stanford, CA 94305-5025, USA*

† The HTML version of this article has been enhanced with colour images.

made by density functional theory (DFT) models of these systems in order to gain insight into trends in catalyst activity.

It has been found that ΔG_H, the hydrogen binding energy to a given surface, is a good descriptor for identifying electrocatalyst materials with high exchange current densities.[1,7,8] A recent study[5] using DFT showed that the active sites on nitrogenase and hydrogenase bind hydrogen weakly, similar to Pt. It was also found that the overpotential of carbon supported MoS_2 is comparable to the DFT calculated hydrogen binding energy on the edge of the nanoparticles. In another study, MoS_2 nanoparticles on Au(111) were synthesized under UHV conditions, characterized by STM and examined for HER activity.[6] This study showed direct evidence that the active site of the MoS_2 nanoparticles is indeed the edge. The exchange current density was also found to be in agreement with the volcano relation between the HER exchange current density and the DFT calculated values for ΔG_H proposed by Nørskov et al.[7] By having identified the active site of MoS_2 particles, the next step is to modify that edge such that its ΔG_H approaches even closer to zero where the HER volcano curve has its maximum, and this is a major aim of the work presented herein.

Bulk MoS_2 consists of stacked S–Mo–S layers, and MoS_2 nanoparticles can be synthesized as single layer hexagonal structures exposing two different kinds of edges, the so-called Mo-edge and the S-edge.[9] It has been shown that the structure of nanoparticulate MoS_2 is a single layered truncated triangle primarily exposing the Mo-edge when supported on Au(111),[6,9] highly ordered pyrolytic graphite (HOPG)[10] or graphitic carbon.[11] Brorson et al.[11,12] also found truncated triangles by means of HAADF-STEM (high-angle annular dark-field-scanning transmission electron microscopy) in their investigation of MoS_2, WS_2 and cobalt-promoted MoS_2.

Estimating the number of active sites on a nanoparticulate catalyst is not trivial. One approach is to measure activity on well defined model systems characterized by STM, for example UHV-deposited nanoparticles[6,13] or physi- or chemisorbed molecular clusters.[14] Another option is to use a well established method to measure electrochemically active surface area such that used with Pt based on the adsorption–desorption behavior of underpotentially deposited hydrogen, H_{upd}.[15] We note, however, that this method still relies upon the assumption that the sites active for H_{upd} are the same as those active for the HER. As we are studying metal sulfides where no such method exists, the irreversible oxidation of metal sulfides will be investigated as a measure of their surface area and edge sites.

In the following, we will show our investigation of MoS_2, WS_2 and cobalt promoted WS_2 (Co–W–S) and MoS_2 (Co–Mo–S), prepared similarly to the ones imaged by Brorson et al.[11] and supported on Toray carbon paper. The electrochemical measurements will be discussed in relation to DFT calculations of ΔG_H for each of the metal sulfates investigated in order to identify structure–composition–activity relationships for these systems.

Results and discussion

Synthesis and electrochemical characterization of MoS_2

MoS_2 particles on Toray carbon paper were prepared by dropping 25 μL of an aqueous ammonia heptamolybdate (1mM Mo) solution onto 1 cm^2 of Toray paper. The sample was dried in air at 140 °C followed by sulfidation in 10% H_2S in H_2 at 450 °C for 4 hours, and subsequently cooled in that same gas stream. This preparation method would typically give the highest HER current on a per gram basis; higher loadings usually led to lower HER currents.

HER activity was measured (see experimental details) and the results are plotted as Tafel (log i–E) and polarization curves (i–E) in Fig. 1. The Tafel plot exhibits a slope of 120 mV dec^{-1} and an exchange current density of 4.6×10^{-6} A $cm^{-2}_{geometric}$. Samples prepared by different methods have often yielded different Tafel slopes,

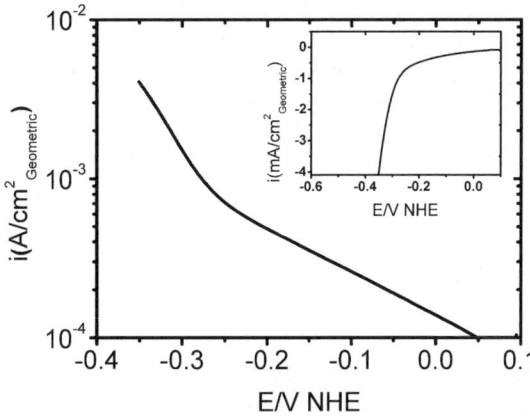

Fig. 1 Tafel plot (main) and polarization curve (inset) in the cathodic potential range of MoS_2 supported on Toray paper. The scan rate is 5 mV s^{-1} and the Tafel slope in the HER region is found to be 120 mV dec^{-1}.

ranging between 110 mV dec^{-1} and several hundred mV dec^{-1}. We attribute this to transport limitations through the fibrous, porous network characteristic of Toray carbon paper. Although sample/substrate preparation could potentially be optimized further, the consistent results achieved using the preparation method described above allows for accurate cross-comparisons among different catalyst materials. It should be noted that hydrogen evolution is taking off at around −0.2 V *vs.* NHE just as we have previously seen on MoS_2.[5,6]

The current measured from approx. +0.1 V *vs.* NHE to −0.15 V is most likely not due to the HER but rather oxygen reduction at the interface between the electrolyte and the electrode. Finally, it should be noted that sweeps between −0.35 and +0.1 V *vs.* NHE showed negligible change over time, apart from the effects of bubble formation on the electrode.

Fig. 2 shows a cyclic voltammogram of MoS_2/C where the potential is cycled between −0.3 and +1.05 V *vs.* NHE. At approx. +0.6 V *vs.* NHE an irreversible

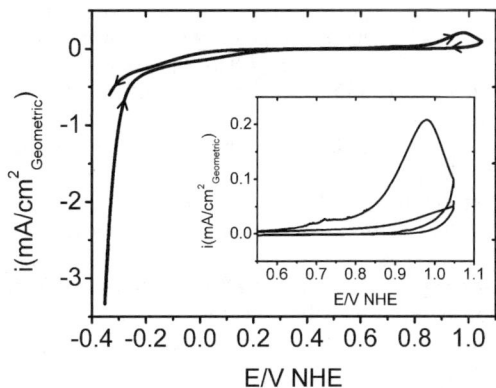

Fig. 2 Cyclic voltammogram of the oxidation and subsequent deactivation of the MoS_2 sample. Scan rate 2 mV s^{-1}. Main: the deactivation of the sample showing one sweep from −0.35 V *vs.* NHE to 1.05 V *vs.* NHE and back to −0.35 V *vs.* NHE. On the 1st anodic sweep an irreversible oxidation peak occurs at 0.6 V *vs.* NHE and is followed by a subsequent decrease in current at cathodic potentials (−0.35 V *vs.* NHE), indicating a deactivation of the active sites. Inset: the first and second sweep at anodic potentials showing a significant decrease in the oxidation peak.

oxidation begins to occur with a maximum at +0.98 V vs. NHE. On the subsequent cathodic sweep a significant drop in HER activity is noticed. On the ensuing anodic sweep seen in the inset of Fig. 2, the oxidation peak is no longer present. Thus, the loss of HER activity is attributed to irreversible MoS_2 oxidization. In subsequent studies, fresh samples were subjected to CVs in which an initially narrow potential window was widened gradually to more positive (anodic) potentials. It was found that the HER activity of MoS_2/C remained stable with every sweep as long as the anodic potential was limited to ≤+0.6 V vs. NHE.

MoS_2 electro-oxidation

To our knowledge, the electrochemical oxidation of nanoparticulate MoS_2 is not covered in the literature, which instead focuses on the corrosion of bulk MoS_2. Kautek and Gerischer[16] found that the bulk system preferentially oxidized at the (10$\bar{1}$1) face and that it did not corrode at the (0001) basal plane. On the nanoparticles this would correspond to corrosion of the particle edges. Closer examination of the insert of Fig. 2 reveals two distinct oxidation peaks. The major peak has its maximum at approx. +0.98 V vs. NHE whereas the minor peak has its maximum at approx. +0.7 V vs. NHE. As the edges of MoS_2 nanoparticles are expected to be more readily oxidized than the basal plane,[16] we interpret the two distinct oxidation peaks to correspond to the edges (minor peak, +0.7 V vs. NHE) and the basal planes (major peak, +0.98 V vs. NHE) of the particles. While only one cycle to +1.05 V vs. NHE will completely deactivate the sample for the HER, it takes several cycles to +0.7 V vs. NHE to achieve the same effect. This implies that not all edge sites are oxidized with a single sweep to +0.7 V vs. NHE. Had the sample been deactivated for the HER after a single sweep to +0.7 V vs. NHE, we could definitely have used this peak as a measure of the concentration of edge sites. But as this is not the case, we will use the major peak at +0.98 V vs. NHE to determine the total surface area of MoS_2/C. We have however attempted to use the weak feature at +0.7 V vs. NHE to get an estimate of our particle size. At low sweep rates (2 mV s^{-1}), the feature is typically not dominated by the major feature at +0.98 V vs. NHE. The area of the edge feature is approx. 8% of the major peak. If the particles are triangular, this corresponds to an edge length of around 25 nm, consistent with the particle sizes observed by Brorson et al.[11]

XPS was also employed in this investigation to study the MoS_2/C at three stages of its life: freshly prepared, after HER in H_2SO_4 and after oxidation in H_2SO_4 at high anodic potentials (see experimental details). To obtain a reasonable signal to noise ratio for the XPS studies the Toray paper was dip coated in a 0.14 M Mo solution instead of dropping a known amount of solution on the surface, resulting in a higher loading of Mo than previously described (a factor of 5–10 according to the charge of the oxidation peak). The survey spectra of the different samples showed no contaminants on the freshly prepared samples. On the samples that had been submerged in H_2SO_4, peaks corresponding to sulfate were seen and a peak corresponding to N 1s was also seen. The N 1s peak is most likely caused by trace amounts of NH_3 present in air absorbed by H_2SO_4 as $(NH_4)_2SO_4$ with a N 1s binding energy of 401.3 eV.[17]

The XPS data, see Fig. 3. reveals that the freshly prepared sample (no. 1) of MoS_2 is similar to previously reported MoS_2 spectra.[18–20] The XPS data from a similarly prepared sample that was tested for the HER (sample no. 2a) by sweeping the potential between +0.1 V and −0.45 V vs. NHE, showed an increase in the SO_4^{2-} peak which is to be expected as the sample had been submerged in H_2SO_4. Apart from the increase in the SO_4^{2-} peak, no significant changes were found compared to the freshly prepared sample, indicating that MoS_2/C does not change significantly during the HER. After XPS analysis of sample no. 2a, it was examined for the HER again, then cycled between −0.4 V vs. NHE and +1.4. V vs. NHE and removed from the solution at −0.32 V vs. NHE (sample no. 2b in the XPS spectra). A significant decrease of the Mo 3d, Mo 3p, S 2s and S 2p peaks was observed and there was

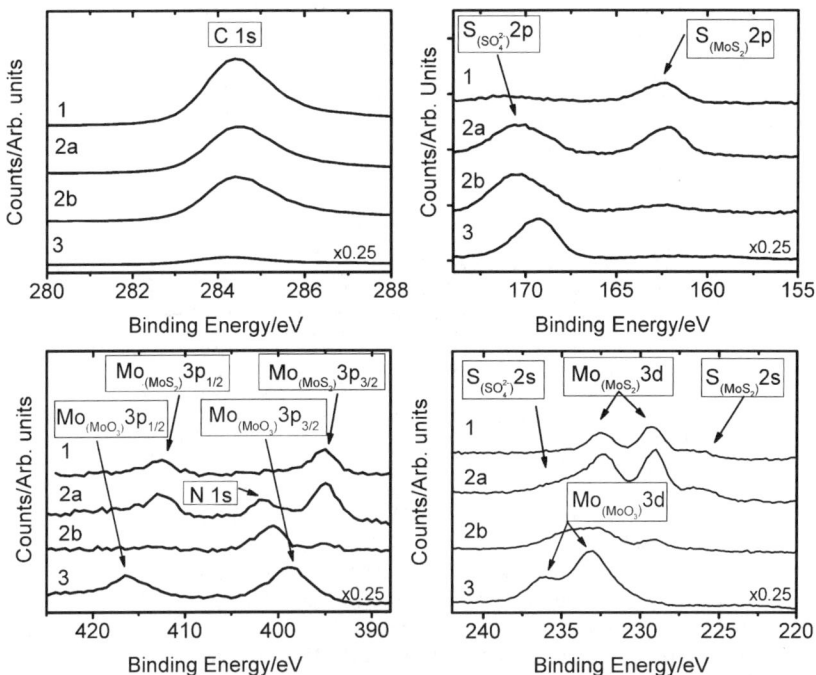

Fig. 3 XPS spectra of MoS$_2$ on Toray paper recorded at different stages of its life. (1) As prepared after sulfidation. (2a) After initial activity measurements of the HER (CVs between +0.1 and −0.4 V *vs.* NHE). (2b) Sample 2a after measurements of the HER and subsequent oxidation/deactivation (CVs between +1.4 and −0.4 V *vs.* NHE) and removal from the electrolyte at −0.32 V *vs.* NHE. (3) After measurements of the HER and subsequent oxidation/deactivation (CVs between +1.4 and −0.4 V *vs.* NHE) and removal from the electrolyte at 0.4 V *vs.* NHE.

no XPS signal corresponding to MoO$_3$. Thus, although the amount of surface Mo decreased significantly it still maintained its Mo^{4+} character (as in MoS$_2$). There are several possible explanations for the lack of Mo on the surface: (1) the MoS$_2$ desorbs from the surface at high anodic potentials, (2) the oxidation product of MoS$_2$, MoO$_3$, dissolves,[21] (3) that MoO$_3$ is reduced to Mo^{3+} at −0.32 V *vs.* NHE and subsequently dissolves.[21] To answer this question, sample no. 3 was subjected to the same oxidation treatment as sample no. 2b but in this case the sample was pulled out of solution at a higher potential (+0.4 V *vs.* NHE), where MoO$_3$ is thermodynamically stable according to the Pourbaix diagrams.[21] XPS reveals a shift of the Mo 3d and Mo 3p towards higher binding energies just as expected for MoO$_3$. Thus, it is unlikely that MoS$_2$ dissolves at anodic potentials.

We note that the Mo peaks of the MoO$_3$ were significantly greater than the Mo peaks observed on the other samples and at the same time the C 1s peak was significantly smaller. The increase in intensity could be due to a higher loading on this specific sample but we only found a factor of 2 larger oxidation peak on sample 3 than on sample 2a/b. This leads us to believe that there could be surface enrichment of Mo species on the outermost exposed surface of the Toray paper after repeated dissolution–redeposition cycles during each potential sweep.

Having established that the MoS$_2$ is in fact being oxidized at high anodic potentials we will now elaborate on possible reaction mechanisms. The reaction mechanism will enable us to use the irreversible oxidation peaks to determine the amount of MoS$_2$ present on the surface. A plot of the correlation between the amount of Mo used during synthesis and the charge of the irreversible oxidation

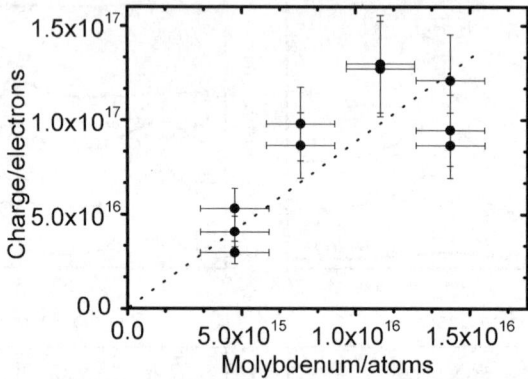

Fig. 4 The charge of the irreversible oxidation peak as a function of the amount of Mo used during the synthesis of MoS_2.

peak is shown on Fig. 4. In a corrosion study by Jaegermann and Schmeisser,[18] bulk MoS_2 was electrochemically oxidized in KNO_3 and examined by XPS. A shift toward higher binding energies was observed for the S 2p and Mo 3d peaks and a broadening was observed in the S 2p line. This was interpreted as MoS_2 degradation to SO_4^{2-}, S_2^{2-} and MoO_3. We can, with this knowledge, consider how many electrons we expect to use to oxidize one Mo atom. If we consider one extreme where the carbon supported MoS_2 is decomposed into MoO_3 and SO_4^{2-} the following reaction would take place, where 18 electrons are transferred per Mo atom:

$$MoS_2 + 11H_2O \rightarrow MoO_3 + 2SO_4^{2-} + 22H^+ + 18e^- \qquad (1)$$

The other extreme would be that MoS_2 is decomposed into MoO_3 and S_2^{2-} where 4 electrons are needed:

$$MoS_2 + 3H_2O \rightarrow MoO_3 + S_2^{2-} + 6H^+ + 4e^- \qquad (2)$$

According to Fig. 4, the correlation between the oxidation peak and the deposited amount of Mo yields 8.9 ($r^2 = 0.55$) electrons per Mo atom used in the deposition. This number is in between the two extremes mentioned above. Revisiting the XPS data we can not see whether we have produced excess SO_4^{2-} due to the background of H_2SO_4. We are, however, also not seeing any significant amounts S_2^{2-} after cyclic voltammetry. While the samples have been subject to a high anodic (1.4 vs. NHE) potential where S_2^{2-} can be oxidized to SO_4^{2-}, the subsequent high cathodic (−0.4 vs. NHE) potential can reduce the S_2^{2-} to H_2S.[21] We can not conclusively determine the exact nature of the oxidation reaction. But our measurements indicate that the sulfur in the MoS_2 is only partially oxidized during anodic sweeps, resulting in the following proposed reaction mechanism:

$$MoS_2 + 7H_2O \rightarrow MoO_3 + SO_4^{2-} + \tfrac{1}{2} S_2^{2-} + 14 H^+ + 11e^- \qquad (3)$$

HER activity of MoS_2/C

In order to determine the activity of the MoS_2/C system per active site, we start with the irreversible oxidation to estimate the total surface area of MoS_2 on the Toray paper. The irreversible oxidation peak of the sample, shown in Fig. 1 and Fig. 2, has a charge of 0.014 C. If we assume that 11 electrons are involved in the oxidation of MoS_2, as presented in the previous section, the surface area will be 3.8 cm² of single layered MoS_2 giving an exchange current density (i_0) of 1.2×10^{-6} A cm^{-2} (and a Tafel slope of 120mV dec^{-1}). We have previously shown that the active sites of Au(111) supported MoS_2 nanoparticles are situated on the edge ($i_0 = 7.9 \times 10^{-6}$ A cm^{-2}).[6]

The exchange current density on a per active site basis will clearly be higher than the exchange current densities reported above, since few of the MoS_2 sites are on the edge. Thus, the values above constitute a lower bound for activity. If we incorporate the fact that the MoS_2 nanoparticles are triangular with an edge length of 25 nm, approx. 8% of the atoms will be situated at the edge of the particle. This would lead to a 12-fold increase in exchange current density per active site.

Electrochemical characterization of WS_2

WS_2 exhibits a layered structure similar to MoS_2,[11,12] forming the same triangular shape as MoS_2 when prepared under similar conditions. WS_2 supported on SiO_2 has previously been proposed as a catalyst for the hydrogen evolution reaction.[22]

The WS_2 was studied on Toray paper and the preparation method was similar to that of the MoS_2 samples (see experimental details). In Fig. 5, the results of the electrochemical measurements are shown. Fig. 5A shows a Tafel plot (log i–E) and a polarization curve (i–E) in the region where we have previously observed hydrogen evolution on MoS_2 samples. On the polarization curve (inset of Fig. 5A) the cathodic current increases at potentials more negative than −0.2 V vs. NHE ascribed to HER activity. The Tafel slope on this sample is found to be 135 mV dec^{-1}, indicating that the current could be transport limited. Fig. 5b shows the deactivation of the sample at positive potentials. As with the MoS_2/C sample, an oxidation feature is observed with a peak at approx. 1 V vs. NHE, and on consecutive sweeps the peak disappears concurrent with a drop in the HER current. This is the same behavior as we have seen on the MoS_2/C sample except that the WS_2 sample required two sweeps towards highly anodic potentials before the HER current was affected. This behavior was also observed with high loadings of MoS_2/C samples that surely had formed multilayers. In this case, the outer layer could be passivated by a sulfur/oxide layer, thus requiring several oxidation/reduction steps to completely dissolve the metal sulfide. Apart from the potential formation of multilayers the oxidation of the WS_2 is similar to that of MoS_2/C, and it is assumed that the oxidation process is similar to that of MoS_2/C.

Cobalt promoted MoS_2 and WS_2

Cobalt is often used to promote WS_2 and MoS_2 in catalyzing the hydrodesulfurization reaction. Both the structural and the catalytic effect of adding cobalt has been extensively studied.[2] It is widely accepted that the cobalt is located at the edge of MoS_2, more specifically the so called S-edge ($\bar{1}110$). Cobalt promotion of MoS_2 has also been shown to change the morphology significantly.[11] Cobalt promoted

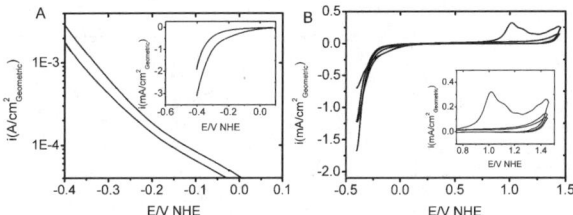

Fig. 5 A: Tafel and polarization curve (inset) of WS_2/C, scan rate 5 mV s^{-1}, both the initial and the final stable scan is shown. B: CV of WS_2/C showing the deactivation of WS_2.

MoS_2 is usually found as truncated triangles exposing the S-edge ($\bar{1}110$) predominantly, unlike the unpromoted MoS_2, in which the triangles are less truncated and primarily expose their Mo-edge ($\bar{1}010$).[23] In the following we will show data for sulfided Co and Co promoted WS_2 and MoS_2.

Electrochemical characterization of cobalt sulfide (CoS_x)

The first step in testing the promotion by cobalt is the test of sulfided cobalt itself. We have used Co(acetate) as the Co precursor as described in ref. 11. The precursor was sulfided under the same conditions as the MoS_2 and the WS_2 samples (see experimental details). The Co is expected to be in the form of Co_8S_9 immediately after sulfidation, but as this form is not stable in air,[24] our Co sulfide is most likely partially sulfided (CoS_x).

Fig. 6A shows the initial and the stable Tafel (log i–E) and polarization curves (i–E) within a narrow potential window (maximum +0.1 V vs. NHE). Initially the activity is high, but unlike MoS_2 and WS_2, subsequent sweeps within this potential window show a significant decrease in activity. The decrease is most likely due to the CoS_x instability in sulfuric acid, introducing ambiguity into the interpretation of the current at cathodic potentials as the HER competes with cathodic desorption or dissolution of CoS_x. In Fig. 6B a wide sweep is exhibited. The CoS_x exhibits similar oxidation features as we have seen on the MoS_2 and WS_2, but in this case the oxidation peak is shifted towards a higher potential (1.14 V vs. NHE). After, oxidation the HER activity drops just as with MoS_2 and WS_2, again indicating oxidation of the material.

Cobalt promoted MoS_2(Co–Mo–S) and WS_2(Co–W–S)

The Co promoted WS_2 and MoS_2 was prepared by co-impregnation of the Mo/W and the Co precursor (see experimental details). Fig. 7A and C shows the Tafel (log i–E) and the polarization (inset) curve (i–E) within a narrow potential window (maximum +0.1 V vs. NHE). The HER current diminishes just as on the pure CoS_x sample: it is initially high and after subsequent sweeps the current decreases noticeably, but unlike the case of pure CoS_x, remains stable at a fairly high level. This indicates that some of the Co promoter is in the state of CoS_x, but as the current stabilizes at a higher level than pure CoS_x, MoS_2 or WS_2, the remaining Co must have a promotion effect. The Tafel slopes are also in the expected region (Co–W–S 132 mV dec^{-1}, Co–Mo–S 101 mV dec^{-1}). In Fig. 7B and D, the CVs within a wide potential window are shown, and just as on the other metal sulfides we observe an irreversible oxidation peak followed by a significant decrease in HER activity. The peak maximum, however, seems to be shifted to a more anodic potential (approx. 1.1 V vs. NHE) than those corresponding to the unpromoted MoS_2 and WS_2.

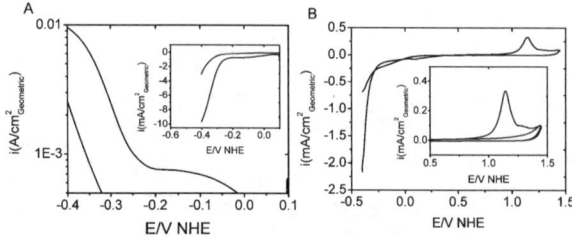

Fig. 6 A: Tafel and polarization curve (inset) of CoS_x/C, both the initial and the final stable scan is shown. B: CV of CoS_x/C showing the oxidation/deactivation of CoS_x.

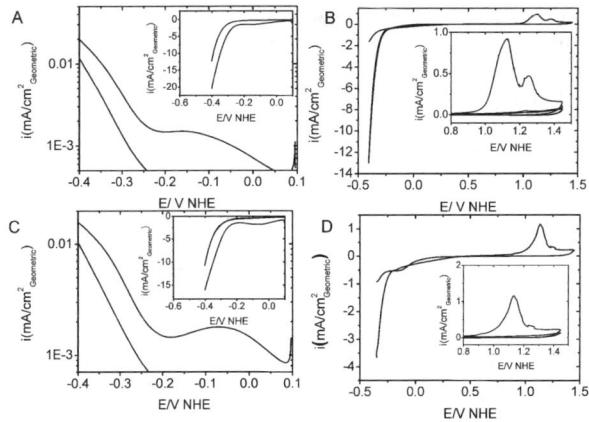

Fig. 7 A,C: Tafel and polarization curve (inset) of Co–Mo–S (A) and Co–WS (C), the scan rate is 5 mV s^{-1} both the initial and the final stable scan is shown. B,D: CV of Co–Mo–S (B) and Co–W–S (D) showing the deactivation of Co–Mo–S and Co–W–S.

DFT calculations on WS$_2$, MoS$_2$, Co–Mo–S and Co–W–S

We have calculated ΔG_H at the S-edge ($\bar{1}010$) and the Mo/W-edge ($1\bar{0}10$) of WS$_2$ and MoS$_2$ and on the Co promoted S-edge ($\bar{1}010$) edge of WS$_2$ and MoS$_2$ over a wide range of S coverage and H coverage. The choice of the relevant edge configurations have been based on the chemical potential of hydrogen and sulfur at the experimental sulfiding conditions using a thermodynamic model similar to the one presented in ref. 25. The structure and the differential free energies of H adsorption for these structures can be seen in Fig. 8. The results indicate that non-promoted WS$_2$ and MoS$_2$ nanoparticles should be reasonably good hydrogen evolution catalysts, since both edges on both systems have free energies of adsorption close to zero. Hydrogen evolution on MoS$_2$ is expected to take place predominantly at the Mo-edge ($\Delta G_H = 0.08$ eV) rather than the S-edge ($\Delta G_H = 0.18$ eV), while for WS$_2$ both edges are equally good ($\Delta G_H = 0.22$ eV). Given these values for ΔG_H, non-promoted MoS$_2$ is predicted to be a better hydrogen evolution catalyst than WS$_2$.

The incorporation of cobalt into the edge structures of both WS$_2$ and MoS$_2$ is expected to have a promotion effect. The cobalt only incorporates itself into the S-edge of both cases, so ΔG_H values at the Mo/W-edge remain unaffected. At the S-edge, however, ΔG_H is reduced to 0.10 eV and 0.07 eV for MoS$_2$ and WS$_2$, respectively (down from 0.18 eV and 0.22 eV). We note that the free energy of hydrogen adsorption at the cobalt-promoted S-edge of MoS$_2$ is very similar to the free energy

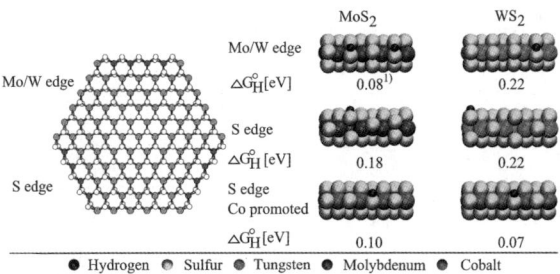

Fig. 8 Left: ball model of a Mo/WS$_2$ particle exposing both S-edge and Mo/W-edge. Right: differential free energies of hydrogen adsorption. 1 from ref. 5.

of hydrogen adsorption on the Mo-edge of MoS_2. Therefore, for MoS_2 the effect of promotion is the increase in the number of sites with high activity. On WS_2, the effect of cobalt promotion is the creation of new sites with higher activity than that prior to promotion.

In comparing all catalyst systems, DFT calculations suggest that cobalt-promoted MoS_2 (Co–Mo–S) should be a better catalyst than Co promoted WS_2 (Co–W–S) because it would have active sites on both edges and therefore a higher total number of active sites.

Linking catalyst structure and composition to HER activity

Calculated DFT values are best compared to experimental data where the activity has been normalized with respect to the number of active sites on the catalyst, in this case the number and type of edge sites on the different metal sulfides. We accomplish this normalization by using the irreversible oxidation features of each sulfide.

Fig. 9 exhibits normalized polarization curves (E–i) pertaining to each of the different samples. There is an apparent promotion effect of Co on both the MoS_2 and the WS_2 samples. The promotion effect on the WS_2 sample can be explained by the DFT calculations predicting that the Co promotion should decrease the free energy of hydrogen adsorption from 0.22 eV to 0.07 eV on the S-edge, and thus effectively increasing the activity of the active site. MoS_2 is a slightly different case. It has previously been found that the Mo-edge of MoS_2, which has a ΔG_H of 0.08 eV, is the major edge exposed, and that this edge does not adsorb cobalt.[23,26,27] However, the inhomogenous nature of these nanoparticulate catalysts suggests that both the Mo-edge and the S-edge will be present in significant fractions. Thus, the cobalt on the S-edge of MoS_2 promotes the HER, as its free energy of hydrogen adsorption is decreased from 0.18 eV to 0.10 eV. In other words, the number of active sites is increased since the normally less active S-edges becomes more active in the presence of cobalt.

Experimental and calculation details

Toray carbon paper was used as support material because it is inert, of high purity, has high conductivity and because it has adsorption sites/defects that will anchor the metal sulfide particles. The Toray paper was cut into strips that were 1 cm wide and 5 cm long. The Toray paper was loaded with catalyst by wetness impregnation with an aqueous solution of $(NH_4)_6Mo_7O_{24} \cdot 4H_2O$ in the case of MoS_2 and an aqueous solution of $H_{24}N_6O_{39}W_{12} \cdot xH_2O$ in the case of WS_2. In the case of the sulfided Co, $C_4H_4CoO_4 \cdot 4H_2O$ in an aqueous solution was used. The promoted WS_2 and MoS_2 were made by co-impregnation of Co and Mo/W. The impregnation of the pure sulfides was done by dropping a 25 µL aliquot (0.3–1 mM for Mo, 0.8 mM for W, 4mM for Co). The co-impregnation of the promoted sulfides was done by

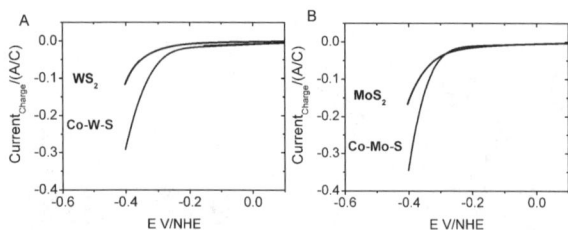

Fig. 9 Polarization curves where the currents of the different metal sulfides have been normalized with respect to the charge of the irreversible oxidation peak. A: Polarization curve of the HER on WS_2 and cobalt promoted WS_2(Co–W–S). B: Polarization curve of the HER on MoS_2 and cobalt promoted MoS_2(Co–Mo–S).

adding a 25 μL aliquot of Mo (0.7 mM) or W (0.8 mM) solution followed by a 25 μL aliquot of Co (4 mM) solution. A different sample preparation was used for the MoS_2 sample for XPS analysis where a the Toray paper was dip coated in the Mo solution (0.14 M) to obtain a more uniform impregnation.

The samples were dried at 140 °C and afterwards sulfided in a tube furnace under 10% H_2S in H_2 at 450 °C for 4 hours. The samples were cooled down in the same gas stream.

The electrochemical measurements where performed in N_2 purged 0.5 M H_2SO_4 (pH 0.4). To avoid contamination from the SCE reference electrode, a salt bridge was used. A Pt mesh was used as the counter electrode.

The XPS data was recorded using a Perkin-Elmer surface analysis system (Physical Electronics Industries Inc., USA) with a chamber base pressure of 10^{-10} Torr. Al-Kα radiation (1486.6 eV) was used for excitation. The XPS scans on Fig. 4 were measured with a pass energy of 100 eV, a step size of 1 eV, and 250 ms step^{-1}.

DFT calculations

An infinite stripe model, which has previously been proven successful to investigate MoS_2 based systems,[5,28–30] is used to investigate the edges of MoS_2. The infinite stripe exposes both the ($\bar{1}$010) Mo-edge and the ($\bar{1}$1$\bar{1}$0) S-edge. The supercell has 4 Mo atoms in the x-direction and 4 Mo atoms in the y-direction, in order to allow for important reconstructions with a period of 2 in the x-direction and to allow decoupling of the Mo-edge and the S-edge in the y-direction. The stripes are separated by 14.8 Å in the z-direction and 9 Å in the y-direction.

The plane wave density functional theory code DACAPO[31,32] is used to perform the DFT calculations. The Brillouin zone is sampled using a Monkhorst–Pack k-point set[33] containing 4 k-points in the x-direction and 1 k-point in the y- and z-direction. The calculated equilibrium lattice constant is 3.235 Å and 3.214 Å for MoS_2 and WS_2, respectively. A plane-wave cutoff of 30 Rydberg and a density wave cutoff of 45 Rydberg are employed using the double-grid technique.[34] Ultrasoft pseudopotentials are used except for sulfur, where a soft pseudopotential is employed.[35,36] A Fermi temperature of $k_BT = 0.1$ eV is used for all calculations and energies are extrapolated to zero electronic temperature. The exchange correlation functional RPBE is used. The convergence criterion for the atomic relaxation is that the norm of the total force should be smaller than 0.15 eV Å^{-1}, which corresponds approximately to a max. force on one atom below 0.05 eV Å^{-1}. Figures of atomic structures have been made using VMD.[37]

The differential free energies are calculated as described in ref. 5, where 0.29 eV is added to the pure DFT energy of adsorption in order to take zero point energy and entropy into account.

Conclusions

We have studied the hydrogen evolution on Co promoted and unpromoted nanoparticulate MoS_2 and WS_2 structures. Cyclic voltammetry revealed that they are irreversibly oxidized at high anodic potentials. We have used the irreversible oxidation features to determine the surface area of MoS_2 and proposed a possible oxidation mechanism of MoS_2. XPS analysis showed no change in the oxidation state of MoS_2 after HER measurements; but after oxidation at potentials above 0.6 V vs. NHE, MoS_2 was oxidized. We found that the activity of the carbon supported MoS_2 is comparable to that of our previously published results on Au(111) supported MoS_2. WS_2 has a similar structure and was also investigated in this study. It was found to irreversibly oxidize at high anodic potentials, just like MoS_2, and was found to be almost as active. Tests of cobalt promoted MoS_2 and WS_2 samples were also performed and Co is indeed promoting the HER in both cases. The findings are corroborated by DFT calculations showing that the activity of the different samples should be $WS_2 < MoS_2 = Co-Mo-S < Co-W-S$.

Acknowledgements

J.B. acknowledges support from the Danish Strategic Research Council. T.F.J. acknowledge H. C. Ørsted Postdoctoral Fellowships from the Technical University of Denmark. The Center for Atomic-scale Materials Design is supported by the Lundbeck Foundation. We thank the Danish Center for Scientific Computing for computer time. The Center for Individual Nanoparticle Functionality is supported by the Danish National Research Foundation.

References

1 J. Greeley, T. F. Jaramillo, J. Bonde, I. B. Chorkendorff and J. K. Nørskov, *Nat. Mater.*, 2006, **5**, 909–913.
2 H. Topsøe, B. S. Clausen and F. E. Massoth, *Hydrotreating Catalysis*, Springer-Verlag, Berlin, 1996.
3 H. Tributsch, *Z. Naturforsch., A: Phys., Phys. Chem., Kosmophys.*, 1977, **32**, 972–985.
4 J. P. Wilcoxon, *J. Phys. Chem. B*, 2000, **104**, 7334–7343.
5 B. Hinnemann, P. G. Moses, J. Bonde, K. P. Jorgensen, J. H. Nielsen, S. Horch, I. Chorkendorff and J. K. Nørskov, *J. Am. Chem. Soc.*, 2005, **127**, 5308–5309.
6 T. F. Jaramillo, K. P. Jørgensen, J. Bonde, J. H. Nielsen, S. Horch and I. Chorkendorff, *Science*, 2007, **316**, 100–101.
7 J. K. Nørskov, T. Bligaard, A. Logadottir, J. R. Kitchin, J. G. Chen, S. Pandelov and U. Stimming, *J. Electrochem. Soc.*, 2005, **152**, J23.
8 J. Greeley, J. K. Nørskov, L. A. Kibler, A. M. El-Aziz and D. M. Kolb, *ChemPhysChem*, 2006, **7**, 1032–1035.
9 S. Helveg, J. V. Lauritsen, E. Lægsgaard, I. Stensgaard, J. K. Nørskov, B. S. Clausen, H. Topsøe and F. Besenbacher, *Phys. Rev. Lett.*, 2000, **84**, 951–954.
10 J. Kibsgaard, J. V. Lauritsen, E. Laegsgaard, B. S. Clausen, H. Topsoe and F. Besenbacher, *J. Am. Chem. Soc.*, 2006, **128**, 13950–13958.
11 M. Brorson, A. Carlsson and H. Topsøe, *Catal. Today*, 2007, **123**, 31–36.
12 A. Carlsson, M. Brorson and H. Topsøe, *J. Catal.*, 2004, **227**, 530–536.
13 J. Meier, K. A. Friedrich and U. Stimming, *Faraday Discuss.*, 2002, **121**, 365–372.
14 T. F. Jaramillo, J. Zhang, B. L. Ooi, J. Bonde, K. Andersson, J. Ulstrup, J. K. Nørskov and I. Chorkendorff, *J. Phys. Chem. C*, 2008, in press.
15 N. M. Markovic and P. N. Ross, *Surf. Sci. Rep.*, 2002, **45**, 121–229.
16 W. Kautek and H. Gerischer, *Surf. Sci.*, 1982, **119**, 46–60.
17 C. D. Wagner, A. V. Naumkin, A. Kraut-Vass, J. W. Allison, C. J. Powerll and J. R. Rumble, Jr, NIST X-Ray Photoelectron Spectroscopy Database, Standard Reference Database 20, Version 3.4, 2008, http://srdata.nist.gov/xps.
18 W. Jaegermann and D. Schmeisser, *Surf. Sci.*, 1986, **165**, 143–160.
19 T. Weber, J. C. Muijsers, H. J. M. C. van Wolput, C. P. J. Verhagen and J. W. Niemantsverdriet, *J. Phys. Chem.*, 1996, **100**, 14144–14150.
20 J. H. Nielsen, K. P. Jørgensen, J. Bonde, K. Nielsen, L. Bech, Y. Tison, S. Horch, T. F. Jaramillo and I. Chorkendorff, 2008, in preparation .
21 M. Pourbaix, Atlas of Electrochemical Equilibria, 1966.
22 A. Sobczynski, A. Yildiz, A. J. Bard, A. Campion, M. A. Fox, T. Mallouk, S. E. Webber and J. M. White, *J. Phys. Chem.*, 1988, **92**, 2311–2315.
23 J. V. Lauritsen, J. Kibsgaard, G. H. Olesen, P. G. Moses, B. Hinnemann, S. Helveg, J. K. Nørskov, B. S. Clausen, H. Topsøe, E. Lægsgaard and F. Besenbacher, *J. Catal.*, 2007, **249**, 220–233.
24 I. Alstrup, I. Chorkendorff, R. Candia, B. S. Clausen and H. Topsøe, *J. Catal.*, 1982, **77**, 397–409.
25 M. V. Bollinger, K. W. Jacobsen and J. K. Nørskov, *Phys. Rev. B: Condens. Matter Mater. Phys.*, 2003, **67**, 085410.
26 P. Raybaud, J. Hafner, G. Kresse, S. Kasztelan and H. Toulhoat, *J. Catal.*, 2000, **190**, 128–143.
27 H. Schweiger, P. Raybaud and H. Toulhoat, *J. Catal.*, 2002, **212**, 33–38.
28 B. Hinnemann, J. K. Nørskov and H. Topsøe, *J. Phys. Chem. B*, 2005, **109**, 2245–2253.

29 J. V. Lauritsen, M. Nyberg, R. T. Vang, M. V. Bollinger, B. S. Clausen, H. Topsøe, K. W. Jacobsen, E. Lægsgaard, J. K. Nørskov and F. Besenbacher, *Nanotechnology*, 2003, **14**, 385–389.
30 L. S. Byskov, J. K. Nørskov, B. S. Clausen and H. Topsøe, *J. Catal.*, 1999, **187**, 109–122.
31 S. R. Bahn and K. W. Jacobsen, *Comput. Sci. Eng.*, 2002, **4**, 56–66.
32 B. Hammer, L. B. Hansen and J. K. Nørskov, *Phys. Rev. B: Condens. Matter Mater. Phys.*, 1999, **59**, 7413–7421.
33 H. J. Monkhorst and J. D. Pack, *Phys. Rev. B: Solid State*, 1976, **13**, 5188–5192.
34 K. Laasonen, A. Pasquarello, R. Car, C. Lee and D. Vanderblit, *Phys. Rev. B: Condens. Matter Mater. Phys.*, 1993, **47**, 10142–10153.
35 N. Troullier and J. L. Martins, *Phys. Rev. B: Condens. Matter Mater. Phys.*, 1991, **43**, 1993–2006.
36 D. Vanderbilt, *Phys. Rev. B: Condens. Matter Mater. Phys.*, 1990, **41**, 7892–7895.
37 W. Humphrey, A. Dalke and K. Schulten, *J. Mol. Graphics*, 1996, **14**, 33.

Influence of water on elementary reaction steps in electrocatalysis†

Yoshihiro Gohda,‡ Sebastian Schnur and Axel Groß

Received 11th February 2008, Accepted 9th April 2008
First published as an Advance Article on the web 8th August 2008
DOI: 10.1039/b802270d

We studied simple reaction pathways of molecules interacting with Pt(111) in the presence of water and ions using density functional theory within the generalized gradient approximation. We particularly focus on the dissociation of H_2 and O_2 on Pt(111) which represent important reaction steps in the hydrogen evolution/oxidation reaction and the oxygen reduction reaction, respectively. Because of the weak interaction of water with Pt(111), the electronic structure of the Pt electrode is hardly perturbed by the presence of water. Consequently, processes that occur directly at the electrode surface, such as specific adsorption or the dissociation of oxygen from the chemisorbed molecular oxygen state, are only weakly influenced by water. In contrast, processes that occur further away from the electrode, such as the dissociation of H_2, can be modified by the water environment through direct molecule–water interaction.

I. Introduction

The details of many simple reactions at the solid–gas interface relevant for heterogeneous catalysis have in recent years been clarified by a close collaboration between theory and experiment.[1] In electrocatalysis, reactions occur at the solid–liquid interface in the presence of ions and an external field. This adds considerable complexity to the reaction pathways and thus makes the elucidation of microscopic reaction steps much harder. Consequently, relatively little is known about the elementary reaction steps occurring in such seemingly simple reactions, such as the hydrogen evolution/oxidation reaction and the oxygen reduction reaction in electrocatalysis,[2] in spite of considerable advances in the experimental microscopic characterization of structures and processes at the solid–liquid interface.[3] For example, the exact microscopic structure of water at the solid–liquid interface is still a subject of debate (see, *e.g.*, ref. 4–6). Whereas, for single water layers on electrode surfaces several surface science techniques with microscopic resolution can be applied,[7] molecular scale studies for thicker water layers are scarce (for a discussion see, for example, ref. 8–12). Thus, it is still not clear whether water assumes an ice-like crystalline structure or rather a disordered liquid-like structure directly at the water–metal interface. The situation becomes even more complex if, additionally, the specifc adsorption of ions is considered.

Nowadays, first-principles calculations based on density functional theory (DFT) can shed light on microscopic structures and processes at interfaces.[13] The interaction of molecules with electrode surfaces in the presence of water layers has

---

Institute for Theoretical Chemistry, Ulm University, Ulm, D-89069, Germany

† The HTML version of this article has been enhanced with colour images.
‡ New Address: Department of Applied Physics, The University of Tokyo, 7-3-1 Hongo, Bunkyo-ku, Tokyo 113-8656, Japan

already been addressed by several DFT studies.[14–23] Some of these studies indicate that the chemical bonding in specific adsorption is hardly influenced by the presence of water because of the weak water–metal interaction.[17] However, this is not necessarily true for reaction barriers in electrocatalytic reactions. It has been found that water has a promoting effect in the CO oxidation on metal surfaces,[15,16] whereas the activation barrier for the Tafel reaction $2H_{ad} \rightarrow H_2$ on Pt(111) seems to be not significantly changed by the presence of water.[22] On the other hand, the significant overpotential required to measure currents in the oxygen reduction reaction[2] indicates that the oxygen dissociation might be hindered by the presence of water.

Hence, it is certainly fair to say that our understanding of the role of water in electrocatalytic reactions is far from being complete. In this contribution, we address the influence of water on simple, but still electrocatalytically relevant reactions by periodic density functional theory calculations. This work has a modest, but rather fundamental goal. By determining the barrier for simple reactions in the presence of water in different structures we want to contribute to the elucidation of the structure–reactivity relationship for water. In order to concentrate on this fundamental issue, we neglect the influence of the electrode potential.

We particularly focus on the dissociation of H_2 and O_2 on Pt(111) which represent important reaction steps in the hydrogen evolution/oxidation reaction and the oxygen reduction reaction, respectively. By comparing the reaction paths with and without the presence of water and analyzing the underlying electronic structure, we are able to elucidate the microscopic factors determining the influence of water on these reactions. In addition, we will discuss the role of anions in electrocatalytic reactions which is also an issue of particular importance in electrocatalysis.

II. Computational details

All DFT calculations were performed using the VASP code[24,25] with the exchange–correlation effects described with the generalized gradient approximation (GGA) by the functional of Perdew, Burke and Ernzerhof (PBE).[26] This functional is well-suited for the present study because it gives a rather good description of the hydrogen bonding with respect to bulk water,[27] but also for small water clusters[28] and hydrogen bonding networks in supramolecular assemblies.[29] The ionic cores were represented by projected augmented wave (PAW) potentials.[30,31] The Kohn–Sham states were expanded in a plane wave basis with an energy cutoff of at least 400 eV. The Pt(111) substrate was modeled by a four-layer slab with the bottom layers kept fixed at their bulk positions with a lattice constant of 3.97 Å, whereas all other atomic positions were fully relaxed. In the case of the O_2 dissociation, we employed a 2×2 surface unit cell and a $6 \times 6 \times 1$ Monkhorst–Pack grid to sample the **k**-points for the integration over the first Brillouin zone, whereas the H_2 dissociation was described within a $\sqrt{3} \times \sqrt{3}R\ 30°$ surface unit cell and a 5×5 **k**-point sampling.

Note that all binding energies to Pt(111) listed in this work are given with respect to the corresponding molecules in the gas phase. For the energy balance appropriate for electrocatalytic reactions, the solvation energy of the molecules in the corresponding solution has to be taken into account. However, here we focus on details of elementary reaction steps occurring directly at the electrode surface in the presence of water where solvation effects in the bulk liquid do not play any direct role.

III. Interaction of water with Pt(111)

As a first step, we addressed the structure of water on a Pt(111) surface. In order to model the metal–water interface, we considered ice-like water structures, as shown in Fig. 1. *Ab initio* molecular dynamics simulations of water–metal interfaces at room temperature[32,33] indicate that the water molecules of the first layer at the interface

Fig. 1 Illustration of the most stable water structures on Pt(111) according to the DFT calculations. (a) H-down bilayer structure, (b) H-down double bilayer structure.

remain rather localized. Hence the energy minimum structure might also be stable at finite temperatures.

As for the bilayer structure shown in Fig. 1a corresponding to a water coverage of $\theta_{H_2O} = 2/3$, there are two undissociated structures, a H-down and a H-up structure, where one of the hydrogen atoms of every second water molecule is oriented either towards or away from the surface. In agreement with previous studies,[4] we find that the H-down structure is slightly more stable than the H-up structure. Our calculated binding energies per H_2O molecule of 481 meV and 464 meV for the H-down and H-up structure are slightly less than the previous results,[4] most probably due to the different treatment of the ionic cores.

We added another water bilayer in order to study the influence of a thicker water environment. It is not easy to unambiguously determine the minimum energy structure of the second bilayer. The upper water layer can be shifted almost arbitrarily on top of the lower water layer, and, in addition, different combinations of H-up and H-down water layers are possible. Many of these possible structures exhibit similar binding energies per water molecule in the range between 464 meV and 482 meV. In order to gain insight into the principle mechanisms, we focus mainly on the structure that can be seen in Fig. 1b. This structure has a H_2O binding energy of 482 meV per molecule which is almost the same as in the case of a single water layer with a binding energy of 481 meV per molecule.

In order to determine the modification of the Pt electrode atoms upon the adsorption of water, we compare in Fig. 2 the local density of states (LDOS) of the Pt(111) substrate atoms for the clean surface with those of water-covered surfaces. For the water bilayer shown in Fig. 1a, there are three inequivalent Pt surface atoms per surface unit cell, either non-covered or covered by a water molecule bound *via* an oxygen atom or a hydrogen atom. Fig. 2 demonstrates that the LDOS of all these three Pt atoms hardly differs from the LDOS of the Pt atoms at the clean Pt(111) surface. In addition, we have also considered the adsorption of a single water molecule to Pt(111), which in its most favorable position lies almost flat on the surface with the oxygen atom above a Pt atom, as already identified in previous studies.[18,34]

Although the binding energy of the single water molecule of about 350 meV is actually smaller than the binding energy per water molecule in the ice-like bilayer structure, the single water molecule is interacting more strongly with the Pt substrate than the water molecules in the bilayer structure where the energy gain upon adsorption is mostly due to the hydrogen bonding within the water network.[18] And indeed we find that the electronic structure of the Pt atom below the water monomer is modified to a larger extent than those of the Pt atoms covered by a water bilayer. The increased LDOS for the monomer adsorption at about −4.5 eV is due to the

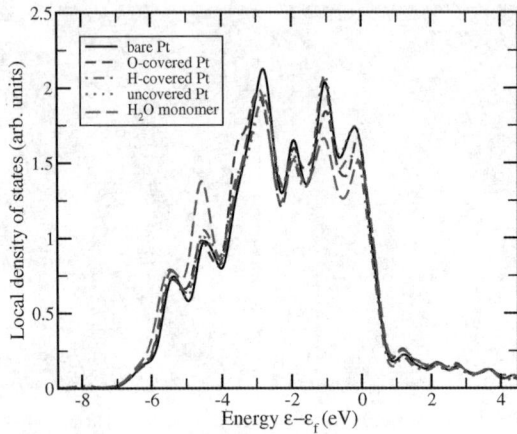

Fig. 2 Local density of states (LDOS) of the Pt(111) surface atoms without and with the presence of water. For the adsorption of a single water molecule, only the LDOS of the Pt atom directly below the water molecule is plotted, whereas for the H-down water bilayer the LDOS of the three inequivalent Pt atoms within the surface unit cell is shown.

hybridization of the water $1b_1$ orbital with the Pt d-band.[34] Still, the peak positions and the width of the Pt d-band are hardly modified by the presence of water indicating that the interaction of water with late transition metals is rather weak. This explains why the chemical bonding of specifically adsorbed species to late transition metal electrode surfaces is only weakly influenced by the presence of water.[17,18]

IV. Oxygen dissociation on Pt(111)

O_2 adsorbs on clean Pt(111) in two molecular chemisorption states, the non-magnetic peroxo state and the magnetic superoxo state[35–38] which can be spontaneously accessed from the gas phase,[36–40] *i.e.*, without encountering any adsorption barrier. Here, we focus on the dissociation of O_2 from these molecular chemisorption states. The peroxo state corresponds to a top–fcc-hollow–bridge configuration, *i.e.*, the O_2 center of mass is located above the fcc hollow position, whereas the two oxygen atoms are oriented towards the top and bridge site, respectively. In contrast, in the superoxo state the O_2 molecule is adsorbed in a top–bridge–top configuration.

The reaction paths for the O_2 dissociation from the two molecular states were determined using the nudged elastic band (NEB) method.[41] Fig. 3 shows the change in the total energy as well as the magnetic moment during the O_2 dissociation. The results agree well with those of previous DFT studies.[37] Obviously, the dissociation barriers from both molecular O_2 states are rather similar and in the order of ≥ 0.7 eV.

One remarkable finding is that the magnetization along the dissociation path from the nonmagnetic peroxo molecular adsorption state rises to a value of $\mu = 2\mu_B$ at the transition state. In order to analyze this change in the magnetic moment, we projected the wave functions to atomic orbitals so that the magnetic moments of the individual atoms could be determined. Thus, we find that most of the magnetic moment at the transition state originates from the magnetic polarization of the 5d electrons of the Pt substrate atoms.

Next, we included the effect of water on the O_2 dissociation barrier by adding one water molecule. In fact, one isolated water molecule should lead to a stronger local perturbation than a water bilayer, as discussed in the previous section. We fully relaxed the atomic structure, including the H_2O molecule, and monitored the oxygen

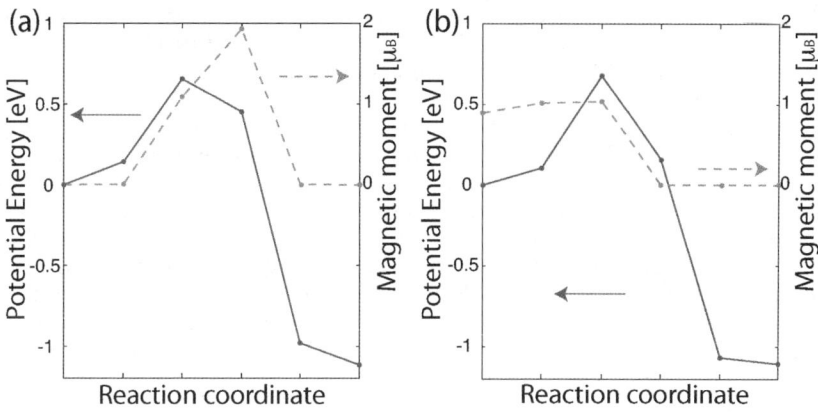

Fig. 3 The total energy relative to the initial molecular state (solid line) and the magnetic moment (dashed line) during the O_2 dissociation for (a) peroxo and (b) superoxo adsorption states on the Pt(111) surface.

dissociation process. Water modifies the dissociation barrier and the magnetization along the dissociation paths rather modestly. In the case of the dissociation of the peroxo precursor state, the presence of water makes the dissociation barrier slightly smaller and narrower, as seen in Fig. 4. This suggests that it is not the O_2 dissociation itself in the presence of water that causes the significant overpotential required for the oxygen reduction reaction, but rather other processes such as the H_2O_2 or H_2O formation that are part of the oxygen reduction reaction schemes,[2] or more complex reorganization effects of the water around the oxygen molecule.

Furthermore, we explored the influence of the additional presence of anions on the electrode surface. As a prototype, we chose Cl in the −1 charge state. Since we considered a situation where the cations are supposed to be far away from the

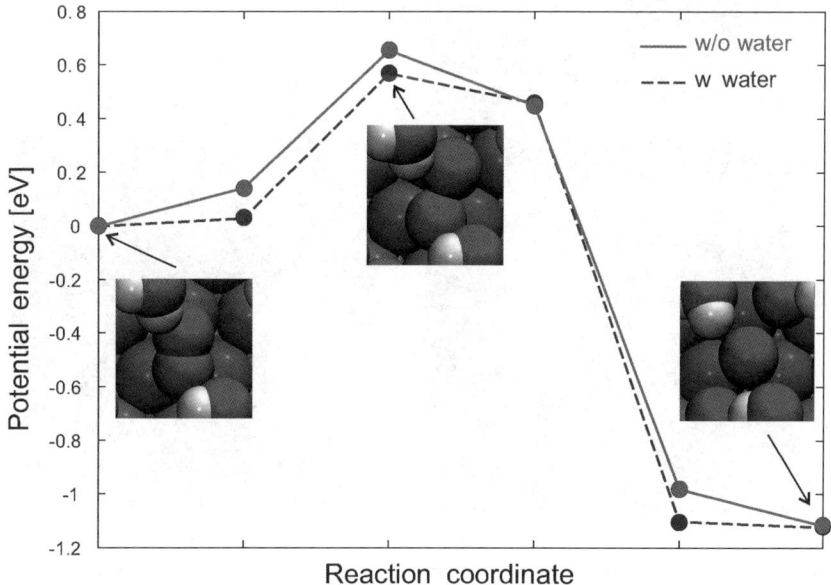

Fig. 4 Potential energy along the O_2 dissociation path on Pt(111) with and without water starting from the nonmagnetic molecular precursor state of O_2.

Pt surface, they are not explicitly included in our model. Instead, the calculations for the negative charge state were performed with a compensating positive uniform background charge to avoid the divergence for charged supercells in the reciprocal space representation.

For the Cl^- ion, the most stable adsorption site on the clean Pt(111) surface is the top site, whereas it is the bridge site for the neutral charge state. To model the O_2 dissociation process in the presence of chloride and water (see Fig. 5), we put Cl^- at the top site and optimized the atomic structure. The Cl^- ion remains at the top site for the initial O_2 superoxo state (Fig. 5c), whereas it is slightly displaced to the top–fcc-hollow site due to the close presence of oxygen atoms in the final state of the atomic oxygen adsorption (Fig. 5d). To find the transition state, we used a denser sampling in the reaction coordinate along the dissociation path, since a more complex pathway is expected due to the larger number of relevant degrees of freedom in the presence of both water and chloride in addition to the oxygen molecule.

The calculated atomic structure for the transition state of this reaction is shown in Fig. 5a. The oxygen molecule is almost in a top–hcp-hollow–bridge configuration. As the comparison of Fig. 5b with Fig. 4 clearly shows, the presence of Cl^- affects the transition barrier and the magnetization remarkably during the dissociation, in spite of the fact that the anion is neither directly involved in the reaction nor

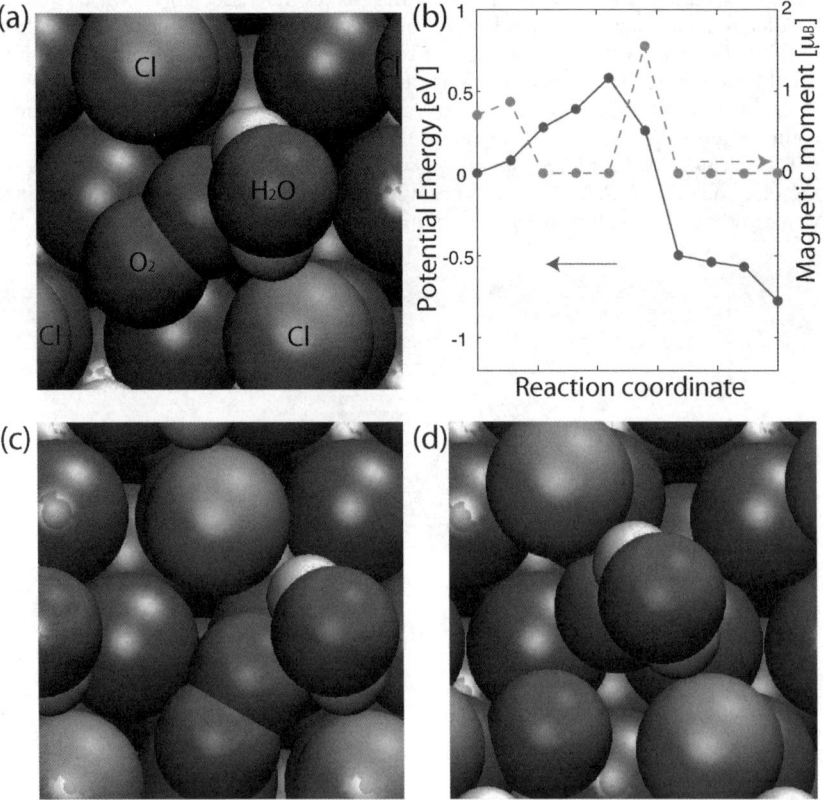

Fig. 5 (a) Atomic structure of the transition state for the dissociation of O_2 superoxo adsorption state on the Pt(111) surface with the presence of H_2O and Cl^-. (b) Potential energy relative to the initial molecular state (solid line) and the magnetic moment (dashed line) during the O_2 dissociation. (c) The initial-state configuration with the O_2 superoxo state. (d) The final-state configuration with atomic O adsorption.

magnetic. The calculated barrier height is slightly lowered due to the presence of chloride and water.

In contrast to the case without Cl⁻ where the magnetization is kept almost constant before overcoming the dissociation barrier, it decreases in the early stage of the dissociation process in the presence of Cl⁻ and then vanishes. Still, the system becomes magnetic again near the transition state. This behavior is due to the fact that the dissociation pathway proceeds through the top–hcp-hollow–bridge configuration, which is a nonmagnetic metastable precursor state. This dissociation pathway corresponds to the rotation of the oxygen molecule, which is caused by the fact that one of the oxygen atoms in O_2 is anchored by the water molecule that is bound to chloride through hydrogen bonding. It should be noted that the presence of water without chloride cannot anchor the oxygen atom, because the water molecule follows the motion of oxygen resulting in the nearly unaffected dissociation pathway. This anchor effect is removed after overcoming the dissociation barrier, because atomic oxygen strongly prefers to be at the fcc-hollow site. Indeed, atomic adsorption of oxygen at the Pt(111) top site is unstable.[37]

Although the spin polarization of the oxygen molecule plays an essential role in the magnetization for the initial state of the superoxo molecular adsorption, the drastic increase in the magnetic moment in the vicinity of the transition state is mainly attributed to 5d electrons of the Pt substrate as mentioned in the discussion of the dissociation of the peroxo molecular oxygen. Since the ground state configuration of 5d electrons of Pt surfaces is nonmagnetic, this remarkable change in the electronic states indicates the strong involvement of the Pt electrons in the dissociation process, explaining the high catalytic activity of Pt surfaces.

V. H_2 dissociation on Pt(111)

In contrast to O_2, for H_2 there are no chemisorbed molecular adsorption states on metal surfaces, except at defect sites such as steps.[42,43] Furthermore, the rather small dissociation barrier of H_2 on clean Pt(111) is located more than 2 Å away from the surface,[44,45] at about the same height as the water bilayer. Therefore we included the whole water bilayer structure in order to model the H_2 dissociation on water-covered Pt(111).

To characterize the reaction path, we determined two-dimensional cuts, so-called elbow plots,[46] through the potential energy surface of the hydrogen dissociation reaction as a function of the center of mass distance of H_2 from the surface and the intramolecular H–H distance. The dissociation path corresponds to a fcc-hollow–top–hcp-hollow (h–t–h) geometry with the H–H-bond axis parallel to the surface, as shown in the inset of Fig. 6b.

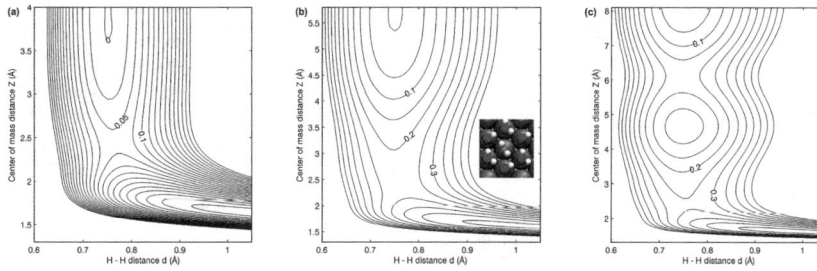

Fig. 6 Two-dimensional cuts through the potential energy surface of the interaction of H_2 with clean Pt(111) (a) and with Pt(111) covered by one (b) and two (c) water bilayers. The potential energy is plotted as a function of the H–H distance d and the H_2 center of mass distance Z from the surface. The lateral position and orientation of the H_2 molecule correspond to a fcc-hollow–top–hcp-hollow configuration, as indicated in the inset. The contour spacing in (a) is 25 meV, while it is 50 meV in (b) and (c). Note the different scales of the Z axis.

On clean Pt(111), there is only a small dissociation barrier of 54 meV for the hollow–top–hollow dissociation of H_2, in good agreement with previous calculations.[44,45] The barrier is at 2.5 Å above the Pt(111) surface plane (Fig. 6a). This barrier is a so-called early barrier,[47] *i.e.*, the barrier is located before the curved region of the PES. The adsorption energy of atomic hydrogen in the hollow position on the three layer Pt(111) slab is calculated to be 468 meV with respect to the free H_2 molecule.

In the presence of the water bilayer, we expected the minimum energy path of the H_2 dissociation to go through the middle of the water bilayer above an uncovered Pt atom as shown in the inset of Fig. 6b. All water molecules and the first Pt layer were completely relaxed in the absence of H_2, but for the determination of the PES, all positions were kept fixed. The potential energy is given relative to the energy of a H_2 molecule 9 Å above the Pt surface and a H–H bond length of 0.75 Å.

The water bilayer affects the H_2 dissociation considerably (see Fig. 6b). While the position of the dissociation barrier at 2.5 Å above the surface plane is hardly modified, the barrier height is increased by 167 meV to 221 meV. In order to figure out the origin of this increase in the barrier height, we removed the Pt slab and calculated the PES with the fixed atomic positions of the water molecules. Thus, we find a barrier for the propagation of H_2 through the water bilayer of 165 meV but at a position that would correspond to a distance of 3.1 Å above the Pt surface (see Table 1). This indicates that there is a direct repulsive interaction between the water bilayer and the H_2 molecule, so that the modification of the barrier is not the result of the (weak) water-induced modification of the electronic structure of Pt(111).

Addressing the problem of different reaction pathways and dissociation sites of hydrogen, we also performed a nudged elastic band calculation[41] to check our preliminary assumptions. As a result, we found that the reaction barrier changes by less than 10 meV in the NEB calculations, which indicates that the barrier that we derived from the PES is very close to the true one. Hence, the assumed reaction pathway is also very close to the result of the NEB calculation as well. Regarding the other inequivalent h–t–h configuration, we rotated the H_2 molecule by 60° around the z-axis onto the corresponding h–t–h dissociation path. We found a small rise in the dissociation barrier of 20 meV. Comparing the two different ice-like H-down and H-up structures, the H-up structure exhibits a slightly increased dissociation barrier compared to the H-down structure by 39 meV from 221 meV to 260 meV. All these data are also collected in Table 1.

Our result that taking a water bilayer into account increases the barrier for the H_2 dissociation on Pt(111) by about 170 meV seems to be at variance with recent DFT calculations by Skúlason *et al.* that found that the barrier for the reverse reaction,

Table 1 Binding energies of water in meV molecule^{-1} and atomic hydrogen binding energies in meV atom^{-1} on Pt(111) with respect to the free water molecule and H_2 molecule, respectively, H_2 dissociation barriers ΔE in meV and distance z of the dissociation barriers from the surface in Å for different water structures on Pt(111). In addition, results for free-standing water layers without the Pt substrate are included. (a) refers to the H-down and (b) to the H-up structure of the water bilayer

θ_{H_2O}	$E_b^{H_2O}$	E_b^H	ΔE_{lower}	z_{lower}	ΔE_{upper}	z_{upper}
2/3 (a)	481	399	221	2.5	—	—
2/3 (b)	464	395	260	2.4	—	—
4/3 (a)	482	378	253	2.4	178	6.1
4/3 (b)	476	352	277	2.3	191	6.1
0	—	468	54	2.5	—	—
2/3 (a) without Pt	451	—	165	3.1	—	—
4/3 (a) without Pt	457	—	183	3.1	175	6.1

the Tafel reaction $2H_{ad} \rightarrow H_2$, is not changed significantly by the presence of water.[22] However, one has to consider that the presence of water also leads to a reduction of the atomic hydrogen binding energies from 468 meV at the clean Pt(111) surface to 395 meV per hydrogen atom (see Table 1). This reduction of the atomic binding energies compensates the increase in the barrier height, so that the difference, which is the barrier for hydrogen desorption, remains almost unchanged. In passing, we note that the activation barrier for the Tafel reaction determined by Skúlason et al. is smaller than the one determined in our calculations. This is due to the fact that Skúlason et al., instead of the PBE functional, used the RPBE functional[48] which is known to yield lower binding energies of adsorbates at surfaces.

So far we have neglected the influence of further water layers. To study the influence of surrounding water molecules on the reaction pathway we studied the dissociative adsorption of H_2 through two water bilayers. We calculated the PES in the same way we did in the case of only one water bilayer. As a result, we get an additional barrier of 175 meV located 6.1 Å above the surface in the middle of the upper water layer (see (Fig. 6c). This confirms that there is Pauli repulsion between the close-shell H_2 molecule and the close-shell water molecules. The lower barrier is weakly influenced by the additional water layer and rises from 221 meV to 253 meV. If we neglect the Pt surface and consider the two water bilayers alone, we find a barrier of 183 meV for the upper layer and 175 meV for the lower layer, again at positions corresponding to a height of 3.1 Å and 6.1 Å above the surface, respectively. The position of the second upper barrier is the same with and without the Pt slab, only the position of the barrier in the first lower barrier is again moved towards the surface due to the presence of the Pt substrate. This indicates again that the total barrier can be regarded as a superposition of the H_2/Pt(111) dissociation barrier and the barrier for the parallel propagation of the H_2 molecule through the water bilayer. Considering differences between H-up and H-down double layer structures, we find slightly higher barriers for the H-up structure as already observed in the single bilayer calculations (see Table 1).

Interestingly enough, the upper barrier depends only very weakly on the molecular orientation. The barrier decreases by less than 10 meV if the molecule is turned in an upright orientation perpendicular to the surface. This can be understood considering the fact that the charge distribution of the free H_2 molecule that is dominated by the σ_g state is almost spherically symmetric so that the repulsion between the H_2 molecule and the water molecules hardly depends on the orientation of the molecule. We performed *ab initio* molecular dynamics runs of H_2 molecules approaching the Pt substrate through the holes in the water bilayer for different initial kinetic energies. We find that the H_2 molecules do not change their orientation significantly when they propagate across the upper barrier demonstrating the potential is indeed almost isotropic. The molecular dynamics simulations also showed that approaching H_2 molecules can become trapped in the potential minimum between first and second water bilayer and then rather return instead of adsorbing. Note that the energy associated with this minimum in Fig. 6c can also serve as a rough indicator for the small solvation energy of H_2 in water, although water relaxation effects are not included.

Finally, we also studied the influence of disorder in the water structures on the dissociation barriers. Naturally, there is a vast number of possible arrangements of the water layer which might only slightly differ from each other. In order to have a systematic approach, we decided to successively remove up to six molecules alternately from the first and the second layer. Some of these arrangements are certainly not very realistic and might hardly occur at a metal–water interface. Still, this approach allows us to get insight in the basic dependence of the dissociation barriers on the surrounding water environment. A $2\sqrt{3} \times 2\sqrt{3}$ unit cell consisting of 16 water molecules was chosen to reasonably describe the disordered water system. All excluded molecules are direct neighbors of the considered H_2 molecule, and all water structures were relaxed before the interaction of H_2 with these structures was evaluated. The PES was calculated in the same manner as we did in the case of the two complete water layers, *i.e.*, the center of mass of the H_2 molecule was fixed at the top site of the uncovered Pt atom and the H_2 orientation was also frozen. To determine the true dissociation path, a nudged elastic band calculation would be required, in particular considering the fact that the symmetry of the water layers is reduced by extracting water molecules; however, the computational cost for such calculations would be exceedingly high. Still, trends in the height of the barriers can also be derived from the PES calculations with the center of mass of the H_2 molecule at the center of the hexagonal ring.

To speed up the calculations, we considered only a three-layer Pt slab, however, we checked that the calculated barrier is hardly influenced by the reduced slab thickness, as a comparison of the third line of Table 1 with the first column of Table 2 shows, where the barrier heights for the considered disordered structures are listed.

Interestingly enough, we find that removing one molecule from the lower layer leads to a small increase of the dissociation barrier. Even if we exclude a second molecule from the lower layer, the barrier remains high and is even raised when a third molecule is excluded from the upper layer. This particular water structure is illustrated in Fig. 7. These findings are surprising because, naively, one would expect that the removal of water molecules would lead to a more open structure through which it is easier to propagate. However, one has to consider that the lattice constant of the water bilayer on Pt(111) and in hexagonal ice differ: the water layers on Pt(111) in the regular hexagonal structure are expanded by about 7%.[6] If the

Table 2 Dependence of the dissociation barriers E_b in meV in the lower and the upper water layer on the number of molecules removed from the water layers (#)

Lower layer							
#	0	1	1	2	2	2	3
E_b	270	296	284	276	275	323	189
Upper layer							
#	0	0	1	1	2	3	3
E_b	173	172	162	139	97	9	0

Fig. 7 Illustration of a disordered water structure considered in the calculations with two water molecules extracted from the lower water layer and three from the upper.

ordered structure is destroyed, the water structure relaxes and the distances between the water molecules can become reduced. In addition, water molecules from the upper layer relax downward, reducing the available space for the H_2 molecule propagation. Altogether, this leads to the increase in the barrier height. Only the exclusion of a third water molecule in the lower layer leads to a significant decrease of the barrier down to 189 meV. In contrast, removing molecules from the upper layer leads to a continuous decrease of the barrier for the propagation through the upper layer. This decrease is again small when only one water molecule is removed, but the barrier even vanishes when three molecules are excluded in the upper layer. These results indicate that disordered water structures at the solid–liquid interface will result in a broadened distribution of barriers for the H_2 dissociation which can be larger as well as smaller than the barrier for the ordered hexagonal bilayer structure.

VI. Conclusions

We studied simple reactions on Pt(111) in the presence of water by density functional theory calculations. The adsorption of an ice-like water bilayer on Pt(111) perturbs the electronic structure of the Pt atoms only weakly. Hence, processes that occur directly at the surfaces are only weakly influenced by the presence of water. Such a process is the dissociation of O_2 from the molecular O_2 chemisorption state, which was addressed using the nudged elastic band method. We considered both the non-magnetic peroxo state and the magnetic superoxo state as initial states. We calculated the dissociation barrier of the O_2 molecule on clean Pt(111) and in the presence of a water molecule alone and additional adsorbed Cl anions. The O_2 molecule becomes magnetically polarized at the transition state from the nonmagnetic peroxo state to the nonmagnetic atomic state with the Pt d states playing an important role. The dissociation barrier and adsorption energies, as well as the magnetization, are hardly affected by the presence of one H_2O molecule. In contrast, they are remarkably changed in the presence of Cl^- ions. In this case, water plays the role of an anchor for oxygen with the water bound to chloride through a hydrogen bond.

In the case of the H_2 dissociation on Pt(111), the water bilayer leads to an increase in the H_2 dissociation barrier height. This barrier can be regarded as a superposition of the H_2 dissociation barrier on clean Pt(111) with the barrier for the H_2 propagation through an ice-like hexagonal water layer. The H_2 dissociation barrier further increases, but only slightly, when a second water bilayer is taken into account. The barrier for the reverse reaction, the Tafel reaction, on the other hand, is not changed significantly by the presence of water, since the increase in the dissociation barrier is compensated by reduced atomic hydrogen binding energies to Pt(111). The barrier for the H_2 propagation through the second bilayer depends only very weakly on the H_2 orientation. Disordered water structures result in a broadened distribution in H_2 dissociation barrier heights.

In the future, we plan to perform *ab initio* molecular dynamics simulations in order to study dynamical details of simple reactions at electrode surfaces in the presence of water.

Acknowledgements

Support by the German Science Foundation (DFG) through contract GR 1503/18-1, the Alexander von Humboldt Foundation and the Konrad-Adenauer-Stiftung is gratefully acknowledged.

References

1 A. Groß, *Surf. Sci.*, 2002, **500**, 347.
2 N. M. Marković and P. N. Ross, Jr, *Surf. Sci. Rep.*, 2002, **45**, 117.

3 D. M. Kolb, *Surf. Sci.*, 2002, **500**, 722.
4 S. Meng, L. F. Xu, E. G. Wang and S. W. Gao, *Phys. Rev. Lett.*, 2002, **89**, 176104.
5 P. J. Feibelman, *Phys. Rev. Lett.*, 2003, **91**, 059601.
6 S. Meng, L. F. Xu, E. G. Wang and S. W. Gao, *Phys. Rev. Lett.*, 2003, **91**, 059602.
7 A. Michaelides, *Appl. Phys. A*, 2006, **85**, 415.
8 P. Vassilev, R. A. van Santen and M. T. M. Koper, *J. Chem. Phys.*, 2005, **122**, 054701.
9 P. Jungwirth, B. J. Finlayson-Pitts and D. J.Tobias, *Chem. Rev.*, 2006, **106**, 1137.
10 I. Benjamin, *Chem. Rev.*, 2006, **106**, 1212.
11 A. Verdaguer, G. M. Sacha, H. Bluhm and M. Salmeron, *Chem. Rev.*, 2006, **106**, 1478.
12 J. S. Filhol and M.-L. Bocquet, *Chem. Phys. Lett.*, 2007, **238**, 203.
13 R. Guidelli and W. Schmickler, *Electrochim. Acta*, 2000, **45**, 2317.
14 S. K. Desai, V. Pallassana and M. Neurock, *J. Phys. Chem. B*, 2001, **105**, 9171.
15 S. K. Desai and M. Neurock, *Electrochim. Acta*, 2003, **48**, 3759.
16 X.-Q. Gong, P. Hu and R. Raval, *J. Chem. Phys.*, 2003, **119**, 6324.
17 A. Roudgar and A. Groß, *Chem. Phys. Lett.*, 2005, **409**, 157.
18 A. Roudgar and A. Groß, *Surf. Sci.*, 2005, **597**, 42.
19 C. Hartnig and E. Spohr, *Chem. Phys.*, 2005, **319**, 185.
20 J. S. Filhol and M. Neurock, *Angew. Chem., Int. Ed.*, 2006, **45**, 402.
21 C. D. Taylor, S. A. Wasileski, J.-S. Filhol and M. Neurock, *Phys. Rev. B: Condens. Matter Mater. Phys.*, 2006, **73**, 165402.
22 E. Skúlason, G. S. Karlberg, J. Rossmeisl, T. Bligaard, J. Greeley, H. Jónsson and J. K. Nørskov, *Phys. Chem. Chem. Phys.*, 2007, **9**, 3241.
23 P. Vassilev and M. T. M. Koper, *J. Phys. Chem. C*, 2007, **111**, 2607.
24 G. Kresse and J. Furthmüller, *Phys. Rev. B: Condens. Matter Mater. Phys.*, 1996, **54**, 11169.
25 G. Kresse and J. Furthmüller, *Comput. Mater. Sci.*, 1996, **6**, 15.
26 J. P. Perdew, K. Burke and M. Ernzerhof, *Phys. Rev. Lett.*, 1996, **77**, 3865.
27 D. R. Hamann, *Phys. Rev. B: Condens. Matter Mater. Phys.*, 1997, **55**, R10157.
28 B. Santra, A. Michaelides and M. Scheffler, *J. Chem. Phys.*, 2007, **127**, 184104.
29 M. Alves-Santos, L. Y. A. Dávila, H. M. Petrilli, R. B. Capaz and M. J. Caldas, *J. Comput. Chem.*, 2006, **27**, 217.
30 P. E. Blöchl, *Phys. Rev. B: Condens. Matter Mater. Phys.*, 1994, **50**, 17953.
31 G. Kresse and D. Joubert, *Phys. Rev. B: Condens. Matter Mater. Phys.*, 1999, **59**, 1758.
32 S. Izvekov, A. Mazzolo, K. Van Opdorp and G. A. Voth, *J. Chem. Phys.*, 2001, **114**, 3284.
33 S. Izvekov and G. A. Voth, *J. Chem. Phys.*, 2001, **115**, 7196.
34 A. Michaelides, V. A. Ranea, P. L. de Andres and D. A. King, *Phys. Rev. Lett.*, 2003, **90**, 216102.
35 C. Puglia, A. Nilsson, B. Hernnäs, O. Karis, P. Bennich and N. Mårtensson, *Surf. Sci.*, 1995, **342**, 119.
36 A. Eichler and J. Hafner, *Phys. Rev. Lett.*, 1997, **79**, 4481.
37 A. Eichler, F. Mittendorfer and J. Hafner, *Phys. Rev. B: Condens. Matter Mater. Phys.*, 2000, **62**, 4744.
38 M. Lischka, C. Mosch and A. Groß, *Electrochim. Acta*, 2007, **52**, 2219.
39 A. Groß, A. Eichler, J. Hafner, M. J. Mehl and D. A. Papaconstantopoulos, *Surf. Sci.*, 2003, **539**, L542.
40 A. Groß, A. Eichler, J. Hafner, M. J. Mehl and D. A. Papaconstantopoulos, *J. Chem. Phys.*, 2006, **124**, 174713.
41 G. Mills, H. Jónsson and G. K. Schenter, *Surf. Sci.*, 1995, **324**, 305.
42 P. K. Schmidt, K. Christmann, G. Kresse, J. Hafner, M. Lischka and A. Groß, *Phys. Rev. Lett.*, 2001, **87**, 096103.
43 M. Lischka and A. Groß, *Phys. Rev. B: Condens. Matter Mater. Phys.*, 2002, **65**, 075420.
44 R. A. Olsen, G. J. Kroes and E. J. Baerends, *J. Chem. Phys.*, 1999, **111**, 11155.
45 R. A. Olsen, H. F. Busnengo, A. Salin, M. F. Somers, G. J. Kroes and E. J. Baerends, *J. Chem. Phys.*, 2002, **116**, 3841.
46 A. Groß, *Surf. Sci. Rep.*, 1998, **32**, 291.
47 A. Dianat, S. Sakong and A. Groß, *Eur. Phys. J. B*, 2005, **45**, 425.
48 B. Hammer, L. B. Hansen and J. K. Nørskov, *Phys. Rev. B: Condens. Matter Mater. Phys.*, 1999, **59**, 7413.

PAPER | www.rsc.org/faraday_d | Faraday Discussions

Co-adsorbtion of Cu and Keggin type polytungstates on polycrystalline Pt: interplay of atomic and molecular UPD

Galina Tsirlina,*[a] Elena Mishina,[bc] Elena Timofeeva,[a]
Nobuko Tanimura,[c] Nataliya Sherstyuk,[b] Marina Borzenko,[a]
Seiichiro Nakabayashi[c] and Oleg Petrii[a]

Received 14th February 2008, Accepted 28th April 2008
First published as an Advance Article on the web 28th August 2008
DOI: 10.1039/b802556h

Second harmonic generation (SHG), electrochemical quartz microbalance (EQCM), and cyclic voltammetry are applied to clarify the structure and properties of Cu adlayers formed in the presence of Keggin polytungstate anions. For 0.02–10 mM $CuSO_4$ solutions, no pronounced suppression of underpotential copper deposition (Cu UPD) by 0.1–10 mM $H_3PW_{12}O_{40}$ (PW12) or $H_4SiW_{12}O_{40}$ (SiW12) is observed in electrochemical experiments. Moreover, coadsorption with polyanions results in an increase of charge in the Cu UPD region. EQCM data demonstrate high surface coverage with polytungstate in the overall potential range and their pronounced co-adsorption with Cu^{2+} cations under open circuit. The unusual potential dependence of EQCM response of polytungstates is discovered and discussed in terms of anion interactions with adsorbed hydrogen. The SHG response of Cu UPD demonstrates a non-linear dependence on Cu surface coverage, which is interpreted in terms of discontinuous submonolayers consisting of 2D Cu islands. The additives of PW12 or SiW12 decrease copper SHG response at low and high $CuSO_4$ concentrations, with minor effect for a mid range of concentrations. In all mixed solutions, the potential dependence of the SHG response remains typical for Cu UPD, not for polytungstates. SHG transients measured under potential step mode demonstrate that the initial non-steady-state SHG behavior of the adlayer is more close to the behavior of polytungstates, but typical copper features appear at longer wavelength. These facts favor the hypothesis of Cu adatom penetration through anionic adlayers and formation of a metal submonolayer at the vacant areas between large quasi-spherical polyanions, with subsequent transformation into a Pt/Cu/polytungstate layered structure.

Introduction

The effect of anions on foreign metal underpotential deposition (UPD) is one of the corner stones of electrochemical surface science. It is usually discussed in terms of adlayer lattice, and there are still some problems with the links between this geometrical consideration, interfacial thermodynamics, and physical properties of

[a]*Department of Electrochemistry, Moscow State University, Leninskie Gory 1-str.3, 119992 Moscow, Russia. E-mail: tsir@elch.chem.msu.ru*
[b]*Moscow Institute of Radioengineering, Electronics and Automation, prosp. Vernadskogo, 78, 117454 Moscow, Russia*
[c]*Department of Chemistry, Faculty of Science, Saitama University, Saitama, 338-8570, Japan*

metal-adsorbate system resulting from partial (or complete) charge transfer. Polytungstate adsorption at Pt in combination with Cu UPD presents an example of pronounced co-adsorption with charge transfer from (to) both species coexisting in adlayer. For the former adsorbate, this charge transfer is evidenced by optical observation of tungstate reduction (appearance of "blue color").[1] We considered it as molecular UPD.

Cu UPD on polycrystalline Pt demonstrates rather complex voltammetric behavior,[2-10] which is usually interpreted as the superposition of adsorption at various low index single crystal surfaces.[11,12] The role of adsorbed oxygen (or hydroxide) in formation of active cites for Cu UPD is now exactly recognized for both poly- and single crystals of platinum.[13,14] The important role of halide ions consists of suppression of oxygen adsorption.[15] These anions never inhibit, but accelerate the formation of Cu adatoms because of creation of new active sites for Cu adatoms due to electrostatic attraction.[8]

The structure of mixed adlayers (Cu co-adsorbed with anions or polar molecules) was intensively studied by electrochemical and various spectroscopic and probe techniques (see ref. 8, 11 and 14–23, for example). At least two basic configurations were proposed for adlayers of sulfate and halide anions co-adsorbed with Cu. For Cu coverage below half monolayer, metal adatoms are assumed to be surrounded by anions; at higher coverage they are believed to form salt-like lattices with Cu located closer to the support surface, as compared to anion. It is a general problem to determine the quantities of co-adsorbed species from coulometry, and an even more hopeless problem to estimate the degree of charge transfer.

Optical second harmonic generation (SHG) is widely used for investigation of the adsorption processes in electrochemical environments,[24-26] including quasi-equilibrium UPD[27,28] and its kinetics.[29] Efficiency of SHG techniques arises owing to its high sensitivity to the interface symmetry and charge transfer. Additionally, SHG is an instrument to evidence the formation of metal nanoclusters, as its intensity is enhanced up to several orders of magnitude when these clusters appear.[30]

This paper presents the unusual type of Cu coadsorption with inorganic polyanions of Keggin type, namely phosphododecatungstate (PW12) and silicododecatungstate (SiW12). Recently, we reported SHG and the electrochemical behavior of these anions at polycrystalline platinum,[1] and concluded that they combine typical anionic behavior induced by electrostatic interaction with a charged interface with typical cationic behavior resulting from reductive chemisorption. Hydrogen adsorption on platinum was assumed to be very important for the latter feature, and chemical reduction of polytungstate adsorbate with hydrogen adatoms was indirectly evidenced in ref. 1.

We report below the experimental results which argue in favor of permeability of polytungstate adlayers for low-molecular species, as well as of their mobility in adlayers. Permeability is important for electrocatalytic phenomena known for polytungstate-modified electrodes.[31-33]

Experimental

The majority of SHG and electrochemistry experimental details, including reactants and solution preparation, can be found in ref. 1. Additionally, $CuSO_4 \cdot 5H_2O$ (Wako, 99.5%) was used. To check possible interactions of Cu^{2+} with polytungstate anions in solution bulk, UV-vis absorption spectra and DC polarography on dropping mercury electrode were applied. All spectra and polarograms demonstrated more or less additive behavior of signals in solutions with both additives, which can hardly be the case if ionic association is strong (innersphere). All specific features of polytungstate reduction on mercury mentioned in ref. 1 remain unchanged after subtraction of the Cu^{2+} reduction wave from polarographic currents obtained for mixed solutions. This fact can be interpreted as a preferential participation of non-associated anions in the interfacial cathodic reaction on positively charged mercury. Of course, all these tests do not exclude the existence of outersphere weakly bonded Cu^{2+}·polytungstate ion pairs in equilibria with free ions. Note that strong Cu^{2+}–polytungstate complexes, like those studied in ref. 34, are usually formed by lacunar or at least less symmetric polyanions.

Along with SHG under potential cycling mode (SHG-CV), SHG transient measurements were performed with averaging of optical signal over 1 s. In order to compare slow and fast behavior, the transient data for certain time intervals (quasi-CVs) are plotted below at some graphs jointly with "normal" SHG-CVs. The procedure of data averaging for quasi-CVs was as follows: for each time presented in the graph (10, 25 s, *etc.*), the intensity was averaged over three adjusting points. For instance, for $t = 10$, $I_{SHG} = [I_{SHG}(t = 9) + I_{SHG}(t = 10) + I_{SHG}(t = 11)]/3$.

For EQCM-based estimates of adsorbate coverage and its potential dependence, a HQ-101B EQCM controller in combination with a HZ-3000 (HAG-1510 m) electrochemical measurement system was applied. A T-cut quartz crystal (6 MHz) was used with evaporated polycrystalline platinum. The roughness of this platinum film determined from the H UPD charge was close to unity within the accuracy of 2–3%, which corresponds to usual coulometric accuracy. A limited stability of evaporated Pt films (300 nm thick) with Ti underlayer (20 nm) prevented long experiments in acidic solution. Much better stability was observed in experiments with 0.1 M H_2SO_4 solution as compared to the usual 0.5 M background. We report below the data for this more diluted supporting electrolyte, which can not be compared quantitatively with SHG and electrochemical data for 0.5 M background. So EQCM data presentation is aimed exclusively at comparison of adsorbate quantities for various solution compositions and under various modes. For EQCM calibration, Cu UPD was applied (complete monolayer weight was 7.8 ng cm^{-2}).

All potential values are reported below in RHE scale. These values are recalculated from directly measured values referred to saturated calomel electrode with a salt bridge (saturated KCl).

Cu UPD in sulfuric acid medium

As far as we know, all previously reported SHG studies of Cu UPD are limited to gold supports (see references in ref. 27, 29, 35 and 36). The dependence of SHG intensity on the electrode potential of polycrystalline gold is presented only in ref. 27 and 37. The SHG intensity decrease induced by copper UPD is explained by copper d-electron delocalization in the UPD gold–copper–oxygen configuration. In ref. 35, azimuthal dependences of SHG intensity are measured for copper UPD at gold single crystalline electrode. It is found that at Au(111) SHG intensity at 1064 nm is less sensitive to CuUPD as compared to Tl UPD. On the other hand, it is known for various materials, including Cu,[38] that SHG intensity *vs.* potential dependencies are very sensitive to the angle of incidence and wavelength, therefore any comparison of results as well as interpretation are valid only for identical experimental conditions. In ref. 29, the kinetics of copper UPD at gold single crystalline electrode was studied, and it was shown that isotropic component of the SHG field is described by the first-order exponential dependence with time constant of several tenths of ms (the anisotropic component is time-independent in this time scale).

Dealing with SHG in the context of our previous data[1] and the currently discussed problem of co-adsorption we addressed mostly the intensity *versus* potential dependence (Fig. 1a). The important qualitative feature is the presence of SHG maxima in the region 0.4–0.6 V. For the studied concentration range x of $CuSO_4$ additives ($x = 0.02–5$ mM), the dependence on concentration (Fig. 1b) was found only in the vicinity of the cathodic limit (corresponding to incomplete copper monolayer for low x; this limit was chosen in order to avoid complications with formation of 3D nuclei in the closest vicinity of Cu^{2+}/Cu equilibrium potential).[20]

Cyclic voltammograms (Fig. 1c) confirm very slow copper adsorption in the most dilute solutions. The resulting surface coverage attained at the cathodic potential

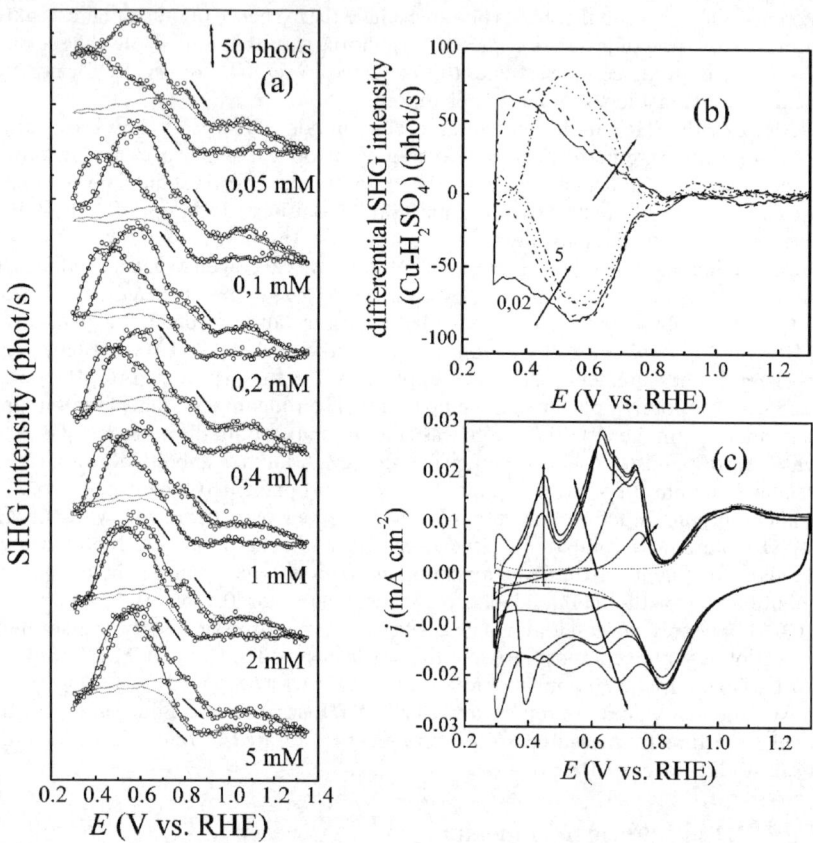

Fig. 1 (a) Concentration dependence of SHG intensity *versus* potential curves for 0.5 M H_2SO_4 + x mM $CuSO_4$ solutions, concentrations are indicated in the graph, background curves are dashed; (b) the same data for various concentrations (x values are indicated in the figure) after background correction; (c) representative voltammograms measured under the same conditions in supporting electrolyte (dashed) and in solutions with x = 0.2; 0.4; 1; 2; 5 (solid). Arrows indicate the effect of x increase. Scan rate: 10 mV s^{-1}.

limit (as estimated from desorption charge)[4] remains below 0.2, even for x = 0.2 mM. When x = 0.4 mM, surface coverage is already close to 0.4, and subsequent increase of x induces only minor growth of coverage up to final concentration-independent values. In general, electrochemical responses (Fig. 1c) look more complex than SHG *vs.* potential dependencies, as they contain a number of peaks at different potentials.

Surface coverage appears to be close for all x only at potentials more positive than 0.7–0.8 V, and in this region the corrected SHG curves (Fig. 1b) also demonstrate no difference. In this region of rather low copper coverage, we observe a systematic decrease of SHG intensity with further decrease of coverage.

SHG behavior in the region of potentials more negative than 0.4 V is more complex. Actually, for lower coverage (attained in more dilute solutions) we observe higher intensities, and SHG hysteresis loops (these loops are wider for lower x) do not follow hysteresis of charge.

As in our previous study of tungstate/platinum systems,[1] we present a simple analysis of nonlinear optical response in the adsorbate/substrate interface. Nonlinear polarization of the metal is modified by a total charge transfer σ_{tot} from a metal to adsorbate, and in the framework of a jellium model can be written as

$$P^{Pt}(2\omega) \propto a_0^{Pt} + \Delta a(\sigma_{tot}) \quad (1)$$

where the Rudnick–Stern parameter $a_0 < 0$, $\Delta a = 4n\theta^{Cu}\delta\mu_2^{Cu\rightarrow Pt}(\sigma_{tot})$, where n is electron density, and $\delta\mu_2^{Cu\rightarrow Pt}$ is the second-order dipole, which is formed at the interface owing to the charge transfer.[39]

Nonlinear polarization of adsorbate $P^{ad}(2\omega)$ can be described as

$$P_i^{Cu}(2\omega) = \gamma^{Cu}\theta^{Cu}E_j(\omega)E_k(\omega) \quad (2)$$

where θ^{Cu} is surface coverage, and γ^{Cu} is the first hyperpolarizability of Cu adatom. The value of γ^{Cu} (amplitude and phase) may reflect the resonance (wavelength dependent) properties of the effective surface dipole.[38]

It is interesting to compare hydrogen[1] and Cu (Fig. 1) adsorption data. For the former, SHG intensity increases with coverage in the H-UPD region and reach the highest value for a monolayer (in the vicinity of zero potential). Oppositely, when we decrease the potential in H_2SO_4 + $CuSO_4$ solutions, the SHG intensity first increases in the beginning of the Cu UPD region, and then drastically decreases, being rather low for the copper adsorbate saturation. Thus, the maximum of SHG intensity is observed for some intermediate coverage (dependent on $CuSO_4$ concentration).

There are several possible reasons for non-monotonous SHG behavior. First, the σ_{tot}-dependent term in eqn (1) may induce this behavior, since the total charge transfer from adatom to the surface is, in general, coverage-dependent. The second reason can be some interplay of coverage effects on the quantities presented by eqn (1) and (2), if these effects differ in sign. Finally, it is expected that under some circumstances Cu islands can give much higher contribution to intensity as compared to more lengthy areas covered with Cu adatoms. A size effect of this type was reported, for example, in ref. 40. The enhancement of the SHG field in metal nanoparticles was already discussed earlier in relation to non-monotonous dependence of SHG intensity on surface coverage for other surfaces undergoing submonolayer deposition of foreign metals.[37,41] Two effects should be mentioned regarding this enhancement: surface plasmon resonance and the lightning-rod effect,[37] and for both the degree of enhancement depends on the island shape (ellipticity) and wavelength.

Indirect evidence of island formation in Cu adlayers at polycrystalline platinum was mentioned earlier in ref. 2 and 4–6 in relation to various non-stationary electrochemical features and assigned to pronounced structural sensitivity of copper adatoms (*i.e.* to their tendency to adsorb at some crystallographically homogeneous areas of the inhomogeneous surface). In ref. 42, this effect was found to be striking for strongly inhomogeneous columnar polycrystalline platinum. 2D metal islands can be considered as the limiting case of oblate spheroid. According to the model considered in ref. 37, the lightning-rod effect is negligible in this case, when surface plasmon resonance (SPR) should play a dominant role. For 3D copper nanoparticles, the impressive examples of size-dependent SPR can be found in ref. 43 and 44.

To summarize, the SHG response for the system under study contains three contributions: from Pt support, copper 2D islands and copper 2D films. For high values of x, only two contributions are important, with nonlinear susceptibility of copper film being of the opposite sign with respect to the nonlinear susceptibility of platinum. This can explain the observed SHG intensity decrease with the copper deposition. For low x, formation of small 2D clusters is expected. Growth of these clusters induces the changes in their size and ellipticity, with unavoidable change of SPR position, and corresponding nonlinear susceptibility change. We even can assume that the SPR position passes the fundamental wave energy of 1.6 eV, as we observe a change of the copper islands contribution sign resulting in the increase of a total SHG intensity for low x (low coverage). In fact, if the sign does not change, we would observe the enhancement of copper contribution in an even deeper SHG signal decrease.

We assume that the hysteresis of curves in Fig. 1a, which is more pronounced for lower concentrations, reflects the difference in kinetics of island formation/destruction in the course of anodic and cathodic scans. The difference in non-steady-state behavior of SHG responses was observed for Cu UPD on gold,[29] as well as the difference in deposition and desorption kinetics.

At this stage, we use the specific SHG behavior of Cu adatoms simply as a test for its presence at a platinum surface, in order to consider the interplay of Cu and polytungstate contribution in the course of coadsorption. For polytungstates, a tendency to systematic increase of SHG signal with concentration was observed,[1] in contrast to Cu UPD.

Cu UPD in polytungstate-containing solutions

Fig. 2 and 3 present the comparison of SHG data for polytungstates under study (open squares†) with currently obtained data for the same polytungstate concentrations in solutions containing $CuSO_4$ additives (solid circles in Fig. 2 and 3). The curves for tungstate-free $CuSO_4$ solutions are also plotted (open triangles).

Concentrations of co-adsorbing species were chosen in order to check the effect of PW12 (SiW12) and Cu^{2+} ratio for both lower and higher concentrations of each component (corresponding to lower and higher surface coverage).

The observed tendencies look as follows:

(i) for any ratio of bulk concentrations the shape of SHG *versus* potential dependence is more close to the shape typical for Cu UPD in polytungstate-free solutions;

(ii) the higher the $CuSO_4$ concentration, the lower the SHG response;

(iii) the increase of PW12(SiW12) concentration also decreases SHG response, but the effect is weaker than mentioned in point (ii).

SHG intensities obtained in mixed solutions demonstrate no additivity. These values never exceed the intensities measured in any single-component solution of the same concentration. Moreover, typically the response of mixed solution appears to be even lower than any of the "individual" SHG responses of Cu and PW12 (SiW12).

The decrease of Cu response can be induced by suppression of its adsorption by polytungstate and/or the effect of polytungstate on characteristic geometry of copper islands.

The decrease of polytungstate response in mixed solutions can be induced by suppression of hydrogen adsorption with Cu adatoms (as just adsorbed hydrogen is assumed to reduce polytungstate, and just reduced polyanions are responsible for SHG intensity). However, this simple hypothesis is correct only if polytungstates are separated from platinum by a Cu interlayer. The possibility of this separation and the mechanism of copper penetration through the tungstate adlayer will be the key points of further discussion.

For pure polytungstates solution, the nonlinear optical response of tungstate adsorbate can be written as

$$P_i^W(2\omega) = \gamma^W \theta^W E_j(\omega) E_k(\omega) \tag{3}$$

and modification of the support response, as in eqn (1), is determined by

$$\Delta a = 4n\theta^W \delta\mu_2^{W \to Pt} (\sigma_{tot}) \tag{4}$$

---

† To avoid Cu OPD we could not use the same cathodic potential limit as in ref. 1 (0.05 V). To check the effect of cathodic limit, we made test measurements with 0.25–0.3 V potential limit for a number of copper-free solutions and found that deviations from the curves reported in ref. 1 are within the accuracy of SHG signal averaging.

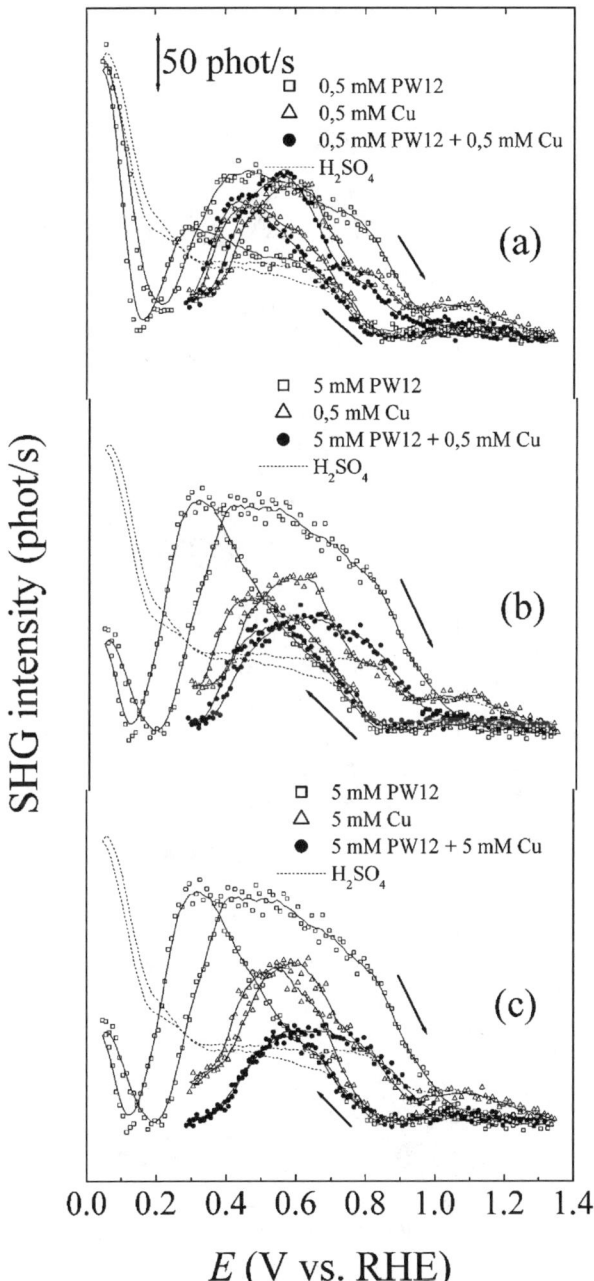

Fig. 2 SHG intensity *versus* potential curves for various ratios of single (open symbols) and co-existing (solid symbols) additives in 0.5 M H_2SO_4 + x mM $CuSO_4$ + y mM PW12 solutions. In all graphs dashed line corresponds to background signal in 0.5 M H_2SO_4; $x = 0$ (no copper)—squares, $y = 0$ (no PW12)—triangles, coexisting additives—circles. (a) Both concentrations are low, $x = 0.5$, $y = 0.5$; (b) excess of PW12, $x = 0.5$, $y = 5$; (c) both concentrations are high, $x = 5$, $y = 5$. Scan rate: 10 mV s^{-1}.

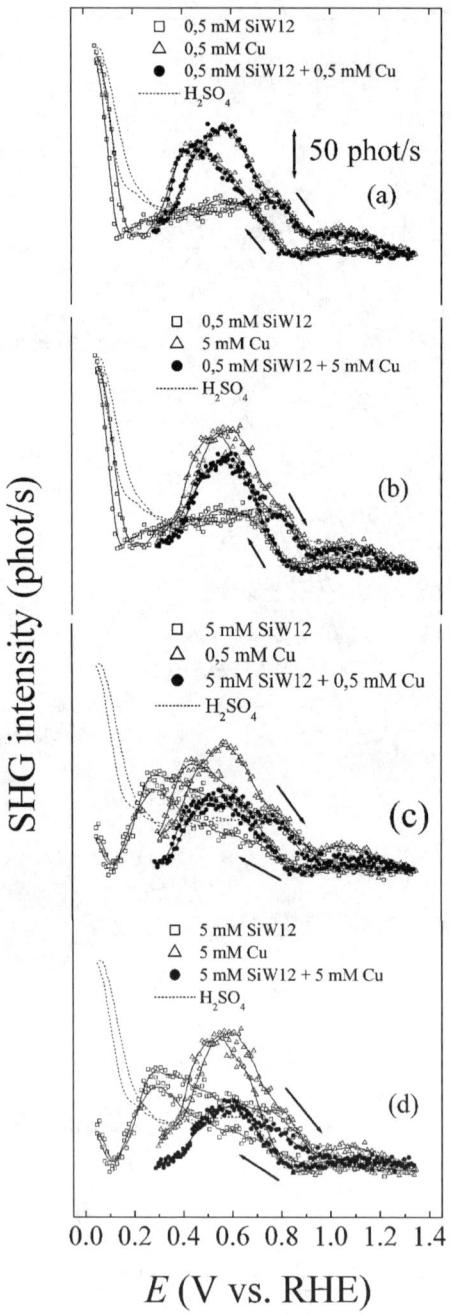

Fig. 3 SHG intensity *versus* potential curves for various ratios of single (open symbols) and coexisting (solid symbols) additives in 0.5 M H_2SO_4 + x mM $CuSO_4$ + y M SiW12 solutions. In all graphs dashed line corresponds to background signal in 0.5 M H_2SO_4; $x = 0$ (no copper)—squares, $y = 0$ (no SiW12)—triangles, coexisting additives—circles. (a) Both concentrations are low, $x = 0.5$, $y = 0.5$; (b) excess of $CuSO_4$, $x = 5$, $y = 0.5$; (c) excess of SiW12, $x = 0.5$, $y = 5$; (d) both concentrations are high, $x = 5$, $y = 5$. Scan rate: 10 mV s^{-1}.

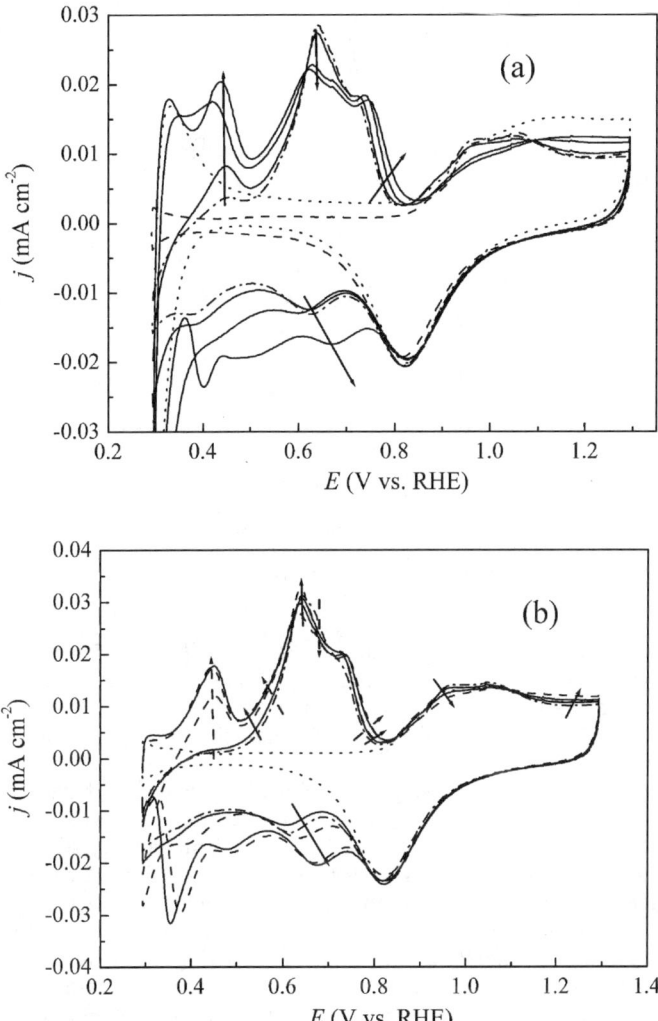

Fig. 4 Cyclic voltammograms related to SHG data in Fig. 2(a) and in Fig. 3(b). (a) Dashed curve corresponds to supporting electrolyte, dotted curve is for PW12, dash dots are for $CuSO_4$; three solid curves are plotted for mixed solutions, with arrows indicating the sequence (a)–(b)–(c) in Fig. 2. (b) Dotted curve corresponds to supporting electrolyte (both with and without SiW12). Solid curves demonstrate the effect of the increasing SiW12 concentration at fixed 0.5 mM $CuSO_4$ concentration (related to (a) and (b) in Fig. 3). Dashed curves present the effect of increasing Cu concentration at fixed 10 mM SiW12 concentration (related to (c) and (d) in Fig. 3). Solid and dashed arrows indicate corresponding tendencies. Scan rate: 10 mV s^{-1}.

In contrast to individual Cu adsorption, SHG vs. potential dependence for individual polytungstate adsorption is closer to coverage dependence on electrode potential.‡ This type of behavior means the nonlinear susceptibility of adsorbed anion gives a major contribution to the optical response.

For SHG intensity under co-adsorption conditions, contributions of all four terms presented by eqn (1)–(4) should be taken into account.

Electrochemical responses measured in mixed solutions (Fig. 4) demonstrate the increase of total charge in the Cu UPD region in the presence of polytungstates.

‡ Comparison of SHG and EQCM data in the next section confirms this statement.

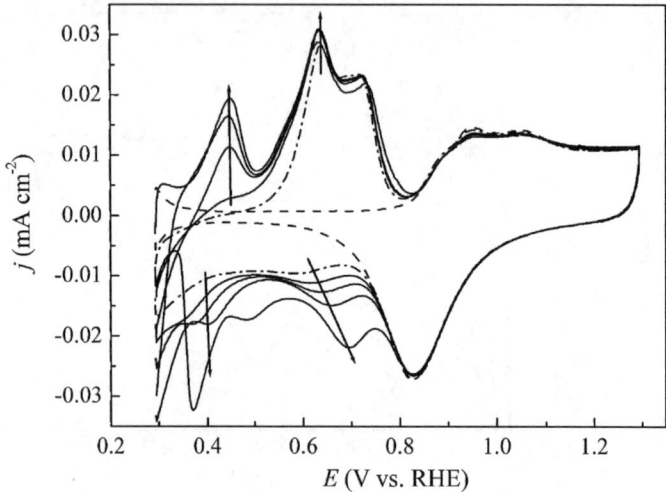

Fig. 5 Cyclic voltammograms measured in 0.5 M H_2SO_4 + x mM $CuSO_4$ + 3 mM SiW12 (solid curves), x = 0.1, 0.25, 2, 4. Arrows indicate the effect of x increase. Dash dots correspond to 0.5 M H_2SO_4 + 0.5 mM $CuSO_4$. Scan rate: 10 mV s^{-1}.

Comparison with the curve for 0.5 mM $CuSO_4$ (Fig. 1c) also demonstrates some enhancement of Cu adsorption induced by polytungstate additives. However, the electrochemical behavior is scan-rate dependent, and even a qualitative difference appears at high scan rates, manifesting rather complex adlayer dynamics. We shall touch upon this point below.

To clarify the interplay of responses induced by two adsorbates, we studied systematically the effects of each component concentration (at fixed concentration of the second component). An example of CVs transformation in a series: no mutual suppression of coadsorbing species is observed from the increasing total charge, when SHG tends to decrease as compared to both "individual" SHG responses.

From comparison of Fig 6a, 6c and 5, one can conclude that the effect of Cu concentration at fixed polytungstate concentration (Fig. 6a) is more pronounced than for the fixed $CuSO_4$ concentration and polytungstate concentration varied in a wide range (Fig. 6b). For the former series, the evident non-monotonous behavior of SHG intensity is observed, *i.e.* the pronounced decrease of optical signal at low and high $CuSO_4$ concentrations. At the same time coulometric data (Fig. 7) behave monotonously in both series of mixed solutions.

It is difficult to separate the contributions of Cu and Keggin anions into this charge value. Some chemical interaction in the adlayer can not be completely excluded (like for Cu coadsorbed with some organic molecules),[14] and this can lead to spillover-like manifestations. For further understanding of this complex behavior we attracted a number of evaluative EQCM experiments.

Composition of adlayers: comparative EQCM estimates

For the EQCM study limited by Pt film stability and thus requiring lower acid concentration, we have chosen SiW12, as EQCM has been already applied to characterize its adsorption on gold[45] (the difference in calibration constants in ref. 45 and our study does not exceed 10%). For more exact comparison with these data, an attempt was made to determine the mass change induced by injection of polytungstate additive into the supporting electrolyte solution under open circuit conditions. This procedure is rather disputable because the resonant frequency can be affected by perturbations induced by injection, but we rely on parallel blank experiments

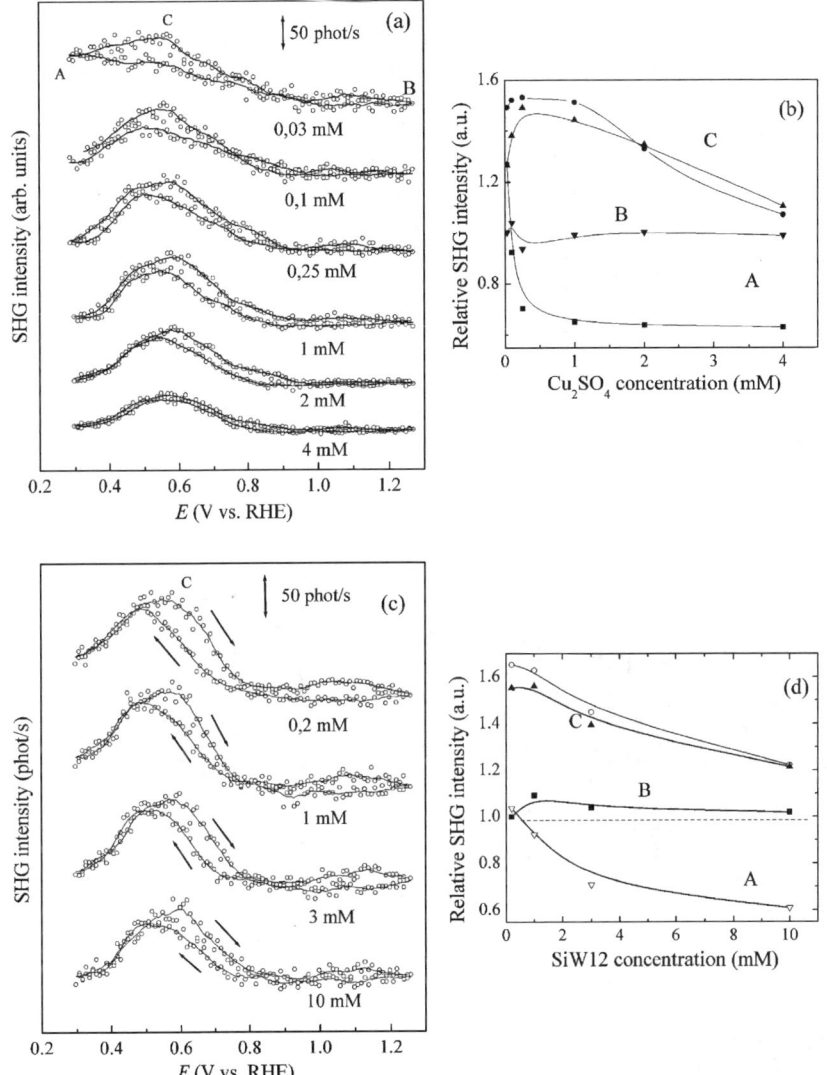

Fig. 6 SHG intensity *versus* potential curves measured in (a) 0.5 M H_2SO_4 + x mM $CuSO_4$ + 3 mM SiW12 and (c) 0.5 M H_2SO_4 + 2 mM $CuSO_4$ + y mM SiW12 (y = 0.2, 1, 3, 10). (b) and (d) present corresponding dependences of SHG intensity on x (b) and y (d).

(with perturbation induced by adding the same supporting 0.1 M H_2SO_4 solution). For blank experiments, the initial frequency decrease of *ca.* 20 Hz (having no relation to any mass increase) continued for few seconds, see open squares in Fig. 8. Relaxation (slow frequency increase) started after 5 s.

A much stronger effect was observed after SiW12 injection (solid squares in Fig. 8), and no relaxation was observed for at least 200 s. The steady-state frequency values were achieved after *ca.* 30 s, in good agreement with the characteristic time found in ref. 45 for 5 mM SiW12 solution on gold.

The accuracy of the coverage value determined from the EQCM response is rather low because of the uncertainty of the steady-state blank value. The range of possible values (assuming the highest and the lowest contribution from mass-independent

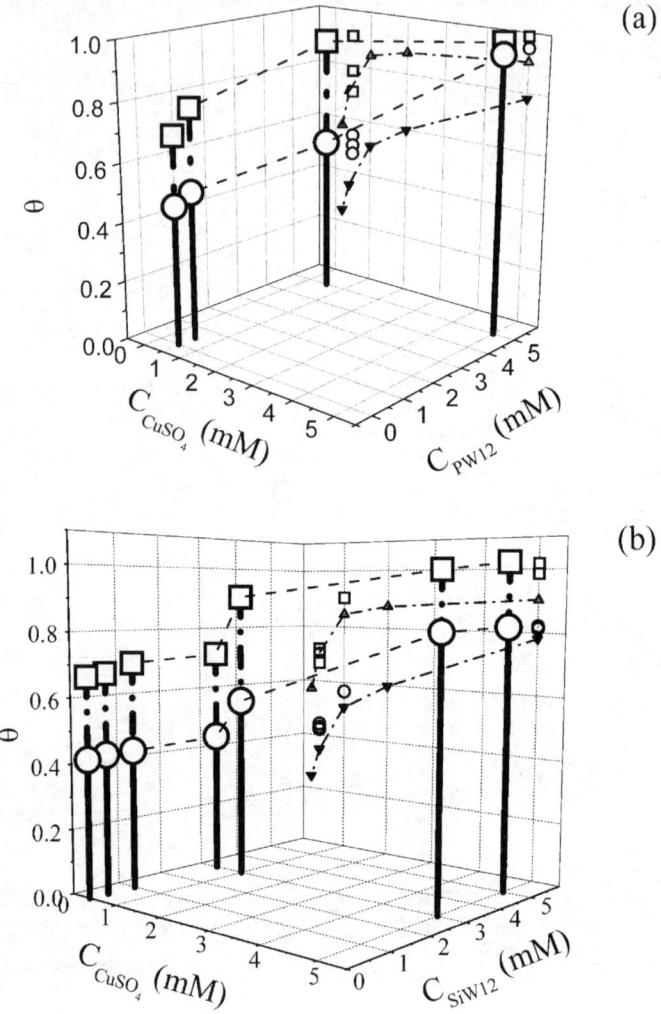

Fig. 7 General 3D plots demonstrating the mutual effect of coadsorbing species on the total charge for Cu-PW12 (a) and Cu-SiW12 (b). Instead of charge, the total surface coverage θ is plotted which is formally calculated for Cu UPD under the assumption of 420 μC cm^{-2} charge of monolayer formation. θ vs. concentration dependencies can be seen as 2D graphs (triangles). In the presence of both PW12 and SiW12 (up triangles) the charge (coverage) is always higher than for polytungstate-free solutions (down triangles). All data correspond to scan rate 10 mV s^{-1}.

perturbation) is 80–110% of monolayer.§ The two times higher value reported in ref. 45 is evidently overestimated, because the authors assumed the monolayer to consist of close-packed balls of 0.5 nm radius (real molecular size is at least 20% larger). Taking into account these details, one can conclude that there is no dramatic difference in the surface coverage of gold and platinum with SiW12 under open circuit conditions.

Injection of SiW12 into 0.1 M H$_2$SO$_4$ + 0.5 M CuSO$_4$ solution induced a stronger frequency decrease (solid triangles in Fig. 8). The additional mass (as compared to

---

§ To estimate the mass of dense monolayer, we considered the geometry of SiW12 adlattice on Ag single crystal.[46]

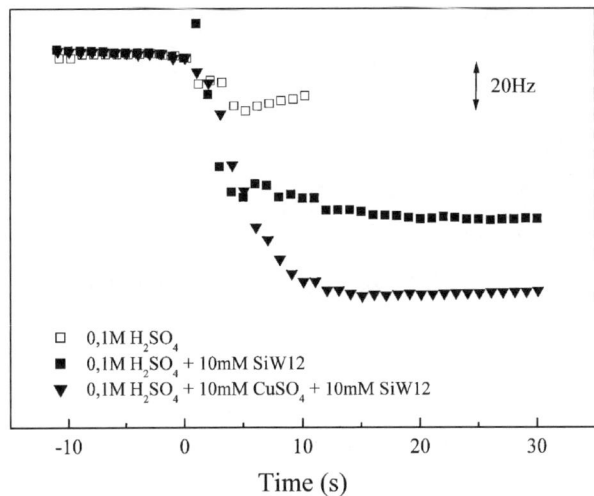

Fig. 8 EQCM transients measured under open circuit conditions. Zero time corresponds to injection of 10 mM SiW12 additive into 0.5 M H$_2$SO$_4$ solution (solid squares) and into 0.5 M H$_2$SO$_4$ + 10 mM Cu SO$_4$ solution (solid triangles). Open squares correspond to blank experiments imitation EQCM perturbation induced by injection.

the previous experiment) was very close to the mass of the Cu adatoms monolayer. As the open-circuit potential (OCP) of *ca.* 0.9–1.0 V corresponds to complete desorption of Cu adatoms (Fig. 4 and 5), the observed difference of curves (solid squares and triangles in Fig. 8) should be attributed to interfacial ionic association, not to Cu UPD. Probably, it is a sort of some innersphere complex formation, as mentioned in ref. 16 for copper–chloride systems.

Formally, the effect demonstrates some similarity to the increase of anion adsorption induced by cation (adatom). This phenomenon was discussed, for example, for the copper sulfate system,[47] but the interpretation proposed in ref. 47 (shift of the potential of zero charge resulting from Cu adatoms formation) can be hardly applied to the system under study, at least at OCP.

EQCM data obtained under potential cycling provided much better accuracy. For the supporting electrolyte (open squares in Fig. 9a), a small but rather typical[48–50] effect of anions adsorption and surface oxide formation was observed. For SiW12-containing solution, a complex figure-of-eight shape response was observed, with the total mass change below 10–15% of monolayer. This finding is very important because it confirms directly that there is no potential of complete desorption inside the region where the platinum electrode works as perfectly polarizable (in agreement with our previous assumption[1] based on SHG data).

Despite some difference in the cycling mode (no excursions to O UPD region were done in ref. 45), characteristic regions of the EQCM hysteresis obtained in this work (Fig. 9a) and in ref. 45 are qualitatively similar. In Cu^{2+}-containing solution (Fig. 9b), this specific loop disappears completely. The total mass change (solid squares in Fig. 9b) remains practically the same as in SiW12-free solution (open squares). The preliminary conclusion agrees with the previous assumption about the absence of direct contact between polytungstate and platinum and the Cu adatoms location under the tungstate layer. An alternative explanation consists of complete desorption of Keggin anions, but it is easily disproved by electrochemical data reported above, as well as by pronounced difference of SHG behavior of Cu adatoms in the absence/presence of SiW12 (PW12).

To summarize, we observe a sort of surface complexation,[51] for which more weak coverage *versus* potential dependence is expected as compared to usual competitive

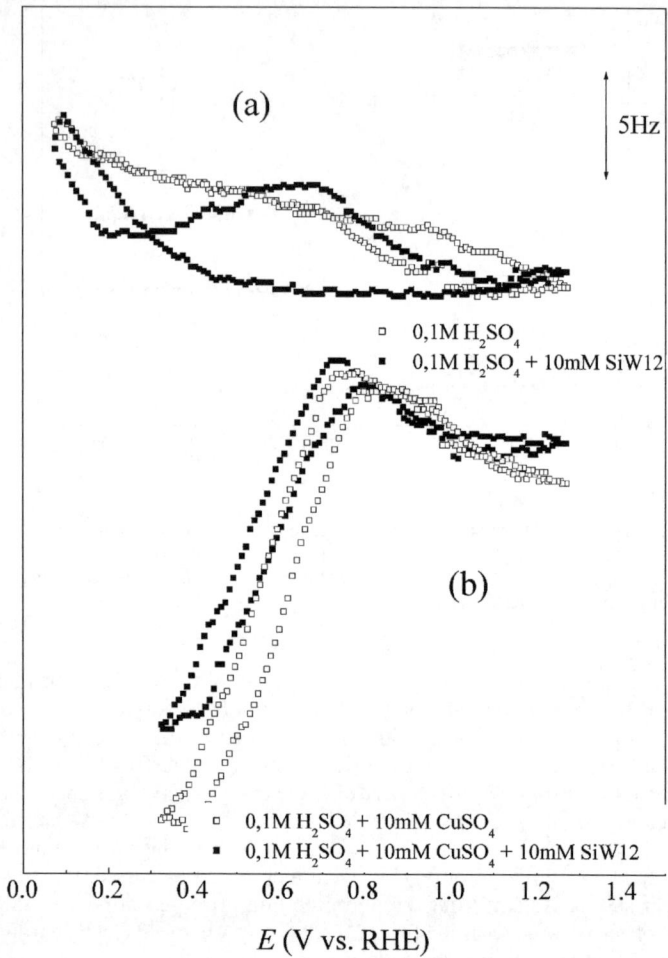

Fig. 9 EQCM data obtained under potential cycling mode for supporting electrolyte (a, open squares), with 10 mM SiW12 additive (a, solid squares), with 10 mM CuSO$_4$ additive (b, open squares), and with both additives present simultaneously (b, solid squares). Scan rate: 10 mV s^{-1}.

adsorption. EQCM estimates confirm that Keggin polytungstates adsorption on platinum results in high surface coverage, as for other previously studied electrode materials.[45,52–58] The specific features of adsorption on Pt come to hydrogen-induced changes in the adsorbate state, which are probably also responsible for some events in mixed Cu-polytungstate adlayers.

Events in mixed adlayer: present-day hypothesis

Quantitative separation of various contributions to SHG remains a serious problem, but it is not crucial for further discussion based on the data of several techniques. At this stage we can discuss the role of polytungstate surface coverage in generation of this signal, dealing with the width of hysteresis loops, not with intensity. Note, that outside the oxygen adsorption region hysteresis is absent in background experiments (dashed curves in Fig. 1a, 2 and 3), that is why we assign this phenomenon to polytungstates (with the assumption of possible secondary contributions from free metal surface, water, adsorbed hydrogen, *etc.*).

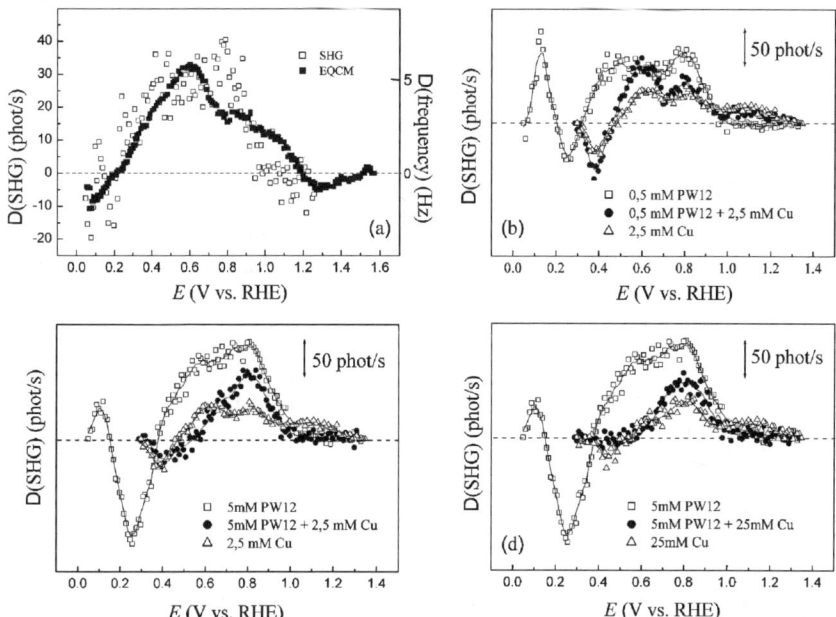

Fig. 10 (a) Correlation of SHG (open squares) and EQCM (solid squares) hysteresis effects for 0.5 M H$_2$SO$_4$ + 10 mM SiW12 solution; (b–d) comparison of hysteresis effects for Cu, PW12, and Cu–PW12 systems. Concentrations of PW12 and symbols in b, c and d are the same as in Fig. 2 a, b and c respectively.

Comparison of typical SHG hystereses with EQCM hysteresis in Fig. 9a demonstrates general similarity of the loops: intersection of forward and reverse branches in H UPD region; appearance of maxima in the double layer region; hysteresis disappearance in O UPD region.¶ To check whether the predominating contribution to SHG is proportional to polytungstate coverage, we put the difference of SHG at anodic and cathodic scans and the corresponding difference of EQCM frequency in one graph (Fig. 10a).∥ It is easy to see that hysteresis behavior of the SHG and EQCM frequency is very similar, especially for the ascending branch at lower potentials. For this region we can come to a rather reliable conclusion on the predominating role of coverage-dependent contribution in the mechanism of SHG generation.

In a narrow region of 0.6–0.8 V the relative width of SHG hysteresis is slightly higher. Just for this region, we assumed[1] the complete disappearance of W (V), which is responsible for the dye-like behavior of partly reduced polytungstates. Now we can tell solidly that the reduction is slower in the course of the cathodic scan, giving a lower SHG contribution (as normalized per coverage).

In contrast, for the most positive potentials above 1 V, the relative width of EQCM frequency hysteresis exceeds the relative width of SHG hysteresis. The most transparent explanation of this tendency is the interplay with oxygen adsorption (*i.e.* its stronger contribution to the mass changes than to SHG intensity). To imagine oxygen adsorption on platinum covered by tungstate monolayer at the same potentials as on bare platinum one should assume the permeability of

---

¶ To avoid misunderstandings: quite in contrary, background curves demonstrate hysteresis in O UPD region,[1] but it disappears in the presence of tungstates.
∥ This comparison is correct within the accuracy of the possible effect of different H$_2$SO$_4$ concentrations (0.5 M for SHG and 0.1 M for EQCM), which can in its turn affect the polytungstate surface coverage. However we rely on the minor sensitivity of the differential values to this difference.

a complete monolayer, at least for water and hydronium ions, which is rather natural because of the existence of empty space between closely packed polyanions.

The important consequence of our analysis is the dependence of surface coverage on polytungstate concentration in the overall region studied in ref. 1 and in this paper (up to 10 mM). This conclusion follows unambiguously from the coverage dependence of SHG, so the dependence of SHG intensity on concentration (curve C in Fig. 6b in ref. 1) can be considered as a portion of an "adsorption isotherm". Its shape is typical for mid and high coverages of the adsorbates with relatively weak lateral interactions. As very strong electrostatic repulsion is unavoidable for the anionic adlayer, this experimental fact can be only explained by co-adsorption of cationic species screening the repulsion.

In the absence of copper ions (or other metal cations), this screening can be provided only by hydronium ions and formally corresponds to a lower pK_a for adsorbed anion as compared to bulk pK_a. This hypothetic phenomenon is rather intriguing and probably typical for adsorption of multicharged anions. In fact, predomination of adsorbed bisulfate at the Pt/solution interface with sulfate species predominating in the bulk[59] presents exactly the same phenomenon. From an electrostatic point of view protonation-induced stabilization should be weaker than that assisted by Cu^{2+} cations. In the potential region of copper UPD, association and adatoms formation are expected to take place in parallel, and the total amount of copper accumulated at the interface should be higher than the submonolayer formed in the absence of polytungstate. This qualitative analysis explains the difference in the EQCM curves in Fig. 9b.

As it follows from the previous section, under OCP conditions the quantity of copper species estimated from adlayer mass increase is close to the value for (sub)-monolayer of Cu adatoms. In the absence of charge transfer from Cu, the only possible explanation can be attachment of Cu^{2+} ions to adsorbed polytungstates. Formation of any type of the salt-like 2D lattice should start from the Pt/PW12(SiW12)/Cu^{2+} configuration, as Cu^{2+} attachment can start only from the solution side. At the same time, there are no doubts that sooner or later (in the course of the UPD process or even before it starts) copper species penetrate closer to platinum surface. This hypothesis is confirmed by copper-like SHG behavior found in mixed solutions (Fig. 2, 3 and 6). Some specific features of this penetration process can be judged from hysteresis behavior demonstrated in Fig. 10b, c and d in terms of differential SHG curves (differences of anodic and cathodic scans).

For comparative study of adsorption dynamics, some useful information can be extracted from comparison of SHG *vs.* potential curves measured under cycling mode and similar curves constructed from SHG transients (Fig. 11). The meaning of points obtained from the set of transients at a certain time is the same as that of the SHG intensity values at the cathodic branch of the usual SHG-CVs, as the potential steps were done from 1.2 V to more negative potentials.

SHG transients usually demonstrate a maximum in the initial region. As for anodic transients (steps from H UPD or double layer region potentials to O UPD region), they are rather monotonous and reach the same SHG intensity values as in usual CVs, in 1–3 s.

Details of SHG transients for Cu UPD in polytungstate-free solutions (Fig. 11a and b, right side) depend on solution concentration, signal stabilization after the maximum being faster in more concentrated solution (Fig. 11b). For Cu-free solutions of polytungstate (Fig. 11c, right side), the time necessary to reach a steady-state SHG value is longer than for Cu-containing solutions.

The opposite tendencies are observed for potential regions corresponding to reduced and oxidized polytungstate (growth and decrease with time, respectively). Comparison in Fig. 11c (left side) demonstrates that potential cycling at 10 mV s^{-1} does not correspond to a stationary adlayer, as further slow decrease of signal is observed for several hundreds of seconds. The same tendency is found in background experiments: transients measured in 0.5 M H_2SO_4 solution also demonstrate

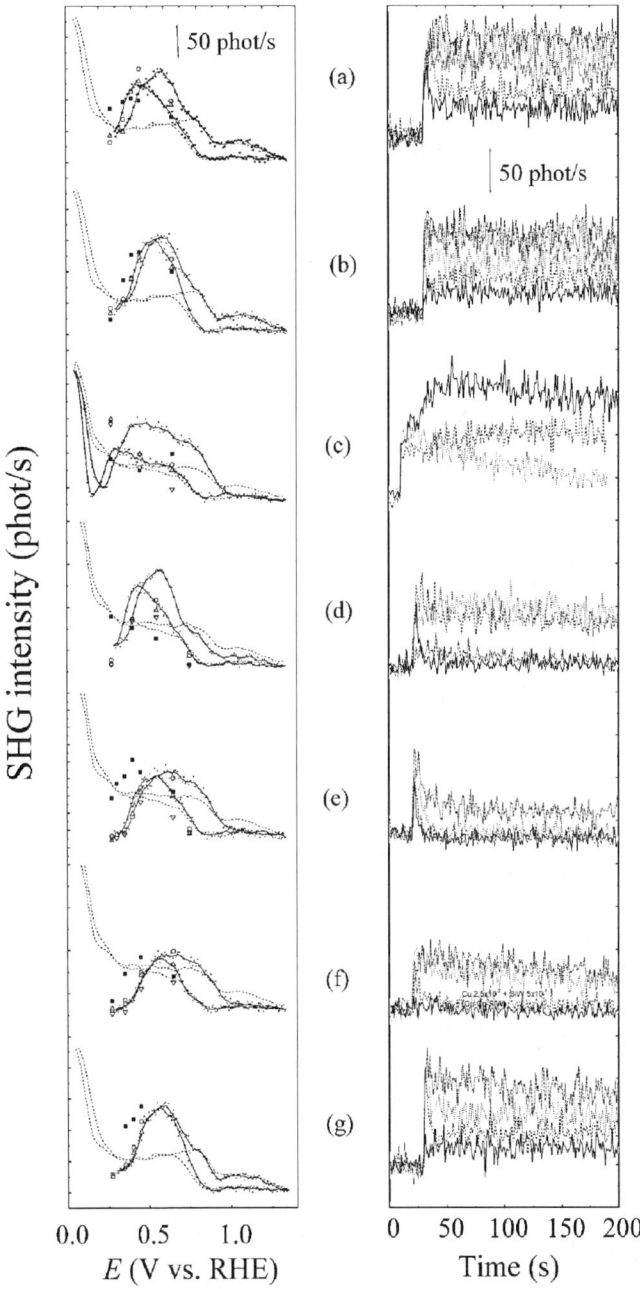

Fig. 11 Sets of SHG intensity transients measured in the same solutions as compared in Fig. 2 and 3, and comparison of SHG quasi-CVs for various fixed times with usual SHG CVs measured at 10 mV s^{-1} scan rate. (a, b): 0.5 M H_2SO_4 + x mM Cu SO_4, x = 0.5 (a) and 5 (b); (c–f): 0.5 M H_2SO_4 + x mM Cu SO_4 + y mM PW12, x = 0, y = 0.5 (c); x = 0.5, y = 0.5 (d); x = 0.5, y = 5 (e); x = 5, y = 5 (f); (g) 0.5 M H_2SO_4 + 5 mM Cu SO_4 + 0.5 mM SiW12. All transients were measured after the potential step from 1.0 V to more negative potentials (0.26…0.74 V). Characteristic times: 10 (solid squares), 25 (open circles), 60 (open triangles), and 200 s (open inverted triangles).

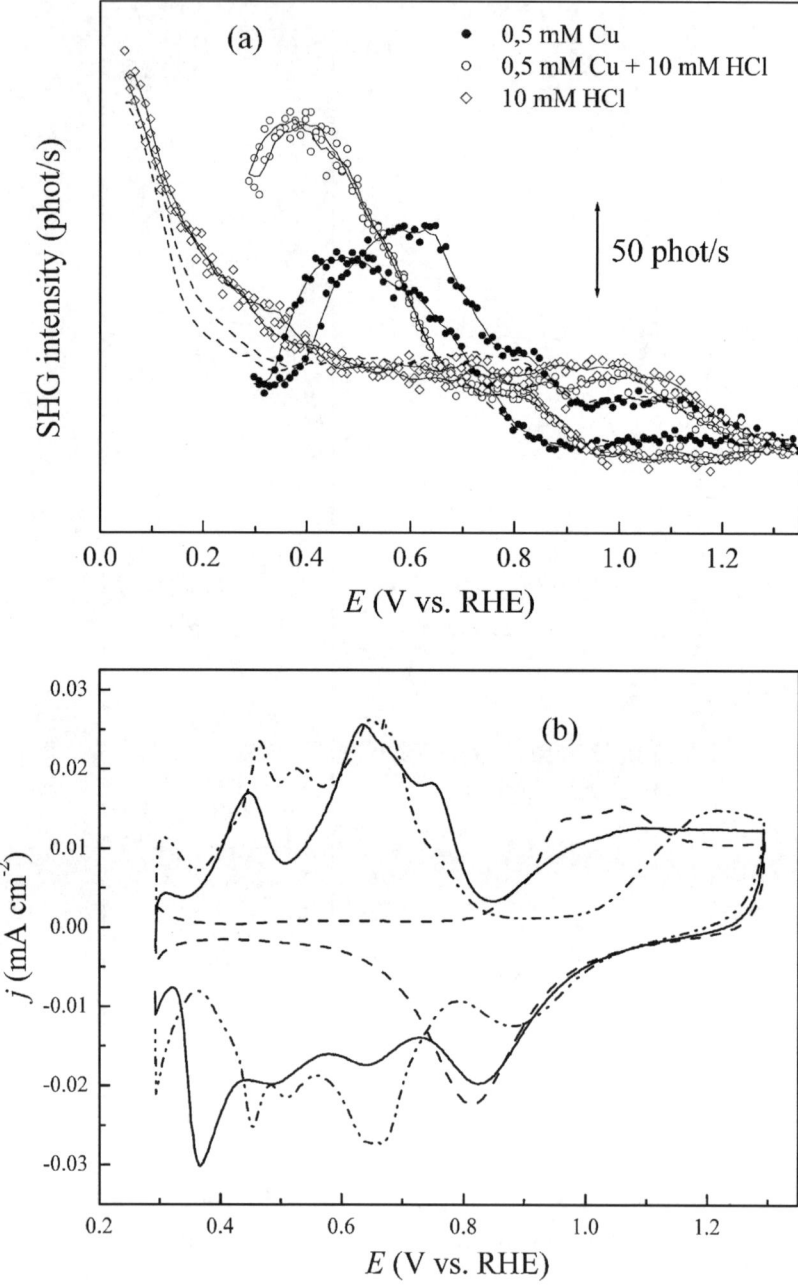

Fig. 12 (a) SHG intensity *versus* potential curves measured in 0.5 M H_2SO_4 (dashed), 0.5 M H_2SO_4 + 10 mM HCl (open diamonds), 0.5 M H_2SO_4 + 0.5 mM Cu SO_4 (solid circles), and 0.5 M H_2SO_4 + 0.5 mM Cu SO_4 + 10 mM HCl, both with and without 3 mM SiW12 (open circles). (b) Cyclic voltammograms measured in 0.5 M H_2SO_4 (dashed), 0.5 M H_2SO_4 + 0.5 mM Cu SO_4 + 3 mM SiW12 (solid) and in the same solution after addition of 10 mM HCl (dot dash). Scan rate: 10 mV s^{-1}.

a slight decrease of signal at certain potentials, especially in the region of adsorption of the oxygen-containing species. The extremely slow equilibration of oxygen on platinum is a well-known fact, and corresponding SHG manifestations are already discussed in ref. 1. We see that the same is typical also for polytungstates.

For mixed dilute solutions (Fig. 11d), transients behavior is close to what we observed for Cu UPD, and the increase of PW12 (Fig. 11e) concentration results

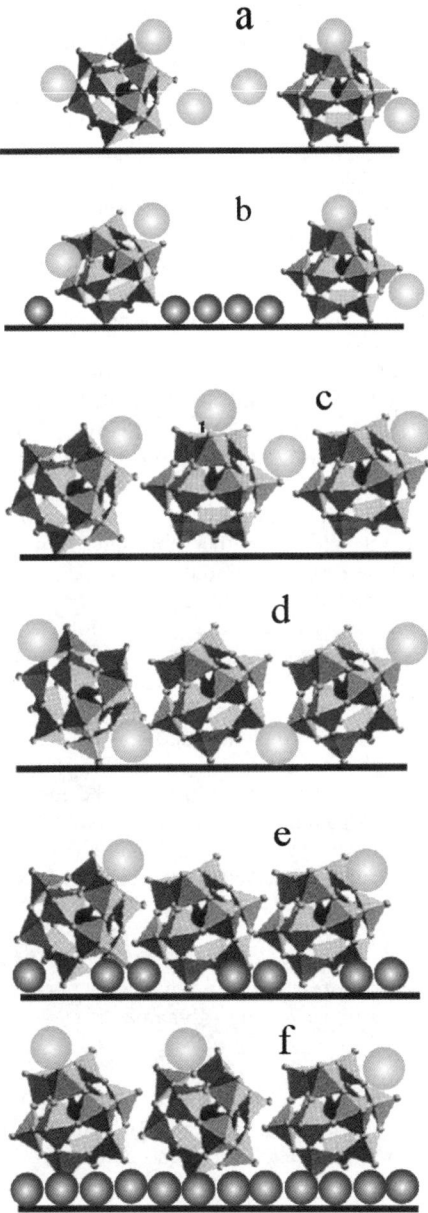

Fig. 13 General scheme of events in mixed adlayers for low (a,b) and high (c–f) surface coverage with polytungstate, see description in the text. Bright spheres represent copper ions, darker spheres correspond to copper adatoms.

in even better similarity. The competition of 'metallic' and 'anionic' behavior is more pronounced when we further increase $CuSO_4$ concentration (Fig. 11f and g for PW12 and SiW12 respectively). It is not surprising because in these experiments we reach a concentration region of SHG signal decreasing with increase of copper content in mixed solutions (see Fig. 6a).

SHG transients confirm again that steady-state SHG is predominantly determined by Cu adatoms, and it takes dozens of seconds to form Cu adlayers in the presence of polytungstate, *i.e.* to penetrate through the polytungstate adlayer. Some additional support of this hypothesis can be obtained from comparison with the chloride effect on SHG behavior of Cu UPD (Fig. 12). The ability of chloride to accelerate Cu^{2+} discharge on Pt is a well-known fact, which finds additional confirmation in complete disappearance of SHG hysteresis. The shift of the UPD region towards more negative potentials induced by chloride is also clearly seen from Fig. 12a. There is no difference of SHG CVs for 10 mM HCl and (10 mM HCl + 3 mM SiW12) solutions, and from the first glance this result can be understood as complete replacement of SiW12 with chloride. However, experiments with copper lead to another conclusion. When comparing Fig 12a and b as well as typical effects of chloride on Cu UPD response,[6,16] we conclude that polytungstate 'protects' a portion of adsorbed Cu from chloride effect, but the main SHG response goes from another portion, for which the state of adatoms is already changed by coadsorbed chloride. 'Protective' action of polytungstate can be easily understood if it is located on the top of Cu adlayer fragments.

The events in mixed Cu-polytungstate adlayers can be schematically represented by the following scheme (Fig. 13). At low PW12(SiW12) concentrations the amount of free space for Cu UPD is high enough (Fig. 13ab), and polytungstate can only affect the geometry of Cu islands. Data from Fig. 2, 3 and 6 confirm that this effect is more pronounced for higher Cu^{2+} concentrations. The structure of the resulting adlayer can not be judged unambiguously and requires additional studies in more dilute solutions in order to check whether tungstates remain at Pt (Fig. 13b) or move to find themselves on top of Cu islands.

This situation looks more comprehensible for higher polytungstate concentrations (coverage), for which Cu penetration through the monolayer of anions is evidenced. We can assume that Cu^{2+} ions association with adsorbed anions (Fig. 13c) take place in the movement/rotation of adsorbed species, and polyanion works like a molecular pinion, which is scrolling the cation. Finally, some cations appear between coadsorbed anions (in Fig. 13d), with better chances to be discharged, and form very small islands in the vicinity of any anion (Fig. 13e). The geometric feature important for this configuration is a half-open cavity between the edge of anion and support. As the polyanion is now surrounded by Cu adatoms, and its interaction with Cu is strong enough, it moves to newly formed islands and gives a chance of subsequently discharging Cu^{2+} to occupy Pt atoms previously bonded with polytungstate (Fig. 13f).

Schematic picture Fig. 13d reflects the distribution of Cu^{2+} and polytingstate anions in a 2D lattice of a complex salt $[Cu_4(H_2O)_2(P_2W_{15}O_{56})_2]$ on pyrographite.[34] However copper–polytungstate interaction with lacunar $P_2W_{15}O_{56}$ fragment is expected to be stronger, leading to slower dynamics.

Concluding remarks

This paper grounds the hypothesis of easy penetration of various species through polytungstate adlayers and considers spatial and dynamical aspects of the problem in the context of chemosorption and electrostatic factors. The former factor is not specific for inorganic polyanions: for example, two types of coadsorption behavior (organic adsorbate under and upon Cu adlayer) were assumed for some organic additives.[60] Formation of a Cu underlayer is known for alkanethiols,[61] as their interaction with Cu adatoms is stronger than with gold support. Very strong adsorption of SiW12 on copper can be concluded from the data on templated Cu deposition.[62]

According to ref. 56, Keggin anions bonding with silver is stronger than with gold. For silver, polytungstate adsorption was considered irreversible. We can not exclude the possibility that a slow transition from reversible to irreversible adsorption takes place on platinum as well, but in general there is no exact interrelation of adsorption energy and reversibility. For example, polytungstate adsorption on mercury is extremely strong,[51,57,58] but completely reversible in a short-time scale, as follows from an electrocapillary study.[58,63]

Local electrostatic effects in coadsorption processes are poorly understood, but are attracting more and more attention. In this context, the analogies with other types of 2D nanostructures are of interest, like with very high stability of self-assembled thiol-modified alkanoic acids[64] and alkanephonic acids[65] on Cu adatoms (as compared to bare gold). In contrast to alkanethiols, the molecules studied in ref. 64 and 65 (if deprotonated) can undergo association with cations. The anionic nature of molecules makes these systems closer to the polytungstates under study, *i.e.* favors formation of the salt-like layers. If the cations participating in such a process appear to be electroactive, their discharge takes place despite high surface coverage with large anions. Remember that on mercury no barrier properties in respect to discharge of cations were found for a large group of polytungstates, despite very strong manifestations of their adsorption.[58] In this aspect, no qualitative difference of adsorption of Keggin anions on platinum and mercury is concluded, at least for the already studied timescale. At the same time, a thermodynamic difference is evident, because adsorption on platinum takes place with participation of co-adsorbed hydrogen, and it is also accompanied by pronounced charge transfer. This situation requires a thermodynamic approach specific for perfectly (not ideally) polarizable electrodes.[66–68]

We believe that electrostatic effects in formation of mixed adlayers should be interpreted by taking into account the basic quantity of the free zero charge.[66] Modification of platinum and gold with Cu adatoms shifts the point of zero free charge (pzfc) towards more negative potentials. As the presence of adsorbed anions is the main reason for cation coadsorption at positive electrode charges, this pzfc shift should only slightly affect the cations adsorption. So the total effect should be the additional stabilization of the adlayer. This reasoning is true for any adlayer based on chemosorbed anions.

The unique property of polytungstates is their partial reduction in an adlayer (molecular UPD). From our coulometric analysis, we can not exclude that this process also occurs (probably even to a higher reduction degree) in the presence of Cu adatoms. The resulting combination of atomic and molecular UPDs taking place with different rates presents the unique chance to construct exactly layered heterostructures, and provides a unique model system for further studies of interfacial charge distribution at a molecular level. An important advantage of the system is the possibility to control partial rates of coupled adsorption processes by changing the ratio of bulk concentrations and electrode potential. Some limitations are possible in an extended timescale, for which slow chemical interactions with supporting metal can contribute to the formation of another type of a salt-like adlayer. However, the already studied timescale of 1–100 s order is not so bad for the aforementioned prospective studies.

Finally, the ability of both co-adsorbing components to generate characteristic optical signals gives a chance to combine thermodynamic studies with the estimate of true charge transfer,[67,69] one of the most intriguing quantities in electrochemistry.

Acknowledgements

This work is supported by the Russian Foundation for Basic Researches (RFBR) and by the Japanese Society for the Promotion of Science.

References

1. E. D. Mishina, G. A. Tsirlina, E. V. Timofeeva, N. E. Sherstyuk, M. I. Borzenko, N. Tanimura, S. Nakabayashi and O. A. Petrii, *J. Phys. Chem. B*, 2004, **108**, 17096.
2. G. W. Tindall and S. Bruckenstein, *Anal. Chem.*, 1968, **40**, 1051.
3. N. Furuya and S. Motoo, *J. Electroanal. Chem.*, 1976, **72**, 165.
4. V. A. Safonov, A. S. Lapa, O. A. Petrii and G. N. Mansurov, *Electrokhimiya*, 1980, **16**, 439.
5. M. W. Breiter, *Electrochim. Acta*, 1989, **34**, 1119.
6. M. E. Martins, R. C. Salvarezza and A. J. Arvia, *Electrochim. Acta*, 1992, **37**, 2203.
7. N. Marcovic and P. N. Ross, *Langmuir*, 1993, **9**, 580.
8. A. I. Danilov, E. B. Molodkina and Yu. M. Polukarov, *Russ. J. Electrochem.*, 1994, **30**, 674.
9. M. Wunsche, H. Meyer and R. Schumacher, *Electrochim. Acta*, 1995, **40**, 629.
10. A. I. Danilov, E. B. Molodkina and Yu. M. Polukarov, *Russ. J. Electrochem.*, 1997, **33**, 288, 295.
11. P. C. Andricacos and P. N. Ross, *J. Electroanal. Chem.*, 1984, **167**, 301.
12. C. Nishihara and H. Nozoye, *J. Electroanal. Chem.*, 1995, **396**, 139, 386, 75.
13. A. I. Danilov, E. B. Molodkina, A. V. Rudnev, Yu. M. Polukarov and J. M. Feliu, *Electrochim. Acta*, 2005, **50**, 5032.
14. C. Alonso, M. J. Pascual and H. D. Abruna, *Electrochim. Acta*, 1997, **42**, 1739.
15. R. Gomez, H. S. Yee, G. M. Bommarito, J. M. Feliu and H. D. Abruna, *Surf. Sci.*, 1995, **335**, 101.
16. N. Marcovic and P. N. Ross, *Langmuir*, 1995, **11**, 4098.
17. C. A. Lucas, N. M. Marcovic, I. M. Tidswell and P. N. Ross, *Phys. B*, 1996, **221**, 245.
18. N. M. Marcovic, C. A. Lucas, H. A. Gasteiger and P. N. Ross, *Surf. Sci.*, 1997, **372**, 239.
19. C. A. Lucas, N. M. Marcovic and P. N. Ross, *Phys. Rev. B: Condens. Matter Mater. Phys.*, 1997, **56**, 3651.
20. A. I. Danilov, E. B. Molodkina, Yu. M. Polukarov, V. Climent and J. M. Feliu, *Electrochim. Acta*, 2001, **46**, 3137.
21. J. Inukai, Y. Osawa, M. Wakisaka, K. Sashikata, Y.-G. Kim and K. Itaya, *J. Phys. Chem. B*, 1998, **102**, 3498.
22. Y. Shingaya, H. Matsumoto, H. Ogasawara and M. Ito, *Surf. Sci.*, 1995, **335**, 23.
23. R. Michaelis, M. S. Zei, R. S. Zhai and D. M. Kolb, *J. Electroanal. Chem.*, 1992, **339**, 299.
24. T. D. Hewitt and D. Roy, *Chem. Phys. Lett.*, 1991, **181**, 407.
25. G. L. Richmond, *Chem. Phys. Lett.*, 1985, **113**, 359.
26. S. Mirwald, B. Petiinger and J. Lipkowski, *Surf. Sci.*, 1995, **335**, 264.
27. M. J. Bennahmias, S. Lakkaraju, B. M. Stone and K. Ashley, *J. Electroanal. Chem.*, 1990, **280**, 429.
28. D. Koos and G. L. Richmond, *J. Electrochem. Soc.*, 1989, **136**, 218C.
29. E. D. Mishina, N. Ohta, Q. K. Yu and S. Nakabayashi, *Surf. Sci.*, 2001, **494**, L748.
30. C. K. Chen, T. F. Heinz, D. Ricard and Y. R. Shen, *Phys. Rev. B: Condens. Matter Mater. Phys.*, 1983, **27**, 1965.
31. Wei Tze, M. I. Borzenko, G. A. Tsirlina and O. A. Petrii, *Russ. J. Electrochem.*, 2002, **38**, 1250.
32. I. Lavrik, P. Staiti, P. Novak and S. Hocevar, *J. Power Sources*, 2001, **96**, 303.
33. P. J. Kulesza, K. Karnicka, K. Miecznikowski, M. Chojak, A. Kolary, P. J. Barczuk, G. Tsirlina and W. Czerwinski, *Electrochim. Acta*, 2005, **50**, 5155.
34. M. S. Kaba, I. K. Song, D. C. Dubcan, C. L. Hill and M. A. Barteau, *Inorg. Chem. B.*, 1998, **37**, 398.
35. D. A. Koos and G. L. Richmond, *J. Phys. Chem.*, 1993, **96**, 3770.
36. S. Lakkaraju, M. J. Bennahmias, J. G. Borges, J. G. Gordon II, M. Lazaga, B. M. Stone and K. Ashley, *Appl. Opt.*, 1990, **29**, 4943.
37. G. T. Boyd, Th. Rasing, J. R. R. Leite and Y. R. Shen, *Phys. Rev. B: Condens. Matter Mater. Phys.*, 1984, **30**, 519.
38. V. L. Shannon, D. A. Koos, S. A. Kellar, P. Huifang and G. L. Richmond, *J. Phys. Chem.*, 1989, **93**, 6434.
39. F. Rebentrost, *Prog. Surf. Sci.*, 1995, **48**, 71.
40. I. Yagi, S. Idojiri, T. Awatani and K. Uosaki, *J. Phys. Chem.*, 2005, **109**, 5021.
41. J. Y. Zhanga, Y. R. Shen, D. S. Soane and S. C. Freilich, *Appl. Phys. Lett.*, 1991, **59**, 1305.
42. D. Margheritis, R. C. Salvarezza, M. C. Giordano and A. J. Arvia, *J. Electroanal. Chem.*, 1987, **229**, 327.
43. G. Celep, E. Cottancin, M. Lermé, M. Pellarin, L. Arnaud, J. R. Huntzinger, J. L. Vialle, M. Broyer, B. Palpant, O. Boisron and P. Mélinon, *Phys. Rev. B: Condens. Matter Mater. Phys.*, 2004, **70**, 165409.
44. J. Zhao, H. Zhang and G. Wang, *J. Phys. Chem. Solids*, 1996, **57**, 225.

45 B. Keita, L. Nadjo, D. Belanger, C. P. Wilde and M. Hilaire, *J. Electroanal. Chem.*, 1995, **384**, 155.
46 L. Lee, J. X. Wang, R. R. Adzic, I. K. Robinson and A. A. Gewirth, *J. Am. Chem. Soc.*, 2001, **123**, 8838.
47 G. Horanyi, *Electrochim. Acta*, 1980, **25**, 43.
48 C. A. Jeffry, W. M. Storr and D. A. Harrington, *J. Electroanal. Chem.*, 2004, **569**, 61.
49 R. T. S. Oliveira, M. C. Santos, L. O. S. Bulhoes and E. C. Pereira, *J. Electroanal. Chem.*, 2004, **569**, 233.
50 M. S. Ureta-Zanartu, C. Yanez and C. Gutierrez, *J. Electroanal. Chem.*, 2004, **569**, 275.
51 E. Guaus, F. Sanz, M. Sluyters-Rehbach and J. H. Sluyters, *J. Electroanal. Chem.*, 1995, **385**, 121.
52 M. Ge, B. Zhong, W. G. Klemperer and A. A. Gewirth, *J. Am. Chem. Soc.*, 1996, **118**, 5812.
53 J. Kim and A. A. Gewirth, *Langmuir*, 2003, **19**, 8934.
54 C. M. Teague, X. Li, M. E. Biggin, L. Lee, J. Kim and A. A. Gewirth, *J. Phys. Chem. B*, 2004, **108**, 1974.
55 D. E. Clinton, D. A. Tryk, I. T. Bae, F. L. Urbach, M. R. Antonio and D. A. Scherson, *J. Phys. Chem.*, 1996, **100**, 18511.
56 L. Lee and A. A. Gewirth, *J. Electroanal. Chem.*, 2002, **522**, 11.
57 C. Rong and F. C. Anson, *Inorg. Chim. Acta*, 1996, **242**, 11; C. Rong and F. C. Anson, *Anal. Chem.*, 1994, **66**, 3124.
58 M. I. Borzenko, G. A. Tsirlina and O. A. Petrii, *Mendeleev Commun.*, 2002, **12**, 126.
59 A. Lachenwitzer, N. Li and J. Lipkowski, *J. Electroanal. Chem.*, 2002, **532**, 85.
60 A. S. Dakkouri, N. Batina and D. M. Kolb, *Electrochim. Acta*, 1993, **38**, 2467.
61 M. Nishikawa, T. Sunagawa and H. Yoneyoma, *Langmuir*, 1997, **13**, 5215.
62 P. Ficoteaux and O. Savadogo, *Electrochim. Acta*, 1999, **44**, 2927.
63 M. I. Borzenko, G. A. Tsirlina and O. A. Petrii, *Russ. J. Electrochem.*, 2000, **36**, 452.
64 S.-Y. Lin, C.-H. Chen, Y.-C. Chan, C.-M. Lin and H.-W. Chen, *J. Phys. Chem. B*, 2001, **105**, 4951.
65 M. V. Baker, G. K. Jennings and P. E. Laibinis, *Langmuir*, 2000, **16**, 3288.
66 A. N. Frumkin and O. A. Petrii, *Electrochim. Acta*, 1970, **15**, 391.
67 A. N. Frumkin, B. B. Damaskin and O. A. Petrii, *J. Electroanal. Chem.*, 1974, **53**, 57.
68 A. N. Frumkin and O. A. Petrii, *Electrochim. Acta*, 1975, **20**, 347.
69 V. E. Kazarinov, A. M. Foontikov and G. A. Tsirlina, *J. Electroanal. Chem.*, 1990, **282**, 253.

PAPER

Aqueous-based synthesis of ruthenium–selenium catalyst for oxygen reduction reaction

Cyril Delacôte,[a] Arman Bonakdarpour,[a] Christina M. Johnston,[b] Piotr Zelenay[b] and Andrzej Wieckowski*[a]

Received 15th April 2008, Accepted 12th May 2008
First published as an Advance Article on the web 20th August 2008
DOI: 10.1039/b806377j

Carbon-supported Se/Ru(Se) catalysts of a broad range of composition were synthesized *via* a reduction procedure in which a mixture of $RuCl_3$, SeO_2 and Black Pearl carbon was treated with $NaBH_4$ in basic media at room temperature. Physical characterization of the catalyst was performed by X-ray diffraction, energy dispersive X-ray spectroscopy and by high resolution transmission electron microscopy. The effect of NaOH addition during the reduction by $NaBH_4$ and the impact of a post-reduction thermal treatment at 500 °C were interrogated. The activity of the catalyst towards the oxygen reduction reaction was studied by the use of a rotating disk electrode. It was found that the half-wave potential for the oxygen reduction reaction was about 0.78 V *vs*. RHE. The Se-to-Ru ratio and metal loading on carbon were optimized for the oxygen reduction reaction and the optimized catalyst was tested at the cathode of a polymer electrolyte fuel cell. The stability of the Se/Ru(Se) catalyst was evaluated by electrochemical cycling and by leaching the catalyst in 0.5 M H_2SO_4 at 80 °C.

Introduction

A significant amount of research is being conducted on fuel cells as a clean, alternative energy source. There are, however, several barriers to their commercialization, including the poor kinetics of the oxygen reduction reaction (ORR) at the cathode.[1,2] Although Pt and Pt alloys have traditionally been considered the best catalysts for oxygen reduction, the high cost of Pt and its sensitivity to the fuel cross-over and atmospheric contaminants have prompted studies of non-Pt based ORR catalysts.[3–7]

Alonso-Vante *et al.* demonstrated that ruthenium-based selenide materials (Ru_xSe_y) showed excellent activity in catalyzing the reduction of oxygen.[8,9] Unlike Pt, these Ru_xSe_y materials are fully tolerant against poisoning by methanol,[10–14] and are therefore suitable to serve as a cathode in the direct methanol fuel cell (DMFC).[4,15–17] Proportionally to these observations, the syntheses of ruthenium-based chalcogenide catalysts have received increased attention.[5] The first chalcogenide materials studied were chevrel phase-type compounds (*e.g.* $Mo_4Ru_2Se_8$) obtained by sintering a mixture of high-purity elements in sealed quartz tubes at 1470 K.[8,9] The synthesis of Ru_xSe_y type materials was later performed under milder conditions by the decomposition of carbonyl precursors in organic solvents.[10,18]

[a]Department of Chemistry, University of Illinois at Urbana-Champaign, Urbana, IL, 61801, USA. E-mail: andrzej@scs.uiuc.edu; Fax: +1 217 244 8068; Tel: +1 217 333 7943
[b]Materials Physics & Applications Division, Los Alamos National Laboratory, Los Alamos, New Mexico, 87545, USA

Improvements on the carbonyl procedure have recently been reported,[15,19] but there are still several drawbacks to preparing Ru_xSe_y type materials by this pathway. In particular, carbonyl compounds are expensive and the yield of the products obtained by this procedure is always less than 100% (~40–60%).[20] Other synthesis procedures of chalcogenide materials have also been proposed. The reduction of ruthenium salts and selenium oxides with different kinds of reducing agents has been used by several groups. For example, the reduction of $RuCl_3$ and SeO_2 has been performed with H_2 on impregnated carbon[21] or with $Li(Et)_3N$ by a colloidal method.[22] More recently, Rao and Viswanathan proposed the reduction of $RuCl_3$ and $HSeO_3$ with $NaBH_4$ by using the reverse microemulsion method.[23] All of these synthetic routes need to be either performed at high temperatures (200–800 °C) or involve the use of organic compounds (*e.g.* solvents and/or templates). Harsh synthetic conditions (high temperature, organic solvent) and/or the difficulties in fully removing the organic compounds are some of the problems in synthesizing ruthenium-based chalcogenides on an industrial scale.

Recently, Campbell proposed an "environment-friendly" aqueous method for preparing active catalysts such as Ru_xSe_y.[24] He suggested reducing $RuCl_3$ and SeO_2 on carbon with $NaBH_4$ in a water/propanol solvent at room temperature. We have adopted a similar procedure without the use of propanol to synthesize carbon supported ruthenium selenide catalyst (Se/Ru(Se)/C), meaning that the Se/Ru is the catalytic surface site but there is still Se in the bulk of the catalyst. In an earlier report, we presented some electrochemical characterizations of this catalyst using the rotating disk electrode (RDE) and rotating ring disk electrode (RRDE).[25] In this paper, we investigate different parameters affecting the synthesis procedure and the materials thus obtained. We focus on the structure of the catalyst and its ability to catalyze oxygen reduction (on RDE). We also discuss the stability of our compound and the role of the metal-on-carbon loading on the ORR activity. The synthetic procedure was optimized to produce the Se/Ru(Se)/C catalyst with the best ORR activity on RDE. The optimized catalyst was tested as a candidate cathode catalyst in a polymer electrolyte fuel cell (PEFC).

Experimental methods

Chemicals and reagents used

All reagents were analytical grade and used as received from the suppliers. Ruthenium trichloride ($RuCl_3 \cdot xH_2O$, Aldrich), selenium dioxide (SeO_2, Alfa Aesar), sodium borohydride ($NaBH_4$, Aldrich), and sodium hydroxide (NaOH, Aldrich) were used. The carbon used was Black Pearl 2000 (particle size 15 nm, specific area 1500 $m^2\ g^{-1}$) from Cabot. The electrolyte was prepared from high purity, double-distilled sulfuric acid (H_2SO_4, GFS Chemicals) and high purity water. In all experiments, high purity water (18 MΩ cm) obtained from a Millipore water purification system was used.

Synthesis of Se/Ru(Se)/C catalysts

The basic principles of the synthesis were inspired by Campbell's method[5,24] and are schematically presented in Fig. 1. Our procedure did not involve propanol in the synthesis; only high purity water was used. First, carbon (BP 2000) was dispersed in water by sonicating the reaction flask for one hour. $RuCl_3$ and SeO_2 salts were added to the carbon and water suspension and sonicated for 10 minutes. The suspension was stirred at 180 rpm and heated at 80 °C for one hour. During this heating stage, Ar was purged through the suspension to completely remove any dissolved O_2 and to help dissolve the salts. After cooling to room temperature (RT), an aqueous solution of $NaBH_4$ (0.1 M) and NaOH (0.2 M) was added dropwise to the solution. NaOH allows the stabilization of borohydride in water ($NaBH_4$ in aqueous environment is reduced by H^+ to H_2 and boron derivatives) and its addition

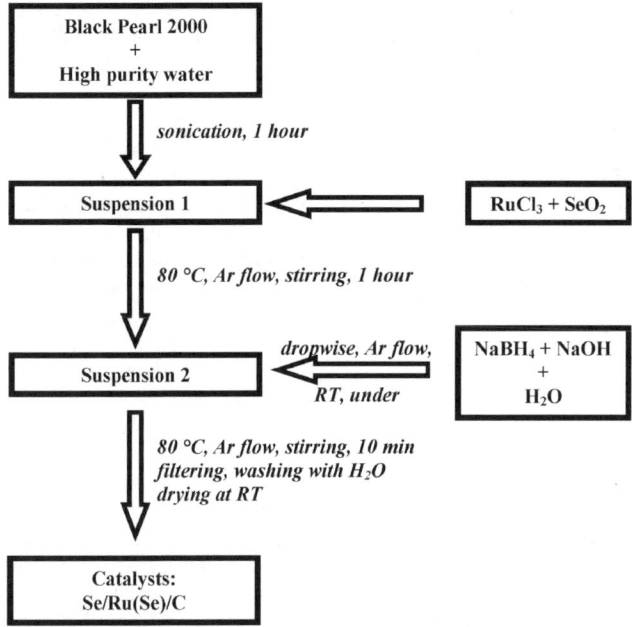

Fig. 1 Schematic representation of the aqueous synthetic procedure used to prepare Se/Ru(Se)/C nanoparticles.

causes the precipitation of the Ru and Se hydroxides. These hydroxides are then reduced to their respective metallic forms by the addition of NaBH$_4$. During the reduction step, the mixture was maintained under bubbling Ar and placed into an ultrasonic bath to enhance the dispersion of the carbon supported nanoparticles. In a final step, the excess borohydride was removed by heating the suspension at 80 °C for 10 min. After cooling, the resulting material was filtered, thoroughly washed with water, and dried in air over 24 h. An additional, post-synthesis annealing (500 °C for 2 h under Ar) could then be applied to the sample. Characterization of the Se/Ru(Se)/C materials was performed on both as-prepared and heat-treated samples.

In this work, the metal (Ru + Se) loading on carbon for each catalyst (metal + carbon) is expressed in weight percent (wt%). Metal composition (Ru or Se) is reported in atomic percent (at%).

Physical characterizations

X-Ray diffraction (XRD) measurements were performed using a Panalytical X'pert MRD system with a Cu Kα radiation (λ = 0.15418 nm) using an X-ray beam collimated to 1 mm and 8 mm dimensions in the directions parallel and perpendicular to the diffraction plane, respectively. The secondary optics consisted of a parallel plate collimator, a flat graphite monochromator and a proportional detector. Conventional $2\theta/\theta$ scans (2θ is the Bragg angle) were performed with a 0.1° step and 15 s counting time per step.

Transmission electron microscopy (TEM) was performed using a high-resolution JEOL 2100 LaB$_6$ microscope operating at 200 kV. For the TEM measurements, the sample was dispersed in ethanol/water mixture by ultrasonic blending, and the suspension was applied onto a lacey carbon-coated copper 200 grid, and dried in air.

The composition of the samples was measured by a JEOL 6060 LV SEM system equipped with an Oxford Link energy dispersive spectrometer (EDS). The beam voltage was 20 kV and a typical area of 50 μm × 50 μm was rastered during the measurements. The characteristic L peaks were used for quantification of Ru ($L_{\alpha 1}$: 2.559 keV) and Se (L_{α}: 1.329 keV).

Electrochemical characterization

Oxygen reduction activity of the samples was evaluated by the rotating disk electrode method. A catalyst ink was prepared by dispersing a precise amount of catalyst in 250 μL of Nafion® (5 wt% solution) and 1.250 mL of Milli-Q H_2O. After homogenizing the ink suspension under ultrasonic conditions, 3 μL of the ink were deposited on the polished glassy carbon (GC) surface of an RDE (3 mm diameter). In the case of a 20 wt% catalyst, 10 mg of the catalyst in the suspension ensured a 56 μg cm^{-2} total metal (Ru and Se) loading on the GC electrode. The electrode tip (from an AutoLab RDE system) was dried at room temperature and in air. An Ag/AgCl (0.280 V *vs*. RHE) and a Pt wire were used as reference and counter electrodes, respectively. The electrochemical measurements were performed by Autolab potentiostat (PGSTAT100) and the associated GPES electrochemical analysis software (Eco Chemie). The electrochemical measurements were carried out in a 0.1 M H_2SO_4 solution and at room temperature (21 °C).

The procedure described below was used for recording ORR activity. Namely, the first conditioning step, consisting of 20 electrochemical cycles between 0.90 and 0.10 V *vs*. RHE in Ar-saturated electrolyte, was applied at 100 mV s^{-1}. A steady-state voltammogram then obtained indicated that the surface was stabilized. The ORR voltammogram was next recorded in O_2-saturated electrolyte by cycling the working electrode between 0.90 and 0.10 V *vs*. RHE at 5 mV s^{-1}. The rotation speed was 1600 rpm. To compare the ORR activity of a series of samples, the half-wave potentials were obtained from the maxima of the derivatives of reduction curves.

To estimate electrochemical durability of the samples, first the ORR activity was recorded. Then 1000 electrochemical cycles were performed from 0.85 to 0.60 V *vs*. RHE at 20 mV s^{-1} in the O_2-saturated H_2SO_4 solution. The final ORR activity was recorded at the end of the cycling, or after waiting overnight in the O_2-saturated electrolyte at open circuit potential.

Fuel cell test

Fuel cell tests of Se/Ru(Se)/C catalyst were performed at Los Alamos National Laboratory. Catalyst was dispersed in a mixture of 5 wt% Nafion® solution, and glycerol to make an ink. The composition was determined to produce a final dry Nafion® volume of approximately 40% (balance Se/Ru(Se)/C). The ink was coated onto a 5 cm^2 area of a Teflon-coated decal to give a nominal catalyst loading of 1.0 mg cm^{-2}. The decal was then hot-pressed onto a Nafion® membrane in sodium-ion form (as opposed to the proton form) for 5 min at 210 °C under 600 lb of force. The anode was prepared by coating 3–6 mg of 5 wt% Nafion® solution onto a 5 cm^2 square of the commercially available platinized carbon cloth (ELAT from BASF), drying it under a heating lamp, then hot-pressing it onto a Nafion® 1135 membrane (in proton-form) for 90 s at 125 °C under 600 lb of force. The MEA was prepared using two Nafion® 1135 membranes, with the cathode applied onto one membrane and the anode onto the other membrane. These so-called "half-MEAs" were pressed together in the fuel cell hardware, as reported in previous publications.[26] This arrangement aids in separately analyzing the anode and cathode after the completion of the experiment.

Before the data were taken, the cell was "conditioned" by running at 0.10 V overnight. The gas flow and temperature conditions for the H_2 were as follows: 300 sccm flow, 30 psi backpressure, and 105 °C humidification temperature. For the air, the parameters were: 500 sccm flow, 30 psi backpressure, and 90 °C humidification

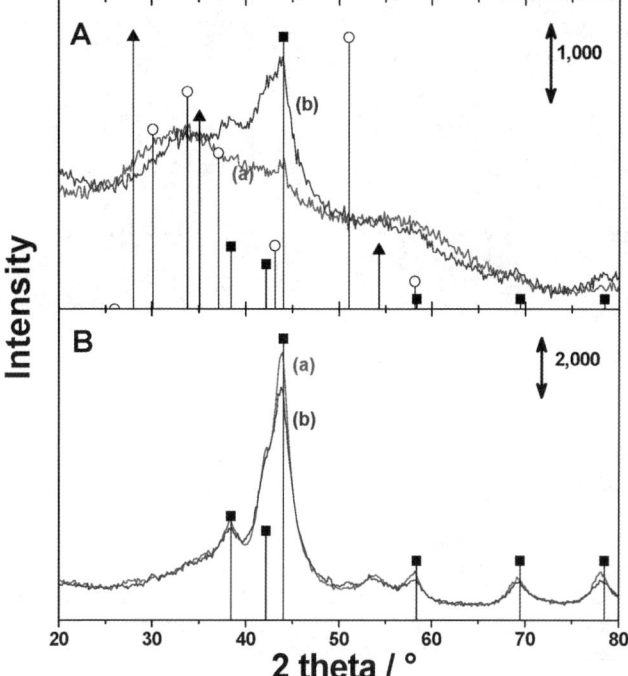

Fig. 2 X-Ray diffraction patterns of Se/Ru(Se)/C (70 wt%) synthesized in the presence of NaOH (curve (a)) and in the absence of NaOH (curve (b)). Panel A and B show the as-prepared and heat-treated (500 °C for 2 h, under Ar flow) samples, respectively. Powder diffraction cards (PDFs-JCPDS)† of RuO_2 (triangles (▲); No. 00-040-1290) and $RuSe_2$ (circles (○); No. 00-080-0670) and Ru (squares (■); No. 00-006-0663) phases are also shown.

temperature. The operating temperature of the cell was 80 °C during the measurements.

Results and discussion

Physical characterization

To evaluate the impact of the NaOH addition during the synthesis, two Se/Ru(Se)/C samples (70 wt% metal loading) were prepared: one with and one without NaOH in the reducing solution. (When NaOH is not present in the reducing solution, $NaBH_4$ is not stable and slowly decomposes to H_2 and boron hydroxides.) The reducing solution must be freshly prepared and the use of a large excess of hydride ensures a complete reduction of the metallic salts. This was confirmed by a total discoloration of the suspension. All of these samples were characterized by EDS, XRD and HRTEM before and after heat-treatment (500 °C for 2 h, under Ar flow). Bulk composition of all four catalysts, determined by EDS, showed an average composition of 64 at% Ru and 36 at% Se. The measured composition was in agreement with the nominal composition assumed from the amount of the starting materials. The metal salts were completely reduced whether NaOH was used or not. No significant changes in the catalysts compositions were observed upon heat-treatment.

† Copyright © JCPDS-ICDD 2002 JCPDS—International Centre for Diffraction Data, 12 Campus Boulevard, Newtown Square, PA 19073-3273 USA; www.icdd.com

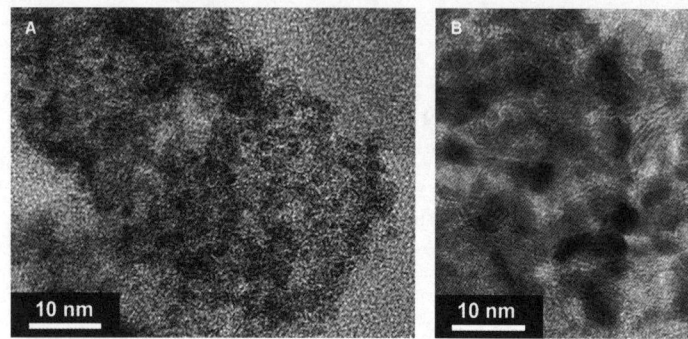

Fig. 3 TEM images of Se/Ru(Se)/C nanoparticles for the as-prepared (A) and heat-treated (500 °C for 2 h, under Ar flow) samples (B).

XRD patterns of all four samples are shown in Fig. 2. The as-prepared samples (Fig. 2A) show broad Bragg reflections indicative of the formation of amorphous material. When NaOH is used during the synthesis, one can observe a mixture of $RuSe_2$ and RuO_2 phases. When NaOH is not used, a broad peak, centered about $2\theta = 43.9°$, is observed. This peak is attributed to the presence of small Ru crystallites. The presence of NaOH in the reduction medium allows for the precipitation of Ru hydroxides on the carbon and seems to favor the formation of more homogenous Se/Ru(Se) nanoparticles. XRD patterns of heat-treated samples are shown in Fig. 2B. The two samples, synthesized with or without NaOH, respectively, show similar diffraction patterns. In both cases hexagonal-closed-packed (hcp) Ru crystallites can be identified. It is interesting to mention that Alonso-Vante *et al.* have observed similar patterns in the case of Ru_xSe_y clusters obtained by decomposition of carbonyl complexes. They have provided evidence that the core of particles is essentially constituted of Ru atoms surrounded by Se atoms.[11] We can expect that our catalyst exhibits the same structure but more analyses are necessary to confirm this conjecture.

When comparing Fig. 2A and B, there is evidence that heat-treatment induces crystallite growth. This observation is confirmed by TEM images of the samples synthesized with NaOH, as shown in Fig. 3. The images were acquired using both as-prepared and heat-treated samples. For the as-prepared samples, the average Se/Ru(Se) particle size is 2 nm. After annealing, the nanoparticles appeared bigger, around 6–8 nm. Both XRD and TEM observations showed growth of nanoparticles upon heat-treatment. Fiechter *et al.* have reported similar results and they have observed a particle growth from 4 to 10 nm upon annealing at 900 °C.[27]

TEM images of the catalysts synthesized in the absence of NaOH (not shown here) were similar to those presented in Fig. 3. The use of NaOH did not influence the morphology or size attributes of the catalysts. The presence of NaOH in the reduction solution allows both the stabilization of $NaBH_4$ in a fully aqueous environment and also the formation of homogenous nanoparticles by precipitation of the metallic hydroxides as shown by Fig. 2A.

Electrochemical characterization using RDE

Voltammetric cycles measured in a de-aerated electrolyte are shown in Fig. 4. Reduction of Se oxides, observed only during the first cycle, occurs between 0.40 and 0.20 V *vs.* RHE.[28] The presence of Se in the vicinity of Ru atoms prevents their further reoxidation,[28,29] and no other redox peaks can be observed upon the subsequent cycles. The reduction current of the Se oxide, observed upon the first sweep, is more intense when the sample was synthesized with NaOH. This may be due to the presence of higher amounts of Se oxide on the surface of the nanoparticles. We note

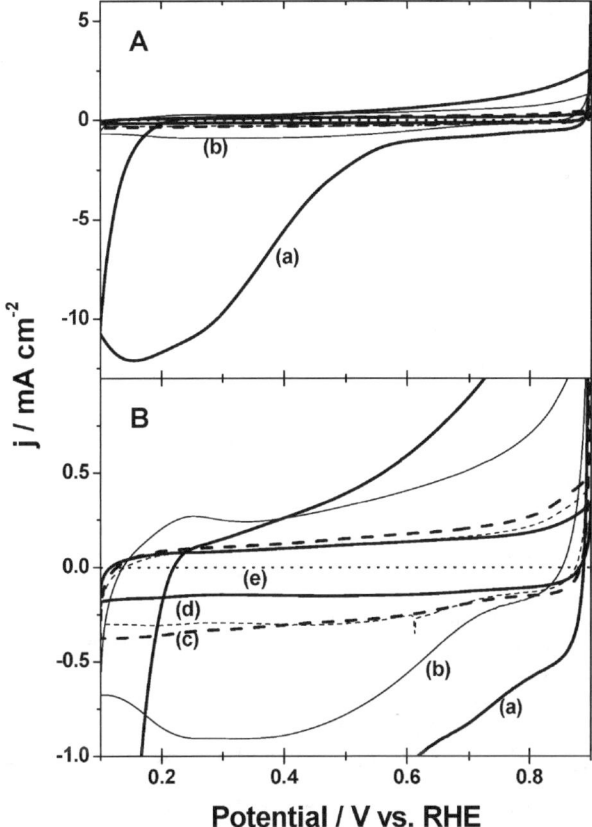

Fig. 4 Voltammetric characterization of Se/Ru(Se)/C catalysts. Except for (e) the data show first electrochemical cycles recorded in de-aerated 0.1 M H_2SO_4 electrolyte from 0.90 to 0.10 V vs. RHE at 20 mV s^{-1}. Panel B shows the data from panel A in more detail. Samples were synthesized in the presence of NaOH (a, c) and in the absence of NaOH (b, d). Samples (a) and (b) are as-prepared Se/Ru(Se)/C, whereas samples (c) and (d) underwent a post-synthesis heat-treatment at 500 °C. The 10th voltammogram in panel B (e), which is similar for all samples, shows that the cyclic voltammogram is at steady-state.

also that the heat-treated samples do not exhibit a strong reduction current and showed similar electrochemical behavior, whether they were synthesized with or without NaOH in the reduction medium. Because the total Se content is unchanged after the annealing step, one can conclude that Se diffuses from the surface to the core of the nanoparticles. Alternatively, heat treatment could help to complete the reduction of surface Se, and affect its electrochemical activity. Under these conditions, there would be less oxidized Se in the first voltammogram. Steady-state voltammograms were achieved after 10 cyclic voltammograms (see Fig. 4).

The cathodic oxygen reduction waves of the samples studied (RDE) are presented in Fig. 5. ORR commenced at an onset potential of ∼0.85 V vs. RHE and a "pseudo" diffusion-limited region was attained at ∼0.60 V vs. RHE. We call this region "pseudo" diffusion-limited, because no flat plateau-like current was observed during the ORR measurements; instead, a gradually increasing reduction current was recorded. A similar phenomenon, i.e. a non-plateau current region, was also observed in other Se/Ru(Se)-based catalysts[12,13,30–36] and, more generally, with Ru-based catalysts.[6,7,37] Under such conditions, it is difficult to determine the value of the diffusion-limited currents. Therefore, we have evaluated the half-wave potentials from the maxima of the derivatives of the reduction currents. Although the

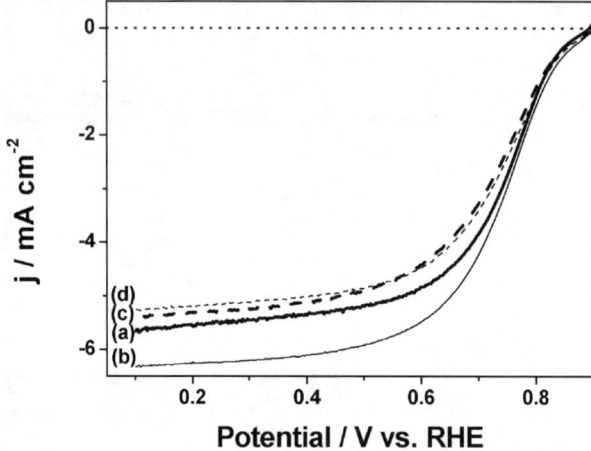

Fig. 5 ORR of Se/Ru(Se)/C catalysts in the O_2-saturated solution recorded at 5 mV s^{-1}, 1600 rpm and after 20 cycles in the de-aerated solution at 100 mV s^{-1}. Samples were synthesized in the presence (a, c) and in the absence (b, d) of NaOH, respectively. Samples (a, b) were as-prepared, whereas samples (c, d) were heat-treated (post-synthesis) at 500 °C.

half-wave potentials may not be rigorously defined here, this allows us to compare the activity of our samples on a relative basis. The half-wave potential measured in Fig. 5 for the as-prepared samples is about 0.78 V *vs.* RHE whether the catalyst was synthesized with or without NaOH. However, the material synthesized in the absence of NaOH showed higher diffusion-limited currents. Half-wave potentials measured on heat-treated samples are close to those observed on as-prepared samples (*i.e.* no significant shift is observed).

Fig. 6 shows ORR curves of the as-prepared catalysts synthesized with NaOH recorded before and after different treatments. The half-wave potential, initially about 0.78 V *vs.* RHE, decreases to 0.77 V *vs.* RHE after 1000 electrochemical cycles.

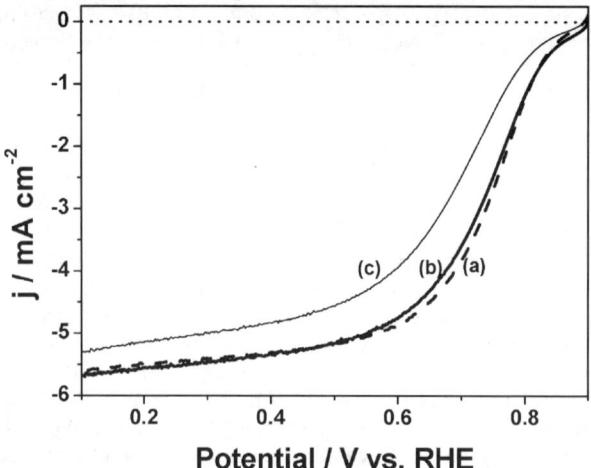

Fig. 6 Laboratory-scale durability testing of the Se/Ru(Se)/C catalyst. ORR measurements recorded in the O_2-saturated solution (0.5 M H_2SO_4) at 5 mV s^{-1}, 1600 rpm after different electrochemical conditionings: after 20 cycles (at 100 mV s^{-1}) in de-aerated solution (a), after conditioning (a) plus 1000 electrochemical cycles (at 20 mV s^{-1}) between 0.85 and 0.60 V *vs.* RHE in the O_2-saturated solution (b), and after conditioning (b) and after an overnight exposure to O_2-saturated electrolyte at open circuit potential (c).

The half-wave potential decreased further to 0.74 V *vs.* RHE after an overnight exposure to the oxygen saturated solution at the open-circuit potential (OCP), following 1000 electrochemical cycles. The OCP measured at the beginning of the experiment was 0.85 V *vs.* RHE and, at the end of night, was over 1.10 V *vs.* RHE. This high OCP value indicates a mixed potential, set by oxygen reduction and oxidation of the catalyst.[28] These results demonstrate that the impact of the immersion in acid and maintaining the equilibrium (under OCP conditions) on the degradation of the catalyst is greater than that of the electrochemical cycling between 0.85 and 0.60 V. The durability test (1000 electrochemical cycles and an overnight exposure to oxygen at the open-circuit potential) was also performed on a sample synthesized in the absence of NaOH, and on heat-treated samples. No differences were observed between the samples prepared with or without NaOH. For the heat-treated catalysts, the shift in the half-wave potential is about 20 mV, as opposed to 40 mV observed with the as-prepared sample. The heat-treatment apparently leads to an improvement in the catalyst stability. This may be attributed to an increase in the catalyst particle size at elevated temperatures.

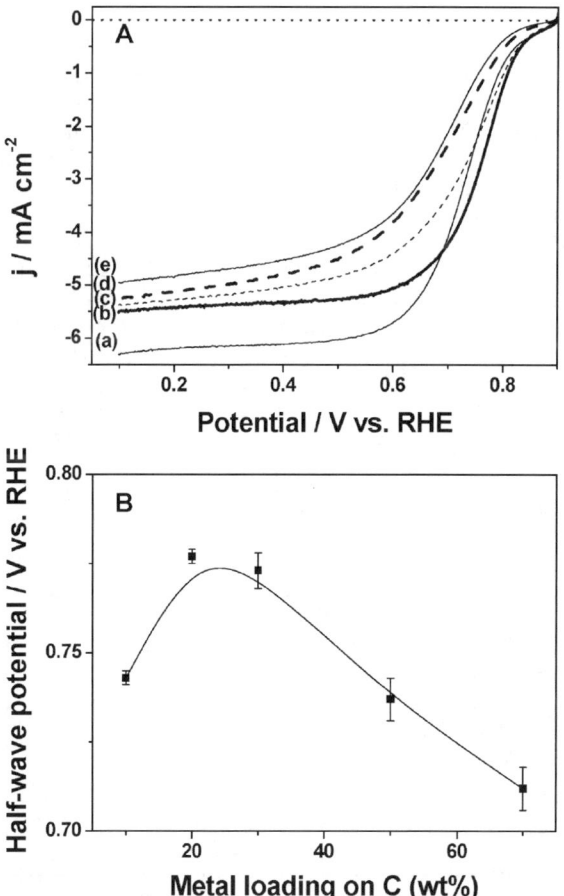

Fig. 7 Optimization of the metal loading in Se/Ru(Se)/C catalyst. (A) ORR curves recorded in the O_2-saturated solution (0.5 M H_2SO_4) at 5 mV s^{-1}, 1600 rpm. The metal loading on the electrode was maintained at 56 µg cm^{-2}. Samples were characterized by 10% (a), 20% (b), 30% (c), 50% (d) and 70% (e) Ru + Se on the carbon support (wt%). (B) Values of half-wave potential obtained from the data in (A) as a function of metal loading on carbon (BP2000).

To evaluate the catalyst stability, *via* mimicking the polymer electrolyte fuel cell environment, "acid-leaching" experiments were performed. Namely, the as-prepared samples were dispersed in acid (0.5 M H_2SO_4) and kept at 80 °C for one week, followed by a one week exposure of the catalysts to the solution at room temperature. The initial composition of 64 : 36 (Ru : Se) changed to 67 : 33 after the leaching experiments, namely, no significant amount of Se was lost from the bulk of the particles. Se was mainly lost from the surface of the particles and this affected the oxygen reduction activity.

Catalyst optimization and fuel cell measurements

Before measuring the fuel cell performance of Se/Ru(Se)/C catalyst synthesized by the aqueous procedure, the catalyst composition and the metal-to-carbon ratio were optimized for the best ORR activity on RDE.

A series of Se/Ru(Se)/C samples with different metal (Ru + Se) loading on the carbon support (BP 2000) was synthesized. The oxygen reduction activity of these samples was compared by maintaining the 56 μg_{metal} cm^{-2} catalyst loading on the GC tip of RDE. (In our previous study, we reported that at least 56 μg_{metal} cm^{-2} of Se/Ru(Se)/C on the RDE should be used to obtain diffusion-limited currents at low H_2O_2 release.[25]) The ORR voltammograms and half-wave potentials for various metal loading on the carbon support are shown in Fig. 7. The half-wave potential reaches a maximum when the metal-on-carbon loading is about 20 wt%. Increasing the metal loading from 20 to 70 wt% leads to a drop in half-wave potential. This drop coincides with a higher H_2O_2 yield, as shown in the earlier study done using a rotating ring-disk electrode.[25] In contrast, decreasing the metal loading below 20 wt% leads to a shift of the half-wave potential to lower values. At this stage, we have no data that could explain this observation.

The impact of Se content on the ORR catalytic activity was also investigated. Fig. 8 shows half-wave potentials (for ORR) *vs.* the Se content for a series of Se/Ru(Se)/C catalysts. An optimum ORR activity is observed when the bulk Se content in the sample is about 37 at%. It is worth mentioning that the optimal values close to 30 to 35 at% Se have been reported for Se/Ru(Se)/C catalysts prepared by other synthetic routes.[10,23] For the catalysts prepared by modifying Ru by Se, the optimal Se content for the sample has been somewhat lower—about 30 at%.[17,32,33] In this latter case, the Se atoms are present on the catalyst surface only.

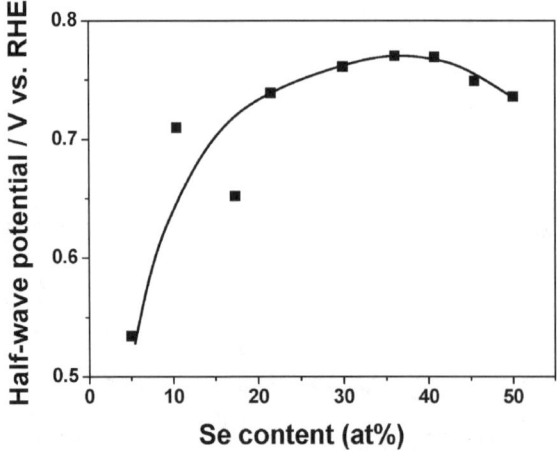

Fig. 8 Optimization of Se content in the Se/Ru(Se)/C catalyst. Half-wave potentials of ORR curves (recorded in the O_2-saturated solution 0.5 M H_2SO_4 at 5 mV s^{-1}, 1600 rpm) are presented as a function of Se content (in atomic percentage).

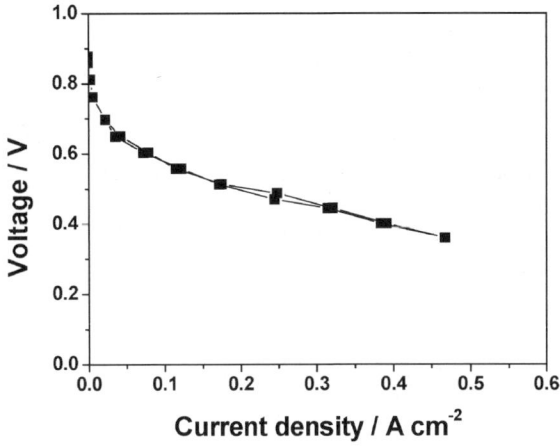

Fig. 9 Voltage–current density plot for MEA with Se/Ru(Se)/C 20 at% measured after conditioning overnight at 0.10 V. Cathode: 1.1 mg cm^{-2} in 20 wt% Se/Ru(Se)/C; anode: 0.25 mg cm^{-2} Pt (catalyzed ELAT carbon cloth). H$_2$: 300 sccm, 30 psi backpressure, 105 °C humidification temperature. Air: 500 sccm, 30 psi backpressure, 90 °C humidification temperature. Cell temperature was 80 °C.

As shown above, the optimized Se/Ru(Se)/C catalyst based on the RDE data had 20 wt% metal loading on carbon with the Ru : Se composition of 64 : 36 at%. This composition was selected for further testing as the cathode of a PEMFC. The nominal loading of this catalyst (Se/Ru(Se)/C) on the MEA was 1.1 mg cm^{-2}, corresponding to a Ru loading of 0.15 mg cm^{-2}. Fig. 9 shows the fuel cell polarization curve. The Nafion® 1135 sandwich used here had a relatively high resistance of 0.18 Ω cm^2, so the data were corrected by 0.10 Ω cm^2 to the resistance value expected for a single, standard Nafion® 112/212 membrane. The cell current densities measured at 0.40 and 0.50 V were 0.32 and 0.17 A cm^{-2}, respectively. These results are consistent with those reported for Se/Ru(Se)/C catalysts but still lower than those reported for Pt.[15,38,39] Wipperman *et al.* have compared the polarization curves obtained in DMFC using Se/Ru(Se)/C catalyst with different carbon support.[16] They reported 53 and 45 mA cm^{-2} at 0.40 V *vs.* RHE for Vulcan XC72 and BP2000, respectively. Thus, the performance of Se/Ru(Se)/C catalysts appears to depend on the choice of carbon support. In future work, some other carbon supports *vs.* that used in this study will be screened for potentially improved ORR activity of the catalyst.

Conclusion

An aqueous route for the synthesis of Se/Ru(Se)/C catalysts was explored, using NaBH$_4$ as a reducing agent for RuCl$_3$ and SeO$_2$ in basic media. NaBH$_4$ was stabilized in NaOH solution, allowing for the formation of homogenous nanoparticles as precipitates of the metallic hydroxides. The mild synthetic conditions at low temperatures (RT or 80 °C) led to a product with a particle size of around 2 nm. During heat-treatment at 500 °C, however, particle grain growth occurred resulting in larger 6–8 nm crystallites. The heat-treatment and the presence and/or absence of NaOH in the synthesis bath did not affect the catalytic activity of the catalysts, as measured by the ORR half-wave potential using RDE. Both the as-prepared and heat-treated Se/Ru(Se)/C catalysts exhibited high oxygen reduction activity with a half-wave potential close to 0.78 V *vs.* RHE. PEFC performance of our catalyst is comparable to that observed for other Ru-based chalcogenide catalysts. Further work is currently underway to optimize the catalyst for fuel cell applications.

Acknowledgements

The support from Los Alamos National Laboratory *via* the US Department of Energy (DOE) Office of Hydrogen, Fuel Cells & Infrastructure Technologies is gratefully acknowledged. AB acknowledges the National Science and Engineering Research Council of Canada for a postdoctoral fellowship. XRD, TEM and EDS analyses were carried out in the Frederick Seitz Materials Research Laboratory Central Facilities, University of Illinois, which is partially supported by the US Department of Energy under grants DE-FG02-07ER46453 and DE-FG02-07ER46471.

References

1. D. M. Bernardi and M. W. Verbrugge, *J. Electrochem. Soc.*, 1992, **139**, 2477–2491.
2. H. A. Gasteiger, S. S. Kocha, B. Sompalli and F. T. Wagner, *Appl. Catal., B*, 2005, **56**, 9–35.
3. K. Wiesener, D. Ohms, V. Neumann and R. Franke, *Mater. Chem. Phys.*, 1989, **22**, 457–475.
4. N. Alonso-Vante, *Fuel Cells*, 2006, **6**, 182–189.
5. L. Zhang, J. Zhang, D. P. Wilkinson and H. Wang, *J. Power Sources*, 2006, **156**, 171–182.
6. L. Liu, J.-W. Lee and B. N. Popov, *J. Power Sources*, 2006, **162**, 1099–1103.
7. L. Liu, H. Kim, J.-W. Lee and B. N. Popov, *J. Electrochem. Soc.*, 2007, **154**, A123–A128.
8. N. Alonso-Vante and H. Tributsch, *Nature*, 1986, **323**, 431–432.
9. N. Alonso-Vante, W. Jaegermann, H. Tributsch, W. Hoenle and K. Yvon, *J. Am. Chem. Soc.*, 1987, **109**, 3251–3257.
10. V. Le Rhun and N. Alonso-Vante, *J. New Mat. Electrochem. Syst.*, 2000, **3**, 333–338.
11. N. Alonso-Vante, in *Catalysis and electrocatalysis at nanoparticle surfaces.*, ed. A. Wieckowski, E. R. Savinova and V. G. Constantinos, Marcel Dekker, Inc., New York, Basel, 2003, pp. 931–958.
12. D. Cao, A. Wieckowski, J. Inukai and N. Alonso-Vante, *J. Electrochem. Soc.*, 2006, **153**, A869–A874.
13. L. Colmenares, Z. Jusys and R. J. Behm, *J. Phys. Chem. C*, 2007, **111**, 1273–1283.
14. J. M. Ziegelbauer, A. F. Gulla, C. O'Laoire, C. Urgeghe, R. J. Allen and S. Mukerjee, *Electrochim. Acta*, 2007, **52**, 6282–6294.
15. C. Cremers, M. Scholz, W. Seliger, A. Racz, W. Knechtel, J. Rittmayr, F. Grafwallner, H. Peller and U. Stimming, *Fuel Cells*, 2007, **7**, 21–31.
16. K. Wippermann, B. Richter, K. Klafki, J. Mergel, G. Zehl, I. Dorbandt, P. Bogdanoff, S. Fiechter and S. Kaytakoglu, *J. Appl. Electrochem.*, 2007, **37**, 1399–1411.
17. G. Zehl, G. Schmithals, A. Hoell, S. Haas, C. Hartnig, I. Dorbandt, P. Bogdanoff and S. Fiechter, *Angew. Chem., Int. Ed.*, 2007, **46**, 7311–7314.
18. O. Solorza-Feria, K. Ellmer, M. Giersig and N. Alonso-Vante, *Electrochim. Acta*, 1994, **39**, 1647–1653.
19. H. Cheng, W. Yuan and K. Scott, *Electrochim. Acta*, 2006, **52**, 466–473.
20. N. Alonso-Vante, I. V. Malakhov, S. G. Nikitenko, E. R. Savinova and D. I. Kochubey, *Electrochim. Acta*, 2002, **47**, 3807–3814.
21. K. Nagabhushana, E. Dinjus, H. Bönnemann, V. Zaikovskii, C. Hartnig, G. Zehl, I. Dorbandt, S. Fiechter and P. Bogdanoff, *J. Appl. Electrochem.*, 2007, **37**, 1515–1522.
22. V. I. Zaikovskii, K. S. Nagabhushana, V. V. Kriventsov, K. N. Loponov, S. V. Cherepanova, R. I. Kvon, H. Bonnemann, D. I. Kochubey and E. R. Savinova, *J. Phys. Chem. B*, 2006, **110**, 6881–6890.
23. C. V. Rao and B. Viswanathan, *J. Phys. Chem. C*, 2007, **111**, 16538–16543.
24. S. A. Campbell, *US Patent*, 2004, US 2004/0096728 A0096721.
25. A. Bonakdarpour, C. Delacote, R. Yang, A. Wieckowski and J. R. Dahn, *Electrochem. Commun.*, 2008, **10**, 611–615.
26. P. Piela, C. Eickes, E. Brosha, F. Garzon and P. Zelenay, *J. Electrochem. Soc.*, 2004, **151**, A2053–A2059.
27. S. Fiechter, I. Dorbandt, P. Bogdanoff, G. Zehl, H. Schulenburg, H. Tributsch, M. Bron, J. Radnik and M. Fieber-Erdmann, *J. Phys. Chem. C*, 2007, **111**, 477–487.
28. A. Lewera, J. Inukai, W. P. Zhou, D. Cao, H. T. Duong, N. Alonso-Vante and A. Wieckowski, *Electrochim. Acta*, 2007, **52**, 5759–5765.
29. F. Dassenoy, W. Vogel and N. Alonso-Vante, *J. Phys. Chem. B*, 2002, **106**, 12152–12157.
30. N. Alonso-Vante, H. Tributsch and O. Solorza-Feria, *Electrochim. Acta*, 1995, **40**, 567–576.
31. T. J. Schmidt, U. A. Paulus, H. A. Gasteiger, N. Alonso-Vante and R. J. Behm, *J. Electrochem. Soc.*, 2000, **147**, 2620–2624.

32 L. Colmenares, Z. Jusys and R. J. Behm, *Langmuir*, 2006, **22**, 10437–10445.
33 J. Inukai, D. Cao, A. Wieckowski, K. C. Chang, A. Menzel, V. Komanicky and H. You, *J. Phys. Chem. C*, 2007, **111**, 16889–16894.
34 P. Kulesza, K. Miecznikowski, B. Baranowska, M. Skunik, A. Kolary-Zurowska, A. Lewera, K. Karnicka, M. Chojak, I. Rutkowska, S. Fiechter, P. Bogdanoff, I. Dorbandt, G. Zehl, R. Hiesgen, E. Dirk, K. Nagabhushana and H. Boennemann, *J. Appl. Electrochem.*, 2007, **37**, 1439–1446.
35 N. Bogolowski, T. Nagel, B. Lanova, S. Ernst, H. Baltruschat, K. Nagabhushana and H. Boennemann, *J. Appl. Electrochem.*, 2007, **37**, 1485–1494.
36 D. C. Papageorgopoulos, F. Liu and O. Conrad, *Electrochim. Acta*, 2007, **53**, 1037–1041.
37 K. Suarez-Alcantara, A. Rodriguez-Castellanos, R. Dante and O. Solorza-Feria, *J. Power Sources*, 2006, **157**, 114–120.
38 C. Christenn, G. Steinhilber, M. Schulze and K. Friedrich, *J. Appl. Electrochem.*, 2007, **37**, 1463–1474.
39 A. Kolary-Zurowska, A. Zieleniak, K. Miecznikowski, B. Baranowska, A. Lewera, S. Fiechter, P. Bogdanoff, I. Dorbandt, R. Marassi and P. Kulesza, *J. Solid State Electrochem.*, 2007, **11**, 915–921.

PAPER

Size and composition distribution dynamics of alloy nanoparticle electrocatalysts probed by anomalous small angle X-ray scattering (ASAXS)†

Chengfei Yu,[a] Shirlaine Koh,[a] Jennifer E. Leisch,[b] Michael F. Toney[b] and Peter Strasser*[a]

Received 30th January 2008, Accepted 20th March 2008
First published as an Advance Article on the web 20th August 2008
DOI: 10.1039/b801586d

Anomalous small angle X-ray scattering (ASAXS) is shown to be an ideal technique to investigate the particle size and particle composition dynamics of carbon-supported alloy nanoparticle electrocatalysts at the atomic scale. In this technique, SAXS data are obtained at different X-ray energies close to a metal absorption edge, where the metal scattering strength changes, providing element specificity. ASAXS is used to, first, establish relationships between annealing temperature and the resulting particle size distribution for $Pt_{25}Cu_{75}$ alloy nanoparticle electrocatalyst precursors. The Pt specific ASAXS profiles were fitted with log-normal distributions. High annealing temperatures during alloy synthesis caused a significant shift in the alloy particle size distribution towards larger particle diameters.
Second, ASAXS was used to characterize electrochemical Cu dissolution and dealloying processes of a carbon-supported $Pt_{25}Cu_{75}$ electrocatalyst precursor in acidic electrolytes. By performing ASAXS at both the Pt and Cu absorption edges, the unique power of this technique is demonstrated for probing composition dynamics at the atomic scale. These ASAXS measurements provided detailed information on the changes in the size distribution function of the Pt atoms and Cu atoms. A shift in the Cu scattering profile towards larger scattering vectors indicated the removal of Cu atoms from the alloy particle surface suggesting the formation of a Pt enriched Pt shell surrounding a Pt–Cu core.
Together with XRD and TEM, ASAXS is proposed to play an increasingly important role in the mechanistic study of degradation phenomena of alloy nanoparticle electrocatalysts at the atomic scale.

Introduction

Metal nanoparticles are of fundamental scientific and technological interest. Their nanoscale dimensions give rise to fundamentally new optical, electronic, magnetic or catalytic properties, which, for the most part, are sensitively dependent on their

[a] *Department of Chemical and Biomolecular Engineering, University of Houston, Houston, TX, 77204-4004, USA*
[b] *Stanford Synchrotron Radiation Laboratory, Stanford Linear Accelerator Center, Menlo Park, CA, 94025, USA*

† The HTML version of this article has been enhanced with colour images.

structural characteristics on the nanoscale, such as their particle size and shape.[1-3] Metal nanoparticle ensembles rarely have well defined monodisperse size or shape, but instead, are generally characterized by a particle size distribution.

In the field of surface catalysis, metal nanoparticle ensembles play a tremendously important role as active catalysts. Their nanoscale dimensions result in a very high reactive surface area per unit mass, which can significantly improve the product yield of nanoparticle catalysts on a per mass basis.[4] For instance, in fuel cell electrocatalysis, a significant improvement in Pt-mass based current density can be achieved by the use of Pt nanoparticles supported on highly porous carbons compared to unsupported Pt black catalysts.[5] Furthermore, nanoscale metal nanoparticles exhibit a large fraction of low-coordinated atoms on surface edges and kinks. These atoms often exhibit different chemisorption energies of reaction intermediates compared to higher-coordinated atoms on smooth metal surfaces. Particle size and shape, together with the metal crystal structure, also determine which single crystal facets are exposed and the ratio of these exposed faces.[6] As a result of this, particle size may become strongly correlated with the observed catalytic surface reaction rate (particle size effect).[6-10]

Just like with smooth (*e.g.* single crystal) metal surfaces, surface catalytic reactivity of metal nanoparticles can be tuned by alloying two or more metals. Alloying results in electronic (ligand), geometric, or ensemble effects which modify the surface catalytic activity. Pt nanoparticle electrocatalysts show improved reactivity for the electrooxidation of CO when alloyed with Ru due to a bifunctional mechanism.[11-13] Similarly, when alloyed with transition metals, Pt nanoparticles exhibit more favorable electrocatalytic rates for the electroreduction of molecular oxygen (ORR), likely due to a combination of geometric and ligand effects.[14-18] Due to synthetic imperfections, the particle molar composition of alloy nanoparticle ensembles, just like their size, is characterized by a distribution rather than by a single value. Knowledge of this size–composition distribution is important for characterizing and understanding the function of the nanoparticle catalysts.

Metal alloy nanoparticle electrocatalysts, with all their advantages with respect to surface reactivity, come with serious challenges: they are often unstable under catalytic reaction conditions and tend to change their size and composition distributions.[19-22] Particle size distribution changes are caused by a number of factors. Metal nanoparticles show a reduced melting point compared to the bulk, show pronounced surface diffusion, and suffer from thermodynamic instabilities in the presence of larger particles (Ostwald ripening). Ostwald ripening is the dissolution of the smaller particles to become even smaller and the growth of the larger particles, while diffusion leads to particle coalescence. In addition, in an electrocatalytic system, metal dissolution occurs at high electrode potentials resulting in the loss of active metal surface atoms. As a consequence of these three mechanisms, the mean particle size changes during the course of an electrocatalytic reaction, typically resulting in a reduced specific surface area. All three particle instability mechanisms have been experimentally observed for Pt and Pt alloy electrocatalysts using voltammetric and microscopic techniques,[19,20,22-24] and simple models have been developed to capture the dynamics of particle size changes.[25,26] Much research is currently being dedicated to mitigation strategies to stabilize the size and composition of alloy nanoparticle ensembles.[27-32]

Very little work to date has been dedicated to the understanding of the dynamics of composition distributions of alloy nanoparticle electrocatalysts. In principle, one is looking at time-resolved measurements of alloy particle composition and alloy particle size. Another way to approach this question is through metal-specific particle size distributions. Fig. 1 schematically illustrates atom-specific size distributions for a binary Pt–M alloy nanoparticle ensemble. The solid distribution curve represents the number fraction of Pt atoms as function of particle radius. Small particles consist entirely of Pt atoms. The dashed curve represents the number fraction of a base metal M. Large particles are enriched in M, hence the distribution

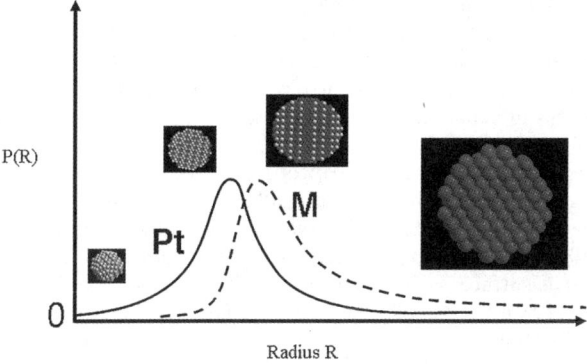

Fig. 1 Schematic illustration of atom-specific distribution functions for Pt atoms and M atoms for a binary Pt–M alloy nanoparticle ensemble with non-uniform composition distribution. Large particles are assumed to be M rich, while small particles are Pt rich.

curve of M lies above that of Pt. If both curves coincide, this would indicate a perfectly alloyed particle ensemble where small and large particles have identical composition. The distribution curves in Fig. 1 represent the compositional picture at some given time. Probing composition dynamics requires repeated measurement of metal atom specific distribution curves at different times.

The most common experimental methods to probe nanometre scale particle size and composition distributions are transmission electron microscopy (TEM) and X-ray scattering techniques.[33,34] While the former can provide real space images, the latter are reciprocal-space techniques and measure the Fourier transform of the electron densities. Suitable scattering techniques comprise the wide angle scattering (X-ray diffraction, XRD) as well as the small angle X-ray scattering (SAXS).[35–37]

Each technique has its advantages and limitations. TEM offers the appeal of images of nanoparticles and (under favorable conditions) individual atoms, yet based on the relatively small particle population analyzed (at best in the thousands) TEM often yields unreliable statistics. TEM particle histograms often suffer from subjective choices that the experimenter makes during the analysis process; for instance, overlooking very small particles and therefore skewing distribution functions towards larger sizes. Based on the interaction of X-ray radiation with atom electron densities, X-ray diffraction offers excellent statistics based on a larger sampling volume (typically more than billions of particles). Mean size estimates are available from analysis of the diffraction peak full width at half maximum (FWHM). However, the mean size estimate refers to crystallite size, which is not necessarily the same as the particle size; this can result in discrepancies between TEM and XRD. Also, being a volume based sampling method, XRD provides volume-averaged sizes, not number averaged sizes as for TEM. SAXS has been used in the past to follow the growth dynamics during the synthesis of Pt nanoparticles.[35] It has also been applied to study the surface oxidation processes of pure Pt electrocatalyst during voltammetric studies.[36] Virtually no work exists on the use of SAXS to study the structural dynamics of alloy nanoparticles.[38]

In this contribution, we aim to highlight the power of SAXS techniques for an *ex situ* and *in situ* probing of the structural and compositional dynamics of Pt alloy nanoparticle electrocatalysts. SAXS is ideally suited for scattering objects in the 1–100 nm range. Performed in the "anomalous" scattering mode (ASAXS), it combines the advantages of diffraction in terms of statistical quality with the power of atom-specific size distributions, according to Fig. 1. In the anomalous scattering mode, the SAXS is measured at several energies near a metal absorption edge where the scattering strength of the metal changes.

We will illustrate the power of SAXS for probing particle structure and composition using a class of Cu-rich Pt–Cu alloy electrocatalysts which were recently reported to offer unprecedented catalytic activity for the electroreduction of oxygen in acidic media.[39,40] We investigated two aspects about this catalyst class.

In a first series of experiments, we correlate alloy particle size with the synthesis conditions of the alloy catalysts.

In a second set of experiments, we probe changes in the alloy particle size and composition during an electrochemical dealloying procedure. The active phase of this particular catalyst system is believed to be formed during electrochemical dealloying of Cu atoms prior to catalytic testing: selective Cu atom dissolution from the particle surfaces is assumed to form a nanoparticle with a Pt rich surface region surrounding a Cu rich alloy particle core (core-shell particle).[40–42] We demonstrate that ASAXS can help clarify the formation dynamics of such structures.

Continued base metal dissolution from Pt alloy particle electrocatalysts under operating conditions[21,43–45] has also often been linked to activity degradation. The present work, therefore, also aims to showcase ASAXS as a suitable experimental technique for a fundamental characterization of alloy particle electrocatalyst degradation behavior during reaction conditions. An atomic-scale understanding of the dynamics of particle size and composition distributions of alloy electrocatalyst degradation would be of tremendous importance for progress in the design and development of corrosion stable electrocatalyst systems.

Experimental

Catalyst synthesis

$Pt_{25}Cu_{75}$ alloy nanoparticle electrocatalyst precursors were synthesized *via* the liquid salt precursor impregnation method. Measured amounts of a Cu nitrate $(Cu(NO_3)_2 \cdot 3H_2O$, Aldrich) were dissolved in de-ionized water (>18.2 MΩ, Milli-Q® gradient system, Millipore Inc.). The precursor solutions were then added to a weighted amount of commercial carbon-supported Pt nanoparticle electrocatalyst (Tanaka Kikinzoku Inc.) with a Pt weight loading of about 30 wt%. The mixtures were then ultrasonicated with a sonifier horn (Branson) for 1 min and then frozen in liquid nitrogen for 5 min. Subsequently, the frozen samples were freeze-dried (Labconco) for 24 h and annealed in flowing hydrogen/argon mixtures in a flow furnace (Lindberg Blue) to a temperature between 600 °C and 950 °C.

Catalyst film electrode preparation

Alloy nanoparticle electrocatalyst precursor electrode films were prepared by depositing a dilute suspension of the catalyst powder synthesized as described above onto inch long carbon tape (3M). The catalyst suspension was made by mixing catalyst powder (2.5 mg) with 100 μL de-ionized water (>18.2 MΩ, Millipore Gradient system) containing 5 wt% Nafion® solution (Sigma, #274704). The suspension was ultrasonicated for 1 minute before being deposited onto the carbon tape and then air dried for 3 hours.

Electrochemical leaching

Size and compositional changes of the catalytic particles were probed *ex situ* after voltammetric protocols using anomalous SAXS. The measurements were carried out on the catalyst film electrodes before and after the following electrochemical protocol: the catalyst films were employed as working electrode in a three electrode half cell set up. A Pt wire served as counter, and a Ag/AgCl microelectrode as reference electrode. Potentials are referenced to the reversible hydrogen electrode (RHE). A constant electrode potential of +1.2 V *vs.* RHE was applied for 5 hours.

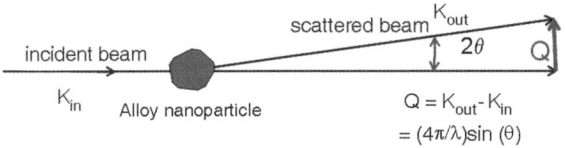

Fig. 2 Schematic experimental set up of a SAXS experiment.

ASAXS

SAXS is a technique to probe structural features of colloidal size. In a SAXS experiment, the X-ray scattering intensity is measured in the small angle regime; here the scattering arises from electron density inhomogeneities over length scales of 1–100 nm[33,46] (see Fig. 2). SAXS, in combination with model-based data analysis, enables the study of size distributions of nanoparticle ensembles as well as temporal changes thereof. If scattering profiles are collected at different X-ray energies, the method is referred to as anomalous small angle X-ray scattering (ASAXS). ASAXS measurements were performed in transmission at the Stanford Synchrotron Radiation Laboratory (SSRL) beam line 4–2 using a MAR165 CCD detector. To obtain information on the platinum in the catalyst samples, SAXS measurements were conducted at two energies, one that is close to the Pt L_3 absorption edge ($E_{L3} = 11564$ eV) at $E_1 = 11551$ eV (wavelength (λ) of 0.1074 nm) and a second energy far from this absorption edge at $E_2 = 11450$ eV ($\lambda = 0.1083$ nm). Similarly, for copper, SAXS measurements were conducted near Cu K absorption edge ($E_K = 8979$ eV) at $E_1 = 8975$ eV ($\lambda = 0.1382$ nm) and far away at $E_2 = 8880$ eV ($\lambda = 0.1397$ nm).

"Fit2D" software[47,48] was used to reduce the raw scattering data. TIFF images obtained from the ASAXS measurements were integrated in azimuthal angle, resulting in intensity-scattering vector ($I(Q)$, with $Q = (4\pi/\lambda) \sin \theta$, see Fig. 2) plots. Thereafter, the dark signal was subtracted from these and the data corrected for sample thickness and incident photon flux changes during the measurements.

XRD

X-ray measurements are performed on a Siemens D5000 ($\theta/2\theta$) Diffractometer equipped with a Cu target and a Braun position sensitive detector (PSD) with an angular range of 8°. The X-ray generator operated at a potential of 35 kV and a current of 30 mA. The data were collected from 21° to 70° 2θ, using step scans of 0.02° per step and a holding time of 10 s per step.

Results and discussion

The ASAXS experiments described here focused on the class of Cu-rich $Pt_{25}Cu_{75}$ alloy nanoparticle electrocatalysts, which were recently reported to be highly efficient catalysts for the electroreduction of molecular oxygen in acidic environments.[39–42,49–53] The active phase of the catalysts is formed by means of a voltammetric surface dealloying process during which a large fraction of the Cu surface atoms are dissolved leaving a multi-layer Pt enriched particle shell behind that surrounds an Pt–Cu alloy particle core.[40]

Synthesis–structure relationships of Pt alloy nanoparticle ensembles

In a first series of scattering experiments, the dependence of the particle size distribution on the synthesis conditions was investigated. In particular, the effect of the annealing temperature on the resulting particle size and structure was studied. Fig. 3 shows X-ray diffraction profiles of the three $Pt_{25}Cu_{75}$ catalysts considered here. Annealing at 600 °C resulted in incomplete alloying of Cu and Pt atoms. An excess pure face-centered cubic (fcc) Cu phase with sharp reflections is discernible

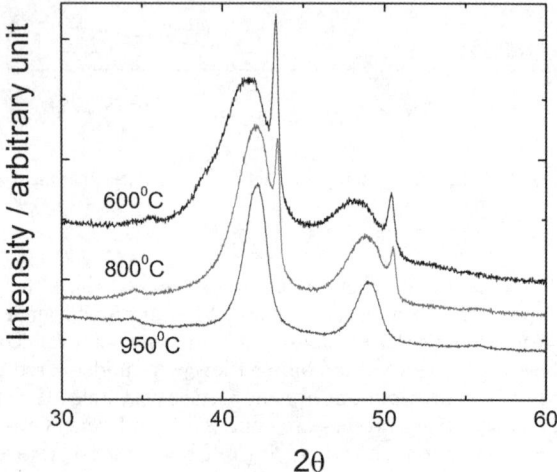

Fig. 3 X-ray diffraction patterns of three $Pt_{25}Cu_{75}$ nanoparticle electrocatalyst precursors. Annealing temperatures of each catalyst are given in the graph.

in Fig. 3 at $2\theta = 43.3°$. The Pt–Cu alloy phase formed in this synthesis process has an estimated composition of $Pt_{55}Cu_{45}$ (see Table 1) as evaluated from the alloy (111) reflection position using Vegard's relation.[54,55] Increased annealing temperatures result in more complete alloying, as shown in Fig. 3, with the 950 °C sample showing no signs of a pure Cu phase. Accordingly, the (111) fundamental reflection of the alloy phase is shifted to larger 2θ, indicating a smaller fcc unit cell parameter (Table 1).

In the ASAXS measurements the scattering intensity, $I_{Pt}(Q,E)$, of a monodisperse ensemble of randomly oriented spherical Pt particles of radius R in vacuum at an incident X-ray energy E is given as[33,35,36,46,56–58]

$$I_{Pt}(Q,E) = N_p \, (n_{Pt} \, f_{Pt}(E))^2 V^2(R) \, F^2(Q,R) \qquad (1)$$

With Q, N_p, n_{Pt}, $f_{Pt}(E)$, $V(R)$, and $F(Q,R)$ denoting the modulus of the scattering vector, the particle density in the sample, the Pt atomic density in the particles, the atomic form factor (scattering factor) of Pt atoms, the volume of any one particle, and the particle form factor given as

$$F(Q,R) = \frac{3[\sin QR - QR\cos QR]}{[QR]^3} \qquad (2)$$

If the Pt particles are embedded in a matrix, for instance a solid polymer, the scattering signal is a result of the difference in the atomic form factor of Pt atoms and the

Table 1 Nanoparticle catalyst compositions, synthesis conditions, alloy phase composition estimates, mean crystallite and mean particle diameter data of $Pt_{25}Cu_{75}$ alloy particles as well as the R^2 factors of the fitting in Fig. 4a

Overall catalyst composition (%)	Annealing temperature/°C	XRD Lattice Parameter $d(111)$/Å	Vegard alloy phase composition (%)	XRD mean crystallite size/nm	ASAXS mean particle diameter/nm	Fitting R^2
Pt 25 Cu 75	600	2.164	Pt 55 Cu 45	2.82	2.95	0.9839
Pt 25 Cu 75	800	2.146	Pt 45 Cu 55	4.07	3.51	0.9834
Pt 25 Cu 75	950	2.141	Pt 40 Cu 60	5.74	4.65	0.9888

surrounding matrix atoms. The term $(n_{Pt} f_{pt}(E) - n_m f_m(E))^2$ replaces the term $(n_{Pt}f_{Pt})^2$ as the source of the scattering contrast in eqn (1), with n_m and f_m denoting the average atomic density and atomic form factor of the matrix atoms. As long as the scattering power of Pt atoms is different from those of the matrix, the contrast is non-vanishing and a scattering signal will be obtained. Fortunately, the scattering atomic form factor of Pt is many times larger compared to those of lighter elements, such as carbon, oxygen or hydrogen, and so SAXS is strong

If the Pt particles are supported on a porous material, the SAXS from the pores $I_{pores}(Q)$ contributes to the overall scattering as

$$I_{total}(Q,E) = I_{Pt}(Q,E) + I_{pores}(Q) \tag{3}$$

The interpretation of SAXS $I(Q)$ curves becomes non-trivial as the unknown scattering contribution of the support is hard to separate from that of the particles.

ASAXS was used to extract the signal of the Pt particles. Near an absorption edge, the Pt scattering intensity is a strong function of the X-ray energy, and the atomic form factor of Pt $f_{Pt}(E)$ changes by up 20% in magnitude[35] while the scattering factors of the matrix atoms, and hence the scattering signal from the pores, are unchanged. This is how element specificity is achieved. We chose $E_1 = 11.551$ keV and $E_2 = 11.450$ keV in our measurements. Subtracting the scattering profiles $(I(Q,E_2) - I(Q,E_1))$ results in a "net" scattering profile $I_{Pt}(Q)$ which represents only the SAXS of the Pt particles.

Similar measurements were carried out near the Cu absorption edge ($E_1 = 8.975$ keV, $E_2 = 8.880$ keV) to obtain $I_{Cu}(Q)$ associated with the Cu only. The ASAXS method cannot distinguish whether or not the Cu atoms are alloyed. If a non-monodisperse particle ensemble is considered as characterized by a particle size distribution, $P(R)$, the scattering intensity for spherical particles becomes

$$I_{Pt}(Q,E) = N_p (n_{Pt} f_{Pt}(E))^2 \int P(R) V^2(R) F^2(Q,R) \, dR \tag{4}$$

Under these assumptions, we can then extract an element-specific particle size distribution for the Pt atoms and Cu atoms according to the discussion in the Introduction (see Fig. 1). Since the particle loading is small in our samples, there is no inter-particle scattering.

Fig. 4a reports the Pt-specific SAXS of the three carbon-supported $Pt_{25}Cu_{75}$ alloy nanoparticle electrocatalyst precursors measured at the Pt edge. To obtain these measured curves, the scattering $(I-Q)$ profiles for the two separate energies were subtracted from each other. For a direct comparison, the three resulting subtracted profiles are shifted in the y-direction in order to normalize their intensity at low Q. Using the relation

$$Q = \frac{2\pi}{d} \tag{5}$$

the Q range reported in Fig. 4a is estimated to span particle diameters d in the 1.5–10 nm range. Larger or smaller objects do not have significant scattering in the chosen Q range.

Fig. 4a reveals that all three scattering profiles start out with a flat portion at low Q, and show a gradual drop in scattering intensity for larger Q. The $Pt_{25}Cu_{75}$ nanoparticle ensemble annealed at 600 °C exhibits larger scattering intensity than the other two nanocatalysts over the entire Q range. Scattering intensity at large Q is seen to drop in the order 600 °C > 800 °C > 950 °C annealing temperatures. The trends in scattering intensities as a function of scattering vector indicate that the particle ensemble annealed at 600 °C consists of smaller particles compared to the other two particle ensembles. Thus, annealing temperature is correlated with the alloy particle diameter.

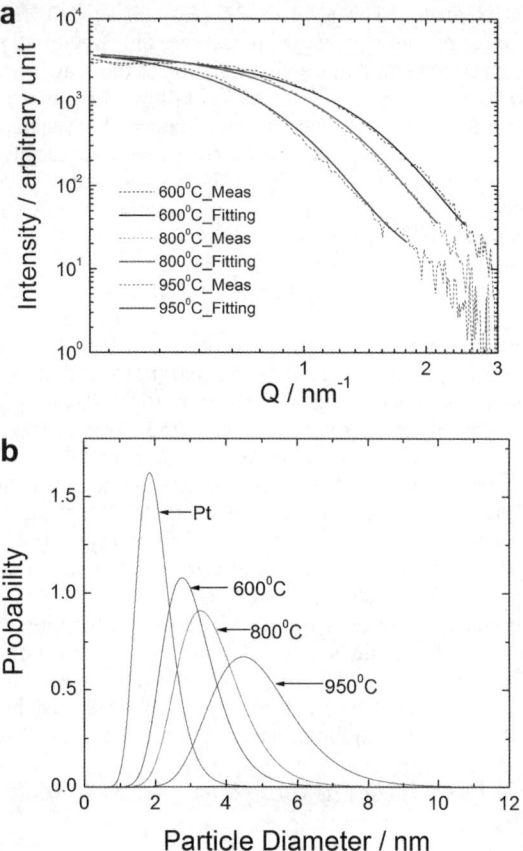

Fig. 4 (a) Measured subtracted ASAXS intensities ($E_2 - E_1$) ("Meas") and intensity fits using log-normal distribution functions ("Fittings") of a $Pt_{25}Cu_{75}$ alloy nanoparticle electrocatalyst precursor prepared at the three different annealing temperatures indicated in the legend. (b) Particle size distribution functions $P(R)$ for each of the Pt–Cu nanoparticle electrocatalyst precursors in (a) in comparison to the pure Pt nanoparticle catalyst used as a precursor in the preparation of the three alloys. The temperatures indicate the annealing conditions of each catalyst.

To more quantitatively determine the nanoparticle size distribution and to determine the mean particle diameter, the scattering profiles of Fig. 4a (dashed lines) were fitted with log-normal distribution function[56,59] for $P(R)$ (see eqn (4)):

$$P(R) = \exp\left[-\frac{1}{2}\frac{\left(\ln\frac{R}{R_0}\right)^2}{\sigma^2}\right]\frac{\exp[-\frac{1}{2}\sigma^2]}{R\sigma\sqrt{2\pi}} \quad (6)$$

where R_0 denotes the mean particle radius of the distribution and σ is the log-normal dispersion or width in size. Fig. 4a compared the measured intensities with the fitted intensities (dashed and solid lines, respectively). With this choice of distribution, the FWHM depends on both R_0 and σ. While there is no *a priori* argument why a log-normal distribution should fit experimental SAXS results better than, say, a Gaussian distribution, for many cases they often empirically provide good fits.

The dispersion of the function $P(R)$ was set to a constant value of $\sigma = 0.25$ during the fitting of the mean radius R_0 in order to increase the accuracy of the fit over the chosen Q range. Typical correlation factors R^2 of the curve fittings were in the range of 0.98–1.0 (see Table 1 last column for fittings in Fig. 4a), and thus indicated a very good fit to the experimental data. Table 1 (second column from right) summarizes the fitted mean particle diameters, and Fig. 4b shows the resulting distribution functions $P(R)$ for the three nanoparticle catalysts. Fig. 4b also includes ASAXS measurements of the carbon supported pure Pt precursor catalyst which was used as starting material in the synthesis of the alloy catalysts. One can see that increasing the annealing temperatures results in a significant shift of the size distribution toward larger diameters. Thus, alloy preparation at higher temperatures resulted in an increase in the resulting mean particle size. The drop in intensity of the distribution maximum is a consequence of the chosen form for $P(R)$ and does not reflect any absolute particle number information. The chosen expression for $P(R)$ causes the distribution FWHM to increase with increasing average size, even though the dispersion σ remained constant for all three fits. $P(R)$ is the number fraction of particles of a given radius (R).

This is one of the first reports where ASAXS has been applied to study size distributions of bimetallic Pt alloy nanoparticle electrocatalysts. Dahn et al.[60,61] and Haubold et al.[36,62] reported SAXS studies on carbon-supported pure Pt nanoparticles, and found that the metal loading had significant impact on the mean particle size. Benedetti et al.[63] investigated carbon-supported Au and Pd particles using ASAXS, yet did not report detailed results on their resulting size distribution functions.

Our experimental and fitting results are in excellent agreement with earlier X-ray diffraction reports on the relationship between mean particle size and thermal annealing.[40,41,43–45,64] Higher annealing temperatures generally cause increased mobility of carbon-supported metal atoms leading to a higher rate of particle coalescence and Ostwald ripening. Given a constant total amount of metal atoms at the beginning of the thermal process, one would expect fewer, but larger, particles for the sample annealed at 950 °C. Previous reports on an exponential size growth with annealing temperatures are not supported by our present data.[64] Increased heat treatment during alloy synthesis resulted in more complete alloying, as shown in Fig. 3. Also, for Pt alloys with first row transition metals, higher annealing temperature resulted in a alloy lattice contraction and a shift of the alloy (111) peak towards larger 2θ values,[64] in agreement with our present results.

A comparison of the mean crystallite sizes obtained from a Debye–Scherrer analysis with the mean particle sizes obtained from ASAXS is given in Table 1. The data suggests that the XRD based crystallite size values are slightly larger than those from ASAXS, especially for higher temperatures. This is consistent with the notion that diffraction data yields volume-weighted size values, while the ASAXS analysis, just like microscopy methods, results in number averages for the particle size. As a result of this, large particles receive increased relative weight in the ensemble average of diffraction data increasing the average slightly.

It should be noted, however, that a number of additional factors may affect the XRD based crystallite size measurements; for instance, strain effects in small particles. Non-uniform strain results in peak broadening of XRD peaks, and therefore may cause an increase in the diffraction peak widths, which, in turn, reduces the mean size estimate.[34]

Another complicating factor in the evaluation of mean crystallite sizes of alloy particles using the Scherrer equation is related to the possible presence of an unknown number of alloy phases with distinct composition. Since diffraction peak positions are sensitive to alloy composition, multiple phases in incompletely alloyed samples result in relatively broad peaks and in an underestimate of the true particle size. This could be the reason why the XRD based crystallite size is smaller than the ASAXS value for the incompletely alloyed 600 °C sample in Table 1.

The present ASAXS experiment highlights the complementary nature of X-ray diffraction and SAXS as well as the advantage of applying SAXS in an anomalous mode. While both techniques result from an interaction of X-rays with atom electron densities, ASAXS provided an accurate correction for background scattering and yielded detailed size distribution information for particles up to about 12 nm (limited by our choice of Q range). In the present study we focused on the scattering contribution from Pt atoms (Pt edge data), yet we will show in the following section that for a uniform bimetallic alloy either metal edge could have been used to determine particle size information.

Electrochemical dealloying of Pt alloy nanoparticle ensembles

In a second series of ASAXS experiments, we investigated the impact of an applied electrode potential and the application of an electrochemical protocol on the size and composition distribution of our Pt–Cu bimetallic catalysts. The nanoparticle size and composition distribution dynamics is of considerable importance for understanding particle degradation phenomena in electrocatalysis. To the best of our knowledge, this is the first time that the changes in the scattering intensity of both Cu atoms and Pt atoms have been investigated for an as-prepared and electrochemically treated (dealloyed) bimetallic Pt alloy nanoparticle electrocatalyst.

We specifically have investigated the size distribution dynamics of $Pt_{25}Cu_{75}$ nanoparticle catalysts annealed at 950 °C. The high annealing temperature ensured that the as-prepared catalyst attains a uniform alloy composition (compare results from previous section). Fig. 5 reports the subtracted Cu and Pt edge ASAXS of the as-prepared pristine alloy nanoparticle catalyst (before electrochemical treatment) and of the dealloyed electrocatalyst (after application of an electrochemical protocol). The electrochemical protocol consisted of a constant potential hold at 1.2 V vs. RHE in 0.1M perchloric acid for 5 hours at room temperature. The scattering data in Fig. 5 reveal that the scattering profiles of both the Cu and Pt ("Cu edge" and " Pt edge") almost completely overlap over the entire Q range, suggesting that the Pt and Cu specific size distribution functions are nearly the same. From our earlier discussion of Fig. 1, this indicates that the alloy nanoparticles possess a uniform composition regardless of their size. This result does not suggest that the alloy composition is $Pt_{50}Cu_{50}$, as the distribution functions are normalized.

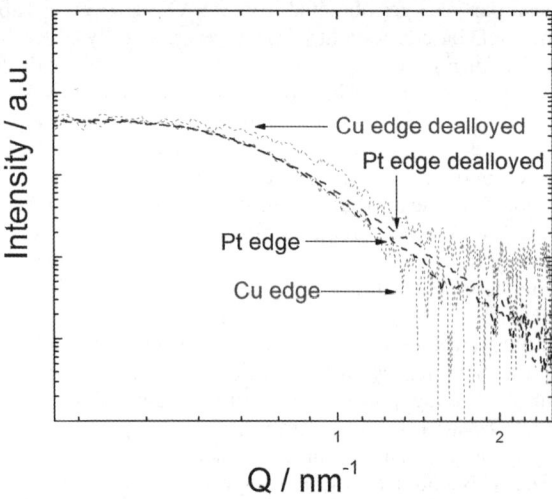

Fig. 5 ASAXS scattering profile of a $Pt_{25}Cu_{75}$ alloy nanoparticle electrocatalyst prepared at 950 °C before and after electrochemical treatment at 1.2 V vs. RHE for 5 hours in oxygen saturated 0.1 M $HClO_4$ electrolyte. "Cu edge" and "Pt edge" denote atom specific scattering profiles of the as prepared catalyst particles. "Cu edge dealloyed" and "Pt edge dealloyed" label the profiles after electrochemical treatment.

Table 2 Metal-specific mean particle diameters determined by fitting the scattering profiles in Fig. 5

	Mean particle diameter Cu edge/nm	Mean particle diameter Pt edge/nm
As prepared	4.77	4.81
After electrocatalysis	4.02	4.67

After application of the electrochemical protocol (profiles marked "Pt edge dealloyed" and "Cu edge dealloyed"), the Pt profile is slightly shifted towards larger values of Q, that is, towards smaller particle sizes, while the Cu atomic scattering profile significantly shifted toward smaller particle diameters (larger Q). Table 2 reports the detailed mean particle diameters of the fit distribution curves for each metal. The difference in size between the Pt and Cu edge for the as prepared catalyst are below the estimated 5–10% experimental error (4.77 nm and 4.81 nm for Cu and Pt, respectively). Their difference is therefore considered to be insignificant.

Similarly, the difference between the Pt particle size before and after application of the electrochemical protocol is in the order of magnitude of the experimental error and is therefore insignificant (4.81 nm to 4.67 nm). Hence, the size of the scattering particles consisting of contiguous Pt atomic density, regardless of the interspersed presence co-alloyed Cu atoms, does not change during the applied potential. The particle size distribution associated with the Cu atoms, however, drops from 4.77 nm to 4.02 nm and represents a statistically significant change in size during the applied potential protocol. These data show that the mean size of scattering particles consisting of contiguous Cu atoms is reduced after electrochemical leaching. The scattering results cannot distinguish between pure Cu particles or Cu atoms co-alloyed with other metals, such as Pt. Since no pure unalloyed Cu particles were present at the beginning of the experiments, the experimental results suggest that a portion of the alloyed Cu atoms were leached out of the initially uniformly alloyed Pt–Cu alloy particles. Fig. 6a schematically illustrates the case before electrochemical dealloying where the Cu and Pt specific particle sizes are identical, while Fig. 6b illustrates the proposed situation after electrochemical leaching of some portion of the Cu atoms from the surface of the alloy particles. Selective dissolution of Cu atoms created a Pt enriched shell surrounding a Pt–Cu alloy core (core-shell nanoparticle structure). In the core-shell structure, the effective particle size of Cu-specific scattering decreased, because only the core of the particles is participating in the scattering profile.

Our present results represent the first report where ASAXS was used to investigate atom-specific size distribution dynamics of alloy nanoparticle electrocatalysts.

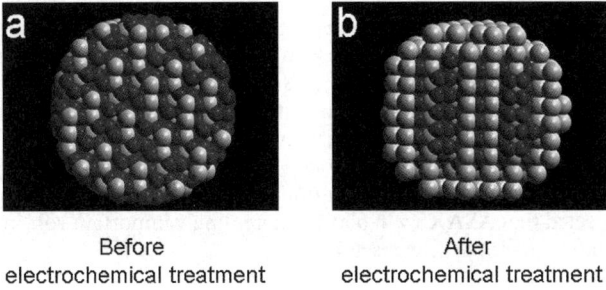

Before electrochemical treatment

After electrochemical treatment

Fig. 6 (a) Schematics of a uniformly alloyed Pt–Cu alloy nanoparticle giving rise to overlapping scattering profiles for the Cu and the Pt edge. (b) Dealloyed Pt–Cu with a Pt enriched shell surrounding a Pt–Cu core giving rise to a Cu specific scattering profile shifted towards smaller mean diameter.

Our findings indicate that ASAXS techniques can resolve structural transformations, such as metal enrichment or metal depletion during electrochemical treatment of alloy nanoparticles. Such subtle intraparticle structural features are usually difficult to detect using regular XRD techniques. Estimates of the resolution of our techniques based on our experimental error give a minimum thickness of the resulting Pt enriched shells of 0.5 nm. Previous XPS studies of dealloyed Pt–Cu nanoparticles, in fact, estimated the resulting Pt enriched shell thickness to be on the order of 1–2 nm.[40] Our results further indicate that ASAXS is insensitive to the degree of alloying of the metal atoms. Hence, the SAXS of unalloyed Cu particles and of Cu atoms inside alloy particles cannot be easily distinguished. This fact underlines the complementary nature of XRD and SAXS, with the former being able to resolve the presence of distinct alloy particle phases, and the latter yielding the detailed atom specific size distribution.

Conclusions

The dynamics of particle size and composition changes is a poorly understood issue of great importance in the area of nanoparticle electrocatalysis. The ability to probe the structural and compositional particle dynamics is pivotal to understanding catalyst degradation behavior, for instance inside fuel cell electrodes, where small alloy particle are exposed to prolonged Faradaic surface electrocatalytic processes, which are known to result in severe decreases in particle surface area.[19,20,22] Monitoring metal dissolution phenomena is also of importance to better understand the activation of alloy catalysts by electrochemical surface dealloying, which was shown to be an effective means to modify surface catalytic activity.[40]

We have demonstrated how ASAXS measurements can be used to obtain atomic-level insight in the size and composition distribution dynamics of Pt–Cu alloy nanoparticle ensembles during alloy particle synthesis and during electrocatalytic treatment of alloy nanoparticles. We found a direct relationship between the resulting mean alloy particle size and the annealing temperature, and have evidenced the loss of Cu atoms from uniform alloy nanoparticles during a constant potential leaching protocol.

ASAXS can resolve shifts in the particle size distribution function for many elements, as long as the X-ray absorption edges are accessible. Hence, it will be of great value for the study of degradation mechanisms of pure Pt alloy catalysts in low temperature fuel cells. Tuning the incident X-ray energy over wider ranges, ASAXS is also capable of providing multiple atom-specific size distribution functions. Scattering contributions from distinct metal atoms in alloy particles can thus be separated. Applied to a series of electrochemically treated samples, ASAXS can resolve the composition dynamics of alloy nanoparticle ensembles, and thus can contribute critical information to the compositional stability of alloy particles.

Outlook

Future efforts in the area of ASAXS applied to the study of degradation studies of alloy nanoparticle ensembles will focus on improved time resolution of ASAXS measurements using *in situ* techniques. *In situ* ASAXS will be able to provide the onset and the details of particle size and composition dynamics under electrochemical conditions. Combined with theoretical models for particle growth,[22,25,26] such *in situ* ASAXS data will be able to test models and provide insight into the molecular mechanisms of nanoparticle transformation and degradation, such as surface area loss. It is expected that ASAXS will play an increasingly important role in the science of electrocatalytic nanoparticle ensembles.

Acknowledgements

This project was supported by the Department of Energy, Office of Basic Energy Sciences (BES), under grant LAB04-20 *via* a subcontract with Stanford Synchrotron

Radiation Laboratory, and by the National Science Foundation, award #0729722. Acknowledgment is made to the Donors of the American Chemical Society Petroleum Research Fund for partial support of this research (grant #44165). Financial support from Houston Area Research Center (HARC) is gratefully acknowledged. Portions of this research were carried out at the Stanford Synchrotron Radiation Laboratory, a national user facility operated by Stanford University on behalf of the U.S. Department of Energy, Office of Basic Energy Sciences.

References

1. D. L. Fedlheim and C. A. Foss, *Metal Nanoparticles: Synthesis Characterization & Applications*, CRC – Taylor Francis Group, London, 2001.
2. G. Schmid. *Nanoparticles: From Theory to Applications*; Wiley-VCH, New York, 2004.
3. D. Astruc. *Nanoparticles and Catalysis*; Wiley-VCH, New York, 2007.
4. M. V. Twigg, *Catalyst Handbook*, Wolfe Publishing Ltd, London, 2nd edn, 1989.
5. *Handbook of Fuel Cells - Fundamentals, Technology, and Applications*, ed. W. Vielstich, A. Lamm and H. Gasteiger, Wiley, Chichester, UK, 2003.
6. K. Kinoshita, *J. Electrochem. Soc.*, 1990, **137**, 845.
7. *Handbook of Heterogeneous Catalysis*, ed. G. Ertl, H. Knözinger and J. Weitkamp, Wiley-VCH, Weinheim, 1997.
8. M.-k. Min, J. Cho, K. Cho and H. Kim, *Electrochim. Acta*, 2000, **45**, 4211–4217.
9. S. Mukerjee, *J. Appl. Electrochem.*, 1990, **20**, 537–548.
10. G. A. Somorjai, *Introduction to Surface Chemistry and Catalysis*, Wiley, New York, 1994.
11. M. Watanabe and S. Motoo, *J. Electroanal. Chem.*, 1975, **60**, 267–273.
12. M. Watanabe, M. Shibata and S. Motoo, *J. Electroanal. Chem.*, 1986, **206**, 197.
13. H. A. Gasteiger, N. Markovic, N. Philip, J. Ross and E. J. Cairns, *J. Phys. Chem.*, 1994, **98**, 617–625.
14. J. Lipkowski; P. N. Ross, *Electrocatalysis*, Wiley-VCH, New York, 1998.
15. S. Mukerjee, S. Srinivasan, M. P. Soriaga and J. McBreen, *J. Electrochem. Soc.*, 1995, **142**, 1409–1422.
16. V. Stamenkovic, B. S. Moon, K. J. Mayerhofer, P. N. Ross, N. Markovic, J. Rossmeisl, J. Greeley and J. K. Norskov, *Angew. Chem., Int. Ed.*, 2006, **45**, 2897–2901.
17. V. Stamenkovic, B. S. Mun, M. Arenz, K. J. J. Mayerhofer, C. A. Lucas, G. Wang, P. N. Ross and N. Markovic, *Nat. Mater.*, 2007, **6**, 241.
18. V. R. Stamenkovic, B. Fowler, B. S. Mun, G. Wang, P. N. Ross, C. A. Lucas and N. M. Markovic, *Science*, 2007, **315**, 493.
19. P. J. Ferreira, G. J. la O', Y. Shao-Horn, D. Morgan, R. Makharia, S. Kocha and H. Gasteiger, *J. Electrochem. Soc.*, 2005, **152**, A2256–2271.
20. J. Xie, D. L. Wood, K. L. More, P. Atanassov and R. L. Borup, *J. Electrochem. Soc.*, 2005, **152**, A1011–A1020.
21. M. Watanabe, K. Tsurumi, T. Mizukami, T. Nakamura and P. Stonehart, *J. Electrochem. Soc.*, 1994, **141**, 2659–2668.
22. Y. Shao-Horn, W. C. Sheng, S. Chen, P. J. Ferreira, E. F. Holby and D. Morgan, *Top. Catal.*, 2007, **46**, 285–305.
23. B. Merzougui and S. Swathirajan, *J. Electrochem. Soc.*, 2006, **153**, A2220–A2226.
24. E. Guilminot, A. Corcella, F. Charlot, F. Maillard and M. Chatenet, *J. Electrochem. Soc.*, 2007, **154**, B96–B105.
25. R. M. Darling and J. P. Meyers, *J. Electrochem. Soc.*, 2003, **150**, A1523.
26. R. M. Darling and J. P. Meyers, *J. Electrochem. Soc.*, 2005, **152**, A242.
27. D. A. Stevens, M. T. Hicks, G. M. Haugen and J. R. Dahn, *J. Electrochem. Soc.*, 2005, **152**, A2309–A2315.
28. A. Seo, J. Lee, K. Han and H. Kim, *Electrochim. Acta*, 2006, **52**, 1603.
29. J. Zhang, K. Sasaki, E. Sutter and R. R. Adzic, *Science*, 2007, **315**, 220.
30. H. A. Gasteiger, S. S. Kocha, B. Sompalli and F. T. Wagner, *Appl. Catal., B*, 2005, **56**, 9–35.
31. E. Antolini, *Appl. Catal., B*, 2007, **74**, 338–351.
32. E. Antolini, J. R. C. Salgado and E. R. Gonzalez, *J. Power Sources*, 2006, **160**, 957–968.
33. *Modern Aspects of Small Angle Scattering*, ed. H. Brumberger, Kluwer Academic Publishers, Dordrecht, Boston and London, 1995.
34. B. D. Cullity; S. R. Stock, *Elements of X ray diffraction*, 3rd edn, Prentice Hall, New York, 2001.
35. H.-G. Haubold, T. Vad, N. Waldoefner and H. Boennemann, *J. Appl. Crystallogr.*, 2003, **36**, 617–620.

36 H.-G. Haubold, X. H. Wang, G. Goerigk and W. Schilling, *J. Appl. Crystallogr.*, 1997, **30**, 653–658.
37 T. Vad, H.-G. Haubold, N. Waldoefner and H. Boennemann, *J. Appl. Crystallogr.*, 2002, **35**, 459–470.
38 P. Fratzl, Y. Toshida, G. Vogl and H. G. Haubold, *Phys. Rev. B: Condens. Matter Mater. Phys.*, 1992, **46**, 11323–11331.
39 P. Mani, S. Srivastava and P. Strasser, *J. Phys. Chem. C*, 2007, DOI: 10.1021/jp0776412.
40 S. Koh and P. Strasser, *J. Am. Chem. Soc.*, 2007, **129**, 12624–12625.
41 P. Strasser, S. Koh and C. Yu, *ECS Trans.*, 2007, **11**, 167–180.
42 S. Koh, C. Yu and P. Strasser, *ECS Trans.*, 2007, **11**, 205–215.
43 S. Koh, J. Leisch, M. F. Toney and P. Strasser, *J. Phys. Chem. C*, 2007, **111**, 3744–3752.
44 S. Koh, C. Yu, P. Mani, R. Srivastava and P. Strasser, *J. Power Sources*, 2007, **172**, 50–56.
45 S. Koh, M. F. Toney and P. Strasser, *Electrochim. Acta*, 2007, **52**, 2765–2774.
46 A. Guinier, G. Fournet, C. B. Walker and K. L. Yudowitch, *Small Angle X Ray Scattering*, Wiley, New York, 1955.
47 A. P. Hammersley, *FIT2D: An Introduction and Overview – ESRF Internal Report, ESRF97HA02T*, European Synchrotron Radiation Facility, Grenoble, 1997.
48 A. P. Hammersley, S. O. Svensson, M. Hanfland, A. N. Fitch and D. Häusermann, *High Pressure Res.*, 1996, **14**, 235–248.
49 Z. Liu, S. Koh, C. Yu and P. Strasser, *J. Electrochem. Soc.*, 2007, **154**, B1192–B1199.
50 P. Mani, R. Srivastava and P. Strasser, *ECS Trans.*, 2007, **11**, 933–940.
51 P. Strasser, *Electrocatalysis of Pt Alloy Nanoparticles at Fuel Cell Cathodes: Correlating Structure, Activity, and Stability using Synchrotron X-ray Scattering –abstract #90*, Electrochemical Society Annual Spring Meeting, Chicago, 2007.
52 S. Koh, M. F. Toney and P. Strasser, *Lattice-strained Pt shell Nanoparticle Catalysts for the Electroreduction of Oxygen at PEMFC cathodes*, AIChE Fall Meeting, San Francisco, 2006.
53 P. Strasser, *Lattice-strained Pt Nanoparticle Catalysts for the Electroreduction of Oxygen at PEMFC Cathodes*, American Chemical Society Annual Fall Meeting, San Francisco, 2006.
54 J. Friedel, *Philos. Mag.*, 1955, **46**, 514.
55 L. Vegard, *Z. Phys.*, 1921, **5**, 2–26.
56 B. Ingham. *Small Angle X-ray Scattering – A manual for curve fitting*, Lower Hutt, New Zealand, 2006.
57 O. Glatter and O. Kratky. *Small Angle X-ray Scattering*, London, 1982.
58 J. Wagner, *J. Crystallogr.*, 2004, **37**, 750–756.
59 The Log-Normal distribution, http://mathworld.wolfram.com/LogNormalDistribution.html, 2007.
60 D. A. Stevens, S. Zhang, Z. Chena and J. R. Dahn, *Carbon*, 2003, **41**, 2769–2777.
61 D. A. Stevens and J. R. Dahn, *J. Electrochem. Soc.*, 2000, **147**, 4428–4431.
62 H.-G. Haubold, P. Hiller, H. Jungbluth and T. Vad, *Jpn. J. Appl. Phys., Part 2*, 1999, **38**, 36–39.
63 A. Benedetti, L. Bertoldo, P. Canton, G. Goerigk, F. Pinna, P. Riello and S. Polizzi, *Catal. Today*, 1999, 485–489.
64 C. W. B. Bezerra, L. Zhang, H. Liu, K. Lee, A. e. L. B. Marques, E. P. Marques, H. Wang and J. Zhang, *J. Power Sources*, 2007, **173**, 891–908.

General Discussion

Professor Parsons opened the discussion of the paper by E Santos, Kay Pötting and Wolfgang Schmickler*:

I would like to make a comment on the history of the "Volcano" curve, before the discussion of the contents of this paper begins. There seems to be some confusion. I believe that the application of the volcano curve to electrochemical problems originates in work by Heinz Gerischer and myself, done independently and simultaneously. I am glad that Prof. Schmickler mentioned Gerischer, although the paper he refers to contains no volcano curves (*Z. Phys. Chem.*, 1956, **8**, 137). Gerischer's work on this subject was presented at a conference in Brussels in 1957 and published in *Bull. Soc. Chim. Belg.*, 1958, **67**, 506. Mine was published in *Trans. Faraday Soc.*, 1958, **54**, 1053. He and I acknowledged each other's work in those papers. My paper went a little further than his, in that I used a Frumkin type of adsorption isotherm which led to a flat-topped volcano and I gave a prediction of the Tafel plots to be expected for electrodes at different points of the volcano, some with two, or even three, different slopes. Both of us used simple kinetic arguments, with, in my case some experimental input using data on platinum electrodes, but essentially it was a theoretical prediction. At the end of the papers each of us discussed the experimental data of hydrogen evolution and hydrogen adsorption then available, but neither of us was confident enough in these data to put points on the volcano curve. This was left to Sergio Trasatti 14 years later (*J. Electroanal. Chem.*, 1972, **39**, 163).

Neither Gerischer nor I mentioned Sabatier's principle, though I guess we must have been aware of it. I came to hear about Balandin's Volcano some time later and I suppose that he deserves the credit for its application in heterogeneous catalysis.

I have a technical comment: while the Gibbs energy of the adsorption of hydrogen from protons changes with the electrode potential this does not mean that the bond strength between the hydrogen atoms and the metal depends on the electrode potential. If there is any change it is quite small.

Professor Schmickler answered:

Many thanks for your clarification on the history of the volcano curve. Concerning your technical comment: The adsorbed hydrogen atom at the fcc hollow site practically carries no excess charge, and the change of the actual chemical bond strength with potential must therefore indeed be small. However, the free energy of adsorption of the proton from the solution changes. So we totally agree on this point.

Dr Ikeshoji commented:

We have observed the similar orbital-electron transfer behavior as Fig. 1 in your paper in our first principles simulation for the Volmer reaction (H adsorption from H_3O^+ with a charge transfer). The electric potential was applied (M. Otani *et al.*, *J. Phys. Soc. Jpn.*, 2008, **77**, 024802). In the unit cell, we have 36 platinum atoms, 31 water molecules, and H_3O^+. Temperature is 80 °C and total simulation time is about 8 ps. When 0.95 electrons was added to the Pt surface, H in the H_3O^+ was adsorbed on Pt after some time (thermal fluctuation). The charge transfer also took place when the Pt–H distance decreased (adsorption) as can be seen in the figure below (Fig. 1). In the figure, density of states (DOS) weighted by the GPOP (Mulliken gross population analysis) projected on H_3O^+ LUMO state is shown. It has a peak (red region) at 10 eV from ε_F before the Volmer reaction. The position of the peak shifts to 2 eV below ε_F after the adsorption. The DOS projected to the topmost Pt atoms simultaneously decreases below ε_F, which consists of the Pt

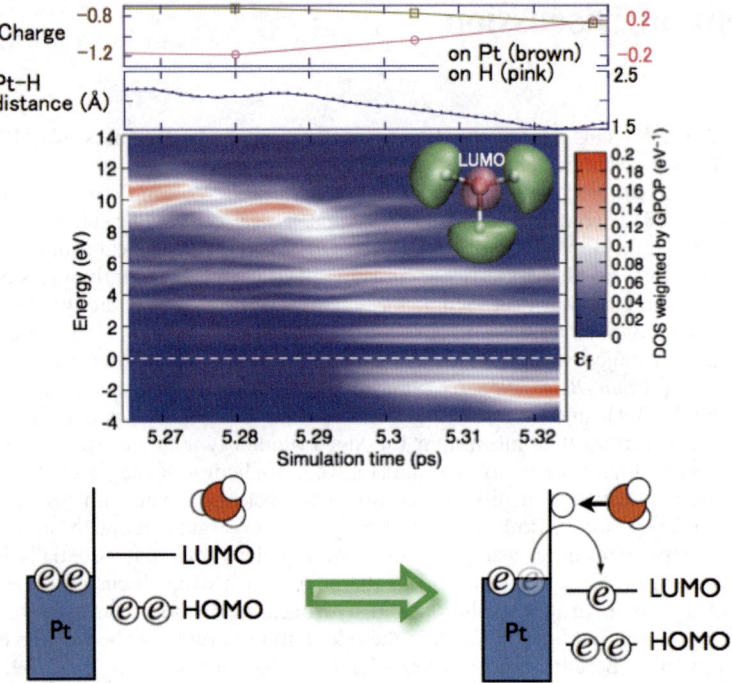

Fig. 1 Time evolution of the DOS weighted by the GPOP during the H-adsorption process. The LUMO state of H_3O^+ cluster (inset) is used for the projection. Charge on surface Pt atoms and the reacting H atom is also shown on the top as well as the Pt–H distance change. This figure was made from two figures in M. Otani et al., J. Phys. Soc. Jpn., 2008, 77, 024802.

5d bands. The electron transfer is characterized by a crossing of the hydronium ion LUMO level and the Pt 5d bands.

Professor Schmickler responded:
I am aware of your valuable work, and though we use rather different approaches, we arrive at the same view for the charge transfer. Your approach should be particularly useful to obtain more detailed information on the role of the solvent, which in our work is just represented by an energy of reorganization.

Professor Wieckowski remarked:
How to predict which metals will be covered by oxides for HER? Are the classical Pourbaix diagrams useful?

Professor Schmickler replied:
With DFT it is possible to calculate Pourbaix diagrams for surface oxides. Our work is not concerned with this problem, but colleagues in the audience, in particular Dr Jacob and Prof. Neurock, have worked on this.

Professor Markovic asked:
HOR is strongly pH dependent, ado you have any suggestion why this is the case? How does H_{upd} affect the HOR rate?

Professor Schmickler responded:
We are presently working on the hydrogen oxidation reaction in the presence of the adsorbed H_{upd}. We have preliminary results, but it is too early to talk about them. The pH dependence may be related to anion adsorption.

Professor Neurock commented:

In response to the question by Professor Wieckowski previously, we have published calculations on the surface Pourbaix diagrams for different metals.[1] Similar work was also recently published by Rossmeisl and Nørskøv.

1. C. D. Taylor, R. G. Kelly and M. Neurock, *J. Electrochem. Soc.*, 2007, **154**(112), F217-F221

Professor Tryk commented:

I would like to point out that Ishikawa and coworkers have found that there is spontaneous formation of a UPD H layer on Pt(111) made up of bridging hydrogens in the 0–0.2 V *vs.* RHE range (*J. Electroanal. Chem.*, 2007, **607**, 37–46; *Molec. Simul.*, 2008, in press). This is different from the result that everybody else is getting, *i.e.*, favored hollow sites, as in vacuum. We feel that this is due to the fact that Ishikawa is including all of the electrons plus relativistic corrections. The energy differences are not large, which is consistent with the fact that adsorbed hydrogen on Pt is highly mobile, but the point is that the hollow sites actually become energy maxima. We need to compare theory and experiment back and forth and improve each until we can achieve convergence. One example is that if the homolytic dissociation is operating, one gets a 30 mV decade^{-1} Tafel slope for HOR on Pt(111), which was not obtained in Markovic's 1997 paper (*J. Phys. Chem. B*, 1997, **101**, 5405–5413). With our proposed mechanism, it may be possible to explain their 74 mV decade^{-1} slope (at 274 K). One has to be careful because the Tafel slope for polycrystalline platinum is close to 30 mV decade^{-1}, due to the overriding contribution of the (110) surface, on which the homolytic (Tafel) step is known to predominate.

Professor Schmickler replied:

I am aware of this work. The differences in adsorption energies for the hollow sites, the bridge and the on-top sites are quite small. Since DFT, after all, contains a semi-empirical components, it is not surprising that different programs may give slightly different answers. Your point about interchange between theory and experiment is well taken.

Professor Schiffrin addressed Dr Santos and Professor Schmickler:

In order to be able to compare experimental results with theoretical calculations, it is necessary to have values of the relevant thermodynamic functions, *i.e.*, Gibbs energies at 298 K. You indicate that the results shown in Fig. 3 in your paper are energy calculations, but those in Fig. 5 in the paper are indicated as free energy surfaces. The use of energy (I guess these values are electronic energy calculations at 0 K, is this correct?) is not very revealing when entropic contributions can be very large, as might be the case for reactions involving a change in the total number of species. In addition, significant zero point energy contributions should be taken into account. The role of the solvent in these calculations appears to have been considered only for the solvent reorganisation energy in the Marcus theory (eqn (8) in the paper). What is the justification for doing the calculations in this way?

Professor Schmickler responded:

The DFT calculations have been performed for 0 K, but these are just used to provide the quantities that we need for our theory. We have taken the entropy of hydrogen and zero point energies into account, and the solvent reorganization is also a free energy. That we represent the solvent only by the energy of reorganization and by one solvent coordinate is simply due to our lack of knowledge of further details. Several groups are performing molecular dynamics for an excess proton in water in front of an electron, and from this kind of work we hope to be able to construct a more detailed model in the future.

Dr Santos said:

The unified model to explain the electrocatalytic effects of different materials presented by Prof. Schmickler can also be applied to understand the differences observed in the electrocatalytic properties of different nanostructures. As an example, I would like to show some results, which predict correctly the tendency on the activity of different surface orientations of silver and copper single crystal electrodes for the hydrogen evolution/oxidation reaction. Experimental results have shown that the activity for silver single crystal electrodes is: Ag(111) > Ag(100) > Ag(110).[1,2] The same trend has been found for copper single crystal electrodes.[3] In the case of gold single crystal electrodes, the opposite tendency has been observed. However, it is well known that gold surfaces reconstruct.[4] The following table (Table 1) shows the adsorption and activation energies obtained for the Volmer reaction, which is postulated as the rate determining step, on the different surfaces employing this theoretical model:

From these results, it can be observed that the first proton transfer occurs uphill in all the cases. There are appreciable differences in the adsorption energies between the different orientations of the same material, and the activation energy values show the tendency observed experimentally.

1. D. Eberhardt, E. Santos and W. Schmickler, *J. Electroanal. Chem.*, 1999, **461**, 164
2. L. Doubova and S. Trasatti, *J. Electroanal. Chem.*, 1999, **467**, 164
3. V.V. Batrakov, Yu. Dittrikh and A.N. Popov, Elektrokhimiya, 1972, **8**, 640
4. A. Hamelin and M. Weaver, *J. Electroanal. Chem.*, 1987, **223**, 171

Professor Schmickler replied:

Many thanks for your valuable addition to our work. It is gratifying to see that our theory can explain the relatively small differences in the rates of the hydrogen reaction on different single-crystal surfaces.

Professor Tsirlina asked:

Does your model assume certain overvoltage, or do these results correspond to zero overvoltage? In the latter case one can use the exchange current densitiy values i_0 for hydronium discharge step obtained from hydrogen evolution data. The selected values for acidic solutions (in the absence of surface oxidation problems) are log i_0 = 10.7–11.8 for Cd and 4.5–6.5 for Au,[1] *i.e.* the difference is 6–7 orders of magnitude. Is it possible to find 2–3 orders more, to be closer to the computed difference? Probably they can go from preexponential term, if the reaction is treated as diabatic?

[1]. O.A.Petrii and G.A.Tsirlina, *Electrochim. Acta*, 1994, **39**, 1739.

Professor Schmickler replied:

The calculations have been performed for zero overvoltage, and the reactions are adiabatic because of the proximity to the surface. Since we do not know the energy of reorganization of the solvent well enough, we cannot calculate absolute values for the activation energies, but can only calculate the trend of the variation. Ours is the first theory that can do this, so we believe that this is an important step forward. It is

Table 1 Adsorption and activation energies obtained for the Volmer reaction

	E_{ads}/eV	E_{act}/eV
Ag(111)	0.42	0.48
Ag(100)	0.52	0.67
Cu(111)	0.05	0.43
Cu(100)	0.17	0.56

true that the pre-exponential term may also vary somewhat between metals, but usually the activation energy should dominate the trend.

Professor Markovic opened the discussion of the paper by Jacob Bonde, Poul G Moses, Thomas F Jaramillo, Jens K Nrskov and Ib Chorkendorff*:
During the anodic sweep direction, in addition to the oxide formation Mo dissolution may take place. Based on this, how reliable is assessment of specific surface area from the oxide formation?

Professor Chorkendorff replied:
The deactivation of the MoS_2 seems to coincide very well with the oxidation feature in the anodic sweep so we have no reason to believe that there is dissolution on Mo prior to the oxidation.

Professor Russell asked:
As you have found that the reaction occurs preferentially at the edges of the MoS_2 islands, have you found any ways to preferentially grow upwards rather than outwards to maximise the amount of such sites?

Professor Chorkendorff replied:
That is a good point, but we have, at present, not systematically studied this. However, we found that when growing MoS_2 at HOPG the nanoparticles are tending to grow in multilayers. This is primarily due to the higher annealing temperature used here in an attempt to get atomic resolution.

Professor Morgan asked:
What do you mean when you say you studied "a range of S coverage"?

Professor Chorkendorff replied:
The question presumably refers to the statement: "We have calculated ΔG_H at the S edge and the Mo/W edge of WS_2 and MoS_2 and on the Co promoted S edge of WS_2 and MoS_2 over a wide range of S coverage and H coverage". Thus, it is the theoretical study where one has to consider the variation of the different termination of the particles, *i.e.* whether it is the metal or the sulfur terminated surfaces.

Professor Savinova commented:
First of all I would like to mention that to my opinion this is a very important contribution, which turns the attention of the electrochemical community to "molecular" catalysts based on non-noble metals.
My question refers to the possibility to design hybrid materials for electrocatalysis of *e.g.* the hydrogen oxidation reaction (HOR), comprising transition metal sulfides and (noble) metal particles. Transition metal sulfides are rather efficient in the hydrogen evolution reaction, as shown by the authors. However, they are not active in the HOR, since they do not catalyze efficiently the H–H bond splitting. Perhaps, addition of a small amount of Pt or other noble metals might help in producing highly efficient novel hybrid materials for the HOR and subsequently for other electrocatalytic reactions?

Professor Chorkendorff replied:
This is a very good point since the MoS_2 nanoparticles are not viable for HOR so adding another metal like Pt could maybe improve this situation. It should, however, be mentioned that this diode effect, *i.e.* that it only works in the direction of HER, may not be undesireable.[1] Imagine that one wants to good catalyst for HER in, for example, photocatalysis, then it would be useful that it is only capable of evolving H_2[2] since the reverse process is not desirable.

Professor Neurock remarked:
Have you tried nickel, iron or NiFe sulfides?

Professor Chorkendorff answered:
We have tried Ni and Fe sulfides but not combinations of those. They seem to oxidize much faster and therefore are not be stable for HER. Further studies of those metals were therefore abandoned.

Professor Wieckowski opened the discussion of the paper by Yoshihiro Gohda, Sebastian Schnur and Axel Gross*:
The role of Cl^- in bridging redox systems in electrocatalysis should be looked into.

Professor Gross responded:
We are aware that theoretical modelling can and should contribute to the understanding of many processes and mechanisms in electrocatalysis such as the inner sphere electron transfer in bridging redox systems. In fact, there is a growing number of theoretical groups addressing rather complex reactions in electrocatalysis from first principles. On the other hand, there are many open questions in electrochemistry and electrocatalysis for seemingly simple but fundamental issues, such as the structure of water at the solid–liquid interface or the role of water in the adsorption at electrode surfaces. Hence, we are convinced that there is still a need for theoretical studies addressing fundamental topics in electrocatalysis. At the same time, it is of course also important to look at more complex processes which we, as others, will certainly continue to do in the future.

Professor Markovic asked:
The effect of Cl^- on the O_2 dissociation barrier is indeed an important issue. However illuminating the effect of oxygenated species (OH^-, O^-) may be of greater importance and should be proven. Cl^- has no effect on the reaction mechanism on Pt(111) at low overpotentials. However, Br^- is strongly affecting the reaction pathway. Comparing the effect of Cl^- with Br^- may help understanding the O_2–halide interactions. Do you have any intention of going in this direction?

Professor Gross responded:
We admit that our present study represents only a first step towards a realistic modelling of the solid–liquid interface that could lead to a better understanding of the role of the electrolyte and in particular the ions in electrochemical adsorption and reaction processes. We are definitely planning to extent these preliminary steps to a systematic study including different anions such as OH^-, O^-, Cl^- and Br^- in a realistic water environment. It is also true that it is necessary to establish chemical trends for, *e.g.*, different halides in order to gain genuine insights.

Professor Janik remarked:
This comment regards the calculation of the effect of Cl^- addition on the O_2 dissociation barrier. In the calculation, a Cl^- ion and compensating background charge are added. This creates spurious interactions of the electronic structure with the background charge, as well as creating a charge separation in the periodic cell, leading to periodic diploe–dipole and multipole interactions. These may differ between the O_2 adsorbed and dissociation transition state and therefore the barrier is not rigorously "defined" in this calculation. Beyond this point, the added charge likely does not localise on the Cl atom, leading to a negative charge in the surface, an interfacial electric field (surface to positive background) which may cause a reduction in dissociation barrier (see electric field included computational work of Koper and Medlin) independent of the addition of Cl atoms. Finally, it should be recognised that this dissociation is occurring at constant charge not electrode potential. Our

recent (Wasileski and Janik, PCCP) paper shows that constant potential can give very different results.

Professor Gross responded:
We agree that the addition of the compensating background charge creates spurious interactions which can influence the determination of the barrier height when the effects are different in the initial and the transition state.

We have made no particular effort to correct for this spurious interactions since we did not find a dramatic effect of the environment on the height of the O_2 dissociation barrier on Pt(111), in agreement with the results at constant charge of your nice and very detailed study.[1]

As far as the localisation of the added charge is concerned, we do not agree that it "likely" does not remain at the chlorine atoms after O_2 dissociation. Within the computational set-up used in our study the addition of the charge means that the number of explicitly considered electrons per unit cell is increased from 175 to 176. Since upon adsorption the 3p-derived levels of chlorine atoms are typically located several eV below the Fermi energy,[2,3] one would assume that the added charge remains close to the chlorine atom or the chlorine–metal bond, irrespective of the presence of extra charges and other electronegative adsorbates such as oxygen atoms. Charge will rather flow from the electron reservoir of the metal electrode to the additional adsorbate. It is true that in our calculations we observe a significant change in the Cl p local density of states (LDOS) after the O_2 dissociation because of the modified Cl–O interaction. However, almost no Cl p LDOS is shifted above the Fermi level so that the charge on the adsorbed Cl atoms remains constant within one per cent upon the O_2 dissociation.

Finally, we also agree that one has to look at the difference between constant charge *vs.* constant potential calculations in more detail, but this was not the issue of our present contribution.

1. S. A. Wasileski and M. J. Janik, *Phys. Chem. Chem. Phys.*, 2008, **10**, 3613.
2. T.A. Baker, C.M. Friend and E. Kaxiras, *J. Am. Chem. Soc.*, 2008, **130**, 3720.
3. N. D. Lang and A. Williams, *Phys. Rev. B*, 1978, **18**, 616.

Professor Ahlberg said:
Single or bilayers of water are commonly used for the theoretical study of solvent effects on adsorption and kinetics of electron transfer reactions. However, for an overall understanding of the system, the bulk electrolyte also needs to be considered. Electrolyte ions, as well as oxygen and hydrogen, structure water in different ways. For example, water forms a cage with a hydrophobic hole where oxygen is located, as illustrated in Fig. 2. The energy required in reconstructing water for adsorption on the electrode surface and electron transfer is about 140 meV.[1] This energy is probably part of the overpotential observed experimentally for oxygen reduction. Can these effects be included in your model?

1. Itai Panas, personal communication

Professor Gross replied:
In order to have a true energy balance, it is of course necessary to realistically take into account solvation and reorganization effects, both in the bulk electrolyte as well as at the electrode surface. There is no fundamental problem including these effects in electronic structure calculations based on density functional theory. However, the realistic modelling of these solvation effects requires a large number of water molecules to be considered in the calculations, which represents a significant computational effort. Still, it can certainly be done, and we will carry out these kinds of calculations in the future. Thus, it should be possible to estimate the energetic cost of the reorganisation of the water structure upon adsorption and reaction at

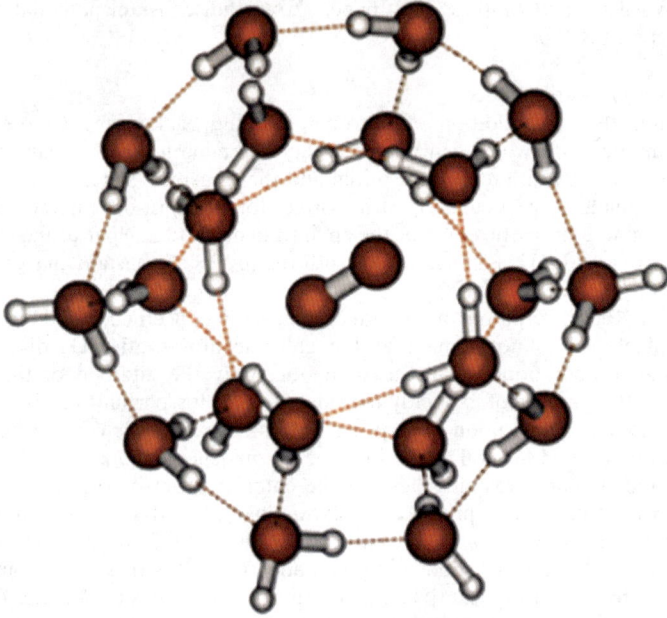

Fig. 2 20 water molecules form a cage around the oxygen molecule. The interior of the cage is hydrophobic which facilitates the dissolution of oxygen molecules in water.

the electrode surface which can contribute, as you rightly suggest, to experimentally observed overpotentials.

Professor Neurock asked:

We have examined H/Pt(111) in multilayers of water to simulate bulk water. We see that the work function changes. In addition we see that the hydrogen transfers into solution as a proton. In your calculations you see that the hydrogen remains bound to the surface. I think that the difference is due to the fact that the proton is not as effectively stabilised on the bilayer as it is in the bulk solution or the multilayers. Can you please comment?

Professor Gross answered:

It is indeed well-known that adsorbed hydrogen layers on metallic surfaces change the work function of the metal.[1,2] In fact, we also find that the work function of the Pt electrode changes significantly, by up to +1.1 eV, upon hydrogen deposition on Pt(111) in the presence of water. Interestingly enough, according to our DFT calculations the work functions of the H-down and the H-up bilayers on Pt(111) differ by more than 2 eV, in agreement with previous results.[3] Yet, both types of bilayers lower the work function of the clean Pt surface by 0.23 eV and 2.27 eV, respectively. This is surprising because naively one would expect that the work function changes induced by these bilayers have opposite signs; it is obviously related to a relatively strong structural and electronic rearrangement of the water bilayers in the presence of the Pt surface. However, we did not observe any transfer of hydrogen atoms from the surface to the solution. On the contrary, when we added hydrogen atoms to the ordered bilayer structures, both to the single as well as to the double layer, the additional hydrogen atoms always relaxed down to the metal substrate. Only for the disordered structures, we observed the formation of H_3O^+ together with OH^- on the Pt surface. Obviously, as Prof. Neurock suggests, the crystalline water structure binds hydrogen less strongly than the Pt substrate.

[1] M. Lischka and A. Gross, *Phys. Rev. B*, 2002, **65**, 075420.
[2] J. Greeley and M. Mavrikakis, *J. Phys. Chem. B*, 2005, **109**, 3460.
[3] J.-S. Filhol and M.-L. Bouquet, *Chem. Phys. Lett.*, 2007, **438**, 203.

Dr Wang asked:
In our combined experimental and DFT calculation studies, we found that the free energy barrier for H_2 dissociation on Pt is 196 meV at the reversible potential for the HOR/HER,[1] which is close to your calculated value using a model with a water bilayer on Pt. We considered that repulsion from underpotentially deposited H adatoms contribute considerably to the barrier. To be more realistic, can you include 2/3 monolayer H_{ad}, in addition to the water layers, in your calculation?

1. J. X. Wang, T. E. Springer, P. Liu, M. Shao and R. R. Adzic, *J. Phys. Chem. C*, 2007, **111**, 12425–12433

Professor Gross answered:
It is indeed gratifying to see that the value of the free energy barrier for H_2 dissociation on Pt(111) that you derive in your combined experimental and DFT study is close to the one that we calculated. However, you determined the barrier height in the presence of an H_{upd} layer, while in our calculations there are no preadsorbed hydrogen atoms. It is no problem at all to include an adsorbed layer of hydrogen atoms in the calculations; in fact, we have already performed a detailed DFT study of the H_2 adsorption on hydrogen-precovered Pd surfaces at the gas-solid interface.[1] We will certainly perform corresponding studies on Pt for the liquid-solid interface. At the moment, one can only speculate on the results. One would expect that the presence of the hydrogen upd layer would lead to an increase in the height of the dissociation barrier because of the repulsive interaction between the adsorbed hydrogen atoms and the impinging H_2 molecule. This would result in a larger discrepancy between your results and our calculations. However, definite answers can only be given once the calculations have been performed.

1. Axel Gross and Arezoo Dianat, *Phys. Rev. Lett.*, 2007, **98**, 206107.

Professor Tryk commented:
I have two points. First, you have not specified a potential, but effectively, there is a potential, which we believe would be somewhere in the double layer region (and might correspond to the potential of zero charge). Assuming this, there should be oxidation to the proton, based on simple thermodynamics. This is in line with what Matt Neurock was saying.

Second, on a somewhat different topic, Ishikawa and coworkers find that the effect of water is to shift the energetic ordering of the various adsorption sites. In a vacuum, everyone agrees that the favoured sites are the hollow sites. In water, suddenly, the bridging sites are favoured. We don't understand this in detail yet but it leads to a spontaneously formed UPDH bridging layer in the proper potential range (see *J. Electroanal. Chem.*, 2007, **607**, 37–46; *Molec. Simul.*, 2008. in press)

Professor Gross responded:
With respect to your first point, I like first to reiterate my reply to the question of Prof. Neurock: it might well be that the ice-like bilayer structure does not support proton transfer from the Pt substrate to the water while more open water structures can stabilise the proton. However, with respect to your remark that the oxidation to the proton follows from simple thermodynamics, I would like to point out that in our study we only looked at total energy minimum structures. As far as thermodynamical aspects are concerned, instead of total energies one rather has to consider free energy differences. Indeed this can be done based on first-principles results using the concept of *ab initio* atomistic thermodynamics[1,2] where the chemical potential of the adsorbate, here hydrogen, enters as the fundamental variable. It might well be

that there are situations where specific adsorption is energetically favorable but thermodynamically the transfer into the solution is stable for entropic reasons. Thus our findings that the hydrogen atoms remains at the surface is not necessarily at variance with thermodynamical considerations.

Turning to your second point, it is important to realize that Pt(111) unlike other transitions metals has the surprising property that the hollow, bridge and top sites exhibit very similar adsorption energies for atomic hydrogen[3,4] resulting in a very small diffusion barrier for hydrogen on Pt(111). This is also reflected in the facile motion of the hydrogen atoms on Pt(111) after H_2 dissociation observed in our *ab initio* molecular dynamics simulations within the first picosecond after the dissociation event. Considering the small differences in the adsorption energies, it is not unlikely that changing the environment of the Pt surface (water deposition, charging of the surface) can modify the energetic ordering of the adsorption sites. It might also well be that the exact nature of the water structure above the Pt(111) electrode and the proper representation of the delocalized nature of the electronic states of the metal play a decisive role in the determination of the most favorable adsorption sites.

1. K. Reuter and M. Scheffler, *Phys. Rev. B*, 2001, **65**, 035406.
2. A. Gross, *J. Comput. Theor. Nanosci.*, 2008, **5**, 894.
3. G. W. Watson, R. P. K. Wells, D. J. Willock, and G. J. Hutchings, *J. Phys. Chem. B*, 2001, **105**, 4889.
4. I. Hamada and Y. Morikawa, *J. Phys. Chem. C*, 2008, **112**, 10889.

Dr Rossmeisl said:

A comment on if H* will become H* + e⁻. This will depend on the area of the simulation cell, since the potential will change from a positive to a negative as the proton is moved to the water layer.

Professor Tryk replied:

This is a valid point, certainly, which everyone has to deal with to some extent. However, Ishikawa is doing cluster calculations with a relatively large (38 atom) Pt(111) cluster, so that the effect of the proton moving from the surface is minimized. Another problem is that the effective pH suddenly goes down. The ultimate answer will involve working with larger systems, for example, with protons already present, so that the effect of producing an additional one will be less.

Dr Ikeshoji asked:

Was H_3O^+ formed from H_2 in your simulation? Where does it come from? If dissociation of a water molecule gives H_3O^+ and OH^-, they will not be stable. They will be recombined soon. At least, such water dissociation is never observed in the first principles MD of bulk water because of small volume and a short time. So, water might be in an unusual circumstance.

Professor Gross answered:

We did in fact observe water dissociation into H_3O^+ and OH^- in our *ab initio* molecular dynamics (AIMD) simulations, and within the run time of 1 ps of the AIMD simulations, which is admittedly still rather short, there was no recombination. However, we found this dissociation only for the strongly disordered water structures with several water molecules removed from the hexagonal ice-like structures. Thus, the water structure favoring dissociation might indeed be unusual. On the other hand, in any molecular dynamics simulation of water there might be some transient dissociation of water due to thermal fluctuations, as you suggest, if the system is sufficiently large and the run time long enough.

Professor Tryk returned to the discussion of the paper by E. Santos, Kay Pötting and Wolfgang Schmickler*:

The hydrogen oxidation reaction (HOR) on the Pt(111) surface has become something of a touchstone in the area of theoretical electrocatalysis, with results appearing recently from several groups,[1-6] including two of the papers in this Discussion.[7,8] The closely related process, hydrogen adsorption, has also been discussed. Much of this work has made use of computational methodologies that have become fairly standard in recent years, and the results have also been somewhat similar. For example, most studies have been in basic agreement with the mechanistic scenario proposed by Gohda *et al.*, involving the hemolytic dissociation (Tafel) reaction:[7]

$H_2 \rightarrow 2H_{ads}$

with the dissociated hydrogens relaxing into hollow sites (3-fold or 4-fold). However, this mechanism is belied by the experimental results of Markovic *et al.*, which indicate a different mechanism, based on a 74 mV decade^{-1} Tafel slope (274 K), rather than a 28 mV decade^{-1} slope, which is found for the more active (110) face, on which the Tafel pathway is known to occur.[9] Recently, using all-electron, scalar-relativistic DFT calculations, together with molecular dynamics, however, Ishikawa and coworkers have found a different scenario, in which the initial step involves heterolytic dissociation with a simultaneous oxidation (Heyrovsky) reaction:[1,6]

$H_2 \rightarrow H_{ad} + H^+$

Moreover, the product H_{ad} is in a bridging site rather than one of the hollow sites, which are indeed found to be energy maxima rather than minima. Thus, we find that the effect of water on the energetics is major, shifting the most stable site from hollow to bridging, whereas neither Gohda *et al.*[7] nor Skulason *et al.*[2] find such a pronounced effect of water, with the latter group finding only a very small effect. Significantly, Gohda *et al.* do not find that the adsorbed hydrogens become oxidized, which should be thermodynamically favored for the uncharged platinum slab.

The computational results thus far are in basic agreement with the experimental results obtained thus far. For example, the underpotentially deposited (UPD) H–Pt bond strengths at zero coverage have been reported for Pt(111) in dilute acid aqueous solution: -262 kJ mol^{-1} [10] and -240 kJ mol^{-1}.[11] Our value is 232.6 kJ mol^{-1} for a Pt(111)$_{38}$ cluster.[1] Interestingly, Markovic *et al.* find the adsorption energy remarkably insensitive to the type of anion or even acid *vs.* alkaline pH.[11] This is consistent with the rather weak interaction we find between the bridging hydrogens (H_{br}) and the closest water molecules.

We have also predicted vibrational frequencies that are consistent with experimental results, although clear results that could distinguish between bridging and hollow sites have yet to be obtained. We are also now able to predict other features for the vibrational spectra, for example, the O–H stretch for water molecules that are interacting with the on-top hydrogen and with the bridging hydrogens. We will report these results elsewhere.

The most striking aspect of the predicted H_{br} is the fact that it is highly energetically favored to form a honeycomb structure with increasing coverage (Fig. 3), which is stable over a potential range from 0.0 to +0.2 V *vs.* the reversible hydrogen electrode (RHE).[6] At 0.0 V *vs.* RHE, the on-top positions are partially filled. In agreement with other groups, we identify these as "overpotentially deposited" (OPD) hydrogen.

In summary, although our results are in agreement with experimental results, the key piece of data that will distinguish bridging from hollow sites is still missing: the former should yield an infrared absorption peak at 1092 cm^{-1}, at low coverages, with peaks in the 1000–1100 and 1300–1400 cm^{-1} ranges appearing at higher coverages[1] (Hamada *et al.* predict 1306 cm^{-1}),[5] while the latter would yield a peak at 1234 cm^{-1}, based on HREEL spectra in ultrahigh vacuum.[12,13]

1. Y. Ishikawa, J. J. Mateo, D. A. Tryk and C. R. Cabrera, *J. Electroanal. Chem.*, 2007, **607**, 37–46.

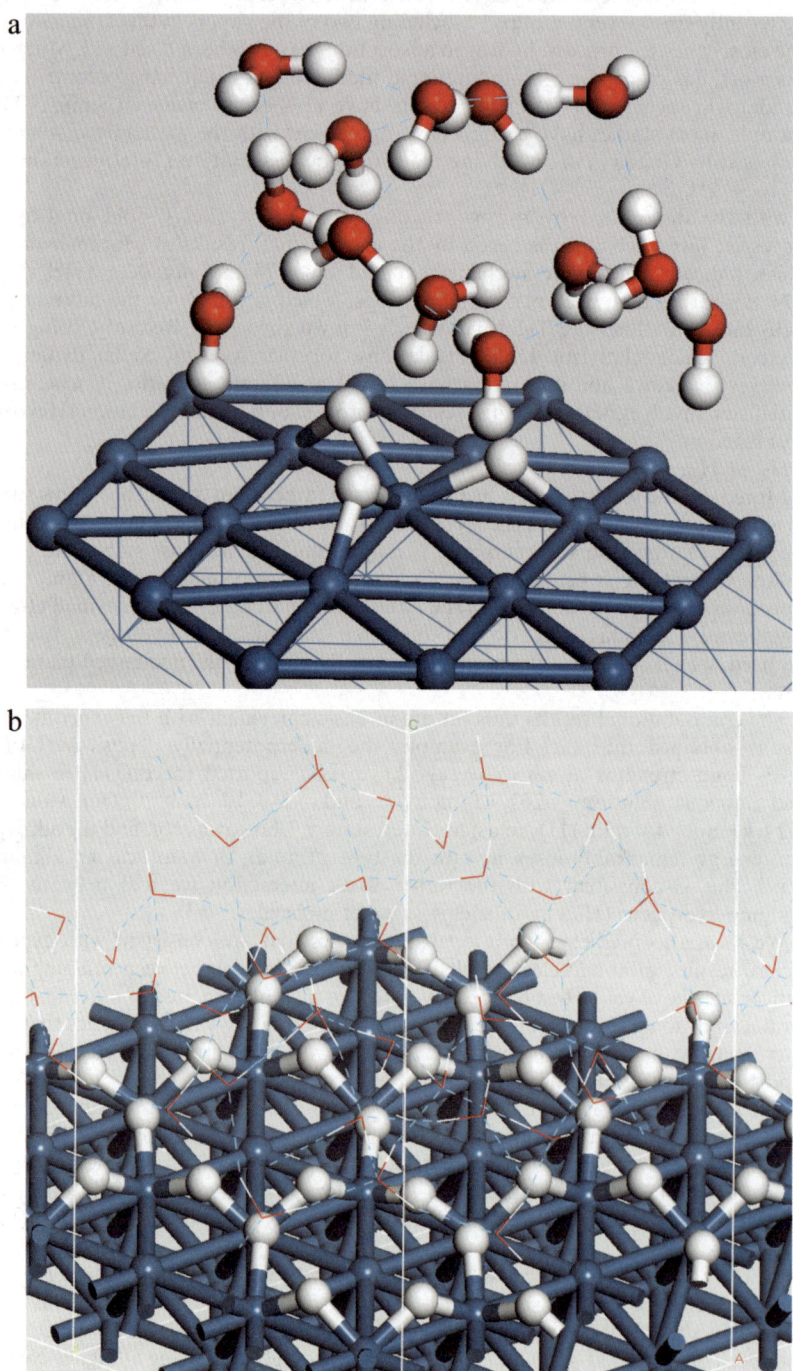

Fig. 3 (a) UPD H_{br} at low coverage over platinum (111) cluster. (b) Self-assembled honeycombs formed by UPD H_{br} at 1 ML coverage.

2. E. Skúlason, G. S. Karlberg, J. Rossmeisl, T. Bligaard, J. Greeley, H. Jónsson and J. K. Nørskov, *Phys. Chem. Chem. Phys.*, 2007, **9**, 3241–3250.
3. J. X. Wang, T. E. Springer, P. Liu, M. Shao and R. R. Adzic, *J. Phys. Chem. C*, 2007, **111**, 12425–12433.
4. M. Otani, I. Hamada, O. Sugino, Y. Morikawa, Y. Okamoto and T. Ikeshoji, *Phys. Chem. Chem. Phys.*, 2008, **10**, 3609–3612.
5. I. Hamada and Y. Morikawa, *J. Phys. Chem. C*, 2008.
6. J. J. Mateo, D. A. Tryk, C. R. Cabrera and Y. Ishikawa, *Molecular Simulation*, 2008, in press.
7. Y. Gohda, S. Schnur and A. Gross, *Faraday Discuss.*, 2008, **140**, DOI:10.1039/b802270d.
8. E. Santos, K. Pötting and W. Schmickler, *Faraday Discuss.*, 2008, **140**, DOI:10.1039/b802253d.
9. N. M. Markovic, B. N. Grgur and P. N. Ross, *J. Phys. Chem. B*, 1997, **101**, 5405–5413.
10. G. Jerkiewicz, *Prog. Surf. Sci.*, 1998, **57**, 137–186.
11. N. M. Markovic, T. J. Schmidt, B. N. Grgur, H. A. Gasteiger, R. J. Behm and P. N. Ross, *J. Phys. Chem. B*, 1999, **103**, 8568–8577.
12. L. J. Richter and W. Ho, *Phys. Rev. B*, 1987, **36**, 9797.
13. S. C. Badescu, K. Jacobi, Y. Wang, K. Bedürftig, G. Ertl, P. Salo, T. Ala-Nissila and S. C. Ying, *Phys. Rev. B*, 2003, **68**, 205401.

Professor Schmickler answered:
Thank you for the additional information on your interesting work.

Professor Gross added:
Your findings that a honeycomb structure of hydrogen adsorbed at bridge sites is formed at positive potentials and increasing coverage is certainly very interesting. As already indicated in my reply to your previous comment, your findings might need to be substantiated by further calculations using different setups.

Professor Gewirth reopened the discussion of the paper by Jacob Bonde, Poul G Moses, Thomas F Jaramillo, Jens K Nrskov and Ib Chorkendorff*:
Your motivation—surely a good one—derives from the active sites of nitrogenase and hydrogenase. Yet MoS_2 and the related materials actually are quite far away from the active sites in these proteins. Are you looking at more closely related complexes?

Professor Chorkendorff replied:
You are completely right, the materials studied here are quite far from the actual co-factor in the enzyme which can be considered as being an assembly of two cubanes containing Mo and Fe connected by S. The MoS_2 nanoparticles was a first attempt to make something simple and at least containing some of the elements. We are now improving on this by, for instance, studying Mo_3S_4 incomplete cubanes and other cubanes with different metal combinations. They also show interesting HER activities (slightly higher than those reported here for MoS_2), but they are on the other hand more difficult to stabilize on the surface. See for instance T. F. Jaramillo, J. Zhang, B. Lean Ooi, J. Bonde, K. Andersson, J. Ulstrup and I. Chorkendorff, Hydrogen Evolution on supported $[Mo_3S_4]^{4+}$ cubane-type electrocatalysts, *J. Phys. Chem.*, 2008, accepted.

Professor Ahlberg reopened the discussion of the paper by E Santos, Kay Pötting and Wolfgang Schmickler*:
You are using DFT calculations to obtain adsorption energies of reaction intermediates in the hydrogen oxidation reaction. In these calculations a free hydrogen atom and a neutral surface form the dissociation limit. However, the ionic product, H^+, and a negative surface are more relevant. Since DFT is unable to treat such excited states what is the relevance of the obtained energies?

Professor Schmickler answered:

Indeed, DFT is not able to treat an electrochemical charge transfer reaction. Therefore, our calculations are not based directly on DFT, but on our own theory of electrocatalysis as presented in ref. 8–10 in our paper and further references therein. We use DFT solely to obtain the system parameters required by our model—interaction constants, densities of states—and to correct for exchange and correlation. So, in Fig. 6 in our paper, the states at $q = -1$ at the left correspond to the H^+, and at the right, at $q = 0$, to the adsorbed hydrogen. To the best of our knowledge, our theory is the only one that can treat such electrochemical electron transfer. So the calculated energies are directly relevant to the electrochemical proton transfer!

Professor Markovic asked:

Your model is applicable for metals with very different activities for the HOR. The question is if you would find differences with the metal with similar I_0 (exchange current)?

Professor Schmickler replied:

As the comment from Dr Santos shows, we are able to reproduce the differences in the exchange current densities for different surfaces of the same metal. We believe we can also explain differences between a bulk metal and an overlayer of the same metal on a foreign substrate. However, in the case of two completely different metals with similar rates, we cannot reliably calculate such small differences. The main difficulty is that we do not know the details of the solvent reorganization well enough. However, the scatter of experimental data between various groups is so large, that I believe that experimentalists cannot distinguish between similar rates on different metals, either.

Professor Scherson opened the discussion of the paper by Galina Tsirlina*, Elena Timofeeva, Nobuko Tanimura, Nataliya Sherstyuk, Marina Borzenko,:

What information could you extract from the SHG data that you could have not obtained from regular linear absorption spectroscopy?

Professor Tsirlina answered:

The advantages of SHG over IR absorption spectroscopy are higher adsorbate/substrate contrast, higher signal-to-noise ratio (100 or 1000 times faster monitoring and hence reliability), and locality (down to 1 µm).

For UPD on ordered surfaces, SHG provides information on a symmetry of adsorbate layer as well. The disadvantage is the insensitivity of the optical signal to molecular vibrations. If to discuss wider and to go to linear techniques in a visible region, reflection and even electroreflection are mostly determined by bulk metal properties, when nonlinear responses are more sensitive to interfacial features, see ref. 1 presenting an example of so-called electroscattering response for platinum.

1. A. M. Brodsky, L. I. Daikhin, A. M. Foontikov, V. E. Kazarinov, G. A. Tsirlina and M. I. Urbakh, *Surf. Sci.*, 1990, **26**, 137.

Professor Gewirth said:

How many monolayers of your POMs were calculated by using QCM? Is there a correlation between the coverage as defined by QCM and SHG measurements?

Professor Tsirlina replied:

Our cautious estimates of POM coverage from EQCM under open circuit conditions gave 0.8–1.1 (some uncertainty results from perturbation induced by POM injection). Under potential cycling the decrease of this open circuit value does not exceed 0.10–0.15, so the lowest possible coverage can be estimated as *ca.* 0.7. We mention some doubts concerning the previous EQCM observations of multilayer POM adsorption on gold. Correlation of EQCM and SHG is specially illustrated in the paper: it is complete within the accuracy of both signals registration.

Professor Markovic asked:

The formation of Cu UPD may go through the formation of Cu^+. Do you have any evidence of the existence of Cu^+?

To asses a true adsorption isotherm for Cu UPD the RRDE is an appropriate method. Do you consider using the RRDE?

Professor Tsirlina answered:

Cu^+ formation complicates Cu UPD study if adsorption potential is close to equilibrium potential. People usually meet this problem when use Cu UPD for true surface area determination, and overcome it either by using RRDE or by other (less precise) means. For the study under discussion, it was less important to have the complete copper monolayer, and we avoided Cu^+-induced complications simply by using rather high cathodic potential limit of 0.3 V RHE. Even for our highest Cu^{2+} concentration it is 0.05 V more positive than the equilibrium potential. This is mentioned briefly in the text (p.3, bottom; ref. 20 is given, which is rather informative).

Professor Ahlberg asked:

What kind of electrocatalytic reactions would be of interest for these type of surfaces?

Professor Tsirlina answered:

Modification of platinum with Keggin anions favors steady-state oxidation of methanol and some other organic fuels at low potentials (the reason is less pronounced self-poisoning). Many electroreduction reactions are catalysed by adsorbed heteropolyoxometalates operating as electron transfer mediators. As for the ternary platinum–copper–Keggin system, it should be considered as a model system at this stage, helpful for understanding the features of Keggin adlayers (partial reduction, coadsorption with cations, and permeability for smaller size adsorbates).

Professor Scherson remarked:

In Fig. 1 in your paper: 50 µM and 10 mVs^{-1} is not enough time for copper to reach the surface. This would explain the hysteresis when the scan is reversed; the coverage of Cu would be higher and thus elicit larger signal. The argument of islands in paper appears incorrect.

Professor Tsirlina answered:

We compared voltammograms and SHG *vs.* potential curves for anodic and cathodic scans separately, with a solid understanding of hysteresis nature in diluted solutions. These solutions were very helpful for obtaining low coverage under the same potentiodynamic mode. It does not matter how certain adsorbate state/configuration (coverage, number and size of islands) was achieved if its electrochemical and optical features are fixed independently.

Interpretation based on island formation explains a well-pronounced effect: independence of SHG signal on copper coverage. This hypothesis (still only a hypothesis!) agrees with all experimental facts available at the moment (for both voltammetric and transient modes), and we failed to find any other consistent hypotheses so far.

Dr Fermin remarked:

I would like to mention a couple of points in relation to the complex potential dependence of the SHG signal as a function of the Cu coverage at the Pt electrode

(Fig. 1 in the paper). The experimental results were qualitatively explained assuming that the non-linear susceptibilities of the Pt and Cu have opposite sign. In addition, it is also proposed that the SHG signal arising from Cu islands at low coverage are enhanced due to a resonance with the collective excitation of free electrons (plasmon). It is difficult to assess from the information given in the paper whether the SHG wavelength will be in resonance with the plasmon signal of Cu. For the sake of clarity in the discussion, the author should quote the value reported (or expected) for the plasmon of the Cu islands and compare it to the wavelength of the incident beam.

The second point is in relation to eqn (2) in the paper. The non-linear polarisation associated with the third order susceptibility tensor ($\chi(3)$) is not explicitly included in the analysis. Taking into account this term, $P(2\omega)$ can be written as:[1] $P(2\omega) = (\chi(2)+\chi(3)E(dc))E(\omega)E(\omega) = \chi\exp(2)E(\omega)E(\omega)$. Consequently, the strong potential dependence of the SHG signal can be interpreted in terms of changes in the local electric field ($E(dc)$) in addition to changes in the Cu coverage. Indeed, these results can provide some fascinating insights into the potential distribution across the metal/electrolyte interface. Can the authors further comment on this issue?

1. A. A. Tamburello-Luca, P. Hébert, P. F. Brevet and H. H. Girault, *J. Electroanal. Chem.*, 1996, **409**, 123.

Professor Tsirlina replied:

1. The majority of SHG and electrochemistry experimental details can be found in Part I (ref. 1 in our currently discussed paper). In particular, the wavelength of our incident beam is 760 nm. Two known values of surface plasmon resonance (SPR) for copper particles are already referred in the text:

(a) 2.15 eV (576 nm) for copper particles of 3 nm diameter (the number of atoms > 2000) embedded in alumina with a well pronounced SPR, Ref. 14; (b) from 2.1 eV (590 nm) to 1.6 eV (775 nm) for copper nanoparticles embedded in lithium fluoride, a blue-shift of SPR with the particle size increase from 2 to 20 nm, ref. 36 in our paper.

Recently published data we can add to this list are as follows:

(c) 690 nm for 2 nm copper nanoparticles embedded in diamond-like carbon (DLC);[1] (d) 760 nm for copper nanorods with the aspect ratio 5, in methanol.[2]

It seems like SPR is size- and shape-dependent, and our wavelength is just in the interval of rather probable resonance manifestations.

2. For material with some fixed configuration of fragments (atoms, molecules, clusters, *etc.*), only one effect of electric field is expected: it can be presented by the term which is the product of $E(dc)$ and third order susceptibility. In the system under study, the fragments (particles) in electric field are mobile. Generally, in order to produce the signal the particle first has to appear, and then it can be subjected to a strong $E(dc)$. The effect of its appearance is the primary effect in this case.

We agree that even in a mobile system something can be fixed at the surface, and then being subjected to a strong $E(dc)$ it can provide the major effect (the second term in the right part is higher in this case). However, comprehensive analysis of both electrochemical and SHG data (including potential and concentration dependencies) resulted in conclusions presented in the paper: (1) the absence of direct contact between polytungstate and platinum, and Cu adatoms location under the tungstate layer; (2) the predominating role of coverage-dependent contribution in SHG generation.

[1]. S. Hussain and A. K. Pal, *Bull. Mater. Sci.*, 2006, **29**, 553.
[2]. A. Azarian, A. Iraji zad, A. Dolati and M. Ghorbani, *J. Phys.: Condens. Matter*, 2007, **19**, 446007.

Professor Sun asked:

The difference in size of Cu atom and keggin type polytungstates is very large. How do you determine the surface coverage of this co-adsorption? Can the coverage of the two species on a polycrystallis electrode be measured precisely?

Professor Tsirlina responded:

Our estimates of partial coverage in mixed adlayer originate from the data of voltammetric, SHG and EQCM techniques and surely involve some assumptions. To make these assumptions more solid we always compared the data of each technique for Cu UPD, Keggin anion adsorption, and coadsorbtion of these species, and also considered correlations of various responses, like those given in Fig.10 (with special attention to deviations from these correlations). What we finally determine is surely enough to support the most important statement: despite of large size of Keggin anions and their rather high coverage (close to monolayer) the quantity of coadsorbed copper is even higher than at Keggin-free surfaces. As for the more general question, coverage can be measured precisely only when direct techniques are applied, like radiotracer or titration techniques. This is the same for single- and polycrystalline surfaces, but sensitivity limitations for less rough surfaces force everybody to apply combinations of indirect techniques (including electrochemical).

Dr Cremers opened the discussion of the paper by Cyril Delacôte, Arman Bonakdarpour, Christina M Johnston, Piotr Zelanay and Andrzej Wieckowski:

Why are such high values of humidification used, much in excess of the due point corresponding to the cell temperature, eventually causing flooding of the cell?

Professor Wieckowski answered:

Tests were run with the cathode temperature set at 80 °C (same as the cell) and 90 °C alternately. There was little performance difference for either voltage–current curves or life tests. In fact, the 90 °C data were slightly better, indicating that flooding was not involved in the measurements (at least not as the main problems of these measurements).

Professor Schiffrin said:

I did a rough calculation of the value of n from the data in Figure 5 and got values close to 4. During the Discussion it was indicated that approximately 10% of the current follows the 2e$^-$ pathway.[1] Since a significant amount of peroxide is produced, how does this affect the long-term chemical stability of this material?

1. A. Bonakdarpour, C. Delacôte, R. Yang, A. Wieckowski and J.R. Dahn, *Electrochem. Commun.* 2008, **10**, 611–615.

Professor Wieckowski responded:

Peroxide formation needs to be avoided for fuel cells applications, especially if a Nafion type membrane is to be used (which is not necessary in laminar and mixed feed fuel cells). Since the Se/Ru materials are relatively inexpensive, the obvious remedy would be to use high load of the catalysts on GDE (or on the membrane). On top, peroxide may contribute to Se dissolution or Ru oxidation (we doubt it, but we do not really know). However, replacing 50% of Ru by Co may entirely prevent the catalyst decomposition, and it is now being investigated.

Dr Buder said:

Decrease of potential at the cathode is related to a mixed potential of the oxygen reduction reaction and the methanol oxidation. Does this not occur on the RuSe electrode also?

Professor Wieckowski answered:

Under certain conditions, the Se/Ru catalyst is completely tolerant to methanol. This would mean no mixed potential for oxygen reduction and methanol oxidation on Se/Ru. The study to document that for the methanol bulk concentration of 0.5 M is: D. Cao and A. Wieckowski, J. Inukai and N. Alonso-Vante, "Oxygen Reduction Reaction on Ruthenium and Rhodium Nanoparticles Modified with Selenium and

Sulfur", *J. Electrochem. Soc.*, 2006, **153**, A869-A874. At higher methanol concentrations and with the use of fuel cell electrodes, the Se/Ru catalyst is not insensitive to methanol crossover over the full potential range. However, the Se/Ru cathode response to methanol is much lower than for Pt over a wide potential range, showing much less methanol reactivity on Se/Ru than on Pt. (Also, at certain lower voltages, the Se/Ru cathode will not be affected by the methanol crossover; but at higher voltages, it may, but much less so than Pt.)

Professor Gewirth asked:
How does the catalyst decompose? Do you ever end up with a Se-free material?

Professor Wieckowski responded:
The Se/Ru catalyst decomposition (and deactivation) occurs due to Se dissolution. This happens at high potential; in fact the beginning of the dissolution is at *ca.* 0.9 (*vs.* RHE). Therefore, as indicated by the room temperature measurements, the Se/Ru catalysts would not be stable at or above 0.9 V. (Fig. 5 in: A. Lewera, J. Inukai, W.P. Zhou, D. Cao, H.T. Duong, N. Alonso-Vante and A. Wieckowski, "Chalcogenide Oxygen Reduction Reaction Catalysis: X-Ray Photoelectron Spectroscopy with Ru, Ru/Se and Ru/S Samples Emersed from Aqueous Media", *Electrochim. Acta*, 2007, **52**, 5759–5765.) After keeping the Se/Ru catalyst on RDE in an electrochemical cell overnight in the presence of oxygen, the OCP is 1.10 V. However, we have never physically tested if any Se remained on the RDE surface. The activity data with the same catalyst (kept in the cell overnight) was lower than the day before. But it was still quite high, which would indicate that there were substantial amounts of Se left on Ru. Bulk composition of the catalyst before and after the one night (or ever a week) exposure to the electrolyte did not show a substantial loss of Se. Evidently, during the overnight experiment at OCP, some amounts of Se were lost from the surface of the Se/Ru particles (deposited on RDE), and this affected the oxygen reduction activity.

Professor Behm asked:
The value of 9–10% of H_2O_2 formation you reported is significantly higher than the yields of 1–3% we determined (at 0.6–0.8 V) on a number of catalysts. Do you have any ideas on possible reasons for this difference?

Professor Wieckowski responded:
We believe that the answer is in catalyst loading (the loading refers to the amount in microgram per cm^2 of the catalyst that is deposited per unit area on the tip of a RDE and/or RRDE). In A. Bonakdarpour, C. Delacôte, R. Yang, A. Wieckowski and J.R. Dahn, Loading of Se/Ru/C Electrocatalyst on a Rotating Ring-Disk Electrode and the Loading Impact on a H_2O_2 Release during Oxygen Reduction Reaction, *Electrochem. Commun.*, 2008, **10**, 611–615 we found the production of H_2O_2 on the level of 1–3 % when the loading was high, *ca.* 90 microgram per cm^2 (at 0.6 V *vs.* RHE). Since we operated at the loading level of *ca.* 50 microgram per cm^2, the amount of the peroxide produced was *ca.* 8 % (at 30 microgram per cm^2, it would indeed be *ca.* 10%). (At 0.8 V, the amount of H_2O_2 formed is much lower for all RDDE curves published, except for very low loading.)

Professor Sun remarked:
The effort to find good Pt-free catalysts is very interesting, not just because of the high price of Pt but also due to the rare reserve of the Pt on the earth. The question is that, except the Se/Ru system, what other kind of Pt free catalysts can be explored in terms of stability and activity that are comparable to Pt-based catalysts? Could the property of Se/Ru be superior to that of Pt-based catalysts?

Professor Wieckowski responded:

Pt is rare, expensive and its delivery is very limited (significant production of Pt is only in South Africa and Russia). While Ru abundance is similar to Pt, that is, is low, Ru is of no interest to the jewellery industry and can be a side product of nuclear fission, (radioactively "cold" after 20 years). The property of Se/Ru may never be superior to that of Pt-based catalysts but may be comparable. Replacement of Pt by Ru may only be a temporary solution. Future alternatives are: (1) the use of ultra-small amounts of Pt, (2) the use of macrocycles or polymers containing some transition metals (mainly Co, Fe and/or Ni), pyrolyzed or not, (3) biocatalysts.

Professor Scherson asked:

Can you comment on the potential hazard of Se, which has been regarded as a problem in other electrochemical applications, and also on the economics of developing technologies based on Ru for which the price has been steadily climbing in recent months?

Professor Wieckowski answered:

"Se is toxic in large amounts, but trace amounts of it are necessary for cellular function in most, if not all, animals, including humans. For instance, exceeding the tolerable upper intake level of 400 micrograms per day can lead to selenosis, a serious illness that needs to be avoided. On the other hand the application of Se in electronic industry is widespread."

The conclusion from this reading (Wikipedia) is that dissolution of Se from Ru should be avoided as it eventually may cause selenium leaking to consumable fluids. However, the Se/Ru catalyst durability is a prerequisite for their applications, and is being taken care of in this research; the promise is that dissolution of Se will be avoided in fuel cell tests or in use in industry.

About the second part of the question, there is a significant incentive to extend the quantitative base of noble metals for fuel cell cathodes. Even assuming that Ru is less active than Pt, the Ru application would increase such a base by *ca.* 50%. The market price of ruthenium at this moment is 1/6 of the market price of platinum. While it may increase amid applications in fuel cells, the price of platinum may also increase (for instance to approach the price of Rh which is *ca.* 5× the current price of Pt).

Professor Markovic remarked:

What is the reaction in mechanisms on Ru/Se and how does Se modify the electronic properties of Ru? A true catalytic effect for the ORR on a high surface area is very difficult to establish without studying well characterised model extended surfaces. Any plans to examine the latter systems?

Professor Wieckowski answered:

1. In the already quoted paper: A. Bonakdarpour, C. Delacôte, R. Yang, A. Wieckowski and J.R. Dahn, "Loading of Se/Ru/C Electrocatalyst on a Rotating Ring-Disk Electrode and the Loading Impact on a H_2O_2 Release during Oxygen Reduction Reaction", *Electrochem. Commun.*, 2008, **10**, 611–615, we concluded that oxygen reduction to H_2O on the Se/Ru/C electrocatalyst occured through the H_2O_2 intermediate. Further work carried out by this group by the use of EC-NMR and synchrotron X-ray shows that Se keeps Ru metallic during the reduction reaction, confirming results of previous studies by Alonso-Vante.

2. Yes, we are planning to develop surface science of Se deposited on ruthenium single crystal surfaces, also including surface motions of Se on nanoparticles of Ru (by EC-NMR).

Professor Schiffrin opened the discussion of the paper by Cyril Delacôte, Arman Bonakdarpour, Christina M Johnston, Piotr Zelanay and Andrzej Wieckowski:

What is the mechanism of oxygen reduction on these materials? Your previous work[1] showed that a decrease in the loading increased the rate of peroxide generated. The reason for this is most likely the same as for the results presented by Behm,[2] *i.e.* the high rate of mass transfer at each separate centre is not compensated by the rate of further peroxide reduction or catalytic decomposition in the rest of the catalyst layer. It is interesting to note, however, that peroxide is indeed produced with this selenide. Hydrogen peroxide is also produced in the reduction of oxygen on pyrite, a disulfide iron mineral, but in much higher yields.[3,4] Do you think that these effects are due to a specific property of the chalcogenides in the material or is the reaction site the metal ion?

1. A. Bonakdarpour, C. Delacôte, R. Yang, A. Wieckowski and J.R. Dahn, *Electrochem. Commun.*, 2008, **10**, 611–615.
2. Y. E. Seidel, A. Schneider, Z. Jusys, B. Wickman, B. Kasemo and R. J. Behm, *Faraday Discuss.*, 2008, **140**, DOI: 10.1039/b806437g.
3. E. Ahlberg and A. E.Broo, *Int. J. Miner. Process.*, 1996, **46**, 73–89
4. E. Ahlberg and A. E.Broo, *J. Electrochem. Soc.*, 1997, **144**, 1281–1286

Professor Wieckowski answered:

All we know about the mechanism of oxygen reduction on Se/Ru is that it that the reduction goes predominantly to peroxide. That is, on isolated Se/Ru sites, the main or sole reaction is the peroxide formation ($n = 2$). However, at high load the peroxide is reacted to water, giving the n value close to 4. So, the answer is yes, "peroxide is indeed produced with this selenide" (*Electrochem. Commun.*, 2008, **10**, 611–615).

I think that Se/Ru sites on the chalcogenides have unique activity to reduce oxygen, apparently first to peroxide and next from peroxide to water (at high load). Reading Nenad's and other people work, it appears that on some metal sites ORR reaction may go to water directly. I am not investigating these reactions myself (as I said at Southampton); I am just sharing with you what I know about Se/Ru and from literature reading. Note that we study supported catalysts and the ratio between metal (Se + Ru) and carbon also affects the peroxide formation process. This will be further investigated.

Miss Hudson opened the discussion of the paper by Chengfei Yu, Shirlaine Koh, Jennifer Leisch, Michael F Toney and Peter Strasser*:

How sensitive is the technique you report for measuring compositional changes?

How much Cu needs to be removed before changes can be observed? Do you feel this technique can be used to characterise core shell particles with different monolayers of Pt?

Professor Strasser answered:

This method is ideal for probing core shell structure. Sensitivity is about 0.1 nm in our experiments. So perhaps 10–20 at% Cu taken from the surface of an alloy particle would suffice to see an effect. This depends on the size of the alloy particle though.

Professor Russell said:

I am very excited by this technique and the idea of combining this with EXAFS to understand particle structure in much more detail and with more confidence. Could you comment on the uniqueness of the result from these measurements, *i.e.* how model dependent is the answer?

I follow this up by asking could you tell the difference between a mixture of small pure Cu particles and layer Pt (pure) particles and small Cu rich regimes inside a larger particle? If not, a combination of techniques will be necessary to give full characterisation of the particles.

Professor Strasser replied:
You would not be able to distinguish between a mix of small Cu particles and large Pt particles and a core-shell arrangement. You need to rely on other knowledge methods to make sure the atoms are intermixed.

Professor Gewirth remarked:
You showed images of dealloyd bulk materials which exhibited internal tortuosity. However, you propose that the particles collapse when they are dealloyed. How big would a particle have to be in order that this internal tortuosity be preserved? Is it possible that a stabiliser could preserve this structure in particles?

Professor Strasser responded:
Roughly it should be on the order of the lengthscale of the filaments of the porous bulk materials, so at least 20–30 nm, that would be my guess.

Professor Chen said:
A question about the electrochemical dealloying of PtCu alloy nanoparticle. Your data show little change in the particle size after dealloying from $PtCu_3$ to form Pt_3Cu. From $PtCu_3$ to Pt_3Cu, about 90% Cu should have been leached off, so the resulted nanoparticle should be highly porous if the particle size doesn't change much. Did you see a significant increase in the active area of nanoparticle catalysts after dealloying? In addition, do you have any idea about the long-term stability of the dealloyed nanoparticle?

Professor Strasser answered:
We see some drop in particle size after dealloying when looking at the Pt edge. We believe we roughen the surface of the particle, which does increase the surface area. After dealloying, the measured ESCA values of the 600 °C annealed Pt-Cu particles are comparable to the pure Pt catalyst we started with in the synthesis, so about 60–70 m^2 g^{-1} ESCA. For Pt-Cu alloy precursors annealed at 950 °C, we measure about 30 m^2 g^{-1} after dealloying, so actually a a drop in surface area.

Professor Behm said:
Comparing with theory or single-crystal experiments, one would expect shifts of the onset for OH(ad) or H(ad) formation in the CV when increasing the thickness of the covering skin layer. Have you achieved such shifts for bimetallic particles with differently thick skin layer?

Professor Strasser replied:
Yes there are shifts in the Pt OH formation in the CV. We didn't see significant shifts in the H ads. We observed that dealloyed particles from Cu richer precursors show more of a shift in the Pt OH onset potential than Pt richer precursors. We are not sure whether that is definitely a function of the resulting shell thickness or of the different strain in the shells due to the different core composition.

Dr Fermin asked:
What are the typical values of particle density required for an acceptable signal-to-noise ratio in these experiments?

Professor Strasser responded:
The method is sensitive to electron density changes. That creates the contrast, not absolute density values. Typically, transitions of a metal to air (in case of particle probing) or carbon and air (pore probing) provide more than enough electron sensitivity contrast. Pt-Pd transition would be harder to probe with ASAXS.

PAPER

Efficient electrocatalytic oxygen reduction by the 'blue' copper oxidase, laccase, directly attached to chemically modified carbons

Christopher F. Blanford, Carina E. Foster, Rachel S. Heath and Fraser A. Armstrong*

Received 27th May 2008, Accepted 29th May 2008
First published as an Advance Article on the web 21st August 2008
DOI: 10.1039/b808939f

This discussion describes efforts to produce a stable, efficient electrocatalyst for four-electron O_2 reduction through the direct attachment of fungal laccase, a 'blue' copper oxidase, to functionalised carbon electrode materials. Commercially available carbons, including fibrous and porous materials, offer important opportunities for achieving high conductivity over high surface areas that can be chemically functionalised. A promising approach for attaching laccase to a carbon surface is to use the diazonium coupling reaction to generate protrusive aromatic functionalities that can bind to hydrophobic residues close to the 'blue' Cu site: this site provides a fast, intramolecular electron relay into the buried trinuclear Cu active site that converts O_2 rapidly and cleanly to H_2O. This enhancement procedure makes possible the stable, direct electrocatalytic reduction of O_2 at high potential with high efficiency in terms of turnover frequency per enzyme active site engaged with the electrode. The absence of electron-transfer mediators and simplicity of electrode system reveals the more inherent characteristics of the electrocatalytic mechanism that are masked in the waveform when a mediator is used. The study includes experiments to assess the effects of methanol and chloride ions on laccase electrocatalysis, complementing studies carried out by other groups, particularly those in which laccase is embedded in an electron-mediating gel.

1. Introduction

Over the past 15 years, we and others have developed a methodology, often known as 'protein film voltammetry', for elucidating the complex catalytic chemistry of redox metalloenzymes.[1-3] The central theme has been that the enzyme is adsorbed on an electrode such that electron transfer in and out of the active sites is facile and full catalytic activity is maintained. One of the outcomes of this research is the realisation that the active sites of enzymes are not only excellent catalysts, as has been long established, but they are also highly efficient '*electrocatalysts*'. When adsorbed on an electrode, many enzymes exhibit the equivalent of a 'high exchange current density' *i.e.* electrocatalysis that responds to application of the smallest of overpotentials.[4] The original data and background to this notion have already been published in several articles, so we devote this paper to enzymes known as 'blue' Cu oxidases, in particular laccases, that do not operate reversibly but nonetheless may well be the best electrocatalysts there are for the four-electron reduction of O_2 to H_2O.

Department of Chemistry, Inorganic Chemistry Laboratory, South Parks Road, Oxford, United Kingdom OX1 3QR. E-mail: fraser.armstrong@chem.ox.ac.uk

Efficient, four-electron interconversion between O_2 and H_2O without releasing intermediates is one of the most important goals in energy-related science.[5] By 'efficient', we mean that the reaction is achieved at a substantial turnover frequency under thermodynamic conditions as close as possible to reversibility. The problem is important for several reasons. First, the O_2 in our atmosphere is biogenic—the result of a single catalyst, a manganese-calcium-oxygen (MnCaO) cluster, contained in the photosynthetic machinery of green plants.[6] This catalyst evolved over two billion years ago to exploit the abundant substance H_2O as the electron donor for solar energy capture, optimising the return on a restricted energy budget from low-energy (2 eV) photons. In this aspect of life on Earth, O_2 is a waste product. Second, in the pursuit of cheap renewable H_2 production, we wish to mimic the efficiency of the MnCaO cluster in devising synthetic catalysts for the water splitting process, the point being that it is much more difficult to produce O_2 than H_2.[5] Thirdly, and most relevant for this paper, O_2 is the oxidant of choice in fuel cells and most of the inefficiency arises from the poor electrochemical kinetics of the reduction of O_2 to H_2O.[7,8]

The catalysts which are widely used for the O_2 reduction reaction are based on platinum (Pt), but in this regard Pt is far from perfect electrochemically, and the economics are ultimately unfavourable because Pt is a limited resource. If instead we look to biology for inspiration, we find that a particular class of enzymes, the so-called 'blue' copper or 'multi-copper' oxidases, are superb catalysts of four-electron O_2 reduction. Already, it has been proposed that laccases and bilirubin oxidase are superior to Pt in terms of the overpotential required.[9,10] The reasoning behind this arises from experiments carried out by Heller and colleagues who have measured the electrocatalytic reduction of O_2 by laccase and bilirubin oxidase, each trapped in a redox hydrogel containing high-potential Os(III/II) pyridyl complexes attached on long linkers.[9,11–13] The Os functions like 'a delivery agent on a bungee cord', providing a semi-mobile electron shuttle between the electrode and the enzyme molecules. The electrode potential at which O_2 is reduced is controlled by the reduction potential of the Os complex and how it is matched to that of the enzyme.

The laccases important for O_2 reduction are of the so-called 'high-potential' variety and they are secreted by tree-decaying white-rot fungi that include *Trametes versicolor* and *Pycnoporus cinnabarinus*.[14,15] These laccases catalyse the oxidation, by O_2, of organic intermediates formed during lignin degradation.[16] Various views of the molecular and active-site structures of *Trametes versicolor* laccase, including a cartoon, are shown in Fig. 1. The cartoon (Fig. 1(a)) emphasises an important feature of the enzyme: that oxidation of the organic substrate (by one-electron transfers to give a radical which reacts further) occurs at a *separate* site to that at which O_2 is reduced to H_2O. For each turnover of O_2, which occurs at a trinuclear Cu active site known as the Type 2/Type 3 Cu site (Fig. 1(b)), four separate electron transfers occur to a remote Cu atom known as the Type 1 or 'blue' Cu, which is located close to the protein's surface.[17,18] Fast, intramolecular electron transfer occurs over a distance up to 14 Å.[19–21] Fig. 1(c) and (d) show areas of surface (both exposed surface and internal channels or clefts) that come within 14 Å of the Type 1 site and Type 2/Type 3 trinuclear centre, respectively, close enough to allow sufficiently fast electron tunnelling from the electrode surface.[22] The Type 1 site lies at the top of a broad cleft approximately 10 Å deep. The Type 2/Type 3 site is buried but accessible to small molecules and ions such as O_2, H_2O and azide *via* two channels (Fig. 1(d)).[23] Fig. 1(e) shows surface regions of hydrophobic character where residues with a hydropathy index > 2.5 (phenylalanine, valine, leucine and isoleucine)[24] are displayed, and Fig. 1(f) shows the specific locations of exposed aromatic phenylalanine, tyrosine and tryptophan side-chains. These regions of the surface are important for considering the design of electrode surface microstructure for interaction with the enzyme. The residues around the Type 1 centre are important because it should be possible to 'hot-wire' the enzyme for O_2 reduction by modifying an

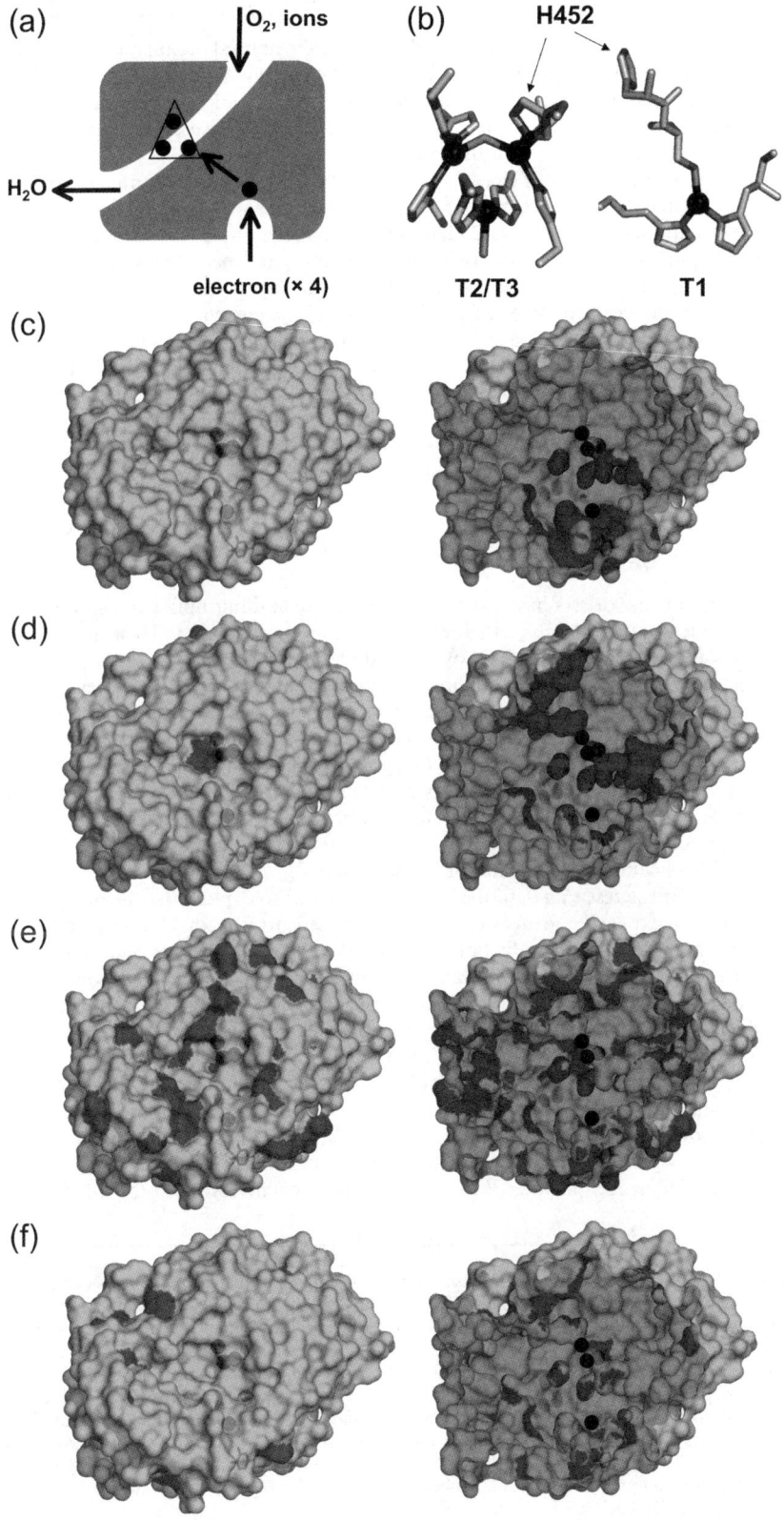

electrode with aromatic functionalities that mimic the natural organic substrates and also bind the enzyme tightly.[25] It is immediately evident that fast, direct electron transfer between an electrode and laccase should be possible provided the enzyme can be held in an orientation that brings the occluded surface close to the Type 1 Cu (the cleft) into contact with the electrode.

Direct electrochemistry of laccases has indeed been reported by several groups,[26–29] so why should there be such interest in using electron-transfer mediators, such as Os complexes or ABTS (2,2′-azinobis(3-ethylbenzothiazoline-6-sulfonate).[13,30] Heller's approach is focused not only on the necessity for mediators to access redox centres in the enzyme regardless of orientation with respect to the electrode surface, but also on the need to increase, greatly, the catalytic current density by engaging a much larger number of laccase molecules than could be adsorbed in a single layer. The redox hydrogel allows a large loading of enzyme, and the Os complexes on swinging linkers allow the apparent diffusion of electrons through a depth equivalent to hundreds of monolayers. In addition to providing a 3D electrode, this also makes an enzyme electrode more durable because a far higher enzyme loading, 'buffered' to loss of catalyst, is employed. High enzyme loading is still possible when electron transfer is direct: Kano and coworkers prepared porous and high-surface area carbon electrodes to adsorb a large amount of laccase which enabled them to obtain and sustain high current density in a fructose/oxygen fuel cell.[34]

Heller and co-workers have proposed that laccase and bilirubin oxidase are more efficient electrocatalysts than Pt for four-electron O_2 reduction. Their evidence is that O_2 is reduced at a smaller overpotential than Pt. In addition the enzymes operate in more neutral aqueous media, suitable for implantable devices.[31,32] An extension of this proposal would be to determine the efficiency based also on the activity per enzyme molecule that is receiving electrons directly from the electrode. Electron mediation by high-potential Os complexes masks the energetics intrinsic to the enzyme. When a very active enzyme is exchanging electrons directly with an electrode it is important also to rotate the electrode at sufficient speed that mass transport of the reactants does not influence the electrochemistry.[33] In practical fuel cell applications this would not be an option.

The aim of the research described in this paper was to explore the chemical modification of inexpensive forms of carbon for the adsorption of laccase to produce a stable and efficient O_2-reduction cathode. We have taken a rational approach, based upon the expectation that laccase should bind strongly at a carbon surface functionalised with polycyclic aromatic molecules targeting the hydrophobic cleft leading to the Type 1 Cu. We have also carried out experiments to assess the performance and stability of laccase in buffered methanol–water solutions and in solutions containing Cl^- ions. Methanol 'crossover' (passage through the electrolyte membrane) is one limitation in fuel cells because it interferes with O_2 reduction at a Pt cathode[35] and the ubiquitous chloride ion adsorbs at Pt.[36] The effect of methanol on catalytic activity has previously been studied with laccase immobilised in a Os redox hydrogel polymer.[37] In our study we have investigated the interference

Fig. 1 Key structural features of laccase. Coppers are shown as black spheres. (a) A simplified view of the laccase structure, highlighting the binding pocket for the electron-donor substrates, the separate gas and water channels and the buried trinuclear copper cluster. (b) Close-up of the coordination of the trinuclear (Type 2/Type 3) cluster and blue (Type 1) copper. For the surfaces shown in (c)–(f), the left-hand image depicts the outer surface and the right-hand image depicts a cut-away through the macromolecule in the same orientation. Note this reveals internal surfaces that form channels. The darker regions correspond to surfaces: (c) within 14 Å of the Type 1 copper, (d) *within* 14 Å of the Type 2/Type 3 copper, (e) with hydrophobic character (phe, ile, val, leu) and (f) with aromatic character (phe, tyr and trp). The copper coordination and protein surfaces are modelled on the crystal structure of *Trametes versicolor* laccase III (PDB 1 KYA).[19]

to direct electrocatalysis resulting from methanol and extended these measurements up to nearly 10 M concentration (33 wt%).

2. Experimental

Buffer solutions for laccase voltammetry

Except where stated, the electrolyte used for all laccase voltammetry was 0.20 M sodium citrate, pH 4.0 at 25 °C, which we refer to more simply as 'citrate buffer'.

Purification of laccase from *Trametes versicolor*

The pH of all buffers containing $(NH_4)_2SO_4$ were re-titrated to the same pH as the buffers free of $(NH_4)_2SO_4$. The column chromatography was carried out at 4–6 °C. The pH of all buffer solutions for chromatography was determined at 5 °C. Crude, lyophilised laccase (Fluka, 22.4 µmol catechol oxidised min^{-1} mg^{-1} at pH 4.5 and 25 °C) was dissolved in 10 mM sodium acetate pH 5.5 at a concentration of 5 g l^{-1}. The mixture was vacuum filtered sequentially through GF/A, 0.45 µm HV and 0.22 µm GV membrane filters (Whatman). The filtrate was loaded on a Q Sepharose Fast Flow strong anion exchange resin (GE Life Sciences, *ca.* 1 g laccase per 10 ml resin) and eluted with a 0–100 mM $(NH_4)_2SO_4$ gradient while monitoring the absorbance at 280 nm and 610 nm. The UV-visible absorbance ratio A_{280}/A_{610} stayed constant for fractions up to approximately 30 mM $(NH_4)_2SO_4$. These fractions were combined then dialysed into 1.65 M $(NH_4)_2SO_4$ in 20 mM sodium acetate pH 4.7 using a 30 kDa polyethersulfone dialysis membrane (Millipore). The dialysate was loaded on a Phenyl Sepharose Fast Flow hydrophobic interaction column (GE Life Sciences) and eluted with a 1.65–0 M $(NH_4)_2SO_4$ gradient. The laccase eluted over the range 0.6–0.9 M $(NH_4)_2SO_4$. Samples with a constant A_{280}/A_{610} ratio were concentrated and dialysed through a 30 kDa membrane into citrate buffer, then frozen rapidly in liquid N_2 and stored at -80 °C.

Isolation and purification of laccase from *Pycnoporous cinnabarinus*

Small squares of *Pycnoporus cinnabarinus* fungus (ATCC 200478) were plated onto malt extract agar and grown at 28 °C for 7 d. Washings from nine plates were used to inoculate 3 l of modified Dodson media. The fungus was grown for 5 d at 28 °C with shaking at >130 rpm; the expression of the *lcc3-1* laccase gene was induced with 2,5-xylidine after 24 h. Extracellular protein was harvested by $(NH_4)_2SO_4$ precipitation (500 g l^{-1}), re-suspended and dialysed overnight into 10 mM pH 4.6 sodium acetate buffer (buffer A). The dialysate was concentrated to approximately 30 ml and loaded onto a DE-52 anion exchange column previously equilibrated with buffer A. Protein was washed with buffer A and eluted with an increasing gradient (0–100 mM) of $(NH_4)_2SO_4$. The laccase-active fractions (determined by ABTS assay) were pooled and dialyzed into 100 mM potassium phosphate, pH 6, before addition of 300 g l^{-1} $(NH_4)_2SO_4$. These fractions were loaded onto a hydrophobic interaction column previously equilibrated with 2.67 M $(NH_4)_2SO_4$, washed, and eluted with a 2.67–0 M $(NH_4)_2SO_4$ gradient. Fractions testing positive for laccase activity by ABTS were concentrated and dialyzed into pH 4.0 citrate buffer, and checked for purity by SDS PAGE. The purest fractions were pooled, rapidly frozen in liquid N_2 and stored at -80 °C.

Pyrolytic graphite "edge" (PGE) rotating disc electrodes (RDEs)

Pyrolytic graphite sheets (Momentive Performance Materials) were machined into 2 mm diameter cylinders with the basal plane oriented normal to the cylinder axis. The graphite was fixed with silver-loaded two-part epoxy (RS Components) to a stainless steel rod inside a nylon casing embedded in epoxy (mixture of

HY1300 and CY 1300, Robnor Resins), then sanded to expose only a circular cross section (area 0.03 cm^2) of the 'edge' surface.

Voltammetry

Electrochemical experiments with PGE RDE working electrodes were carried out in a sealed, 8 ml jacketed cell with a Pt wire counter electrode and a saturated calomel electrode (SCE) maintained at RT and connected to the bottom of the cell by a Luggin capillary. The potential *vs.* SHE is thus 243 ± 1 mV.[38] The potential was controlled and the current was measured by a PGSTAT30 electrochemical analyser (EcoChemie). The RDE was controlled by a Model 636 Ring Disk Electrode System (Princeton Applied Research). Cyclic voltammograms (CVs) of laccase electrocatalysis were recorded between 0.7 V and 0.2 V *vs.* SCE at 5 mV s^{-1}. The potential for chronoamperometry (CA) measurements was set at 0.2 V *vs.* SCE. The rotating disc electrode was rotated at rates ≥2500 rpm for all measurements: a laccase-coated electrode delivering −150 µA cm^{-2} at 0.2 V *vs.* SCE at 2500 rpm produces 98% of the extrapolated current density at infinite rotation rate. For all catalytic voltammetry on RDEs, industrial-grade O$_2$ was flushed through the head space of the cell and out through a water bubbler so that the partial pressure of O$_2$ was maintained close to atmospheric pressure. For all laccase voltammetry on RDEs, 1–2 µl of laccase (typically 1–10 mg ml^{-1}) was applied. The catalytic wave was first monitored through five CV scans at 2 °C, after which the buffer was changed to remove any unbound enzyme and hence eliminate the influence of free enzyme on the catalytic voltammetry. The CV scans were repeated for 3–5 cycles. The difference in the catalytic current between the final scan in each case (enzyme in solution *vs.* no enzyme in solution) gave an initial indication of the strength of the enzyme adsorption: more strongly adsorbed enzyme gave zero or tiny differences.

Aryl diazonium modification of PGE RDEs

Electrode surface modifications were carried out as previously described.[25] Some of the polycyclic amines tested are shown in Fig. 2. Their corresponding diazonium salts are abbreviated thus: anthracene-1-diazonium (A1D), anthracene-2-diazonium (A2D), chrysene-2-diazonium (C2D) and naphthyl-1-diazonium (N1D). Ethanol was added to individual aryl amines to give a concentration of 8.3 mM, then an equal volume of 2 M aqueous HCl was added to the mixture. Similarly, water

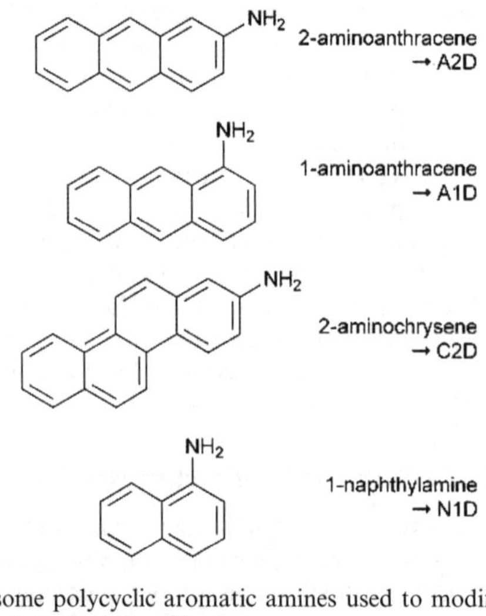

Fig. 2 Structures of some polycyclic aromatic amines used to modify carbon electrodes for enhanced binding and electroactivity of laccase.

was added to sodium nitrite to give a concentration of 15.2 mg ml^{-1}, then an equal volume of ethanol was added. The electrochemical cell solution was 0.1 M HCl in 50 vol% ethanol in water. All solutions were kept at ice temperature. The aryl amines were diazotised by adding 44 μl of the nitrite solution to 956 μl of the aryl amine solution. The solution was allowed to react on ice for 5 min. A known volume of cell solution was added to a jacketed electrochemical cell held at 2–4 °C, then 100 μl of the aryl diazonium solution was added per 1 ml of cell solution. The PGE RDE was rotated at 2500 rpm and the potential was scanned in a single cycle at 50 mV s^{-1} from 0.5 V *vs.* SCE to −0.3 V *vs.* SCE and back. After modification, the electrode was rinsed with ethanol, water and buffer, then excess buffer was removed with paper towel (taking care not to touch the electrode surface), and laccase was immediately applied.

Carbon cloth modification

Plain-weave carbon cloth with no wetproofing (E-TEK, Type A, 116 g m^{-2}, 0.35 mm thick) was cut into rectangles with geometric areas of 4–8 cm^2. Citrate-buffered Nafion solutions were prepared by mixing 0.4 ml of citrate buffer with 0.6 ml of 5.0 wt% tetrabutylammonium bromide–exchanged Nafion in ethanol,[39] then swirled to mix. No precipitate was visible in the final mixture. Two cloths were modified electrochemically in anthracene-2-diazonium (A2D) solution, two were modified by heating with the A2D solution, and two were left unmodified. Cloths to be modified with A2D were soaked in the amine solution for 30–60 min. Cloths left unmodified were soaked in the same acid solution but without 2-aminoanthracene. The vials containing the cloths were sonicated for 15–30 s to nucleate and remove trapped air from the cloths before placing the vials on ice. For electrochemical modification, cloths were placed in sufficient 0.1 M HCl (in 50 vol% ethanol) to cover the cloth but not the metal clip used to make the connection to the working electrode. Cloths were held at −0.3 V *vs.* SCE (−60 mV *vs.* SHE) for 30 s. For thermal modification, cloths were removed from the cold A2D solution, then placed in a 65 °C glassware drying oven for 30–60 min. Before coating with laccase, all the cloths were washed with at least 10 ml of ethanol (until the wash was colourless) then 5–10 ml water and finally 5–10 ml of the aqueous citrate buffer. One of each pair of cloths was coated with laccase in citrate buffer, the other was coated in a mixture of laccase and citrate-buffered Nafion. The loadings of laccase and Nafion were scaled to the geometric area of the cloth: 19 μg of laccase from *Trametes versicolor* (1.73 mg ml^{-1}, determined from the absorbance at *ca.* 610 nm, assuming $\varepsilon = 4900$ cm^{-1} M^{-1})[40] was applied per cm^2 of cloth. Cloths treated with Nafion were coated with 2 μl of buffered Nafion per cm^2 of cloth. A total volume of 15 μl was applied per cm^2 of cloth, with the extra volume made up with citrate buffer and the mixture was thoroughly swirled before use. Cloths were connected to the cathode testing apparatus 5 min after coating with laccase solution.

Cathode testing apparatus

Long-term stability of laccase on modified surfaces was assessed with a custom-built test apparatus. Cathodes for testing were suspended in a plastic storage container (Lock and Lock HPL844, 278 × 115 × 103 mm) containing *ca.* 2 l of 50 mM sodium citrate + 11.13 g l^{-1} MgSO$_4$·7H$_2$O (pH 4.0 at 25 °C). The solution was continually aerated by a Hi Oxygen Koi Pond Airstone (22 × 4 cm) pressurised from air by a Waterlife Ghost 2 air pump. The potential of these working electrodes was maintained at 0.2 V *vs.* SCE (reference electrode placed in the same solution) using a custom-built analogue potentiostat. The current through each working electrode was quantified by measuring the potential drop through a wire-wound resistor (10, 100 or 1000 Ω) and amplifying it 100-fold through a LTC1100 instrument amplifier. The potential drop through the resistor was kept below 50 mV. A 7 mm diameter glass carbon rod was used as the counter electrode to complete the cell circuit. Data acquisition and process control were executed by a LabJack UE9-Pro analogue-to-digital converter/digital-to-analogue converter (ADC/DAC) connected to a PC by a USB interface and controlled by a Visual Basic programme written in-house. Temperature was measured by an externally attached type K thermocouple interfaced to the LabJack UE9-Pro with a LabJack LJ Tick In-Amp (Rev. 2.0). The pH was monitored by a single-junction, gel-filled pH probe (Cole-Parmer) in the buffer, interfaced through a high-impedance amplifier then attached to the same LJ Tick In-Amp. Data were acquired at one-minute intervals. The pH and solution level were controlled by peristaltic pumps interfaced to the LabJack DAC output.

Meso-macro carbon pipe modification and testing

Cylinders of 'meso-macro' porous carbon (2 cm long × 2.5 mm diameter) provided by MAST Carbon (Guildford, Surrey, UK) were modified in a similar fashion to the carbon cloths. The carbon was produced by a proprietary process from epoxy that had been pulverised, extruded and pyrolysed to produce a broad, multimodal pore size distribution of enzyme-inaccessible mesopores with diameters *ca.* 3.5 nm, enzyme-accessible mesopores with modal diameters between 30 and 50 nm, and macropores with diameters of 10s of micrometres. The enzyme-accessible specific surface area of these materials was about 5 m^2 g^{-1}. Each tube weighed about 80 mg. The carbon was soaked in ice-cold 2-aminoanthracene (2AA) mixture (prepared as for the cloths but using 75 vol% ethanol in water) for 90 min. For the first 15 min, the carbon–2AA mixture was sonicated in ice water to help remove trapped air. Ice-cold sodium nitrite (7.6 mg ml^{-1}, in 75 vol% ethanol in water) was added to give a 1.3-fold excess of nitrite in a 4 mM solution of 2-aminoanthracene. After reacting on ice for 5 min, the carbon cylinder was transferred to an electrochemical cell (counter electrode: Pt mesh, reference electrode: SCE) at 4 °C containing 0.1 M HCl in 75 vol% ethanol in water. The cylinder was held for 90 s at −0.3 V *vs.* SCE, and a charge of approximately 0.9 mC was transferred. During electrochemical modifications and measurements, 1 cm of the cylinder was in contact with the rapidly bubbling solution. Over the course of three weeks, the laccase catalytic activity was assessed repeatedly by monitoring the cathodic current at 0.2 V *vs.* SCE at 25 °C. During the chronoamperometry, the cell solution was first purged with argon by continual vigorous bubbling for 30–60 minutes, then the gas was switched to industrial O_2. The laccase activity was taken as the difference between the two cathodic currents extrapolated to infinite time to remove the effects of trapped O_2, electrode resistance and surface charging effects such as double-layer capacitance.

Methanolic buffer preparation

The buffered solvent for the methanol tolerance experiments were prepared as 1.8 wt% (0.54 M), 3.5 wt% (1.1 M), 8.8 wt% (2.7 M), 16.5 wt% (5.0 M) and 33 wt% (9.7 M) mixtures of methanol in water. The molar concentrations were determined using published data.[41] Sodium citrate (Fisher) and citric acid (Acros) were dissolved into each solvent mixture to a concentration of 0.20 M. Proton activity is strongly affected by the presence of the methanol, so that simply adding methanol to aqueous buffer would not only dilute the buffer but would also alter the apparent pH. Thus, each methanol concentration was individually titrated to a measured pH of 4.0.

Methanol inhibition measurements

A film of laccase from *Trametes versicolor* was spotted on a 3 mm^2 PGE electrode that had previously been modified with A2D or C2D. The electrode was rotated at 2500 rpm throughout the experiment, except when changing buffer. Industrial-grade O_2 was sparged through a mixture of 33 wt% methanol in water before flushing into the headspace of a closed, jacketed electrochemical cell. The gas exited the cell through a water bubbler to ensure that the pressure above the cell solution was the same as atmospheric pressure. The catalytic wave was first monitored through five CV scans at 2 °C, after which the buffer was changed to remove any enzyme in solution and thus minimise the influence of free and weakly bound enzyme on the catalytic voltammetry. The CV scans were repeated at 2 °C and the electrode was stored for 18–30 hours at 4–6 °C in buffer with the electrode tip covered in paraffin film but leaving the graphite surface untouched. Subsequent experiments were all carried out at 25 °C. The electrocatalytic activity was monitored by CA followed by three CV scans in each buffer. The start times for each CA measurement were recorded to allow the data to be shown on a single plot. Before and after each measurement in methanol-containing buffer, CV and CA scans were recorded in methanol-free buffer.

Chloride inhibition measurements

The effect of chloride on electrocatalytic O_2 reduction was studied by cyclic voltammetry using *Trametes versicolor* laccase adsorbed on an A2D-modified PGE RDE. Aliquots of 5 M NaCl in citrate buffer were added to a known volume of citrate buffer in the cell. Cyclic voltammograms were recorded until the catalytic wave no longer changed between scans (typically 3–5 cycles for each concentration of NaCl).

3. Results

3.1 Voltammograms for electrocatalytic O_2 reduction on functionalised graphite rotating disc electrodes

In our early attempts to examine the direct, unmediated catalytic electrochemistry of fungal laccases adsorbed on a PGE electrode we initially observed high activity, as reported by others, but even at low temperatures (0 °C) we were unable to obtain a film that lasted for longer than an hour without having excess laccase in the cell solution. Seeking to obtain stable voltammetry without any excess laccase in solution, we explored ways to modify the electrode to introduce aromatic functionalities that we reasoned would target the cleft (a phenylalanine-rich region) leading to the Type 1 Cu site (Fig. 1(f)). Fig. 3 shows voltammograms obtained for O_2 (1 bar) reduction by laccase at pH 4, at a rotating PGE disc electrode, either unmodified or modified by attaching chrysene functionalities *via* the diazonium coupling of 2-aminochrysene. This procedure for modifying electrode surfaces was first described by Savéant and coworkers.[42] The modified electrode showed a higher current density and greatly increased stability, and we have previously shown that no enzyme was required in solution to obtain a stable voltammogram.[25] Interestingly though, we found the voltammetric waveshapes for numerous experiments to be similar regardless of whether and how the graphite surface was functionalised; they vary only in amplitude and stability. All voltammograms comprise a leading edge, covering the potential range at which O_2 is reduced at a low overpotential (down to approximately 0.7 V) followed by a trailing slope at more negative potential.

This looked to be a promising direction and we investigated a selection of different modifiers (Fig. 2). The performances vary, with 2-aminochrysene so far proving to be the best compound for achieving the highest current density and 1-naphthylamine the least effective. Attenuation of the electrocatalytic activity with time occurs in two stages, with an initial steep decrease over hours/days followed by a much slower decrease that lasts for weeks. The enzyme film on the electrode is also stable to electrode rotation (2500 rpm) for at least tens of hours. The enzyme from *Pycnoporus cinnabarinus* gives similar results to the purified, commercially available laccase from *Trametes versicolor* and these enzymes share an 84% amino acid sequence

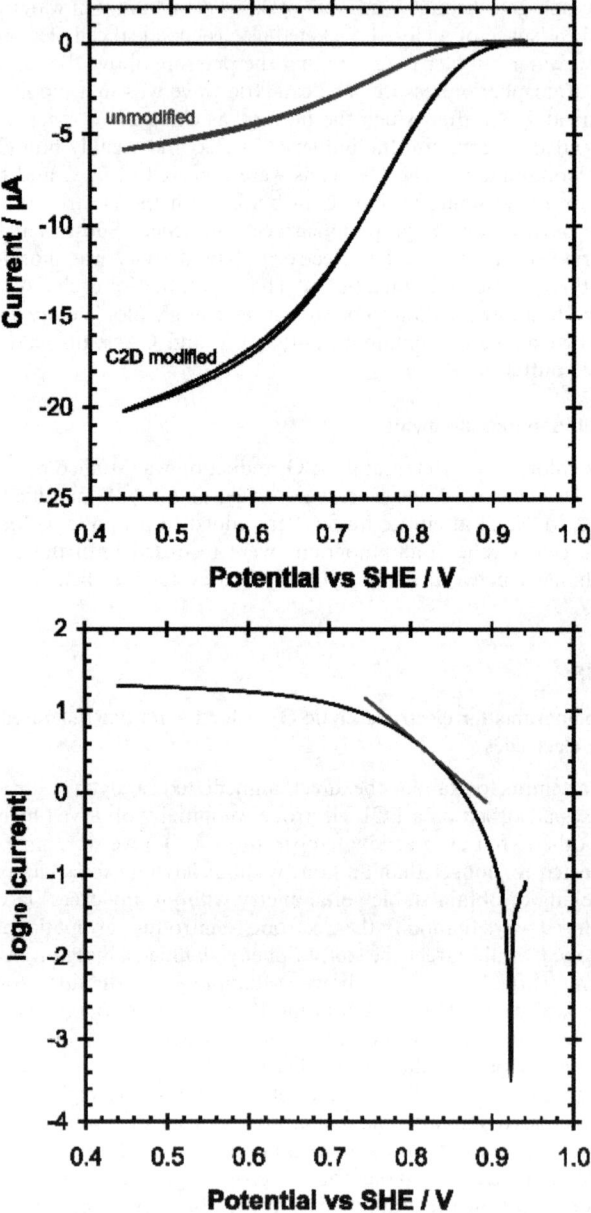

Fig. 3 Top panel: catalytic cyclic voltammetry of laccase from *Trametes versicolor* with and without electrode surface modification by chrysene-2-diazonium (C2D). (Conditions: 1 bar O_2, 3000 rpm, 25 °C, 5 mV s^{-1}, 200 mM sodium citrate pH 4.0, no enzyme in solution) Bottom panel: Tafel plot of the laccase voltammogram on the C2D-modified PGE electrode. The grey line shows the tangent where the Tafel slope equals (118 mV)$^{-1}$.

identity, but the results depicted here have been gained with the latter, for which the crystal structure has been determined.[19] Laccases are glycoproteins and the positioning and homogeneity of glycosylation may have a significant impact on their binding to surfaces *via* non-covalent forces.[43]

The resulting Tafel plot (Fig. 3 lower), shows no linear regions. The Tafel slope corresponding to a one-electron reaction (118 mV)$^{-1}$ lies close to the potential at which the cyclic voltammogram is steepest (*i.e.*, the catalytic potential). This is significant because laccases do not release any intermediate of O_2 reduction such as peroxide or superoxide. As discussed below, this observation is fully consistent with electrocatalysis involving successive one-electron transfers mediated by the Type 1 Cu centre. At potentials less than 800 mV, the slope increases to $(2-3 \text{ V})^{-1}$. The trailing edge is present in all experiments unless the O_2 concentration is low or the electrode is not rotating at sufficient speed to provide adequate transport of O_2. This emphasises the fact that in order to observe the 'true' response of the enzyme it is essential to alleviate effects due to substrate mass transport.

3.2 Longer-term trials using modified carbon materials

We have investigated alternative commercial carbon electrode materials, ranging from carbon cloth to porous carbon. Fig. 4 shows the time dependence of catalytic currents achieved for different carbon materials, modified with A2D then laccase and measured over several weeks. These experiments were carried out with the commercially available enzyme from *Trametes versicolor*. As we have observed with laccase on RDEs (monitored daily rather than continually), the cathodic current decays rapidly over the first few days then greatly stabilises.

In Fig. 4 (top), we show results of experiments in which the laccase-catalysed O_2 reduction current was monitored continuously for several weeks at a potential of 0.44 V *vs.* SHE. The temperature was recorded but not controlled, so some of the plots show oscillations in current density that correlate with changes in room temperature throughout the day. Most strikingly, the plots show a rapid, near total loss in current for cloths with no aryl diazonium modifications. Within 72 hours, the current density from these unmodified cloths dropped to below 10 μA cm^{-2}. Also, as we observed on PGE rotating disk electrodes studied by cyclic voltammetry recorded daily (rather than by continuous CA), the initial cathodic current from A2D-modified materials was about twice that of unmodified materials. Although the plots suggest that the heating modification method provides higher activity, repeated experiments showed that electrochemically diazotised surfaces produce higher catalytic activities more reliably. These experiments also suggested that cloths modified electrochemically have a better long-term stability.

Additionally, we observed that the addition of Nafion to the laccase mixture had no effect on the O_2-reduction catalysis. Nafion has been previously used by Minteer and her group to stabilise multi-copper oxidases in fuel cell applications.[44] We used Nafion to ensure the laccase stays hydrated and has ample proton supply even when it not submerged in buffer (as in a conventional fuel cell stack). We have found that buffered Nafion solutions do not affect laccase voltammetry. Although this is not surprising given that the cloths were immersed in buffered solution, we have found that the same modifications could be adapted to produce a laccase-coated cathode that runs in flowing water-saturated O_2.

3.3 Electrocatalytic O_2 reduction in methanolic solutions

Fig. 5 illustrates how the electrocatalytic response of laccase adsorbed on C2D-modified PGE electrode is influenced by increasing concentrations of methanol. Scans in methanolic buffer were bracketed by scans in methanol-free buffer to differentiate between reversible inhibition of the enzyme and irreversible loss of activity. The top panel of Fig. 5 shows the combined CA scans. The bowed shape of the individual scans results from a convolution of oxygenation of the cell solution and some film loss. Although the electrode used for the plots in Fig. 5 had been 'aged' for about 1.5 d, there appeared to be substantial loss of activity in the first hour of use. The catalytic current in 0.5 M methanol was *higher* than in the methanol-free buffer, which may reflect higher O_2 solubility in methanolic solutions. Subsequent CA measurements did not show any further improvement. Methanol

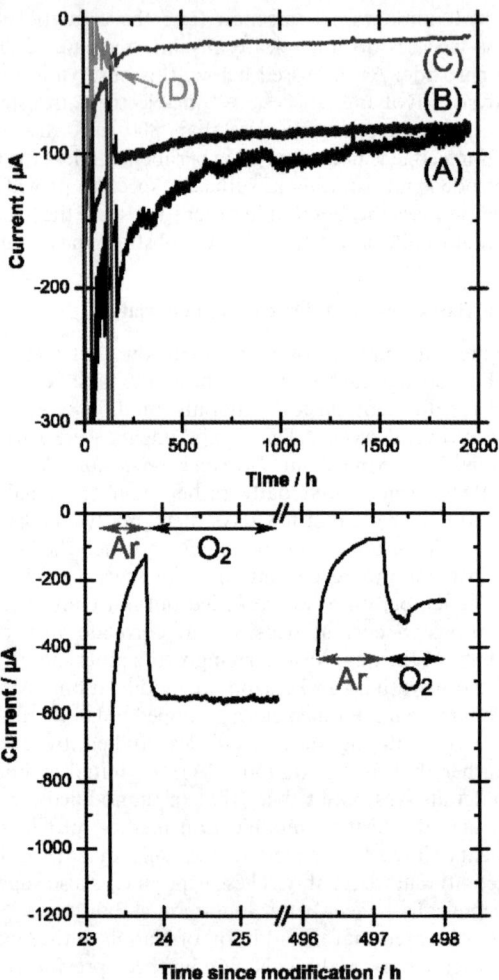

Fig. 4 Extended testing of O_2-reduction catalysis by laccase from *Trametes versicolor* adsorbed on modified carbon materials. Top: O_2-reduction current from laccase adsorbed on carbon cloth (A) thermally modified by A2D and including buffered Nafion, (B) electrochemically modified by A2D and including buffered Nafion, (C) unmodified but including buffered Nafion, and (D) unmodified with no Nafion. The artefacts at 120–140 h were caused by a disconnection in the data-logging hardware. The discontinuity at 170 h was caused by a momentary disconnection of the counter electrode. (Conditions: 0.2 V *vs.* SCE, stationary, 200 mM sodium citrate pH 4, air-sparged, room temperature.) Bottom: O_2-reduction current from laccase adsorbed on porous, A2D-modified carbon from pyrolysed epoxy (0.2 V *vs.* SCE, 25 °C, stationary, 200 mM sodium citrate pH 4, O_2-sparged).

concentrations of 2.7 M and 5.0 M inhibited catalysis (reversibly) by about 10% and 20%, respectively. At the highest methanol concentration (9.7 M), we always observed an irreversible loss of catalytic activity.

The CVs taken after each CA scan are plotted in the middle panel of Fig. 5 and the first derivatives of the reductive sweep of each CV are plotted in the bottom panel. The shape of each CV was similar whether recorded in methanol or not; only the size of the wave changed. Interestingly, for the C2D-modified electrode, we observed a small decrease in the catalytic overpotential between the first CV (in methanol-free buffer) and all subsequent CVs. The catalytic wave shifted about +25 mV.

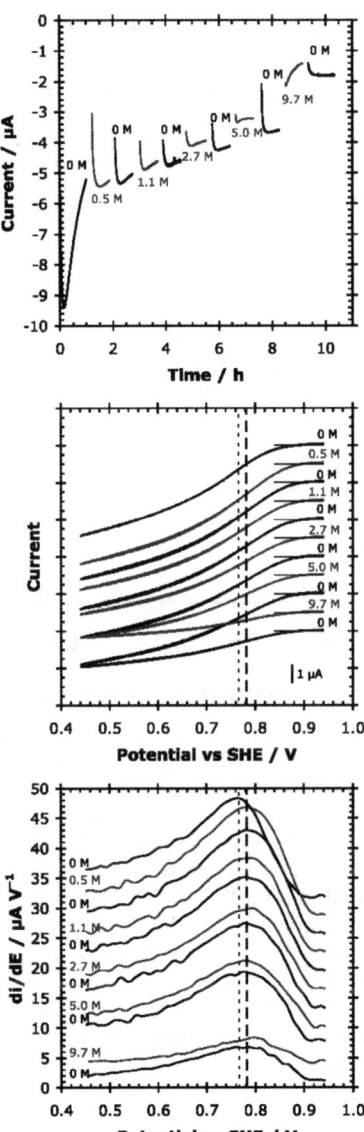

Fig. 5 The effect of methanol on the catalytic voltammetry of laccase from *Trametes versicolor*, with scans in methanol shown in dark grey. Top: Change in reduction current with time and methanol concentration. Middle: Cyclic voltammograms recorded after chronoamperometric measurements shown in the top panel (third scan shown). Horizontal lines show the zero-current point for each scan. Bottom: The first derivative of the cyclic voltammograms shown in the top-right panel. The individual traces in the middle and bottom panels are offset by 1 μA and 3 μA V^{-1}, respectively, with earlier scans shown higher up. The vertical lines show the maximum in the first derivative of the reductive wave of the cyclic voltammograms (dotted: first 0 M CV; dashed: all subsequent scans). Conditions: C2D-modified PGE, 1 bar O$_2$, 3000 rpm, 25 °C, pH* = 4.0, 0.2 V *vs.* SCE (for chronoamperometry), 5 mV s^{-1} (for cyclic voltammetry).

3.4 Electrocatalytic O$_2$ reduction in the presence of chloride

The CVs plotted in Fig. 6 show how NaCl affects electrocatalytic O$_2$ reduction by laccase from *Trametes versicolor* adsorbed on an A2D-modified PGE electrode.

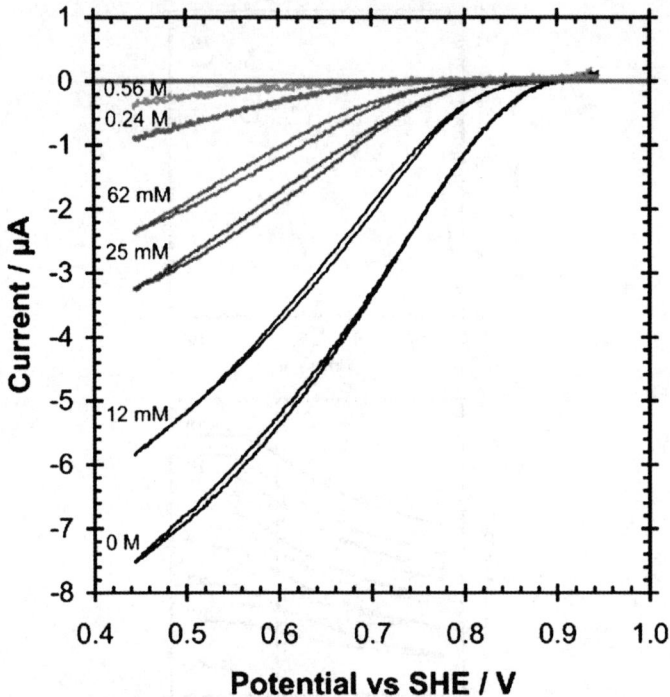

Fig. 6 The effect of chloride on the electrocatalytic reduction of O_2 by laccase from *Trametes versicolor*. Cyclic voltammograms at various chloride concentrations. Conditions: A2D-modified PGE (area 0.03 cm^2), 1 bar O_2, 3000 rpm, 25 °C, pH = 4, 5 mV s^{-1}.

At concentrations of about 1 mM and below, the catalytic potential remains constant. At 0.2 mM chloride, the current measured at 0.2 V (0.45 V *vs.* SHE) was about 10% smaller and in 1 mM chloride, the current was about 20% smaller. These observations are in marked contrast to Pt-catalysed O_2-reduction at similar concentrations (see Discussion).[36] At higher concentrations of chloride, the catalytic wave reproducibly shifted to more reducing potentials, up to −130 mV in 1 M chloride and the leading edge was lost. When subjected to concentrations of chloride ∼1 M, the catalytic activity of a laccase-coated electrode never returned to the same level of electrocatalytic activity.

4. Discussion

The electrocatalytic action of laccase directly adsorbed on carbon electrodes is greatly improved after derivatising the surface with polycyclic aromatic functionalities by diazonium coupling of the corresponding amines. This modification, using different polycyclics, affects stability and current but not (to any great degree) the voltammetric waveshape or overpotential at which O_2 is reduced. Among a selection of polycyclics, C2D provided the highest currents. Two limiting hypotheses for the stabilisation are: (1) that individual polycyclic aromatics act as substrate mimics and the laccase molecules bind in a specific way; (2) that the polycyclic aromatics create much greater local roughness, at the nanometre scale, on the carbon surface. It is likely that polycyclic functionalities protrude naturally from carbon surfaces and the modification procedures simply increase the density of the more strongly binding functionalities. This conceptual model is consistent with our previously published epifluorescence microscopy measurements,[25] our repeated observation that electrocatalytic voltammograms display an initial sharp drop in current followed by a long-period of high stability, and the general similarity in waveshapes regardless of how the electrode has been modified. Comparison may also be drawn with gold, a 'better defined' surface material for electrochemical investigations of proteins. Apart from work by Johnson et al.,[26] the few other examples of laccase catalysis on gold show CVs which appear to be little more than O_2 reduction by gold itself (whether bare or modified with self-assembled monolayers (SAMs) of organic thiols).[45,46]

In studies of laccase electrocatalysis at redox-hydrogel electrodes, the potential at which O_2 is reduced relates strongly to the reduction potential of the mediator (Os). When a mediator is absent, the potential should reflect the properties of the centre controlling electron transfer in the enzyme. Recent work has shown that mutating bilirubin oxidase to lower the reduction potential of the Type 1 Cu results in a corresponding negative shift in the voltammogram for O_2 reduction.[47] An unexpected feature of our work is that the shapes of the catalytic voltammograms vary little with modification or with age of the electrode. In all cases, provided the electrode is rotating, the shape has a leading edge at high potential followed by a trailing residual slope at more negative potential. The Tafel slope is curved with a value corresponding to one electron close to the potential of the Type 1 Cu site, consistent with this centre acting as a one-electron relay between the electrode and the Type 2/Type 3 trinuclear Cu centre. The trailing edge could be due to electrons transferring to the active site of the enzyme inefficiently over a much longer range, i.e. surface regions further away from the Type-1 Cu site. The distance profiles shown in Fig. 1 indeed indicate the feasibility of a slow, direct electron transfer to the Type 2/Type 3 centre *not* involving the Type-1 Cu centre. Given the expected highly efficient coupling between Type 1 and Type 2/Type 3 centres (the cysteine coordinating the Type 1 Cu is adjacent to one of the histidines coordinating Type 3 Cu) this implies that even with our intention of binding the enzyme through substrate-recognition, some enzyme can be bound otherwise. A related proposal has been made by Gorton and coworkers.[48]

The key performance issues for laccase fuel cell cathodes are current density and lifetime,[49] and our investigations have shown that modifying a PGE surface with specific polycyclic aromatics both increases the current density and the lifetime of laccase cathodes.[25] The long-term stable current densities (ca. 50 $\mu A\ cm^{-2}$) measured from A2D-modified cloth are still modest compared to platinum-coated electrodes, but the laccase-coated cathode output should be able to be greatly amplified by optimising surface texture ('roughness'), dimensionality, mass and electrical transport, and laccase loading.

In active direct methanol fuel cells the Pt anode operates typically on methanol concentrations 1–2 M[50] (passive versions work best at 4–5 M)[51,52] but methanol 'crossover' from the anode compartment to the cathode compartment occurs through the polymer electrolyte membrane and this lowers the cell voltage because it interferes with O_2 reduction.[35] A recent study of the effect of methanol on electrocatalysis by laccase in redox hydrogels showed that its activity in methanol was considerably improved over that of Pt. However, the changes in the laccase voltammetry were attributed to the effect of methanol on the structure of the mediator hydrogel, and performance was affected most when the Os complexes were more highly charged.[37] Our studies support the notion that laccase is relatively unperturbed by high concentrations of methanol—in line with a buried Type 2/Type 3 centre to which methanol may have very limited access and a very different mechanism for O_2 reduction.

Oxygen-reduction catalysed by platinum (in H_2SO_4) shows a complex response. The (111) surface of platinum cathode shows no inhibition in the presence of chloride. However, the presence of 1 mM chloride shifts the catalytic wave by over 150 mV for the (110) surface and over 200 mV for the (100) surface.[36] In contrast, laccase catalysis in 1 mM chloride created only a marginal decrease in electrocatalytic oxygen-reduction current. Chloride concentrations more than

1000 times higher are required to produce a similar shift in catalytic potential in laccase compared to Pt(110) and Pt(100). Chloride binding to Pt decreases the number of sites available for O_2 binding and increases the production of peroxide relative to the Pt surfaces in the absence of chloride.

5. Conclusion

We have demonstrated that O_2-reduction cathodes, using laccases as electrocatalysts, without a mediator, can be made by diazotising a variety of carbon forms. These cathodes, which use only small amounts of laccase are far more stable, over a long period of time, than the unmodified materials. They demonstrate high methanol tolerance and a decreased sensitivity to chloride ions compared to several platinum surfaces. Further optimisation of laccase cathodes that use high surface area carbon materials may result in cathodes that can compete with, and perhaps outperform, platinum as the choice catalyst for the oxygen-reduction reaction.

Acknowledgements

The authors thank the Leverhulme Trust (Grant no. F/08 699/C), the UK Engineering and Physical Sciences Research Council (Supergen V), and the BBSRC (studentship no. BBS/S/A/2004/10921) for funding; John Varcoe (University of Surrey) for supplying the tetrabutylammonium bromide-exchanged Nafion solution; Steve Tennison and Tony Rawlinson (MAST Carbon) for samples of porous 'mesomacro' carbon; Greg Wildgoose (Physical and Theoretical Chemistry Laboratory, University of Oxford) for glassy carbon; and Paul Smith (Inorganic Chemistry Laboratory, University of Oxford) for assistance in building the cathode testing apparatus.

References

1 F. A. Armstrong, *J. Chem. Soc., Dalton Trans.*, 2002, 661–671.
2 F. A. Armstrong, *Curr. Opin. Chem. Biol.*, 2005, **9**, 110–117.
3 J. Hirst, *Biochim. Biophys. Acta*, 2006, **1757**, 225–239.
4 K. A. Vincent, A. Parkin and F. A. Armstrong, *Chem. Rev.*, 2007, **107**, 4366–4413.
5 N. S. Lewis and D. G. Nocera, *Proc. Natl. Acad. Sci. U. S. A.*, 2006, **103**, 15729–15735.
6 F. A. Armstrong, *Philos. Trans. R. Soc. London, Ser. B*, 2008, **363**, 1263–1270.
7 T. R. Ralph and M. P. Hogarth, *Platinum Met. Rev.*, 2002, **46**, 3–14.
8 J. Larminie and A. Dicks, *Fuel Cell Systems Explained*, Wiley, Chichester, 2nd edn, 2003.
9 N. Mano, J. L. Fernandez, Y. Kim, W. Shin, A. J. Bard and A. Heller, *J. Am. Chem. Soc.*, 2003, **125**, 15290–15291.
10 V. Soukharev, N. Mano and A. Heller, *J. Am. Chem. Soc.*, 2004, **126**, 8368–8369.
11 A. Heller, *Phys. Chem. Chem. Phys.*, 2004, **6**, 209–216.
12 A. Heller, *Curr. Opin. Chem. Biol.*, 2006, **10**, 664–672.
13 N. Mano, V. Soukharev and A. Heller, *J. Phys. Chem. B*, 2006, **110**, 11180–11187.
14 F. S. Archibald, R. Bourbonnais, L. Jurasek, M. G. Paice and I. D. Reid, *J. Biotechnol.*, 1997, **53**, 215–236.
15 C. Eggert, U. Temp and K.-E. L. Eriksson, *Appl. Environ. Microbiol.*, 1996, **62**, 1151–1158.
16 R. ten-Have and P. J. M. Teunissen, *Chem. Rev.*, 2001, **101**, 3397–3413.
17 R. Lontie, *Copper Proteins and Copper Enzymes*, CRC Press, 1984.
18 L. Quintanar, C. Stoj, A. Taylor, P. Hart, D. Kosman and E. Solomon, *Acc. Chem. Res.*, 2007, **40**, 445–452.
19 T. Bertrand, C. Jolivalt, P. Briozzo, E. Caminade, N. Joly, C. Madzak and C. Mougin, *Biochemistry*, 2002, **41**, 7325–7333.
20 N. Hakulinen, M. Andberg, J. Kallio, A. Koivula, K. Kruus and J. Rouvinen, *J. Struct. Biol.*, 2008, **162**, 29–39.
21 E. I. Solomon, U. M. Sundaram and T. E. Machonkin, *Chem. Rev.*, 1996, **96**, 2563–2605.
22 C. C. Page, C. C. Moser, X. X. Chen and P. L. Dutton, *Nature*, 1999, **402**, 47–52.
23 K. Piontek, M. Antorini and T. Choinowski, *J. Biol. Chem.*, 2002, **277**, 37663–37669.
24 J. Kyte and R. Doolittle, *J. Mol. Biol.*, 1982, **157**, 105–132.
25 C. F. Blanford, R. S. Heath and F. A. Armstrong, *Chem. Commun.*, 2007, 1710–1712.

26 D. L. Johnson, J. L. Thompson, S. M. Brinkmann, K. A. Schuller and L. L. Martin, *Biochemistry*, 2003, **42**, 10229–10237.
27 C.-W. Lee, H. B. Gray, F. C. Anson and B. G. Malmstrom, *J. Electroanal. Chem.*, 1984, **172**, 289–300.
28 S. Shleev, A. Jarosz-Wilkolazka, A. Khalunina, O. Morozova, A. Yaropolov, T. Ruzgas and L. Gorton, *Bioelectrochemistry*, 2005, **67**, 115–124.
29 M. H. Thuesen, O. Farver, B. Reinhammar and J. Ulstrup, *Acta Chem. Scand.*, 1998, **52**, 555–562.
30 G. T. R. Palmore and H.-H. Kim, *J. Electroanal. Chem.*, 1999, **464**, 110–117.
31 H. H. Kim, N. Mano, X. C. Zhang and A. Heller, *J. Electrochem. Soc.*, 2003, **150**, A209–A213.
32 N. Mano, H. H. Kim, Y. C. Zhang and A. Heller, *J. Am. Chem. Soc.*, 2002, **124**, 6480–6486.
33 T. Reda and J. Hirst, *J. Phys. Chem. B*, 2006, **110**, 1394–1404.
34 Y. Kamitaka, S. Tsujimura, N. Setoyama, T. Kajino and K. Kano, *Phys. Chem. Chem. Phys.*, 2007, **9**, 1793–1801.
35 V. Paganin, E. Sitta, T. Iwasita and W. Vielstich, *J. Appl. Electrochem.*, 2005, **35**, 1239–1243.
36 V. Stamenkovic, N. M. Markovic and P. N. Ross, *J. Electroanal. Chem.*, 2001, **500**, 44–51.
37 Y. H. Sun and S. C. Barton, *J. Electroanal. Chem.*, 2006, **590**, 57–65.
38 A. J. Bard and L. R. Faulkner, *Electrochemical Methods: Fundamentals and Applications*, John Wiley, New York, 2001.
39 C. M. Moore, N. L. Akers, A. D. Hill, Z. C. Johnson and S. D. Minteer, *Biomacromolecules*, 2004, **5**, 1241–1247.
40 G. Fahraeus and B. Reinhamm, *Acta Chem. Scand.*, 1967, **21**, 2367.
41 D. R. Lide, H. P. R. Frederikse and Chemical Rubber Company, *CRC handbook of chemistry and physics: a ready-reference book of chemical and physical data*, CRC Press, Boca Raton; London, 1995.
42 M. Delamar, R. Hitmi, J. Pinson and J. M. Savéant, *J. Am. Chem. Soc.*, 1992, **114**, 5883–5884.
43 P. Baldrian, *FEMS Microbiol. Rev.*, 2005, 1–28.
44 S. Topcagic and S. D. Minteer, *Electrochim. Acta*, 2006, **51**, 2168–2172.
45 G. Gupta, V. Rajendran and P. Atanassov, *Electroanalysis*, 2004, **16**, 1182–1185.
46 M. Pita, S. Shleev, T. Ruzgas, V. M. Fernandez, A. I. Yaropolov and L. Gorton, *Electrochem. Commun.*, 2006, **8**, 747–753.
47 Y. Kamitaka, S. Tsujimura, K. Kataoka, T. Sakurai, T. Ikeda and K. Kano, *J. Electroanal. Chem.*, 2007, **601**, 119–124.
48 S. V. Shleev, A. Christenson, V. Serezhenkov, D. Burbaev, A. I. Yaropolov, L. Gorton and T. Ruzgas, *Biochem. J.*, 2005, 385.
49 W. E. Farneth and M. B. D'Amore, *J. Electroanal. Chem.*, 2005, **581**, 197–205.
50 K. Scott, W. M. Taama and P. Argyropoulos, *J. Power Sources*, 1999, **79**, 43–59.
51 J. G. Liu, T. S. Zhao, R. Chen and C. W. Wong, *Electrochem. Commun.*, 2005, **7**, 288–294.
52 G.-G. Park, T.-H. Yang, Y.-G. Yoon, W.-Y. Lee and C.-S. Kim, *Int. J. Hydrogen Energy*, 2003, **28**, 645–650.

Steady state oxygen reduction and cyclic voltammetry

Jan Rossmeisl,[*a] Gustav S. Karlberg,[a] Thomas Jaramillo[b] and Jens K. Nørskov[a]

Received 6th February 2008, Accepted 28th March 2008
First published as an Advance Article on the web 20th August 2008
DOI: 10.1039/b802129e

The catalytic activity of Pt and Pt_3Ni for the oxygen reduction reaction is investigated by applying a Sabatier model based on density functional calculations. We investigate the role of adsorbed OH on the activity, by comparing cyclic voltammetry obtained from theory with previously published experimental results with and without molecular oxygen present. We find that the simple Sabatier model predicts both the potential dependence of the OH coverage and the measured current densities seen in experiments, and that it offers an understanding of the oxygen reduction reaction (ORR) at the atomic level. To investigate kinetic effects we develop a simple kinetic model for ORR. Whereas kinetic corrections only matter close to the volcano top, an interesting outcome of the kinetic model is a first order dependence on the oxygen pressure. Importantly, the conclusion obtained from the simple Sabatier model still persists: an intermediate binding of OH corresponds to the highest catalytic activity, *i.e.* Pt is limited by a too strong OH binding and Pt_3Ni is limited by a too weak OH binding.

Introduction

The largest challenge in proton-exchange membrane fuel cell (PEMFC) catalysis concerns the high overpotentials required to drive the oxygen reduction reaction (ORR).[1] Sluggish ORR kinetics account for the majority of the voltage drop in PEMFCs, limiting state-of-the-art systems to operating voltages of only ∼0.7 V, far from the equilibrium potential of ∼1.2 V. The highest-performance PEMFCs all rely upon Pt to catalyze this reaction despite the high cost of this material, a problem which will be further exacerbated by its scarcity. In recent years, much effort has been devoted to improving the ORR activity of Pt, with secondary goals of improving catalyst stability and reducing catalyst cost. A number of different approaches have been taken in this regard, however three in particular have been shown to be effective: (1) alloying Pt with other transition metals, *e.g.* Co, Ni or Fe,[2–4] (2) modifying Pt's lattice constant *via* Pt (or Pt-alloy) overlayer structures,[5–8] and (3) overlaying other metals, such as Au, onto Pt supports.[9] To date, the most active ORR catalyst was recently reported by Stamenkovic *et al.*,[5] whose experiments revealed an order of magnitude improvement of $Pt_3Ni(111)$ over Pt (111).

The relationship between OH adsorption and the ORR is a critical issue in PEMFC catalysis. In the following work, we investigate this by applying a previously

---

[a] *Center for Atomic-scale Materials Design, Department of Physics, Technical University of Denmark, Lyngby, DK-2800, Denmark*
[b] *Center for Nano-particle Functionality, Department of Physics, Technical University of Denmark, Lyngby, DK-2800, Denmark*

published theoretical model based on first-principles.[2,10] Our general aim is to elucidate the role of adsorbed OH in the ORR catalytic activity. The only inputs to our model are density functional theory (DFT) calculations and standard molecular tables.

The paper is divided into two parts. In the first part we compare the output from our theoretical Sabatier model to recent experimental results for Pt(111) and Pt$_3$Ni(111),[5] as these are well-known "benchmarks" for ORR catalysis. We find that the key features observed in experimental cyclic voltammetry in an O$_2$-free environment are predicted by the simple Sabatier model. In the second part of the paper, we develop a more detailed kinetic model which allows us to compare the calculated and measured steady state current densities in the presence of O$_2$. We find that the ORR current can indeed be expressed as a function of the OH-coverage, noting however that it is the coverage of OH under operating conditions (in the presence of dissolved O$_2$) that is used and not the coverage measured in normal CV experiments.

Cyclic voltammetry

We begin by examining the potential dependence of the OH coverage resulting from water dissociation during cyclic voltammetry in an O$_2$-free environment. The reaction to consider is:

$$H_2O(l) \leftrightarrow OH^* + H^+ + e^- \qquad (1)$$

By introducing a theoretical counterpart to the standard hydrogen electrode,[10] we re-write reaction (1) at zero potential as:

$$H_2O(l) \leftrightarrow OH^* + 1/2H_2(g) \qquad (2)$$

The free energy of this reaction is defined as $\Delta G_{OH} = \Delta E + \Delta ZPE - T\Delta S$, where ΔZPE and $T\Delta S$ account for changes in zero point energy and entropy. ΔE is the reaction energy as calculated using DFT calculations. ΔG_{OH} corresponds to the change in free energy of reaction (1) at zero potential. At finite potentials ($U \neq 0$) the chemical potential of electron in reaction (1) is changed by $-eU$. According to a previously developed method for calculating cyclic voltammograms based on density functional theory,[11] the coverage of OH as a function of potential can be derived from the coverage dependence of ΔG_{OH}.

We will consider the case of adsorbed OH in contact with liquid water and will, therefore, for simplicity assume an ice-like water–hydroxyl layer. Such model systems have been intensively studied in surface science due to their close resemblance to realistic metal-water interfaces.[12] Choosing this model system for the metal–water interface means that the total coverage of OH and H$_2$O is always 2/3 of a monolayer (due to the supply of H$_2$O molecules from the "real" liquid phase).[13] As can be seen from a previously published detailed investigation,[14] there are two different coverage regimes for the OH–H$_2$O interaction for such layers on Pt(111)—up to 1/3 of a monolayer and above 1/3 of a monolayer of OH. The origin of these different regimes is two-fold; first, the OH–H$_2$O interaction is stronger than both the OH–OH and the H$_2$O–H$_2$O interaction, and second, the hydrogen scrambling in these overlayers is very facile,[15] meaning that the relaxation time for finding the most stable overlayer is fast.

For OH coverages less than 1/3 of a monolayer, every OH molecule will have three H$_2$O molecules as nearest neighbors. As the coverage of OH exceeds 1/3 of a monolayer, however, "defects" in the OH/H$_2$O adlayer appear where OH molecules are nearest neighbors. Since the hydrogen bond formed between OH molecules is just half as strong as the hydrogen bond formed between OH and H$_2$O,[14] the latter situation will result in OH molecules less strongly bound to the surface. Looking

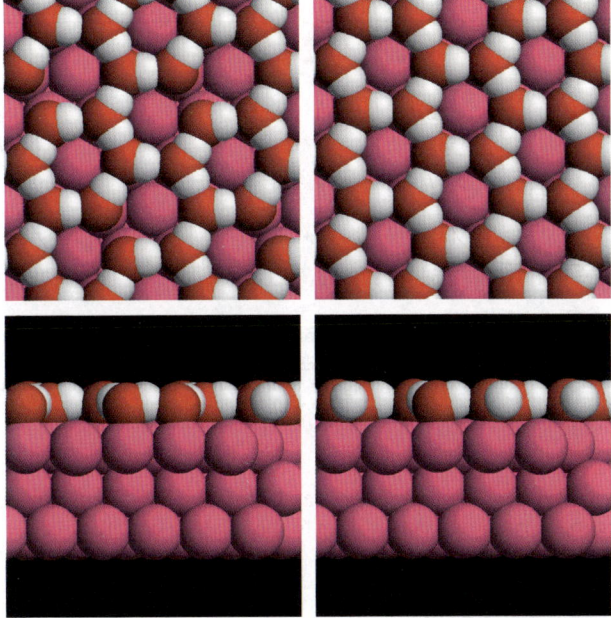

Fig. 1 Top and side view of two configurations with total coverage of OH and H_2O of 2/3 of a monolayer. To the left, 1/3 of a monolayer of OH and H_2O, and to the right, 4/9 of a monolayer of OH and 2/9 of a monolayer of H_2O. The differential adsorption energy for OH is $\Delta E = 0.45$ eV for 1/3 of monolayer of OH and $\Delta E = 0.86$ eV when the OH coverage exceeds 1/3 of a monolayer. For details about the density functional theory calculations see ref 2.

at the reaction energy ΔE of reaction (2) *versus* OH coverage, this corresponds to ΔE being constant for OH coverages less than or equal to 1/3 of a monolayer. When the OH coverage exceeds 1/3, ΔE would become more positive. This jump in ΔE is verified with explicit density functional theory calculations in Fig. 1.

The jump in ΔE is important since it shows that under the potentials of interest in this study the coverage of OH cannot exceed 1/3 of a monolayer. We also note that this value of the coverage compares well to recent experiments (~0.4 ML).[5]

With a weak interaction between hydroxyls in the interesting coverage interval (<1/3 ML), the coverage dependence in ΔG_{OH} will only arise from configurational entropy. Using the configurational entropy of non-interacting particles, $\Delta S = k\ln((1 - \theta_{OH})/\theta_{OH})$, for $0 < \theta_{OH} < 1/3$, we can therefore write the potential and coverage dependence of the reaction free energy of reaction (1) as

$$\Delta G(\theta_{OH}, U) = \Delta G_{OH} - k\ln((1 - \theta_{OH})/\theta_{OH}) - eU \qquad (3)$$

Here ΔG_{OH} is calculated for the standard condition of 1/3 of a monolayer coverage OH and 1/3 of a monolayer H_2O, this structure is depicted to the left in Fig. 1. Assuming that reaction (1) is in equilibrium for all potentials ($\Delta G(U, \theta_{OH}) = 0$), this leads to the following expression for the coverage

$$\Theta_{OH}(U) = (1/3) \, 1/\{1 + \exp[(\Delta G_{OH} - eU)/kT]\} \qquad (4)$$

In the following we will use ΔG_{OH} equal to 0.80 eV and 0.93 eV for Pt and P_3Ni respectively, as was reported in ref. 2. The analytical expression for coverage shown in eqn (4) is determined solely by the equilibrium with water and protons and is the theoretical counterpart to the OH coverage obtained from cyclic voltamograms in

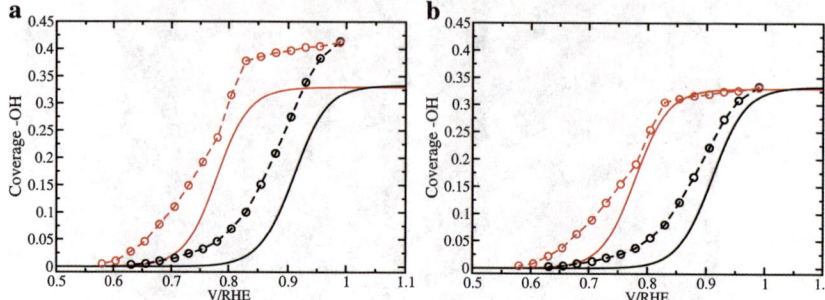

Fig. 2 (a) Experimental (dashed lines with circles) coverage of OH versus potential for Pt$_3$Ni (black) and pure Pt (red) obtained from integration of cyclic voltammograms.[5] The solid lines show the coverage as calculated using eqn (4) with input from density functional theory and standard tables. (b) The same as Fig. 4a but with the experimental coverage scaled so that the max. coverage is 1/3.

an O$_2$-free environment. Computational and experimental[5] results are co-plotted in Fig. 2, revealing a good agreement despite the relative simplicity of the theoretical model.

Steady state current in the presence of molecular oxygen

We consider the following reaction steps for ORR:

$$O_2(g) + 4H^+ + 4e^- \leftrightarrow HOO^* + 3H^+ + 3e^- \leftrightarrow H_2O(l) + O^* + 2H^+ + 2e^- \leftrightarrow H_2O(l) + HO^* + H^+ + e^- \leftrightarrow 2H_2O(l) \quad (5)$$

An alternative to the first reduction step is dissociation of molecular oxygen (O$_2$(g) $\leftrightarrow$ 2O*). However, it has previously been shown that the dissociative mechanism is less important for the metals and potentials of interest in this study.[16]

Based on a detailed theoretical model which includes the effect of the local field in the double layer,[17] we have calculated the potential free energy diagram for the ORR pathway on Pt(111) at 0.9 V vs. NHE, see Fig. 3. From this graph there are only two candidates for the rate-limiting step—formation of OOH and removal of OH. We note that no barriers are included in the model.

Since the adsorption energy of all the intermediates of the ORR can be linearly related to the adsorption energy of oxygen, ΔE_O,[18,19] this energy is a suitable descriptor. The observation that the binding energy of hydroxyl, OH*, scales

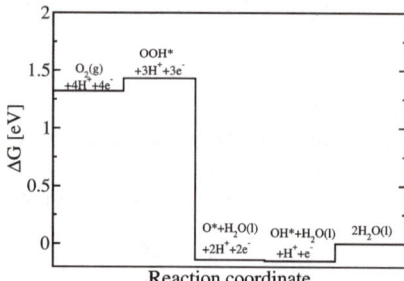

Fig. 3 The free energy diagram for the ORR steps 1–4 at 0.9 V on Pt (111). Notice that there are only two only steps which are uphill: the formation of OOH (step 1) and the formation of H$_2$O (step 4). This allows us to simplify the kinetic model to only consider these two reaction steps.

linearly with the binding energy of oxygen indicates that the binding energy of hydroxyl could just as well have been chosen. In the Sabatier analysis we determine the reaction step related with the most positive change in free energy, and since the rate of a step is proportional to $\exp(-\Delta G)$, this step will be the most difficult along the ORR reaction. The Sabatier volcano for ORR is defined by:

$$\Delta G(U) = \mathrm{Max}[\Delta G_1(\Delta E_O, U), \Delta G_2(\Delta E_O, U), \Delta G_3(\Delta E_O, U), \Delta G_4(\Delta E_O, U)] \quad (6)$$

Where ΔG_{1-4} are the reaction free energies for the four elementary reduction steps of reaction (5) and $\Delta G(U)$ is the most positive change in free energy encountered along the ORR pathway. Fig. 4a shows the relative activities of Pt and Pt$_3$Ni as a function of their DFT-calculated ΔE_O according to the Sabatier model. As shown in Fig. 3, it is either $\Delta G_1(\Delta E_O, U)$ or $\Delta G_4(\Delta E_O, U)$ that defines $\Delta G(U)$.

In order to predict ORR current density computationally, we begin with the Tafel equation

$$j_c = j_0 \exp(\alpha \eta e / kT) \quad (7)$$

where j_c is the cathode current density, j_0 the exchange current, η is the overpotential and α is the transfer coefficient. We will only consider $\alpha = 1$, but the model could be made for other values of α. We use the definition of $\eta \equiv U_0 - U$ in order to rewrite the Tafel equation:

$$j_c = j_0 \exp[\alpha(eU_0 - eU)/kT] = j_{\mathrm{limit}} \exp[\alpha(\Delta G_0 - eU)/kT] \quad (8)$$

where j_{limit} can be written as:

$$j_{\mathrm{limt}} = j_0 \exp[\alpha(eU_0 - \Delta G_0)/kT] \quad (9)$$

The physical meaning of the term j_{limit} is the current density achieved if all surface reactions are exothermic, *i.e.* the highest possible turn-over-frequency per site in an electrochemical cell with minimal diffusion limitations. Clearly, this term is dependent on a number of factors, including the electrode structure and its interface with the liquid electrolyte. In previous work for hydrogen evolution, we found that $j_{\mathrm{limit}} \sim 200$ sites^{-1} s^{-1} (or in terms of surface area, 96 mA cm^{-2}) for Pt(111) fit experimental data well, and we will utilize this value in this work as well.[20] As the

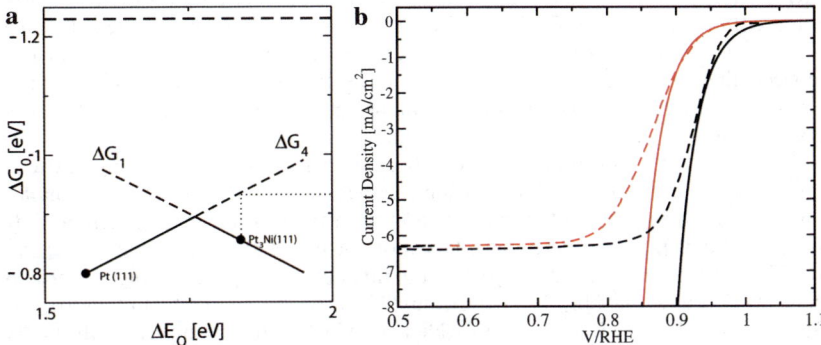

Fig. 4 (a) The volcano at $U = 0$, which defines ΔG_0 as a function of ΔE_O, Either ΔG_1 or ΔG_4 is the limiting step. The vertical dashed line represents zero overpotential. The distance from the solid volcano to the horizontal dashed line is the overpotential at $j = j_{\mathrm{limit}}$. (b) Current density (j_c) from experiment from ref. 5 (dashed lines), and from the model (solid lines). The current density is calculated using eqn (8) with the ΔG_0 for Pt (red lines) and Pt$_3$Ni (black lines) from Fig. 4a as input.

lattice parameters of Pt-alloys change only by a few percent relative to pure Pt, the number of sites cm^{-2} is fairly constant and j_{limit} can be effectively considered material-independent, unlike exchange current density. By incorporating j_{limit} into the analysis, we can thus rewrite the Tafel equation, eqn (8) in such a manner that all material dependence is concentrated within only one computable parameter, ΔG_0. Revisiting eqn (8), ΔG_0 is the largest negative change in free energy along the ORR pathway at $U = 0$, and in Fig. 4a, we show the dependence of ΔG_0 on ΔE_O, simply the Sabatier volcano at $U = 0$. Using the ΔG_0 from Fig. 4a together with eqn (8) we get the polarization curves shown in Fig. 4b co-plotted with the experimentally polarization curves. Since eqn (8) does not include the diffusion limitations specific to the experiments of ref 5, the model and experiments should only be compared in the upper half of Fig. 4b.

Relating ORR activity to OH coverage from water dissociation

In the case of Pt-alloy ORR electrocatalysts, it has become commonplace to explain changes in activity by relating measurements of ORR reaction rates in O_2-saturated solutions to the coverage of OH, measured in O_2-free solutions. Oftentimes in this analysis, it has been rationalized that OH is a "poison" that blocks surface sites from the O_2 reactant, leading to the logical deduction that less OH adsorption on the surface translates directly to improved ORR catalysis.[5,8,21,22]

Consider the measured ORR current densities for Pt (111) and Pt$_3$Ni (111) in Fig. 4b and the OH coverage from water dissociation during cyclic voltammetry in an O_2-free environment shown in Fig. 2. Careful inspection of the experimental data shows that there is no straightforward correlation between ORR current density and the OH coverage determined by the O_2-free CV. In the case of Pt(111), for example, the OH coverage is fairly constant at 0.4 ML in the potential region of $0.82 < V < 1.00$. Still, within this same region, its ORR current density increases by several orders of magnitude from nearly zero to approximately 5 mA cm^{-2}, where the reaction encounters significant transport limitations. A similar phenomenon is observed on the Pt$_3$Ni(111) in the potential region of $0.87 < V < 1.00$, where OH coverage decreases moderately from 0.4 ML to 0.2 ML, and yet ORR current density increases several orders of magnitude from nearly zero to approximately 6 mA cm^{-2} (its transport-controlled limit).

Another noteworthy point concerns different shifts in potential between the Pt(111) and Pt$_3$Ni(111) data in the case of OH coverage (Fig. 2) *versus* that of ORR current density (Fig. 4b). In terms of OH coverage, the data for the two materials are shifted by $\sim$100 mV, while in the case of ORR current density, the half-wave potential is shifted by only $\sim$60 mV. The lack of a one-to-one correspondence between these two shifts also raises doubts as to whether OH is only playing a role as a site-blocker. The theoretical model presented above offers a different interpretation.

Fig. 2 and 4b show that our theoretical model reproduces the key features observed in the experiments. For instance, the calculations reproduce a smaller potential shift between the ORR polarization curves (Fig. 4b) than the potential shift for the OH coverage curves (Fig. 2). Based on the calculations, the physical reason for this observation is that the nature of the rate determining step changes as we go from Pt (111) to Pt$_3$Ni(111). While Pt is limited by too strong of a bond to adsorbed OH, ($\Delta G_4 > \Delta G_1$), this is not true of Pt$_3$Ni(111), which instead is limited by the OOH formation step ($\Delta G_1 > \Delta G_4$). This fundamental difference between the two materials is graphically represented in Fig. 4a, where the two materials are situated on the opposite sides of the volcano. Had both materials been located on the same side of the volcano, one would have expected the ORR current density to shift by approximately the same amount as the coverage $\sim$130 mV, which is indicated with the dotted lines in Fig. 4a.

This analysis shows that the simple Sabatier model accounts for the key features observed in experiments. The higher activity of Pt$_3$Ni(111) is a consequence of a better compromise in bond strength to the OOH and OH intermediates. In other words Pt$_3$Ni is closer to the top of the volcano.

Kinetic volcano

The simple model yields a good agreement with experiments already without a detailed description of the kinetics, indicating that kinetics will play a minor role for the conclusions drawn. In the following we will try to prove this by developing a more detailed kinetic model taking steady-state coverages into account, the scenario encountered under standard fuel cell operating conditions. We start by considering the two steps needed in the kinetic model.

$$O_2(g) + * + H^+ + e^- \rightarrow OOH* \quad \text{Step 1} \tag{10}$$

$$OH* + H^+ + e^- \leftrightarrow H_2O(l) \quad \text{Step 4} \tag{11}$$

From Fig. 3 and Fig. 4a it is seen that, these two steps are the only possible rate-limiting steps. The situation becomes even more simple if we assume Step 1 to be irreversible. This is likely since the OOH* formed rapidly proceeds to OH* (Fig. 3), and implies that the coverage of OOH* always is low.

First we write the rate constants:

$$k_1 = k_1^0 \, \text{Min}(1, \exp[\alpha_1(\Delta G_1 - eU)/kT]) \tag{12}$$

where k_1^0 is the prefactor and ΔG_1 is the change in free energy for Step 1. We introduce an upper limit to all the rate constants, if there is no barrier for a reaction step the rate constant is k_i^0. The expression for k_4 is analogous to k_1, while k_{-4} is given by:

$$k_{-4} = k_{-4}^0 \, \text{Min}(1, \exp[(1-\alpha_4)(eU - \Delta G_4)/kT]) \tag{13}$$

First, at steady state:

$$d\Theta_{OH}/dt = (1 - \Theta_{OH})k_1 - \Theta_{OH}k_4 + (1 - \Theta_{OH})k_{-4} = 0 \tag{14}$$

This means that the OH-coverage can be written as:

$$\Theta_{OH} = (k_1 + k_{-4})/(k_1 + k_4 + k_{-4}) \tag{15}$$

Here Θ_{OH} denotes the OH-coverage in the presence of molecular oxygen. This is not necessarily the same as the coverage obtained in eqn (4), which is the coverage in the absence of molecular oxygen. The steady state current is given by

$$j = j_{\text{limit}}(1 - \Theta_{OH})k_1 = j_{\text{limit}}[\Theta_{OH}k_4 - (1 - \Theta_{OH})k_{-4}] = \\ j_{\text{limit}}[1 - (k_1 + k_{-4})/(k_1 + k_4 + k_{-4})]k_1 = j_{\text{limit}}k_1 k_4/(k_1 + k_4 + k_{-4}) \tag{16}$$

This is the steady state version of eqn (8). Note that the first term is similar to the expression which many authors use to relate the oxygen reduction current to the coverage of OH as given by cyclic voltammetry.[21] However, it is important to note that Θ_{OH} and k_1 are not independent, but related through eqn (15)

In the Sabatier analysis, the change in activity is seen as changes in k_1 and k_4. Looking at the limit where either k_1 or k_4 is large, the expression above eqn (16) approaches:

$$k_1 \gg k_4, k_{-4} \quad j \to = j_{\text{limit}} k_4 \tag{17}$$

or

$$k_4 \gg k_1, k_{-4} \quad j \to = j_{\text{limit}} k_1. \tag{18}$$

This is exactly the result of the Sabatier analysis. This means that in the limit of just one rate limiting step, the Sabatier volcano is obtained. However, when k_1, k_4 and k_{-4} are all of the same size, which is the case at the top of the volcano, the Sabatier analysis is no longer exact. Instead a "kinetic volcano" is obtained. In the following we will estimate this kinetic volcano.

Due to computational limitations we have no means at present to calculate exact values for all the parameters that go into this kinetic modeling based on density functional theory. We will, however, show that for a realistic set of parameters, our kinetic model will give reasonable results. Concerning pre-factors, we assume that k_4^0 and k_{-4}^0 are identical. For the sake of simplicity, for this analysis we will assume k_4^0, k_{-4}^0 and j_{limit} to be 1 in what follows. From the kinetics of the O_2 adsorption step it follows that the pre-factor of Step 1 will contain the partial pressure of oxygen. Hence, varying k_1^0 essentially corresponds to a variation of the oxygen pressure. However, it is reasonable to expect k_1^0 to be less than or equal to 1 (i.e. less than or equal to k_4^0 and k_{-4}^0) since Step 1 includes both adsorption of O_2 and proton transfer to O_2^*, whereas Step 4 only includes the proton transfer between OH^* and H_2O^*. Assuming that the proton transfer in both cases have the same pre-factor, k_1^0 will be smaller than k_4^0 and k_{-4}^0 no matter how big the oxygen partial pressure.

Other unknown parameters are the transfer coefficients α_1 and α_2. They are related to potential dependent barriers, which are very difficult to address with present theoretical methods. Since it is not possible to rigidly include the barriers we will assume that there are no additional barriers for the proton transfer reaction. This is included in our model by the upper limits to the rate constants. We have checked that other models give similar qualitative results, and the conclusions drawn in this paper are robust if we include small potential-dependent barriers.

The steady state current density for oxygen reduction is calculated based on the full kinetics, see Fig. 5. The major effect introduced by the kinetics is observed near the top of the volcano. This can be expected, as in this particular region of the volcano we have relaxed the assumption of a single rate-limiting step.

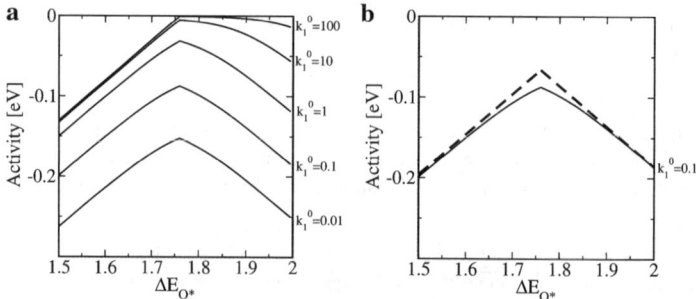

Fig. 5 a. The result of a detailed kinetic analysis. (with $k_1^0 = k_4^0 = 1$). The volcanoes are obtained from the kinetic analysis, the y axis is $kT \times \ln(j)$, j is obtained from eqn (16) with k_4^0, $k_{-4}^0 = 1$ for various values of k_1^0. The most realistic scenario is the case where both k_4^0 and k_{-4}^0 are larger than k_1^0, since k_1^0 includes adsorption of O_2 and proton transfer to O_2^*. k_4^0 and k_{-4}^0 are the prefactors related to the forward and backwards reactions of the last proton transfer step $OH^* + H^+ + e^- \leftrightarrow H_2O^*$, which we consider to be fast. (b) The same as Fig. 5a with only the volcano for $k_0^1 = 0.1$ and with the respective Sabatier volcano indicated by the dashed lines.

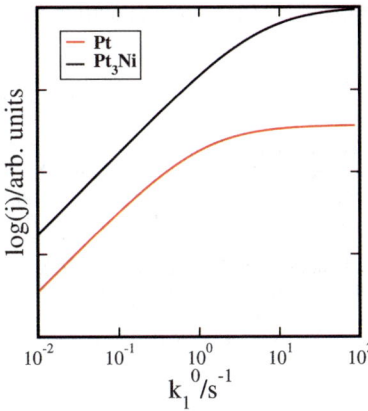

Fig. 6 A log–log plot of the oxygen reduction current density *versus* the pre-factor k_1^0, or equivalently the oxygen partial pressure. The reaction order goes from 1 to zero.

Nevertheless, the activity predictions for Pt and Pt$_3$Ni based on the kinetic model and the reasonable assumption of $k_1^0 < k_4^0$ and k_{-4}^0 are essentially unchanged as compared to the results found by the simple Sabatier model, see Fig. 5b. This confirms the validity of the assumption of a single reaction step being rate determining is valid for both Pt and Pt$_3$Ni, and is the reason why the simple Sabatier approach works.

Another interesting feature to note in Fig. 5 regards the change in current as a function of k_1^0. Obviously, for k_1^0 in the interesting interval in Fig. 5, the change in current is directly proportional to k_1^0. Bearing the connection between k_1^0 and the partial pressure of oxygen in mind, this indicates that the oxygen reduction current is first order with respect to the oxygen partial pressure. We investigate this further in Fig. 6, where the current density is plotted *versus* k_1^0. A first order dependence is clearly seen. For higher values of k_1^0 the reaction order goes to zero, first for Pt and later for Pt$_3$Ni. We don't expect that it is possible to probe the zero-order-regime in experiments since it would require k_1^0 to be larger than one.

Conclusion

We have found that the experimental results of the ORR on Pt$_3$Ni were correctly predicted by the simple Sabatier model. The predictions hold true both for the current density and the OH coverage as functions of the potential. The Sabatier model can be used to predict ORR activity, and is therefore a useful tool for designing new ORR catalysts. Furthermore, we have developed a simple kinetic model of the ORR. The kinetic model reduces to the Sabatier model in the limits where only one reaction step is rate-determining. This kinetic model predicts that the reaction order in oxygen pressure is one in the physically relevant region.

We find that the current density can indeed be written as $(1 - \Theta_{OH})k_1$, as has often been suggested. However, it is very important to note that Θ_{OH} and k_1 are connected by eqn (15), which means that this way of writing the current does not offer insights as to which parameters to change in order to optimize the catalytic activity. The top of the activity volcano is defined by the Sabatier analysis.

Acknowledgements

CAMD is funded by the Lundbeck foundation. This work was supported by the Danish Center for Scientific Computing through grant no. HDW-1103-06. TFJ acknowledges the H. C. Ørsted Postdoctoral Fellowship from the Technical University of Denmark.

References

1 H. A. Gasteiger, S. S. Kocha, B. Sompalli and F. T. Wagner, *Appl. Catal., B*, 2005, **56**, 9–35.
2 V. Stamenkovic, B. S. Mun, K. J. J. Mayrhofer, P. N. Ross, M. N. Markovic, J. Rossmeisl, J. Greeley and J. K. Nørskov, *Angew. Chem., Int. Ed.*, 2006, **45**, 2897–2901.
3 V. Stamenkovic, B. S. Mun, M. Arenz, K. J. J. Mayrhofer, C. A. Lucas, G. Wang, P. N. Ross and N. M. Markovic, *Nat. Mater.*, 2007, **6**, 241–247.
4 S. Koh, J. Leisch, M. F. Toney and P. Strasser, *J. Phys. Chem. C*, 2007, **111**, 3744–3752.
5 V. R. Stamenkovic, B. Fowler, B. S. Mun, G. F. Wang, P. N. Ross, C. A. Lucas and N. M. Markovic, *Science*, 2007, **315**, 493–497.
6 J. Zhang, M. B. Vukmirovic, Y. Xu, M. Mavrikakis and R. R. Adzic, *Angew. Chem., Int. Ed.*, 2005, **44**, 2132–2135.
7 J. Zhang, M. B. Vukmirovic, K. Sasaki, A. U. Nilekar, M. Mavrikakis and R. R. Adzic, *J. Am. Chem. Soc.*, 2005, **127**, 12480–12481.
8 S. Koh and P. Strasser, *J. Am. Chem. Soc.*, 2007, **129**, 12624.
9 J. Zhang, K. Sasaki, E. Sutter and R. R. Adzic, *Science*, 2007, **315**, 220–222.
10 J. K. Nørskov, J. Rossmeisl, A. Logadottir, L. Lindqvist, J. Kitchin, T. Bligaard and H. Jónsson, *J. Phys. Chem. B*, 2004, **108**, 17886–17892.
11 G. S. Karlberg, T. Jaramillo, E. Skulason, J. Rossmeisl, T. Bligaard and J. K. Nørskov, *Phys. Rev. Lett.*, 2007, **99**, 126101.
12 M. A. Henderson, *Surf. Sci. Rep.*, 2002, **46**, 5–308.
13 Here one monolayer corresponds to one adsorbed molecule per Pt surface atom.
14 G. S. Karlberg and G. Wahnström, *J. Chem. Phys.*, 2005, **122**, 194705.
15 A. Michaelides and P. Hu, *J. Am. Chem. Soc.*, 2001, **123**, 4235–4242.
16 G. S. Karlberg, J. Rossmeisl and J. K. Nørskov, *Phys. Chem. Chem. Phys.*, 2007, **9**, 5158–5161.
17 J. Rossmeisl, J. K. Nørskov, C. D. Taylor, M. J. Janik and M. Neurock, *J. Phys. Chem. B*, 2006, **110**, 21833–21839.
18 J. Rossmeisl, A. Logadottir and J. K. Nørskov, *Chem. Phys.*, 2005, **319**, 178–184.
19 F. Abild-Pedersen, J. Greeley, F. Studt, J. Rossmeisl, T. R. Munter, P. G. Moses, E. Skulason, T. Bligaard and J. K. Nørskov, *Phys. Rev. Lett.*, 2007, **99**, 016105.
20 J. K. Nørskov, T. Bligaard, A. Logadottir, J. R. Kitchin, J. G. Chen, S. Pandelov and U. Stimming, *J. Electrochem. Soc.*, 2005, **152**, J23–6.
21 J. X. Wang, N. M. Markovic and R. R. Adzic, *J. Phys. Chem. B*, 2004, **108**, 4127–4133.
22 M. B. Vukmirovic, J. Zhang, K. Sasaki, A. U. Nilekar, F. Uribe, M. Mavrikakis and R. R. Adzic, *Electrochim. Acta*, 2007, **52**, 2257–2263.

PAPER

Intrinsic kinetic equation for oxygen reduction reaction in acidic media: the double Tafel slope and fuel cell applications†

Jia X. Wang,*[a] Francisco A. Uribe,[b] Thomas E. Springer,[b] Junliang Zhang[c] and Radoslav R. Adzic[a]

Received 8th February 2008, Accepted 6th May 2008
First published as an Advance Article on the web 4th September 2008
DOI: 10.1039/b802218f

According to Sergio Trasatti, "A true theory of electrocatalysis will not be available until activity can be calculated *a priori* from some known properties of the materials." Toward this goal, we developed intrinsic kinetic equations for the hydrogen oxidation reaction (HOR) and the oxygen reduction reaction (ORR) using as the kinetic parameters the free energies of adsorption and activation for elementary reactions. Rigorous derivation retained the intrinsic connection between the intermediates' adsorption isotherms and the kinetic equations, affording us an integrated approach for establishing the reaction mechanisms based upon various experimental and theoretical results. Using experimentally deduced free energy diagrams and activity-and-barriers plot for the ORR on Pt(111), we explained why the Tafel slope in the large overpotential region is double that in the small overpotential region. For carbon-supported Pt nanoparticles (Pt/C), the polarization curves measured with thin-film rotating disk electrodes also exhibit the double Tafel slope, albeit Pt(111) is several times more active than the Pt nanoparticles when the current is normalized by real surface area. An analytic method was presented for the polarization curves measured with H_2 in proton exchange membrane fuel cells (PEMFCs). The fit to a typical iR-free polarization curve at 80 °C revealed that the change of the Tafel slope occurs at about 0.77 V that is the reversible potential for the transition between adsorbed O and OH on Pt/C. This is significant because it predicts that the Butler–Volmer equation can only fit the data above this potential, regardless the current density. We also predicted a decrease of the Tafel slope from 70 to 65 mV dec^{-1} at 80 °C with increasing oxygen partial pressure, which is consistent with the observation reported in literature.

1. Introduction

In electrocatalysis, reaction rates are expressed by kinetic equations that describe the current-potential relationship of an electrode reaction in the absence of mass transport limitation. For a single-step redox reaction: O + ne ⇌ R, the kinetic current is formulated by the Butler–Volmer (B–V) equation: $j_k = j_0\,(e^{-\alpha n\eta/kT} - e^{(1-\alpha)n\eta/kT})$.[1] More

---

[a]*Department of Chemistry, Brookhaven National Laboratory, Upton, New York, 11973, USA. E-mail: jia@bnl.gov.gov*
[b]*Los Alamos National Laboratory, Los Alamos, NM 87545, USA*
[c]*General Motors Corporation, Fuel Cell Research Center, 10 Carriage Street, Honeoye Falls, NY, 14472, USA*

† The HTML version of this article has been enhanced with colour images.

than one reaction step occurs in electrocatalyzed reactions because catalysts function through forming adsorbed reaction intermediates. Similar to the rate equations in heterogeneous catalysis, the kinetic equations in electrocatalysis often were formulated by assuming a single rate-determining step (RDS).

Based on this supposition, the modified B–V equations were widely applied with the exchange current and the Tafel slope as semi-empirical parameters, thereby establishing a semi-quantitative connection between the catalysts' properties and their activities. As the Sabatier principle stated, catalytic activity is optimal on a catalytic surface with intermediate binding energy for reactive intermediates.[2] Various volcano plots were generated based on the idea that either adsorption or desorption is involved in the RDS.[3–6] Thus, activity first rises and then falls with increasing binding energy when the RDS switches from adsorption to desorption. As volcano plots recently became widely known for the hydrogen oxidation reaction (HOR) and the oxygen reduction reaction (ORR), the limitation of semi-empirical approach based on a single RDS also became increasingly clear, thereby motivating us to develop the intrinsic kinetic equations.

For the HOR, we derived a dual-pathway kinetic equation,[7] that demonstrated a fast, inverse exponential rising of the HOR kinetic current at small overpotentials through the Tafel or dissociative adsorption (DA) pathway, followed by a more gradual rise at η > 50 mV, mainly via the Heyrovsky or oxidative adsorption (OA) pathway (Fig. 1a). This behavior is dramatically different from the single exponential rise predicted by the B–V equation; it translates into a 20-fold higher activity than previously thought, and explains why the HOR overpotential is negligible when the Pt loading is above 0.05 mg cm^{-2} in H_2-air PEMFCs.

We further elaborated a method to use the free energies of adsorption and activation as the intrinsic parameters for the kinetic equation and adsorption isotherm.[8] Combining experimental data and DFT calculations, we found that the H adsorption energy on Pt is about the same for atop and hollow/bridge (H/B) sites. However, the rates of oxidative adsorption and desorption are considerably higher at atop sites because hydrogen-bond formation with water is much easier for H adatoms at atop than in H/B sites, causing an order-of-magnitude difference in coverage and in the HOR activity (Fig. 1b and c). This example illustrates the important roles of entropy and other factors in determining the catalysts' activities.

Using the method we developed for the HOR, we obtained the kinetic equation and adsorption isotherms for the ORR based on a model containing four essential elementary reactions and two major adsorbed intermediates, O and OH.[9] The free

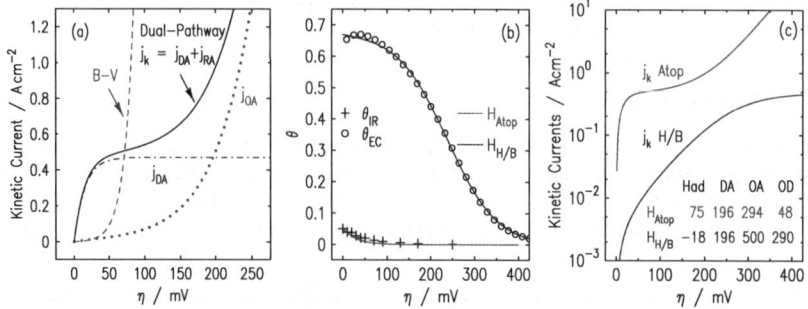

Fig. 1 (a) Comparison of the HOR kinetic current on Pt using the dual-pathway kinetic equation (j_k) and the contributions from the DA and OA pathways with previous results using the B–V equation. (b) H adsorption isotherm measured by IR for the atop-site-adsorbed H and that measured by electrochemical methods, including those adsorbed in hollow and at bridge sites. (c) Kinetic currents corresponding to different adsorption sites. Adapted from previous publications.[7,8]

energies derived from fitting the measured polarization curves for the ORR on Pt(111) are consistent with theoretical calculations, demonstrating through the activity-and-barriers plot how the three comparable activation barriers together determine the ORR kinetic behavior. Understanding this kind of situation is important because lowering the highest activation barrier is how catalytic activities get improved so that the RDS may have a less dominant role in the best catalysts. Our new approach avoids a single-RDS approximation, and thus, opens new ways to resolve controversial issues about mechanism of the ORR reaction.

Polarization curves measured for the ORR on Pt with a rotating disk electrode (RDE) exhibit a lower Tafel slope at high potentials compared to that at low potentials. A common interpretation is that the site-blocking effect by OH or O at high potentials cause the Tafel slope to deviate from its intrinsic value at low potentials.[10–14] The proposed semi-empirical kinetic equations for the ORR assumed that the first electron reduction in the adsorption process was the RDS, and that adsorbed OH and O are the site blockers.[15–19] Combined with the experimentally deduced adsorption isotherm for OH and bisulfate, we fitted the measured polarization curves for Pt(111) in $HClO_4$ and H_2SO_4 solutions.[18] Our results show that the site-blocking effect of OH can lower the Tafel slope at high potentials. However, these methods cannot offer applicable kinetic equations for analyzing polarization curves without independently determining the adsorption isotherms; further, they incorrectly treat the reaction intermediates, OH and O, only as site blockers.

The lack of an adequate kinetic equation able to describe the observed double Tafel slope made it difficult to characterize the ORR activity using kinetic parameters. Thus, catalysts' activities were compared using the numerically converted kinetic currents at certain potentials. Efforts were made to establish activity benchmarks with Pt as the catalysts in PEMFCs;[20] however, this approach is inconvenient because operating conditions often vary significantly in different experiments.

To date, H_2-PEMFC polarization curves have been analyzed *via* the B–V equation with a single Tafel slope for the ORR.[21–26] Most results were interpreted as supporting a single Tafel slope for the ORR kinetics, *i.e.*, $j_k = j_0\, e^{(E_0-V)/b}$; inexplicable features were attributed to possible complications in mass transport or protonic resistance in the catalyst layers. Thus, the existence, or not, of a double Tafel slope for the ORR in PEMFCs is a complex issue that challenges the relevance of fundamental research to fuel cell applications.

In this paper, we focus on resolving the complexity related to the double Tafel slope. Firstly, we discuss the basic concepts of the intrinsic kinetic equation employing the analytical results for the ORR on Pt(111) to illustrate the origin of the double Tafel slope. Secondly, we compare the polarization curves for carbon-supported Pt nanoparticles (Pt/C) measured by the thin-film RDE method with those for the Pt(111) electrode. Then, we present our method and results of analyzing an H_2-PEMFC polarization curve with Pt/C as the catalysts. Further support to the conclusion and testing criteria for the B–V and intrinsic kinetic equations are given by simulating the dependence of polarization curves on Pt loading and oxygen partial pressure.

2. The basic concepts of the intrinsic kinetic equation

A key element of the intrinsic kinetic equations is maintaining the intrinsic connection between the adsorption isotherms of the reaction intermediates and the overall kinetic current through rigorous derivation based on all the elementary reactions in a kinetic model. Chialvo and coworkers used such an approach for the HOR,[27,28] albeit the kinetic parameters were not free energies. Replacing the rate constants and reactants' concentrations by activation and adsorption free energies is another important element, which simplifies the derivation, generates easily comprehensible results, and provides a platform for reconciling various experimental and theoretical results. The new concepts developed are generally applicable for other electrocatalytic reactions.

Below we present the kinetic equation derived from the double-trap kinetic model for the 4e-ORR in acidic media, which includes two strongly adsorbed reaction intermediates and comprises four essential elementary reactions forming two adsorption pathways:

$$1/2 O_2 \leftrightarrows O_{ad} \quad \text{Dissociative Adsorption (DA)} \quad (1)$$

$$1/2 O_2 + H^+ + e^- \leftrightarrows OH_{ad} \quad \text{Reductive Adsorption (RA)} \quad (2)$$

$$O_{ad} + H^+ + e^- \leftrightarrows OH_{ad} \quad \text{Reductive Transition (RT)} \quad (3)$$

$$OH_{ad} + H^+ + e^- \leftrightarrows H_2O \quad \text{Reductive Desorption (RD)} \quad (4)$$

Schematically, they are depicted as

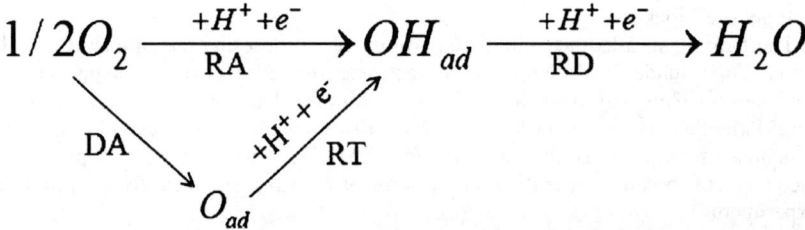

A detailed discussion and derivation can be found in a previous publication.[9]

At the steady state, the net kinetic current can be expressed by the difference between the forward and backward reaction currents of the reductive desorption of OH, which is the last reaction step shared by both RA and DA pathways,

$$j_k = j^* e^{-\Delta G_{RD}/kT} \theta_{OH} - j^* e^{-\Delta G_{-RD}/kT}(1 - \theta_O - \theta_{OH}) \quad (5)$$

Here, $j^* = 1000$ A cm^{-2} is the chosen reference prefactor setting the scale for the activation free energy, just as the normal hydrogen electrode sets the reference for the potential.[8] In most cases, the backward reaction current expressed by the second term in the above equation can be neglected. The adsorption isotherms for the reaction intermediates, O and OH,

$$\theta_{OH} = \frac{g_{DA}(g_{RA} + g_{-RD} - g_{RT}) - (g_{RA} + g_{-RD})(g_{DA} + g_{-DA} + g_{RT})}{(g_{DA} - g_{-RT})(g_{RA} + g_{-RD} - g_{RT}) - (g_{RA} + g_{-RA} + g_{-RT} + g_{RD} + g_{-RD})(g_{DA} + g_{-DA} + g_{RT})} \quad (6)$$

$$\theta_O = \frac{g_{DA}(g_{RA} + g_{-RA} + g_{-RT} + g_{RD} + g_{-RD}) - (g_{RA} + g_{-RD})(g_{DA} - g_{-RT})}{(g_{DA} + g_{-DA} + g_{RT})(g_{RA} + g_{-RA} + g_{-RT} + g_{RD} + g_{-RD}) - (g_{RA} + g_{-RD} - g_{RT})(g_{DA} - g_{-RT})} \quad (7)$$

were obtained by solving the pair of steady-state rate equations for the coverage of O and OH, in which $g_i = e^{-\Delta G_i/kT}$ represents the intrinsic reaction rate for an elementary reaction "i". The minus sign in the subscripts indicates the reaction in backward direction. These intrinsic reaction rates change with potential and are determined by the following equations where β is the electron transfer coefficient,

$$\Delta G^*_{DA} = \Delta G^{*0}_{DA} \tag{8}$$

$$\Delta G^*_{RA} = \Delta G^{*0}_{RA} - \beta e(E^0 - E) \tag{9}$$

$$\Delta G^*_{RT} = \Delta G^{*0}_{RT} - \beta e(E^0 - E) \tag{10}$$

$$\Delta G^*_{RD} = \Delta G^{*0}_{RD} - \beta e(E^0 - E) \tag{11}$$

$$\Delta G^*_{-DA} = \Delta G^{*0}_{DA} - \Delta G^0_O \tag{12}$$

$$\Delta G^*_{-RA} = \Delta G^{*0}_{RA} - \Delta G^0_{OH} + (1-\beta)e(E^0 - E) \tag{13}$$

$$\Delta G^*_{-RT} = \Delta G^{*0}_{RT} - \Delta G^0_{OH} + \Delta G^0_O + (1-\beta)e(E^0 - E) \tag{14}$$

$$\Delta G^*_{-RD} = \Delta G^{*0}_{RD} + \Delta G^0_{OH} + (1-\beta)e(E^0 - E) \tag{15}$$

in which, four activation and two adsorption free energies at the reversible potential (marked by the "0" in superscript) are the adjustable kinetic parameters.

If there were only two elementary reactions, RA and RD, the intrinsic kinetic equation would be simplified to

$$j_k = f^* e^{-\Delta G^*_{RD}/kT}\theta_{OH} - f^* e^{-\Delta G^*_{-RD}/kT}(1-\theta_{OH}) \tag{16}$$

$$\theta_{OH} = \frac{e^{-\Delta G^*_{RA}/kT} + e^{-\Delta G^*_{-RD}/kT}}{e^{-\Delta G^*_{RA}/kT} + e^{-\Delta G^*_{-RA}/kT} + e^{-\Delta G^*_{RD}/kT} + e^{-\Delta G^*_{-RD}/kT}} \tag{17}$$

At zero overpotential, the ratio of OH to empty sites is the sum of the intrinsic rates for the adsorption from oxygen (RA) and water (−RD) *versus* the sum of the intrinsic rates for the oxidative (−RA) and reductive (RD) desorption:

$$\frac{\theta^0_{OH}}{1-\theta^0_{OH}} = \frac{e^{-\Delta G^{*0}_{RA}/kT} + e^{-\Delta G^{*0}_{-RD}/kT}}{e^{-\Delta G^{*0}_{-RA}/kT} + e^{-\Delta G^{*0}_{RD}/kT}} = e^{-\Delta G^0_{OH}} \tag{18}$$

Schematically, we show in Fig. 2, how equilibrium at steady state determines the coverage of OH and the rates of the elementary reactions. At $\eta = 0$, the net reaction rate, $v = v_+ - v_-$, is zero for each of the elementary reactions, which is achieved by the equilibrium coverage of the intermediate. Generally, the higher the activation barrier, the lower is the intrinsic reaction rate, $g_i = e^{-\Delta G^*_i/kT}$. The real reaction rate is the intrinsic rate multiplied by the corresponding coverage: θ_{OH} for the desorption (−RA and RD) reactions, and $1 - \theta_{OH}$ for the adsorption (RA and −RD) reactions. For the case shown in Fig. 2a, the adsorption free energy is below zero so that $\theta_{OH} > 0.5$, and accordingly, $\theta_{OH} > 1 - \theta_{OH}$. Therefore, equal rates for both directions is obtained through a higher OH coverage than that of the empty site, thereby compensating for the higher barrier for −RA relative to RA, and similarly, for RD compared to −RD.

At other overpotentials, the net reaction rate is nonzero, but the same for the RA and RD reactions. In Fig. 2b, we assumed $\beta = 1$ for the RA reaction so that ΔG^*_{RA} decreases by $-e\eta$, while ΔG^*_{-RA} remains constant with decreasing potential

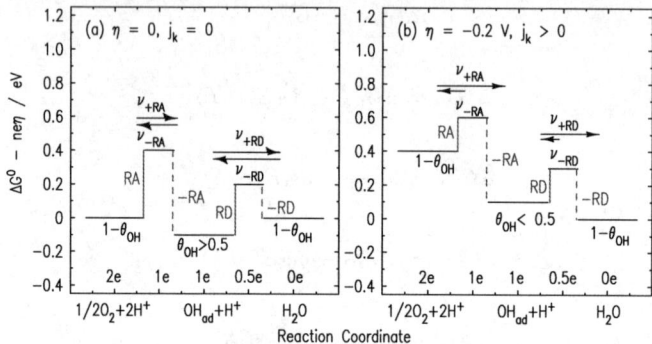

Fig. 2 Free energy diagrams constructed using a single RA pathway model to illustrate the equilibrium at steady state for the ORR at $\eta = 0$ (a) and -0.2 V (b). The lengths of the vertical solid and dashed lines represent the activation free energies for forward and backward reactions, respectively.

(i.e., a negative η). For the RD reaction, we assumed $\beta = 0.5$, that led, respectively, to a decrease and increase of $-0.5e\eta$ for the RD and $-$RD reactions. A new steady-state equilibrium is established in which the OH coverage equalizes the net reaction rate, $\nu = \nu_+ - \nu_-$, for the RA and RD reactions through its dual role as the site-blocker for the adsorption process, and the reactant for the desorption process.

3. The RDE polarization curve for the ORR on Pt(111)

Single crystal electrodes provide a well-defined smooth surface so that the mass-transport limiting current, j_L, is evident in the RDE measurements.[17,29] To better explain the origin of the double Tafel slope, we analyzed a typical ORR polarization curve for Pt(111) measured in an 0.1 M HClO$_4$ solution at 23 °C based on different kinetic models.

The backward reaction rate for the ORR is usually negligible,[9] so that the measured current, j, can be fitted using

$$j = \frac{j_k}{1 + j_k/j_L} \quad (19)$$

or converted to the kinetic current by

$$j_k = \frac{j}{1 - j/j_L} \quad (20)$$

In developing the double-trap kinetic model, we first tested the single RA pathway model. Fig. 3 shows the results of fitting the Pt(111) polarization curve. Imperfect fits, but reasonably good ones can be obtained with the electron-transfer coefficient as 0.5 and 1 for the RA reaction, with that for the RD fixed at 0.5. Although the fitted curves in Fig. 3a overlap, the two corresponding adsorption isotherms in Fig. 3b differ remarkably.

Fig. 3c and 3d depict the activity-and-barriers plots that directly compare the kinetic currents on an equivalent energy scale, $kT\ln(j_k/j^*)$ (black lines) with the potential-dependent activation barriers. The activation free energies are plotted in a negative value, $-\Delta G_i^*$, (grey lines) so that lowering the barrier causes the lines to rise with the increase in activity. The dashed gray lines show the activation free energies for the backward RD reaction that has a slope of $(1 - \beta_{RD}) = 0.5$, while the grey solid lines are for the forward RD reaction with a slope of $-\beta_{RD} = -0.5$. These figures do

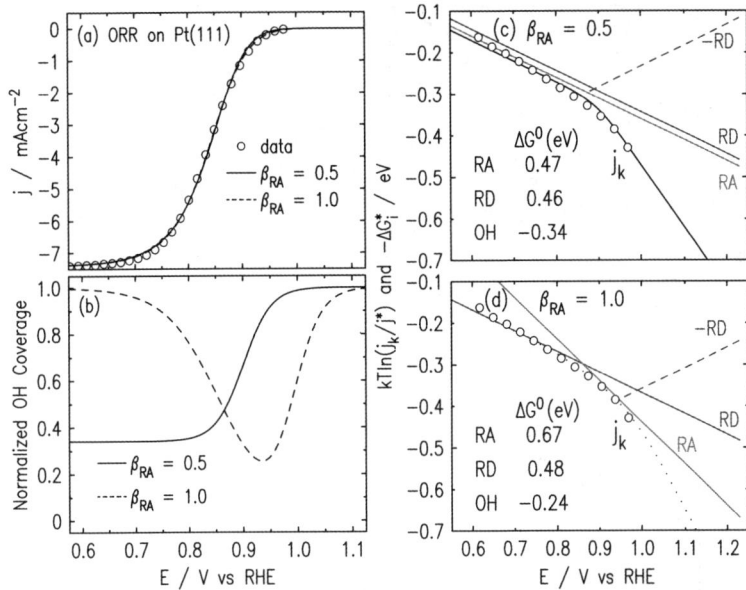

Fig. 3 (a) ORR polarization curve for Pt(111) fitted using eqn (16), (17), and (19) based on the single RA pathway model with the electron transfer coefficient for the RA reaction being 0.5 and 1. (b) The corresponding adsorption isotherms. (c,d) Activity-and-barriers plots made with the given parameters obtained from fitting with $\beta_{RA} = 0.5$, and 1, respectively.

not depict the −RA line because it has much higher barrier (out of the plot's range), and thus a negligible effect on kinetic current.

These two examples show that it is difficult to determine the detailed reaction mechanism, and thus, the cause of double Tafel slope, by simply fitting a polarization curve. To distinguish possible models, we must consider if the corresponding adsorption isotherm is reasonable and whether theoretical calculations support the assumed reaction pathways. Since dissociative adsorption is known to occur for oxygen on Pt at room temperature, and the DFT calculations suggested comparable activation barriers for the DA and RA pathways,[5] a model including both pathways represents the most likely reaction mechanism.

Fig. 4 shows the results obtained using the double-trap kinetic equation. The fit to the polarization curve is nearly perfect (Fig. 4a) yielding three comparable activation free energies for the reductive reactions (Fig. 4b). The corresponding adsorption isotherms reveal that OH coverage increases with decreasing potential as more sites become available for OH adsorption because of the decreasing O coverage (Fig. 4c). Since the kinetic equation can be simplified by omitting the backward reaction current, then

$$j_k = j_f = j^* e^{-\Delta G^*_{RD}/kT} \theta_{OH} \qquad (21)$$

That is, $kT\ln(\theta_{OH})$ determines the difference between the $kT\ln(j_k/j^*)$ and $-\Delta G^*_{RD}$ lines in Fig. 4d. Thus, a straightforward explanation for the variance of the Tafel slope rests on the change in OH coverage as a function of potential. As OH coverage reaches a constant at low potentials, the Tafel slope approaches 118 mV dec^{-1}, a value determined by the symmetric electron-transfer coefficient at room temperature (Fig. 4e).

To elucidate the driving forces behind the changes of the intermediates' coverage, and thus, the Tafel slope, we show the activity-and-barriers plot (Fig. 4f). The DA and RA lines denote that the dominant adsorption pathway switches from the DA to

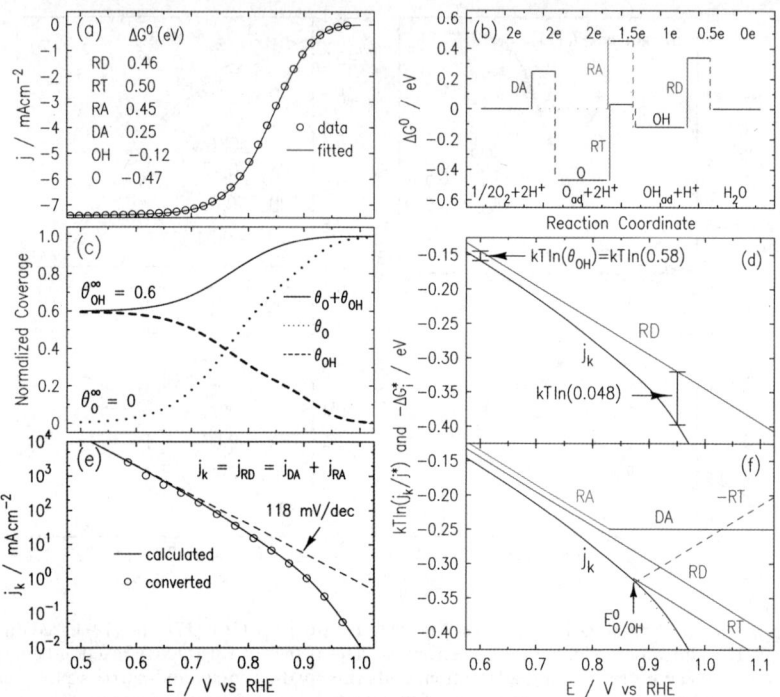

Fig. 4 ORR polarization curve on Pt(111) fitted using eqn (5)–(15), and (19) based on the double-trap kinetic model with the electron transfer coefficient being 0.5 for all the reactions.

RA at the potential where the two lines intercept. At a sufficiently large overpotential, or at low potentials, OH coverage approaches $\theta_{OH}^{\infty} = 1/(1 + e^{(\Delta G_{RA}^{*0} - \Delta G_{RD}^{*0})/kT})$, i.e., 0.5 if $\Delta G_{RA}^{*0} = \Delta G_{RD}^{*0}$. The potential separating the regions of two Tafel slopes is the equilibrium potential for the O and OH transition, viz., at the intercept of the RT and −RT lines (for the forward and backward RT reactions). Thus, when $\Delta G_{RT}^{*} = \Delta G_{-RT}^{*}$ we find, from eqn 10 and 14,

$$E_{O/OH}^{rev} = E^0 - (\Delta G_{OH}^0 - \Delta G_O^0)/e \qquad (22)$$

Above this potential, the slope of the $kT\ln(j_k/j^*)$ curve is close to the difference between the slopes of the highest and lowest barriers for the forward and backward reactions, respectively. That is, $-\beta_{+highest} - (1 - \beta_{-lowest})$, because the determining factors for the net reaction rate are lowering the highest barrier for the forward reactions, and increasing the lowest barrier for the backward reaction. In Fig. 4f, they are the −0.5 for the RT line minus the 0.5 for the −RT lines, so that the slope is −1 (the dotted line). The cases depicted in Fig. 3c and d have a slope of −1 and −1.5 because $\beta_{+highest}$ is 0.5 and 1, respectively. These results demonstrate the origin of the doubling of the electron transfer coefficient for the ORR at high potentials.

In summary, despite some uncertainties in the kinetic parameters due to the lack of an *in situ* technique to measure the O- and OH-specific coverage, we can well explain the double Tafel slope by the intrinsic kinetic equations using activity- and-barrier plots. In general, a lower Tafel slope or a faster increase in current occurs at small overpotentials where the increase in reaction rate is driven not only by the decrease of the highest activation barrier for the forward reactions, but also by the increase of the lowest barrier for the backward reactions. Thus, the rise in the net reaction rate doubles that in the low potential region where the effect of the backward reaction rates vanishes. For the ORR, the RT and −RT have the highest and lowest barriers at zero overpotential for the forward and backward reactions, respectively (Fig. 4b). Therefore, the turning point in a Tafel plot is determined by the reversible potential of the O/OH transition, or the difference between the adsorption free energies for O and OH.

4. The RDE polarization curves for the ORR on Pt/C

To ascertain whether the ORR reaction mechanism is the same on Pt nanoparticles as that on Pt(111), we first analyzed the polarization curves for the Pt/C measured by the thin-film RDE method[30,31] under those same conditions. We mention here a few experimental details that reduced the uncertainties in the results.

Fig. 5a shows the voltammetry curves of Pt/C before and after several potential cycles between 0.05 and 1.2 V, and replacing the electrolyte solution. The high current at the highest potential in the initial voltammetry curve reflects the oxidative removal of organic impurities originating from the residues of surfactants used during synthesis of the catalysts. The ORR activity improves after such cleaning, similar to the effects of initializing process in PEMFC measurements.

The set of voltammetry curves in Fig. 5b show a rise in the reduction current near the zero potential with an increase in the positive potential limit; this phenomenon reflects the slow accumulation of sub-surface oxygen that cannot be removed without reaching very low potentials. Such irreversibility raises the dependence of the Pt/C polarization curve on the potential sweep rate, and on the history of potential sweeps. Similar effects were observed in PEMFC measurements.[32,33]

Suitable Pt loading also is important because the limiting current would be lower should the disk electrode not be fully covered by a thin catalyst layer; conversely, utilization of the catalyst would be lessened if the layers were too thick. In the latter case, the limiting current would be reached very gradually. All these issues might cast doubt upon whether intrinsic kinetic analysis is meaningful for high-surface-area catalysts. Since free energies are the average properties of the entire system, exact

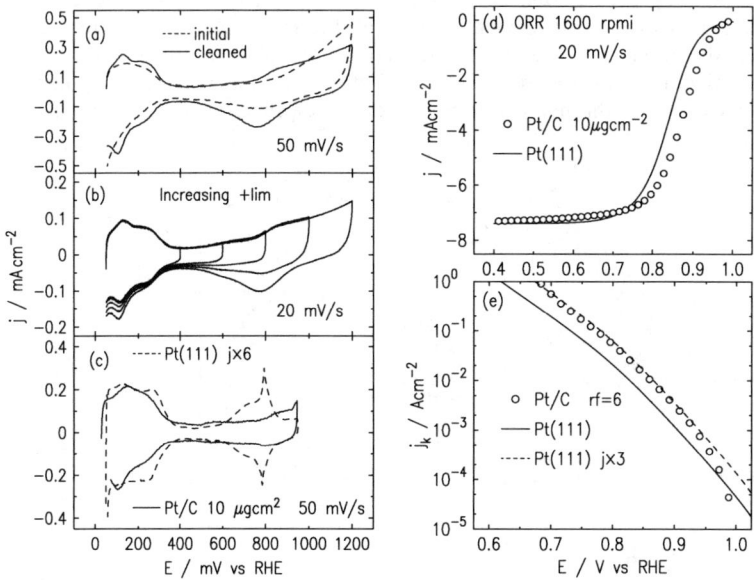

Fig. 5 Linear sweep voltammetry and ORR polarization curves measured for Pt/C (E-TEK, 20%wt Pt, 2–3 nm in diameter) in oxygen-saturated (1 atm) 0.1 M HClO$_4$ solution at 23 °C.

values will vary with the uncertainties in experimental details. Nevertheless, the major features of kinetic behavior should remain intact.

Fig. 5c compares the Pt/C voltammetry curve after subtracting the background measured using only the carbon support (Vulcan XC-72) with the Pt(111) voltammetry curve multiplied by a factor of 6. The similar integrated area in the hydrogen adsorption region shows a roughness factor (*i.e.*, the ratio of electrochemical surface area to electrode's geometric area) of 6 for the Pt/C.

Fig. 5d shows the ORR polarization curves measured at 20 mV s^{-1} during a positive potential sweep. The positive shift in the curve for the Pt/C relative to that for Pt(111) can be attributed to the higher effective surface area of the former. The converted kinetic currents for Pt/C using eqn (20) are shown by the circles in Fig. 5e; they overlap with the Pt(111) curve after multiplying the current by a factor of 3. This fact suggests that the ORR reaction mechanism is the same on Pt/C as on Pt(111), albeit the area-specific activity is reduced by half (*i.e.*, three times the current *versus* six times the surface area).

5. H$_2$-PEMFC polarization curves with Pt/C catalysts

Recently, Neyerlin *et al.* characterized the ORR kinetic behavior in PEMFCs using four semi-empirical parameters: the electron transfer coefficient, the exchange current density, the reaction order for oxygen partial pressure, and the activation energy.[25] They showed that *iR*-free voltage as a function of current density, oxygen pressure, and temperature in the high voltage (>0.79 V) and low current (<0.5 A cm^{-2}) range can be well explained using the B–V equation with a single Tafel slope determined by $b = kT\ln(10)/\beta$, where $\beta = 1$. This finding is encouraging since it demonstrates that well-designed experiments can generate consistent results on ORR kinetic behavior by minimizing many other experimental uncertainties. Below we show that these data are consistent with, and support, the intrinsic kinetic equation through fitting and simulations, despite these authors having interpreted the results using the B–V equation with a simple Tafel slope.

Fig. 6 displays a typical polarization curve measured in an H$_2$-air PEMFC and its corresponding high frequency resistance (HFR). Generally, we can express cell voltage as a function of current density by the reversible potential for the ORR versus the reversible potential for the HOR, E^0, after subtracting voltage losses due to the anode overpotential for the HOR, η_{HOR}, the cathode overpotential for the ORR, η_{ORR}, the mass transport effect, η_{tx}, cell resistance measured by high

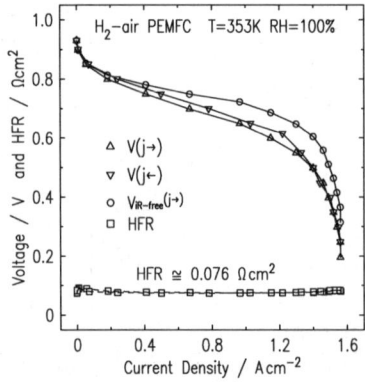

Fig. 6 H$_2$-air PEMFC polarization curve and HFR measured at 80 °C. Anode/cathode catalysts were 20%Pt/C ETEK with 0.19/0.19 mg cm^{-2} Pt loading, corresponding to a roughness factor of 100. H$_2$/air pressures 2/2 atm, flows 160/550 sccm, and 100% humidified. Data measured with a delay of 30 s per point.

frequency impedance, R_{HFI}, and the protonic resistance in the catalyst layers, R_{H^+cl},[25]

$$V(j) = E^0 - \eta_{HOR}(j) - \eta_{ORR}(j) - \eta_{tx}(j) - jR_{HFI} - jR_{H^+cl}(j) \quad (23)$$

Since protonic resistance in the catalyst layer at the electrodes cannot be measured by high frequency impedance, Neyerlin et al. proposed a complicated method to estimate its value.[26] Experimentally, Boyer et al.[34] measured total resistance as function of the composition of a catalyst layer sandwiched between membrane electrolytes. This additional catalyst layer is inert because an electrochemical reaction cannot occur without electronic contact; thus, protons must carry the charge flow through the entire layer. Therefore, Boyer and colleagues calculated the maximum protonic resistance in the inert catalyst layer from the slope of the total resistance *versus* the inert layer loading. For one set of samples, they found 0.100 Ω cm^4 mg^{-1} Pt,[34] implying 0.019 Ω cm^2 for our cathode loading of 0.19 mg cm^{-2}. This value is smaller than the measured HFR of 0.076 Ω cm^2. Including it in the *iR*-correction did not cause the results to differ significantly enough to affect the conclusions. Thus, it is reasonable to omit protonic resistance in the catalyst layer when 100% humidified. Further, the anode overpotential for the HOR is negligible for pure H_2 with sufficient Pt loading (>0.05 mg cm^{-2}).[7,35] Thus, approximately, we have

$$V_{iR\text{-free}}(j) = V(j) + jR_{HFI} = E(j) = E^0 - \eta_{ORR}(j) - \eta_{tx}(j) \quad (24)$$

The circles in Fig. 6 show this so-called *iR*-free polarization curve.

To fit the *iR*-free polarization curve, we assume that the average surface concentration for oxygen in the cathode decreases linearly with increasing current,

$$\frac{C_{O_2}}{C_{O_2}^0} = 1 - \frac{j}{j_L} \quad (25)$$

This assumption yields the equation, $j = j_k/(1 + j_f/j_L)$,[7] which becomes eqn (19) and (20) when the backward reaction rate is negligible, *i.e.*, $j_k = j_f$. In PEMFC measurements, the limiting current, j_L, is less controllable than in RDE measurements. However, it provides a simple first-order approximation to treat the mass-transport effect.

The kinetic equations used for fitting and simulating the dependence on Pt-loading and oxygen-pressure are

(1) For the semi-empirical Butler–Volmer equation

$$j_k = AP_{O_2}^{1/2} j_0 e^{(E-E^0)/b} \quad (26)$$

where A is the roughness factor calculated by the ratio of effective electrochemical area to the electrode's geometric area, and $b = kT\ln(10)/\beta = 70$ mV dec^{-1} for $T = 353$ K and $\beta = 1$.

(2) For the intrinsic double-trap kinetic equation

$$j_k = Aj^* e^{(E-E^0)/kT} e^{-\Delta G_{RD}^*/kT} \theta_{OH} \quad (27)$$

where θ_{OH} is expressed by eqn (6).

Fig. 7a shows the *iR*-free polarization curve, $E(j)$, obtained using a constant HFR of 0.076 Ωcm^2. To analyze this curve, we numerically switched the $E(j)$ to $j(E)$ and applied eqn (19), as we did in analyzing the RDE polarization curve. We then numerically switched back the fitted curve and plotted it in the form of $E(j)$. The solid line is the fit to the *iR*-free polarization curve over the entire voltage range using the double-trap kinetic equation. The fit using the B–V equation with a single Tafel

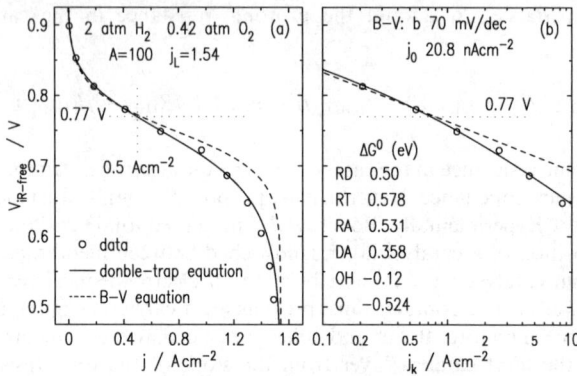

Fig. 7 (a) Fits to the *iR*-free polarization curve using the double-trap (solid lines) and the B–V (dashed lines) kinetic equations with the fitted parameters given in (b). (b) Semi-logarithmic plot of the *iR*-free voltage *versus* kinetic current. The circles are numerically-converted data.

slope (dashed line) is obtained with b fixed at 70 mV dec^{-1} for the data below 0.5 A cm^{-2} (marked by the vertical dotted line). Extending the fitted curve based on the B–V equation to the entire range of current (dashed line) shows that, at large currents or low potentials, it considerably deviates from the measured *iR*-free polarization curve (circles). The difference is the voltage loss unaccounted for by the B–V equation.

Fig. 7b semi-logarithmically plots potential as a function of the kinetic current. Using the fitted adsorption free energies, we obtained the reversible potential for the O/OH transition,

$$E^{rev}_{O/OH} = E^0 - (\Delta G^0_{OH} - \Delta G^0_{O})/e = 1.174 - (-0.120 + 0.524) = 0.77 \text{ V} \quad (28)$$

Below this potential, the data and the curve calculated using the intrinsic kinetic equation diverge from the straight line calculated with the B–V equation with a single Tafel slope, demonstrating the verity of the double Tafel slope for the ORR in PEMFCs. This result is not contradictory with the finding of a single Tafel slope reported by Neyerlin and coworkers[25] because their data were limited to be above 0.79 V.

Further verification can be made from testing the distinct features predicted by simulating the dependence of *iR*-free polarization curves on Pt loading and oxygen pressure. Fig. 8 shows the simulated Pt-loading dependence under the same operational conditions as those for the measured curve in Fig. 6. The #1 and #3 curves are calculated with the same kinetic parameters as those obtained in Fig. 7, except the roughness factor, A, is changed from 100 to 50 and 200, respectively.

In both kinetic equations, the kinetic current is proportional to the roughness factor, so that where the two curves separate changes on the current axis and remains constant on the voltage axis as shown by the dotted lines in Fig. 8. This prediction implies that if the kinetics is the major cause for the B–V curve deviating from the measured *iR*-free polarization curve that can be fully reproduced by the double-trap kinetic equation, it will start at potentials near 0.77 V regardless of current. On the other hand, if the deviation reflects either proton resistance in the catalyst layer, or a more complicated mass-transport effect, these points will be current-dependent. That is, the B–V equation should fit the data below 0.5 A cm^{-2} no matter what the potentials are. This criterion utilizes the data in the middle range, not close to the mass transport limiting current, and thus, to some degree, can tolerate the uncertainties in determining the limiting current.

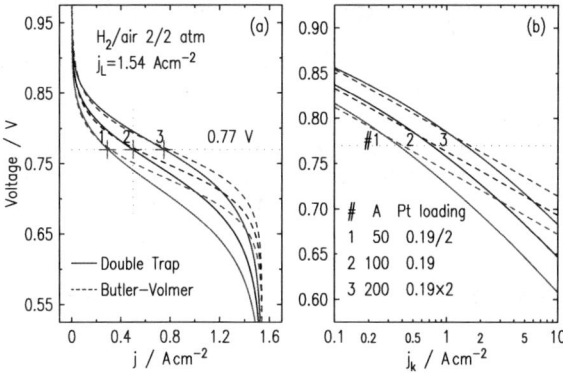

Fig. 8 Simulated Pt-loading dependence of iR-free polarization curves (a), and the voltage versus kinetic current curves (b) based on the parameters obtained in Fig. 7 using the B–V (eqn (26), dashed lines) and the double-trap (eqn (27), solid lines) kinetic equations.

Another way to distinguish different kinetic equations is through studying the dependence of iR-free polarization curves on oxygen pressure. Fig. 9 shows the simulated results based on the changes expected after altering the operating conditions from H_2/air at 2 atm to H_2/O_2 at 2.7 atm and doubling the flow rate for oxygen. The latter condition mimics that employed by Gasteiger et al.[35] for maximizing the mass transport limiting current. We estimated a 15 A cm^{-2} limiting current based on proportionally increasing it by a factor of 10 from 1.54 A cm^{-2}, which is partly from the ratio of oxygen partial pressure, $2.7/(2 \times 0.21) = 6.4$, and partly from the increase of the oxygen flow rate. In the simulations, we also calculated the pressure-dependent reversible potential for the ORR at $T = 353$ K

$$E^0 = 1.170 + kT\ln(P_{H_2}^2 P_{O_2})/4 = 1.170 + 0.0175\log(P_{H_2}^2 P_{O_2}) \quad (29)$$

The higher cell voltage, evidenced by comparing the curves in Fig. 9a, is expected with increasing oxygen pressure partly due to the rise in E^0 from 1.174 to 1.193 V

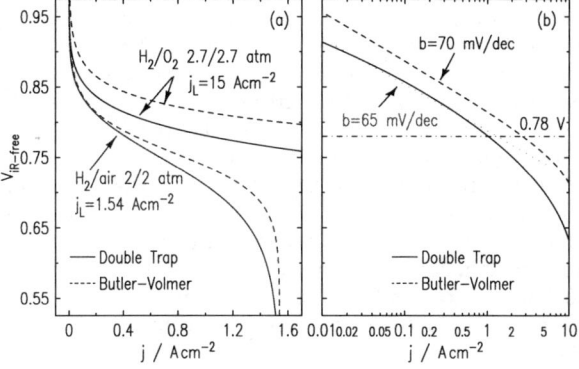

Fig. 9 (a) Simulated iR-free polarization curves for an H_2/O_2 (pure oxygen) fuel cell under higher pressures and higher limiting current (upper part) than those for an H_2/air fuel cell (lower part). (b) Semi-logarithmic plot of the iR-free polarization curves for the H_2/O_2 fuel cell curves in (a). Note that the currents, not the kinetic currents, are plotted, which exhibit nearly linear behavior over a quite large current region because of the high limiting current (~15 Acm^{-2}).

calculated from eqn (29), and partly due to the increase of j_L that reduces voltage loss from mass transport.

More interestingly, different kinetic equations predict the increase from reaction kinetics differently. The B–V equation generates a kinetic current proportional to $P_{O_2}^{1/2}$ and a constant Tafel slope of 70 mV dec^{-1} at 80 °C (dashed lines). For the intrinsic kinetic equation, increasing oxygen partial pressure lowers the activation free energies for adsorption processes, *i.e.*, the DA and RA reactions. The difference can be calculated from the dependence of activation free energies on oxygen concentration,

$$\Delta(\Delta G_{DA}^{*0}) = \Delta(\Delta G_{RA}^{*0}) = -kT\ln(10)\Delta(0.5\log c_{O_2}) = -0.035\Delta\log P_{O_2} \quad (30)$$

Since the activation free energies for the RT and RD reactions are largely unaffected, we expect a smaller gain in activity than that resulting from the B–V equation. Also, activity enhancement by raising oxygen pressure increases with lowering potential because the reaction rate is less predominantly limited by desorption at lower potentials. Accordingly, the Tafel slope slightly declines, from 70 to 65 mV dec^{-1}, at potentials above 0.78 V. This prediction agrees well with an average Tafel slope of 65 mV dec^{-1} at 80 °C for the three curves of different Pt loadings reported by Gasteiger *et al.*[35]

6. Conclusions

The ORR kinetic behavior on Pt in acidic media can be consistently described using the intrinsic kinetic equation based on the double-trap kinetic model. In particular, we elucidated the origin of the double Tafel slope based on the activity-and-barriers plots. Near the reversible potential for the ORR, not only the decrease of activation barrier for the slowest forward reaction step, but also the increase of activation barrier for the fastest backward reaction step, determine the apparent Tafel slope of the overall kinetic current. This statement holds even when the overall backward reaction rate is negligible because the lowest backward reaction barrier can influence profoundly the adsorption isotherms of the reaction intermediates, which, in turn, affects the Tafel slope.

In PEMFC applications, we demonstrated that with 100% humidification, the *iR*-corrected polarization curves could be fitted by the double-trap kinetic equation. An analytic formula for the ORR overpotential as a function of kinetic current is not needed because we can numerically convert an $E(j)$ curve into a $j(E)$ curve for fitting, and then, numerically convert it back again to the $E(j)$ form for plotting. The results show that the Tafel slope changes around 0.77 V on *iR*-corrected voltage; a value consistent with the range of data can be fitted using the B–V equation with a single Tafel slope. Simulations also predict a decrease of the Tafel slope in the high potential region from 70 to 65 mV dec^{-1} at 80 °C with increasing oxygen partial pressure, which agrees well with reported experimental findings.

For fuel cell design, the results suggest that catalyst loading at cathode should be high enough to get the desired current density with the *iR*-free voltage above the turning potential for the double Tafel slope. In catalyst activity tests, reporting the *iR*-corrected polarization curves up to the mass transport limiting current with measured roughness factor can facilitate comparison of kinetic currents over wide potential region either though fitting or numerical converting.

A1 Appendix 1. List of Symbols

b /V decade^{-1}	The Tafel slope
c_{O_2}, c_{H^+} /M or mol cm^{-3}	Concentrations of reactants
E^0/V	Reversible potential for the ORR
$E_{O/OH}^{rev}$ /V	Reversible potential for the O and OH transition

j /A cm^{-2}	Measured current density
j_0 /A cm^{-2}	Exchange current density
j^* /A cm^{-2}	Reference prefactor
j_k /A cm^{-2}	Total net kinetic current density
j_f /A cm^{-2}	Kinetic current for forward reactions
j_L /A cm^{-2}	Mass-transport limiting current density
k /eV K^{-1}	Boltzmann constant, kT(296 K) = 25.51 meV
T /K	Temperature
β	Electron transfer coefficient for reductive reactions
η /V	ORR overpotential
θ_O, θ_{OH}	Fractional coverage of reaction intermediates
v_i /mol s^{-1} cm^{-2}	Reaction rates for i = DA, RA, RT, RD
ΔG_O^0 /eV	Adsorption free energy for O at $\eta = 0$
ΔG_{OH}^0 /eV	Adsorption free energy for OH at $\eta = 0$
ΔG_i^{*0} /eV	Activation free energy of a forward reaction at $\eta=0$
ΔG_i^*, ΔG_{-i}^* /eV	Potential dependent activation free energy
η_{HOR} /V	Anode overpotential for the HOR
η_{ORR} /V	Cathode overpotential for the ORR
η_{tx} /V	The mass transport caused voltage loss
R_{HFI} /Ωcm^2	The cell resistance measured by HFI
R_{H^+cl} /Ωcm^2	The protonic resistance in catalyst layers

Acknowledgements

This work is supported by the U.S. Department of Energy, Divisions of Chemical and Material Sciences, under the Contract No. DE-AC02-98CH1-886.

References

1 A. J. Bard and L. R. Faulkner, in *Electrochemical Methods – Fundamentals and Applications*, John Wiley & Sons, New York, 2001, p. 100.
2 P. Sabatier, *Ber. Dtsch. Chem. Ges.*, 1911, **44**, 1984.
3 R. Parsons, *Trans. Faraday Soc.*, 1958, **54**, 1053.
4 S. Trasatti, in *Handbook of Fuel Cells*, ed. W. Vielstich, A. Lamm and H. A. Gasteiger, Wiley 2003, pp. 88–92.
5 J. K. Nørskov, J. Rossmeisl, A. Logadottir, L. Lindqvist, J. R. Kitchin, T. Bligaard and H. Jonsson, *J. Phys. Chem. B*, 2004, **108**, 17886–17892.
6 J. K. Nørskov, T. Bligaard, A. Logadottir, J. R. Kitchin, J. G. Chen and S. Pandelov, *J. Electrochem. Soc.*, 2005, **152**, J23–J26.
7 J. X. Wang, T. E. Springer and R. R. Adzic, *J. Electrochem. Soc.*, 2006, **153**, A1732.
8 J. X. Wang, T. E. Springer, P. Liu, M. Shao and R. R. Adzic, *J. Phys. Chem. C*, 2007, **111**, 12425–12433.
9 J. X. Wang, J. Zhang and R. R. Adzic, *J. Phys. Chem. A*, 2007, **111**, 12702–12710.
10 A. Damjanovic, in *Electrochemistry in Transition*, ed. O. J. Murphy, S. Srinivasan and B. E. Conway, Plenum Press, New York, 1992, pp. 107–126.
11 R. R. Adzic, in *Electrocatalysis*, ed. J. Lipkowski and P. N. Ross, Wiley-VCH, New York, 1998, pp. 197–242.
12 N. M. Markovic and P. N. Ross, in *Interfacial Electrochemistry – Theory, Experiments and Applications*, ed. A. Wieckowski, Marcel Dekker, New York, 1999, pp. 821–841.
13 U. A. Paulus, T. J. Schmidt, H. A. Gasteiger and R. J. Behm, *J. Electroanal. Chem.*, 2001, **495**, 134–145.
14 M. Gattrell and B. MacDougall, in *Handbook of Fuel Cells*, ed. W. Vielstich, A. Lamm and H. A. Gasteiger, Wiley 2003, pp. 361–367.
15 A. Damjanovic, in *Modern Aspects of Electrochemistry*, ed. J. O. M. Bockris and B. E. Conway, Plenum Press, New York, 1969, pp. 369–483.
16 N. Markovic, H. Gasteiger, B. N. Grgur and P. N. Ross, *J. Electroanal. Chem.*, 1999, **467**, 157.
17 P. N. Ross Jr, in *Handbook of Fuel Cells*, ed. W. Vielstich, A. Lamm and H. A. Gasteiger, Wiley 2003, pp. 465–480.
18 J. X. Wang, N. M. Markovic and R. R. Adzic, *J. Phys. Chem. B*, 2004, **108**, 4127–4133.

19 A. J. Appleby, *J. Electroanal. Chem.*, 1993, **357**, 117–179.
20 H. A. Gasteiger, S. Kocha, B. Sompalli and F. T. Wagner, *Appl. Catal., B*, 2005, **56**, 9.
21 T. E. Springer, M. S. Wilson and S. Gottesfeld, *J. Electrochem. Soc.*, 1993, **140**, 3513–3526.
22 A. Z. Weber and J. Newman, *Chem. Rev.*, 2004, **104**, 4679–4726.
23 M. V. Willians, H. R. Kunz and J. M. Fenton, *J. Electrochem. Soc.*, 2005, **152**, A635–A644.
24 K. C. Neyerlin, H. A. Gasteiger, C. K. Mittelsteadt, J. Jorne and W. B. Gu, *J. Electrochem. Soc.*, 2005, **152**, A1073–A1080.
25 K. C. Neyerlin, W. Gu, J. Jorne and H. A. Gasteiger, *J. Electrochem. Soc.*, 2006, **153**, A1955–A1963.
26 K. C. Neyerlin, W. Gu, J. Jorne, J. Clark A and H. A. Gasteiger, *J. Electrochem. Soc.*, 2007, **154**, B279–B287.
27 M. R. G. d. Chialvo and A. C. Chialvo, *Phys. Chem. Chem. Phys.*, 2004, **6**, 4009–4017.
28 P. M. Quaino, J. L. Fernandez, M. R. G. d. Chialvo and A. C. Chialvo, *J. Mol. Catal. A: Chem.*, 2006, **252**, 156–162.
29 N. M. Markovic, H. A. Gasteiger and P. N. Ross, *J. Phys. Chem.*, 1995, **99**, 3411–3415.
30 T. J. Schmidt and H. A. Gasteiger, in *Handbook of Fuel Cells*, ed. A. L. W. Vielstich, H. A. Gasteiger, John Wiley & Sons, 2003, pp. 316–333.
31 T. J. Schmidt, H. A. Gasteiger, G. D. Stab, P. M. Urban, D. M. Kolb and R. J. Behn, *J. Electrochem. Soc.*, 1998, **145**, 2354–2358.
32 C. H. Paik, T. D. Jarvi and W. E. O'Grady, *Electrochem. Solid-State Lett.*, 2004, **7**, A82–A84.
33 F. A. Uribe and J. T. A. Zawodzinski, *Electrochim. Acta*, 2002, **47**, 3799–3806.
34 C. Boyer, S. Gamburzev, O. Velev, S. Srinivasan and A. J. Appleby, *Electrochim. Acta*, 1999, **43**, 3703–3709.
35 H. A. Gasteiger, J. E. Panels and S. G. Yan, *J. Power Sources*, 2004, **127**, 162–171.

A first principles comparison of the mechanism and site requirements for the electrocatalytic oxidation of methanol and formic acid over Pt

Matthew Neurock,[*a] Michael Janik[b] and Andrzej Wieckowski[c]

Received 17th March 2008, Accepted 13th May 2008
First published as an Advance Article on the web 22nd August 2008
DOI: 10.1039/b804591g

First principles density functional theoretical calculations were carried out to examine and compare the reaction paths and ensembles for the electrocatalytic oxidation of methanol and formic acid in the presence of solution and applied electrochemical potential. Methanol proceeds *via* both direct and indirect pathways which are governed by the initial C–H and O–H bond activation, respectively. The primary path requires an ensemble size of between 3–4 Pt atoms, whereas the secondary path is much less structure sensitive, requiring only 1–2 metal atoms. The CO that forms inhibits the surface at potentials below 0.66 V NHE. The addition of Ru results in bifunctional as well as electronic effects that lower the onset potential for CO oxidation. In comparison, formic acid proceeds *via* direct, indirect and formate pathways. The direct path, which involves the activation of the C–H bond followed by the rapid activation of the O–H bond, was calculated to be the predominant path especially at potentials greater than 0.6 V. The activation of the O–H bond of formic acid has a very low barrier and readily proceeds to form surface formate intermediates as the first step of the indirect formate path. Adsorbed formate, however, was calculated to be very stable, and thus acts as a spectator species. At potentials below 0.6 V NHE, CO, which forms *via* the non-Faradaic hydrolytic splitting of the C–O bond over stepped or defect sites in the indirect path, can build up and poison the surface. The results indicate that the direct path only requires a single Pt atom whereas the indirect path requires a larger surface ensemble and stepped sites. This suggests that alloys will not have the same influence on formic acid oxidation as they do for methanol oxidation.

1. Introduction

The electrocatalytic oxidation of methanol and formic acid has been analyzed in great detail over the past decade as the result of their importance in the development of both methanol and formic acid fuel cells.[1-4] There are a number of elegant studies for methanol as well as formic acid oxidation over well-defined Pt and Pt alloy surfaces that have helped to establish the overall pathways and provide insights into the mechanisms that control these reactions.[4-34] Although much is known about the paths involved, a detailed understanding of the elementary steps in the mechanism as well as the explicit surface ensembles needed to carry out these steps is still

[a] Departments of Chemical Engineering and Chemistry, University of Virginia, Charlottesville, VA, 22904-4741, USA
[b] Department of Chemical Engineering, Pennsylvania State University, University Park, PA, 16802, USA
[c] Department of Chemistry, University of Illinois, Urbana-Champaign, IL, 61801, USA

somewhat speculative and often debated. While theory and simulation have begun to provide mechanistic insights into some of the elementary steps for methanol decomposition in vapor phase over different metals,[35–37] there have been very few studies that have examined the influence of the solution phase or more realistic electrochemical interfaces.[8,38–42]

Methanol oxidation is thought to follow a dual path mechanism over Pt at sufficiently high potentials[4,6–8,18,20,43–45] involving both "indirect" and "direct" pathways. The indirect path proceeds via the formation of CO followed by its subsequent oxidation to CO_2. The direct path proceeds instead through the formation of soluble intermediates, such as formaldehyde and formic acid, which can subsequently oxidize to form CO_2. In the indirect path, methanol weakly adsorbs to platinum and dehydrogenates via a sequence of steps to form adsorbed CO. At lower potentials, CO builds up on the surface and blocks the sites necessary for the activation of water to form surface hydroxyl intermediates that could oxidize it from the surface. At higher potentials, CO can be oxidized and both the indirect and the direct pathways can proceed. The direct path involves the formation of soluble secondary intermediates such as formaldehyde and formic acid[5–7,18,31,46–48] Experiments by Korzeniewski et al.[31,46–48] carried out in different electrolytes suggested that formaldehyde is the predominant secondary product and the path to formaldehyde requires significantly smaller ensemble sizes than that for the primary path. Osawa et al.,[10,21,26] on the other hand, used surface enhanced infrared spectroscopy to suggest that the dual path proceeds through the formation of surface formate intermediates which form on polycrystalline Pt electrodes at potentials greater than 0.7 V RHE and subsequently oxidize to form formic acid rather than formaldehyde. Koper et al.[15,20] showed that formaldehyde and formic acid could react to one another. Formaldehyde, for example, can be hydrolyzed to form methylene glycol which can oxidize over Pt to form formic acid.

The oxidation of formic acid is thought to be significantly easier than methanol since CO_2 is already present intact in the molecule. This would eliminate the need for an external oxidant. This is consistent with the fact that formic acid can oxidize at potentials significantly less than those required to activate water. Three different paths for the oxidation of formic acid to CO_2 have been suggested in the literature.[9,10,24,26,43,44,49,68] The first, which is known as the "direct path", involves the adsorption of formic acid onto the metal followed by the direct removal (dehydrogenation) of both protons to form CO_2. The direct path was clearly identified by carrying out labeling studies in which ^{13}CO was preadsorbed onto Pt before the oxidation of ^{12}C-formic acid. The resulting product was found to be unlabeled, thus demonstrating that CO_2 was formed via a "direct" path from formic acid rather than from the adsorbed ^{13}CO.[25] The second path, which is termed the "indirect path", involves the non-Faradaic dehydration of formic acid to CO and the subsequent oxidation of CO to CO_2. In the final "formate path", the O–H bond of formic acid is cleaved to form a surface formate intermediate which can subsequently react to form CO_2.[9,49]

Osawa et al. carried out the first in-situ internal reflection infrared studies using surface enhanced infrared adsorption spectroscopy (SEIRS) combined with cyclic voltammetry and chronoamperometry to determine the nature of the surface intermediates that form during formic acid oxidation under reaction conditions. They suggest that the reaction proceeds at potentials from 0.35 V–0.9 V NHE through the formation of an adsorbed formate intermediate.[21,25,26] More recent studies by Behm et al.,[9,11,49,24–26] however, indicate that formate decomposition is not involved in the dominant reaction path and plays a very minor role, if any, in the formation of CO_2. They found that the changes in the formate adlayer were similar to those found by acetate and (bi)sulfate and suggest that the adlayer is controlled by a fast adsorption/desorption equilibrium with the electrolyte. The changes in the formate coverage were considered to be the result of this exchange rather than the decomposition of formate to CO_2. The formation of CO_2 was suggested to occur predominantly through the direct activation of the formic acid. The formate that forms was thought to be a spectator species that inhibits the direct decomposition of formic acid to CO_2. They showed, *via* isotopic labeling, that C–H bond activation was involved in the rate determining step for the oxidation of formic acid.

Many of the different rate controlling processes outlined here for both methanol and formic acid oxidation can proceed over different surface ensembles. As such, the development of alloys which lead to enhanced electrocatalytic activity for methanol oxidation may have little influence on formic acid oxidation and *vice-versa*. In order to better understand the mechanisms that control both methanol and formic acid oxidation and the nature of the active sites necessary to carry out the controlling reaction steps, we have carried out first-principle density functional theoretical calculations over the well-defined Pt(111) surface in the presence of solution and over a range of applied potentials. We compare the results for methanol oxidation on different alloys to the results on formic acid in an effort to understand the similarities and differences between the two and how alloying might change the results.

2. Computational methods

Plane wave density functional theoretical calculations were carried out herein using the Vienna *Ab Initio* Software Program (VASP) by Kresse and Hafner[50-52] within the Generalized Gradient Approximation to determine the potential-dependent reaction energies and activation barriers for the oxidation of formic acid over Pt(111) and methanol over Pt(111) and PtRu(111) alloys at the electrified aqueous/metal interface. The metal surface was modeled using a 3×3 super cell comprised of 3 metal layers with 9 metal atoms per layer together with a 10 Å vacuum inserted between the top and bottom surfaces. The bottom layer of the surface was held fixed at the experimental bulk distances. All of the calculations were carried out using the RPBE functional form of the exchange–correlation function,[53-55] ultrasoft pseudopotentials[56] to model the electron ion interactions and a plane-wave basis cutoff energy of 396.0 eV. The first Brillouin zone was sampled using a $3 \times 3 \times 1$ Monkhorst–Pack[57] k-point mesh, which was tested to converge a subset of the relative energies reported herein. Previous tests comparing 3 and 5 layer metal slabs found little change on the structural or potential dependent behavior of water over an Ni(111) surface.[58]

The aqueous solution phase was modeled using 24 explicit solvent molecules chosen to fill up the vacuum region. The height of the background region was chosen so as to match the density of water near a metal surface of 0.86 g cm^{-3} at 0 K.[8,39,40,59,60] The water molecules were initially oriented in a hexagonal bilayer in registry with the (111) surface and subsequently optimized. In order to study the adsorption of the reactant methanol and formic acid molecules and their subsequent reactions, model systems were created containing specific intermediates. One water molecule on the surface was therefore replaced with reactant, intermediate or product species. The surface metal layers, the adsorbates, and the aqueous interface were optimized for all the structures explored.

The potential dependence on the reaction energies and activation barriers were determined using the double-reference method of Filhol and Neurock.[39,59,61] The double-reference method involves calculating two different reference potentials in order to relate the electrified aqueous metal interface to a neutral aqueous system and a vacuum reference potential. The potential is controlled by systematically adding or subtracting fractions of an electron from the unit cell and applying an equal but opposite compensating, homogeneous background charge. The details of the approach and the approximations are described in detail elsewhere.[59]

3. Results

We report below on the results for the oxidation of methanol and formic acid over an ideal Pt(111) surface. The resulting pathways and ensembles are subsequently compared and used to provide an understanding of how alloying may influence each of these systems.

3.1 Methanol oxidation

In a previous communication, we presented DFT-calculated overall reaction energies for methanol decomposition over Pt(111).[8] Herein, we highlight just the salient features of the overall pathways and focus predominantly on the active surface ensembles necessary to carry out methanol decomposition and its subsequent oxidation, as this was not discussed previously. The general pathways involved in the oxidation of methanol are shown in Fig. 1. The decomposition of methanol over Pt proceeds by the adsorption of methanol to the surface. The energies for both the C–H and O–H bond activation paths of methanol were calculated over a range of different potentials. Herein, we summarize the results for an arbitrary potential of 0.5 V NHE. The reaction energies in Fig. 2 indicate that the elementary steps that make up the primary involve:

$$CH_3O^*H \rightarrow {}^*CH_2O^*H + H^+ + e^- \rightarrow {}^*CHOH + H^+ + e^- \rightarrow {}^*COH + H^+ + e^- \rightarrow {}^*CO + H^+ + e^-$$

The elementary steps that make up the secondary reaction path proceeds *via* the initial activation of the O–H bond to form methoxy which then reacts to formaldehyde:

$$CH_3O^*H \rightarrow CH_3O^* + H+ + e^- \rightarrow {}^*CH_2{=}O + H^+ + e^-$$

The structure and specific surface site requirements to carry out both the primary C–H bond activation and the secondary O–H bond activation steps are illustrated in Fig. 3. The primary reaction path requires an ensemble site comprised of at least 4 Pt metal atoms to accommodate the formation of the formyl intermediate that forms. This is consistent with experimental results from Cuesta,[62] who used cyanide to modify the ensembles in the Pt(111) surface and showed that the indirect path for methanol to CO required a surface ensemble with three contiguous Pt atoms arranged triangularly.

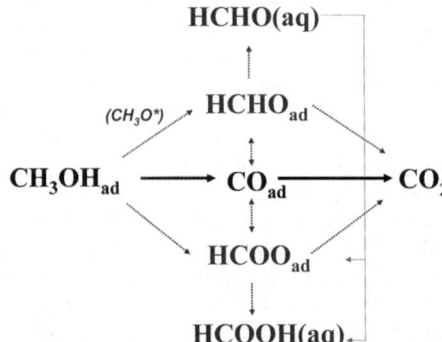

Fig. 1 The initial steps in dual path mechanism for the electrocatalytic oxidation of methanol over Pt to CO_2. The indirect path which proceeds through the formation of CO is shown in the center (in black). The direct paths which lead to the formation of formaldehyde and formic acid are shown at the top and bottom, respectively (in blue). The formaldehyde that forms can also react in solution to form formic acid or adsorbed formate. The interconversion of these steps is sketched in red. A more detailed set of possible elementary steps and how they may interconvert is given by Koper.[15,20]

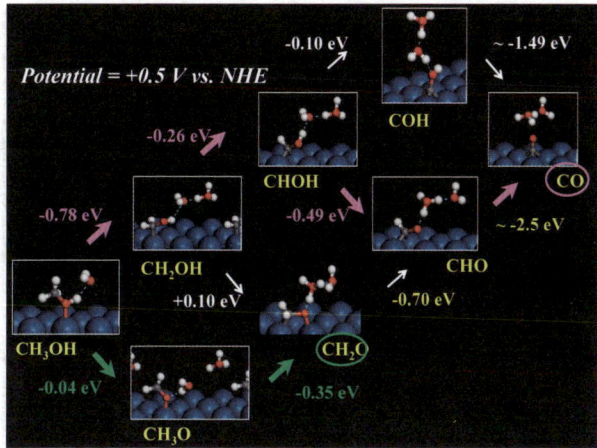

Fig. 2 DFT-calculated reaction energies over Pt(111) for the initial methanol decomposition paths to CO and formaldehyde at a constant potential of 0.5 V. The indirect path through CO shown in purple is favored. The path to formaldehyde is a secondary route that becomes more favorable at potentials slighter greater than 0.5 V.[8]

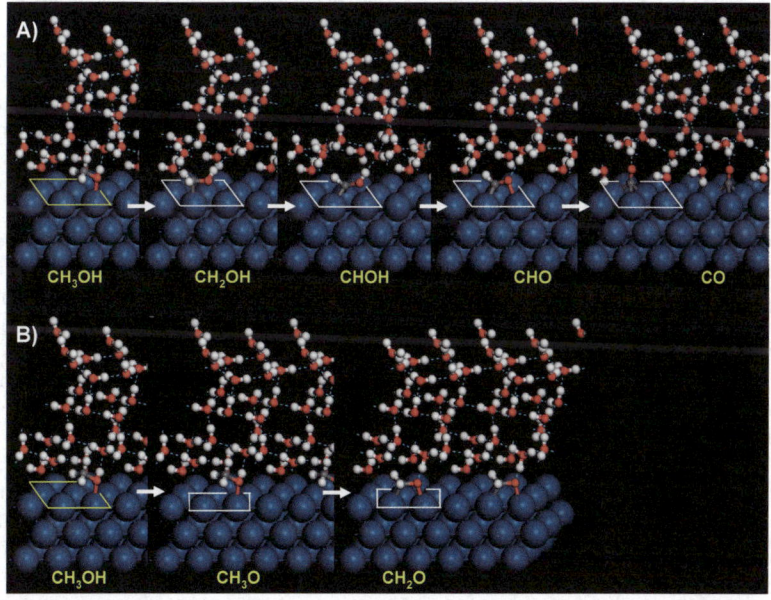

Fig. 3 DFT optimized structures on Pt(111) in solution at 0.5 V for the intermediates in the electrocatalytic oxidation of methanol to: (A) CO *via* the primary path and (B) formaldehyde *via* the secondary path. The primary path requires 3–4 metal atoms whereas the secondary route requires 1–2 metal atoms.

The O–H activation of methanol, as well as the subsequent C–H activation of methoxy to formyl, take place over the top of a single metal atom, as is shown in Fig. 3B and thus require significantly smaller ensemble sizes which consist of at most 1 to 2 metal atoms. This secondary path is much less structure sensitive. This is consistent with the experimental results from Iwasita *et al.*[6,7] and Koper *et al.*[15,20] who demonstrated that the selectivity for the direct path, *i.e.* the formation of formaldehyde and

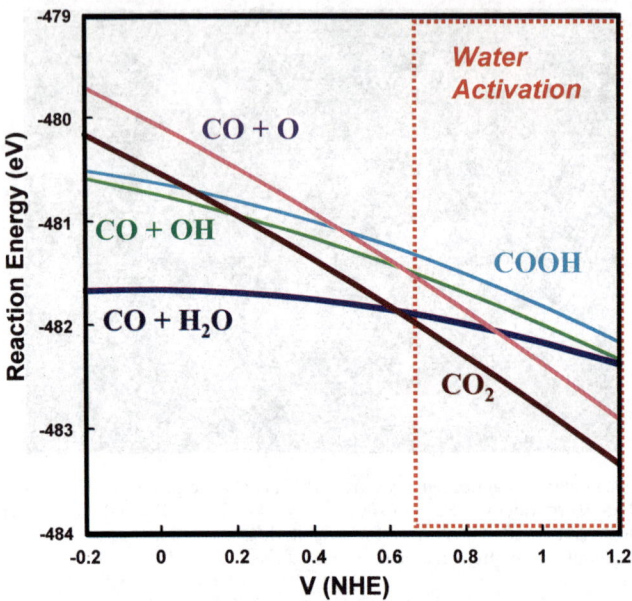

Fig. 4 The DFT-calculated electrochemical surface phase diagrams for the reaction energy *vs.* potential for the reaction of CO and water to CO_2 and COOH over Pt(111).[39]

other soluble intermediates was greatest on the Pt(111) surface in sulfuric acid and lowest in perchloric acid. They demonstrated that the (bi)sulfate strongly adsorbs and inhibits the sites that lead to the indirect path to form CO. Perchlorate, on the other hand, is more weakly held to the surface and requires fewer sites, and therefore does not strongly inhibit the CO formation path for the Pt(111) surface.

In addition to the dehydrogenation pathways depicted in Fig. 2, each of the intermediates that form can subsequently undergo oxygen addition to form CO_2 containing intermediates. We have carried out subsequent DFT calculations to examine most of these paths in the vapor phase. In addition, we have explored, in much greater detail, the potential dependent reaction energies and activation barriers for CO oxidation over Pt(111) in solution, as this is the most difficult and important oxidation step. The overall reaction energies for the oxidation of CO *via* adsorbed atomic oxygen, hydroxyl intermediates, or water to CO_2 and COOH are shown as a function of potential in the phase diagram in Fig. 4. These results indicate that on the Pt(111) surface, hydroxyls are the active oxidants and appear at above 0.6 V NHE as the result of the oxidation of water. CO is oxidized at a just slightly higher potential of 0.66 V NHE. Theory was subsequently used to calculate potential dependent activation barriers for CO oxidation. The calculated activation barriers for CO + OH coupling, suggested to be the rate-limiting step, decreases from about 0.52 eV to 0.38 eV in moving from 0.0 V to 1.0 V NHE, which is consistent with experimental results.[63]

3.2 Influence of ruthenium

The oxidation of methanol is typically carried out by the introduction of Ru which aids in the oxidation of CO to CO_2 *via* both bifunctional and electronic or ligand effects.[29,63,64] In the bifunctional mechanism, water is activated over Ru atoms in the surface to form active hydroxyl intermediates that can then oxidize CO bound to neighboring Pt sites. We used theory to analyze various arrangements of Pt and Ru to better understand these effects. More specifically, we examined the

Table 1 The DFT calculated onset potentials for the activation of water to $OH + H^+ + e^-$ and CO oxidation to CO_2 over model Pt(111) and Ru(0001) alloy surfaces. The higher than expected potentials for the activation of water are due to the fact that activation is carried out directly next to CO. The activation of water on Pt(111) in the absence of CO was calculated to be 0.66 V

SURFACE	H_2O $OH^* + H^+ + e^-$	$CO + H_2O$ $CO_2 + 2 H^+ + 2e^-$
$Pt_{ML}/Ru(0001)$	1.22 V	0.01 V
Pt(111)	1.29 V	0.55 V
Ru dispersed	0.74 V	0.66 V
Ru islands	0.55 V	0.78 V

potential dependent energies for both water activation and CO oxidation steps over Pt, Ru, a well dispersed $Pt_{66.7\%}Ru_{33.3\%}$ surface alloy on Pt, Ru surface ensembles with the Pt(111) surface, and a pseudomorphic Pt overlayer on Ru in order to elucidate the effects on the CO oxidation kinetics. A summary of the DFT calculated results are presented in Table 1 which shows the onset potentials for the activation of water and the subsequent oxidation of adsorbed CO by surface OH groups over each of the alloy structures. The results indicate that the addition of Ru into the surface helps to lower the onset potential for the activation of water, and in addition, subsurface Ru atoms weaken the adsorption of CO. Both effects help promote CO oxidation. There appears to be an optimal balance of the amount of Ru needed in the surface which will enhance the formation of OH without poisoning the surface. This is consistent with other experimental studies.[65–67] While the addition of Ru to the surface decreases the Pt surface ensemble size which is important for methanol decomposition, it also results in a greater number of nearest neighbor Pt-CO and Ru-OH pairs that can enhance the reactivity. Subsurface Ru lowers the barriers for the CO + OH reaction *via* electronic interactions

3.3 Oxidation of formic acid

As was discussed earlier in the introduction, formic acid can be oxidized *via* three different paths. These three paths are illustrated in Fig. 5 (a schematic adopted from Behm[9]). The first path, depicted in the middle of Fig. 5, involves the direct decomposition of formic acid into $CO_2 + 2H^+$ and $2e^-$. The second, depicted at the bottom of Fig. 5, involves an indirect route in which formic acid is first dehydrated to CO. CO is then re-oxidized in a second step to form CO_2. The third path, depicted at the top of Fig. 5, involves the conversion of formic acid into

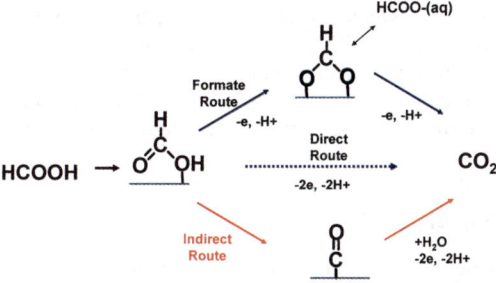

Fig. 5 The reaction paths involved in the electrocatalytic oxidation of formic acid to CO_2. (A) Direct path (center) to CO_2, (B) indirect path through CO (bottom), and (C) formate path though the formation of the formate intermediate (top). Schematic follows that proposed by Behm *et al.*[9,11,49] Experimental evidence also suggests the possible exchange of formate with electrolyte in solution, which is shown.

a surface formate intermediate that can subsequently react to form CO_2. All three of these paths were examined herein using density functional theory together with the double-reference method in order to determine their potential dependence reaction energies and activation barriers. Formic acid can adsorb on Pt(111) in the presence of solution in an upright fashion through both the OH group and the oxygen of the C=O, or lying parallel to the surface via a di-σ interaction with C=O. While the adsorption in the upright mode is more favorable in the gas phase, the two modes of adsorption in solution are nearly equivalent. Formic acid readily reacts from the upright mode to form a surface formate intermediate and proton. The formate intermediate is bound to a bridge site in a di-σ mode through its two oxygen atoms. The molecule is quite rigid in this configuration with the C–H bond oriented along the surface normal, thus making it very difficult to activate the C–H. The dissociation of the C–H bond from this state requires bending the CH group out of the CO_2 plane, which costs over 1 eV in energy. Alternatively, the molecule can reorient itself to be bound through one of the oxygen atoms and the hydrogen. This also costs over 1 eV in energy at positive potentials as it requires breaking the very highly polarized Pt–O bond.

If the C–H bond of formic acid is activated initially instead, the hydroxy carbonyl (COOH) intermediate is formed. At positive potentials, it readily deprotonates to form CO_2 and H^+ with a very low activation barrier. While we did not observe the simultaneous activation of both C–H and O–H bonds, we did find that the O–H bond was very easy to activate subsequent to C–H bond activation. We can not, however, rule out the possible simultaneous activation of both bonds. In either case, the COOH intermediate is unstable and unlikely to be observed, therefore suggesting this path is the direct path of Fig. 5.

Formic acid, formate as well as hydroxy carbonyl intermediates can all potentially react via the activation of the C–O bond, which would result in the formation of adsorbed CO. While we have not calculated this path in solution, we would expect, based on vapor phase calculations for the activation of the RC–OH bond over Pt(111), the barrier to be greater than 1 eV. The C–O bond is quite strong and there is little reason to expect the rate of this non-Faradaic process to be substantially altered by variations in potential. Experimentally, it has been reported that formic acid dehydration can not occur over Pt(111).[69] The reaction instead requires defect sites on the surface. Based on the theoretical estimates and the experimental results, we conclude that CO activation requires step edges or defect sites. Despite the higher activation barrier the calculated overall reaction energy for this step is highly exothermic. The CO intermediates that form are very strongly bound to Pt, and therefore would be expected to be observed at low potentials at which the CO oxidation rate is slow.

The ensemble requirements for both the direct as well as the formate pathways shown along the middle and the top branches in Fig. 6, respectively, are 1 to 2 metal atoms. The indirect path, which is shown along the bottom branch in Fig. 6, on the other hand, requires a significantly larger ensemble size in order to activate the C–O bond. As such, the direct and formate pathways are characteristically different than that for methanol activation, which requires at least 4 metal atoms along its primary path. The indirect path would be similar to methanol since both require the oxidation of CO.

3.4 The effect of potential

We carried out calculations over a range of potentials in order to establish the potential dependence of all three paths, compare with experiment and establish which path ultimately controls the chemistry at different conditions. For the sake of comparison, we report here on the energies at three very different potentials: 0.0, 0.5 and 1.0 V NHE in order to see the changes that can occur as a function of potential.

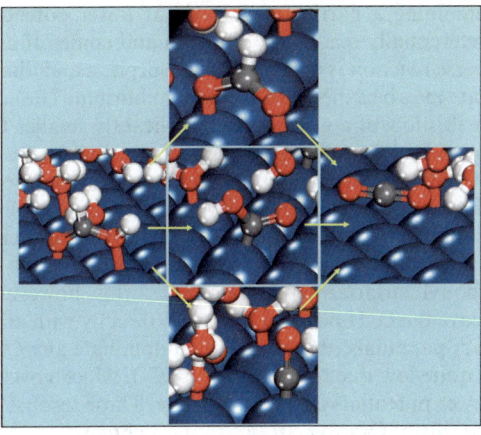

Fig. 6 DFT calculated surface ensembles required for the direct oxidation path (center), the indirect path through CO (bottom) and the formate path (top) for the oxidation of formic acid over Pt(111) to CO_2 at a constant potential of 0.5 V.

The resulting elementary reaction energies calculated at 0 V are shown in Fig. 7. Formic acid can react to form adsorbed CO and water, which is thermodynamically the most favored path, with an overall reaction energy of −2.06 eV. Based on previous gas phase C–OH activation calculations over Pt(111), the activation barrier is speculated to be over 1 eV and will thus require step or defect sites. Any CO that forms on the surface at lower potentials will be strongly bound, build up over time and inhibit the surface. The activation of the O–H bond of formic acid, on the other hand, occurs quite easily in solution with a barrier of less then 0.05 eV and is exothermic at the surface by −0.78 eV. The resulting formate group can either continue on and break its C–H bond to form CO_2 and a proton or undergo the reverse recombination reaction and form formic acid. As was discussed above, the barrier to break the C–H bond of the bound formate intermediate is very difficult

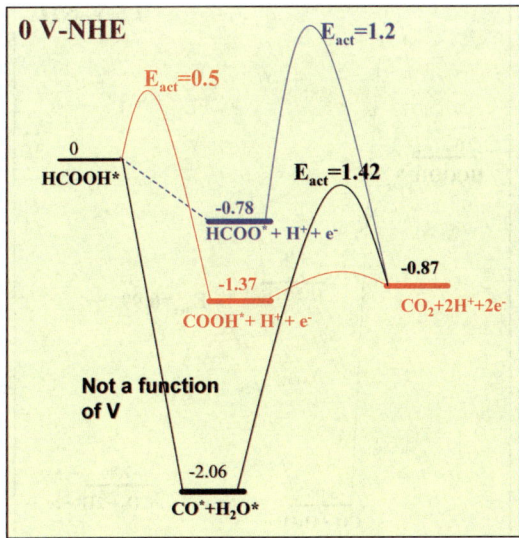

Fig. 7 DFT-calculated potential energy surface for the direct, indirect, and formate paths for the oxidation of formic acid over Pt(111) to CO_2 held at a constant potential of 0 V.

as it requires overcoming a barrier of 1.2 eV. At lower potentials, the adsorbed formate would preferentially react with a proton and come off as formic acid. The barrier for this reverse reaction is 0.78 eV. The adsorption and dissociation of formic acid to form formate may be expected to reach equilibrium. The equilibrium to form surface formate at this low of a potential is significantly smaller than that found at higher potentials, and both surface formate or formic acid species will compete with strongly-bound CO. We would therefore expect a very low surface concentration of formate intermediates and instead a much larger concentration of CO at steady state conditions. The initial activation of the C–H bond of formic acid has a barrier of 0.5 eV, which is greater than that required to break the O–H bond of the acid. The hydroxyl carbonyl surface intermediate that forms, however, is quite reactive and can readily cleave the O–H bond to form CO_2 directly with an activation barrier of 0.52 eV. At lower potentials, this path should dominate any CO_2 production but its activity will be quite low due to inhibition by CO, which covers the surface. The formate path at lower potentials is simply inactive. These results are consistent with the experimental results of Osawa et al.[10,24–26] and Behm et al.,[9–11,49] which show that the surface is highly covered in CO at potentials up to 0.9 V but has a decreasing intensity from 0.3 V to 0.7 V RHE. Formate, which slowly begins to appear at 0.3 V, dominates the surface at 0.9 V RHE.

As the potential is increased to 0.5 V NHE, both the direct path and the formate path become thermodynamically more favorable as both involve electro-oxidation steps (Fig. 8). The formation of surface formate becomes 0.22 eV more exothermic in moving from 0 V to 0.5 V NHE. The subsequent activation barrier for breaking the C–H bond of formate to form CO_2 is reduced by 0.1 eV (from 1.2 to 1.1 eV). The direct activation of formic acid to the hydroxyl carbonyl becomes more exothermic by 0.26 eV and its barrier is lowered by 0.03 eV (from 0.50 to 0.47 eV). There is no change in the thermodynamics or the kinetics for the dehydration of formic acid to CO and water as it is a non-Faradaic reaction. The barrier for the subsequent oxidation of CO to CO_2 is reduced from 1.42 eV at 0 V to 0.99 eV at 0.5 V. While this is a significant drop, there is likely little change in the macroscopic behavior of the system, as CO will still build up and act to poison the surface. The direct path is still the most favorable and remains strongly inhibited by chemisorbed CO on the surface.

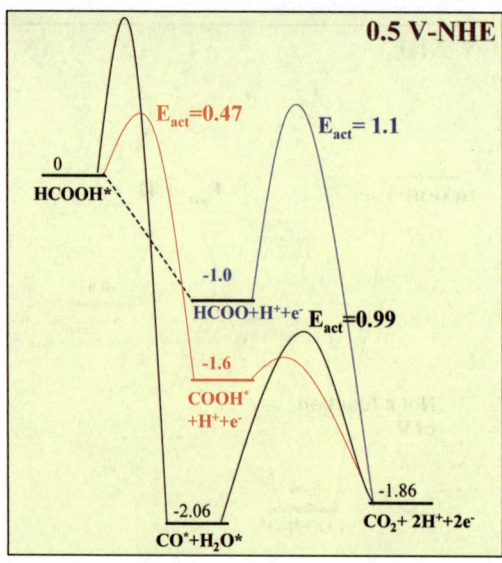

Fig. 8 DFT-calculated potential energy surface for the direct, indirect, and formate paths for the oxidation of formic acid over Pt(111) to CO_2 held at a constant potential of 0.5 V.

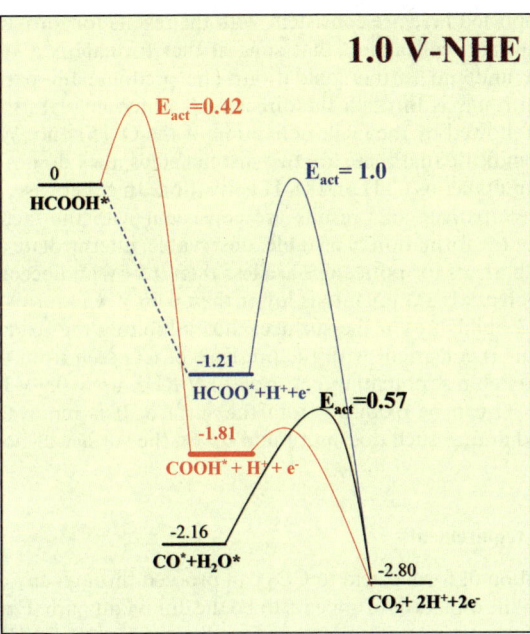

Fig. 9 DFT-calculated potential energy surface for the direct, indirect, and formate paths for the oxidation of formic acid over Pt(111) to CO_2 held at a constant potential of 1.0 V.

As the potential is increased to 1 V NHE more significant changes occur in the importance of the various paths, as seen in Fig. 9. The greatest change is that the indirect route, which produces CO, no longer acts to poison the surface. The activation barrier to oxidize CO is reduced down to 0.5 eV. Earlier, we showed that CO is readily oxidized at potentials greater then 0.66 V. CO therefore no longer acts to inhibit the surface. This is well established experimentally as well, since CO is oxidized from Pt(111) at potentials above 0.6 V RHE.[29,63] Other oxidation reactions also become more favorable as the potential is increased. The reaction energies for the initial steps of both the formate (O–H activation) and direct (C–H activation) paths become more exothermic by 0.21 eV. The C–H activation barrier for the conversion of formic acid to the hydroxyl carbonyl intermediate (*COOH) is lowered down to 0.05 eV.

The barrier for the activation of the C–H bond of adsorbed formate to CO_2 is decreased by 0.1 eV. The change in this barrier is rather small in comparison with the change in the overall thermodynamics for creating the formate intermediate. The activation barrier for the reverse reaction, *i.e.* the reduction of formate back to formic acid is energetically very unfavorable at 1.0 V NHE as the barrier is increased to over 1.2 eV. The result is that formate has a low barrier to form but is energetically a very stable surface intermediate. As such, it remains bound very strongly to the surface and acts as a spectator. The formation of CO_2 is therefore dominated by the direct path. The higher coverage of formate may lead to minor contributions from the indirect path. The results from the calculations suggest that the surface should become covered with formate intermediates at higher potentials as they form faster than they can be removed. They are thermodynamically stable and no longer inhibited from forming by adsorbed CO. The barriers for the C–H as well as the subsequent O–H activation of formic acid *via* the direct path are both lower than 0.5 eV and can readily occur. Formate appears to predominantly be a spectator species.

The results reported here are consistent with the results for formic acid oxidation over Pt by Behm and colleagues[9] that suggest that formate is a spectator species which does not undergo further oxidation. The predominant path to CO_2 over a wide potential range is through the direct path which involves the activation of the C–H bond followed by the facile activation of the O–H bond. While the results reported herein indicate that these are two distinct steps, they do not rule out a path involving the simultaneous C–H and O–H activation. In either case, we refer to this path as the direct path as our results are consistent with the fact that this path proceeds without the formation of a stable, observable, intermediate at any potential of interest. The barriers for both steps are less than 0.5 eV and become more favorable at higher potentials. At potentials lower than 0.66 V, CO forms as the result of dehydration and builds up on the surface, thus inhibiting the oxidation of formic acid to CO_2. This is consistent with the presence of CO seen from the SEIRS data from Osawa and Behm at potentials between 0.1 V RHE up to 0.9 V RHE. At higher potentials, the CO can be oxidized from the surface. It is removed more rapidly than it is formed and as such does not build up on the surface at potentials greater then 0.66 V.

3.5. Ensemble requirements

While the oxidation of formic acid to CO_2 can proceed through any one of the three paths presented, the direct route appears to be the dominant path. For completeness, however, we discuss the site requirements for all three pathways. The direct activation of formic acid to CO_2, which involves the initial activation of the C–H bond, was found to require just one metal atom. The hydroxyl carbonyl intermediate that forms, binds to the surface through the carbon atom in an atop configuration. The subsequent O–H activation of the bound *COOH intermediate occurs quite easily with a low activation barrier and also involves only 1 or 2 metal atoms. The site requirement for the activation of the O–H bond of formic acid to form formate is quite similar, requiring only a single Pt atom. As such, both the direct and the formate paths have much lower ensemble requirements than that for the primary (indirect) path for the activation of methanol over Pt. The final path, which involves the indirect activation of formic acid to CO, requires defect sites as well as a much larger surface ensemble in order to activate the C–O bond. The oxidation of CO that forms is very difficult at lower potentials and CO acts to poison the surface.

The difference in ensemble requirements can lead to significant differences upon alloying. We found that the primary oxidation path for methanol to CO_2 was considerably enhanced by the presence of Ru in the surface. Ru aids in the activation of water to form hydroxyl intermediates which enhances the oxidation of CO to CO_2, *i.e.* the bifunctional mechanism. The most active surfaces would therefore be ones in which Pt and Ru were intimately mixed in the surface in order to enhance their activity. This is the result of both the large surface ensemble requirements as well as the fact that all of the intermediates that might form require an oxidant to ultimately be converted to CO_2. Subsurface Ru significantly enhances the reaction by weakening the adsorption of CO, which is known as the ligand or electronic effect. While the ligand effect is very important, it is not a very strong function of ensemble size and tends to be localized to nearest neighbor surface and subsurface sites.

In the oxidation of formic acid, CO_2 is already explicitly present in the fuel. The site requirements for simple C–H and O–H scission are only 1–2 metal atoms. As such, one might expect small changes in activity as the result of alloying. While the addition of Ru to the surface will help to enhance the oxidation of CO formed *via* the indirect pathway, the results reported here indicate that there is a very small effect, if any, on the overall rates of oxidation as the presence of Ru would not be expected to enhance the activation of the C–H bond of formic acid to initiate the direct path. This is due to the fact that both the direct and the formate pathways do not require the activation of water to produce a sacrificial oxidant as the CO_2 in the reactant remains intact. Both, however, do require the activation of C–H and O–H bonds of the acid. The C–H bond activation in general is preferentially carried out over Pt and should therefore not be explicitly influenced by Ru with exception of weaker electronic influences. The activation of the O–H bond of an acid occurs readily over Pt and therefore does not require Ru either.

4. Summary and conclusions

First principles density functional theoretical calculations help to confirm that methanol oxidation proceeds through both the direct and indirect pathways that are governed by C–H and O–H bond activation, respectively. The indirect path, which proceeds *via* the formation of CO, is the dominant path. The rate, however, is significantly lower at potentials less than 0.6 V NHE over Pt(111) where CO is strongly chemisorbed and inhibits oxidation. The ensemble size required for the indirect and direct paths are characteristically different. The indirect path requires 3–4 Pt atoms in order to accommodate the carbon bound CH_xO fragments that form, which is consistent with experimental findings by Cuesta.[62] The direct path, which leads to the production of formaldehyde and/or formic acid, on the other hand, requires only 1–2 Pt atoms, and as such, is much less influenced by anion adsorption. This is consistent with the experimental results presented by Iwasita *et al.*[6,7] and Koper *et al.*,[15,20] which indicate that bi(sulfate) blocks Pt sites and inhibits the indirect path but has much less or an influence on the direct path.

Formic acid oxidation can also proceed to CO_2 through both direct and indirect pathways. In addition, there is a third path which involves the formation of formate intermediates on the surface. Similar to that for methanol, the direct path proceeds *via* the initial activation of the C–H bond. The barrier for this step is 0.5 eV at 0 V NHE and drops down to 0.4 eV at higher potentials. The barrier to activate the O–H bond of the hydroxyl carbonyl that forms is very low over a range of potentials, and is thus readily activated to form CO_2 directly. Based on the calculated activation barriers, this path appears to be the dominant route to form CO_2 and requires the presence of step or defect sites. At lower potentials, however, the direct path is blocked by CO which covers the surface. The indirect path, which is responsible for CO formation, is controlled by the non-Faradaic hydrolytic activation of the C–O bond of formic acid or formate. At potentials lower than 0.67 V NHE, the CO that forms is thermodynamically quite stable and can not be oxidized from the surface. At potentials greater than 0.6 V, CO is oxidized and the direct path is much less inhibited. The formate path, which involves the activation of the O–H bond of formic acid, initiates quite readily over Pt, thus resulting in the formation of formate intermediates. At lower potentials, this path is blocked due to the presence of CO on the surface. As the potential is increased between 0.5–1.0 V NHE, formate formation becomes preferable. The activation barrier for the subsequent activation of the formate C–H bond, however, is prohibitive. The computational results indicate that formate forms on the surface at these higher potentials but acts as a spectator, which is consistent with experimental results from Behm.[9]

The site specificity for the direct and indirect paths involved in the oxidation of formic acid were found to be quite similar to those found for methanol. The direct path appears to occur over just 1 or 2 Pt atoms. The indirect path, however, requires splitting the C–O bond which requires a significantly greater ensemble size and defect sites. There is one significant difference, however, between the ensemble size effects for methanol and formic acid. In methanol, the indirect path through CO is the major path to CO_2 over a wide range of potentials. In formic acid, however, the direct path is actually the dominant path to CO_2 over nearly all potentials. This helps to explain the significant differences concerning the influence of Ru on methanol and formic acid oxidation. Ru significantly enhances the methanol oxidation but has only a small influence on formic acid oxidation.

Acknowledgements

We gratefully acknowledge the Army Research Office for funding through a MURI grant (DAAD19-03-1-0169) for fuel cell electrocatalysis research as well as the computational time from the U.S. Army Research Laboratory Major Shared High Performance Computing Resource Center, the Molecular Sciences Computing Facility in the William R. Wiley Environmental Molecular Sciences Laboratory at Pacific Northwest Laboratory, and National Center for Computational Sciences at Oak Ridge National Laboratory. AW acknowledges support from the National Science Foundation under grant NSF CHE06-51083. We also kindly acknowledge the very helpful discussions with Sally Wasileski, Christopher Taylor, Jean Sebastian Filhol, and Tom Zawodzinski.

References

1 *Fuel Cell Handbook*, U.S. Department of Energy, 7th edn, 2004.
2 W. Vielstich, A. Lamm and H. A. Gasteiger, *Handbook of Fuel Cells Fundamentals, Technology and Applications*, John Wiley and Sons, 2003.
3 A. Wieckowski, E. R. Savinova and C. G. Vayenas, *Catalysis and Electrocatalysis at Nanoparticle Surfaces*, CRC Press Taylor & Francis Group, 2003.
4 T. D. Jarvi and E. M. Stuve, Fundamental Aspects of Vacuum and Electrocatalytic Reactions of Methanol and Formic Acid on Platinum Surfaces, in *Electrocatalysis*, ed. J. Lipkowski and P. N. Ross, Wiley-VCH, 1998.
5 E. A. Batista and T. Iwasita, Adsorbed intermediates of formaldehyde oxidation and their role in the reaction mechanism, *Langmuir*, 2006, **22**(18), 7912–7916.
6 E. A. Batista, G. R. P. Malpass, A. J. Motheo and T. Iwasita, New insight into the pathways of methanol oxidation, *Electrochem. Commun.*, 2003, **5**(10), 843–846.
7 E. A. Batista, G. R. P. Malpass, A. J. Motheo and T. Iwasita, New mechanistic aspects of methanol oxidation, *J. Electroanal. Chem.*, 2004, **571**(2), 273–282.
8 D. Cao, G. Q. Lu, A. Wieckowski, S. A. Wasileski and M. Neurock, Mechanisms of methanol decomposition on platinum: A combined experimental and ab initio approach, *J. Phys. Chem. B*, 2005, **109**(23), 11622–11633.
9 Y. X. Chen, M. Heinen, Z. Jusys and R. J. Behm, Bridge-bonded formate: Active intermediate or spectator species in formic acid oxidation on a Pt film electrode?, *Langmuir*, 2006, **22**(25), 10399–10408.
10 Y. X. Chen, S. Ye, M. Heinen, Z. Jusys, M. Osawa and R. J. Behm, Application of in-situ attenuated total reflection-Fourier transform infrared spectroscopy for the understanding of complex reaction mechanism and kinetics: Formic acid oxidation on a Pt film electrode at elevated temperatures, *J. Phys. Chem. B*, 2006, **110**(19), 9534–9544.
11 Y. X. Chen, M. Heinen, Z. Jusys and R. J. Behm, Kinetic isotope effects in complex reaction networks: Formic acid electro-oxidation, *ChemPhysChem*, 2007, **8**(3), 380–385.
12 R. B. de Lima, M. P. Massafera, E. A. Batista and T. Iwasita, Catalysis of formaldehyde oxidation by electrodeposits of PtRu, *J. Electroanal. Chem.*, 2007, **603**(1), 142–148.
13 H. Hoster, T. Iwasita, H. Baumgartner and W. Vielstich, Pt-Ru model catalysts for anodic methanol oxidation: Influence of structure and composition on the reactivity, *Phys. Chem. Chem. Phys.*, 2001, **3**(3), 337–346.
14 T. H. M. Housmans and M. T. M. Koper, Methanol oxidation on stepped Pt $n(111) \times (110)$ electrodes: A chronoamperometric study, *J. Phys. Chem. B*, 2003, **107**(33), 8557–8567.
15 T. H. M. Housmans, A. H. Wonders and M. T. M. Koper, Structure sensitivity of methanol electrooxidation pathways on platinum: An on-line electrochemical mass spectrometry study, *J. Phys. Chem. B*, 2006, **110**(20), 10021–10031.
16 T. Iwasita, H. Hoster, A. John-Anacker, W. F. Lin and W. Vielstich, Methanol oxidation on PtRu electrodes. Influence of surface structure and Pt-Ru atom distribution, *Langmuir*, 2000, **16**(2), 522–529.
17 T. Iwasita, Erratum to Electrocatalysis of methanol oxidation [*Electrochimica Acta* **47**(22–23):3663–3674], *Electrochim. Acta*, 2002, **48**(3), 289–289.
18 T. Iwasita, Electrocatalysis of methanol oxidation, *Electrochim. Acta*, 2002, **47**(22–23), 3663–3674.
19 Z. Jusys and R. J. Behm, Methanol oxidation on a carbon-supported Pt fuel cell catalyst - A kinetic and mechanistic study by differential electrochemical mass spectrometry, *J. Phys. Chem. B*, 2001, **105**(44), 10874–10883.

20 S. C. S. Lai, N. P. Lebedeva, T. H. M. Housmans and M. T. M. Koper, Mechanisms of carbon monoxide and methanol oxidation at single-crystal electrodes, *Top. Catal.*, 2007, **46**, 320–333.
21 A. Miki, S. Ye, T. Senzaki and M. Osawa, Surface-enhanced infrared study of catalytic electrooxidation of formaldehyde, methyl formate, and dimethoxymethane on platinum electrodes in acidic solution, *J. Electroanal. Chem.*, 2004, **563**(1), 23–31.
22 H. Miyake, E. Hosono, M. Osawa and T. Okada, Surface-enhanced infrared absorption spectroscopy using chemically deposited Pd thin film electrodes, *Chem. Phys. Lett.*, 2006, **428**(4–6), 451–456.
23 Y. Mukouyama, M. Kikuchi, G. Samjeske, M. Osawa and H. Okamoto, Potential oscillations in galvanostatic electrooxidation of formic acid on platinum: A mathematical modeling and simulation, *J. Phys. Chem. B*, 2006, **110**(24), 11912–11917.
24 G. Samjeske, A. Miki, S. Ye, A. Yamakata, Y. Mukouyama, H. Okamoto and M. Osawa, Potential oscillations in galvanostatic electrooxidation of formic acid on platinum: A time-resolved surface-enhanced infrared study, *J. Phys. Chem. B*, 2005, **109**(49), 23509–23516.
25 G. Samjeske and M. Osawa, Current oscillations during formic acid oxidation on a Pt electrode: Insight into the mechanism by time-resolved IR spectroscopy, *Angew. Chem., Int. Ed.*, 2005, **44**(35), 5694–5698.
26 G. Samjeske, A. Miki, S. Ye and M. Osawa, Mechanistic study of electrocatalytic oxidation of formic acid at platinum in acidic solution by time-resolved surface-enhanced infrared absorption spectroscopy, *J. Phys. Chem. B*, 2006, **110**(33), 16559–16566.
27 H. A. Gasteiger, N. Markovic, P. N. Ross and E. J. Cairns, Methanol Electrooxidation on Well-Characterized Pt-Rn Alloys, *J. Phys. Chem.*, 1993, **97**(46), 12020–12029.
28 A. V. Tripkovic, K. D. Popovic, J. D. Lovic, N. M. Markovic and V. Radmilovic, Formic acid oxidation on Pt/Ru nanoparticles: Temperature effects, *Curr. Res. Adv. Mater. Process.*, 2005, **494**, 223–228.
29 N. M. Markovic and P. N. Ross, Surface science studies of model fuel cell electrocatalysts, *Surf. Sci. Rep.*, 2002, **45**(4–6), 121–229.
30 A. V. Tripkovic, K. D. Popovic, B. N. Grgur, B. Blizanac, P. N. Ross and N. M. Markovic, Methanol electrooxidation on supported Pt and PtRu catalysts in acid and alkaline solutions, *Electrochim. Acta*, 2002, **47**(22–23), 3707–3714.
31 D. Kardash, C. Korzeniewski and N. Markovic, Effects of thermal activation on the oxidation pathways of methanol at bulk Pt-Ru alloy electrodes, *J. Electroanal. Chem.*, 2001, **500**(1–2), 518–523.
32 K. Wang, H. A. Gasteiger, N. M. Markovic and P. N. Ross, On the reaction pathway for methanol and carbon monoxide electrooxidation on Pt-Sn alloy *versus* Pt-Ru alloy surfaces, *Electrochim. Acta*, 1996, **41**(16), 2587–2593.
33 N. M. Markovic, H. A. Gasteiger, P. N. Ross, X. D. Jiang, I. Villegas and M. J. Weaver, Electrooxidation Mechanisms of Methanol and Formic-Acid on Pt-Ru Alloy Surfaces, *Electrochim. Acta*, 1995, **40**(1), 91–98.
34 H. A. Gasteiger, N. Markovic, P. N. Ross and E. J. Cairns, Electrooxidation of Small Organic-Molecules on Well-Characterized Pt-Ru Alloys, *Electrochim. Acta*, 1994, **39**(11–12), 1825–1832.
35 S. K. Desai, M. Neurock and K. Kourtakis, A periodic density functional theory study of the dehydrogenation of methanol over Pt(111), *J. Phys. Chem. B*, 2002, **106**(10), 2559–2568.
36 J. Greeley and M. Mavrikakis, A first-principles study of methanol decomposition on Pt(111), *J. Am. Chem. Soc.*, 2002, **124**(24), 7193–7201.
37 J. Kua and W. A. Goddard, Oxidation of methanol on 2nd and 3rd row Group VIII transition metals (Pt, Ir, Os, Pd, Rh, and Ru): Application to direct methanol fuel cells, *J. Am. Chem. Soc.*, 1999, **121**(47), 10928–10941.
38 S. Desai and M. Neurock, A first principles analysis of CO oxidation over Pt and $Pt_{66.7\%}Ru_{33.3\%}$ (111) surfaces, *Electrochim. Acta*, 2003, **48**(25–26), 3759–3773.
39 M. J. Janik and M. Neurock, A first principles analysis of the electro-oxidation of CO over Pt(111), *Electrochim. Acta*, 2007, **52**(18), 5517–5528.
40 M. J. Janik, C. D. Taylor and M. Neurock, First principles analysis of the electrocatalytic oxidation of methanol and carbon monoxide, *Top. Catal.*, 2007, **46**, 306–319.
41 C. Hartnig, J. Grimminger and E. Spohr, The role of water in the initial steps of methanol oxidation on Pt(211), *Electrochim. Acta*, 2007, **52**(6), 2236–2243.
42 C. Hartnig, J. Grimminger and E. Spohr, Adsorption of formic acid on Pt(111) in the presence of water, *J. Electroanal. Chem.*, 2007, **607**(1–2), 133–139.
43 A. Capon and R. Parsons, *J. Electroanal. Chem.*, 1973, **44**, 1.
44 A. Capon and R. Parsons, *J. Electroanal. Chem.*, 1973, **45**, 205.
45 E. Herrero, W. Chrzanowski and A. Wieckowski, Dual path mechanism in methanol electrooxidation on a platinum electrode, *J. Phys. Chem.*, 1995, **99**, 10423.

46 C. Korzeniewski and C. L. Childers, Formaldehyde yields from methanol electrochemical oxidation on platinum, *J. Phys. Chem. B*, 1998, **102**(3), 489–492.
47 C. L. Childers, H. L. Huang and C. Korzeniewski, Formaldehyde yields from methanol electrochemical oxidation on carbon-supported platinum catalysts, *Langmuir*, 1999, **15**(3), 786–789.
48 M. Islam, R. Basnayake and C. Korzeniewski, A study of formaldehyde formation during methanol oxidation over PtRu bulk alloys and nanometer scale catalyst, *J. Electroanal. Chem.*, 2007, **599**(1), 31–40.
49 Y. X. Chen, M. Heinen, Z. Jusys and R. B. Behm, Kinetics and mechanism of the electrooxidation of formic acid - Spectroelectrochemical studies in a flow cell, *Angew. Chem., Int. Ed.*, 2006, **45**(6), 981–985.
50 G. Kresse and J. Hafner., *Phys. Rev. B: Condens. Matter Mater. Phys.*, 1993, **47**, 558.
51 G. Kresse and J. Furthmüller, *Phys. Rev. B: Condens. Matter Mater. Phys.*, 1996, **54**, 11169.
52 G. Kresse and J. Furthmüller, *Comput. Mater. Sci.*, 1996, **6**, 15.
53 J. P. Perdew, K. Burke and M. Ernzerhof, *Phys. Rev. Lett.*, 1996, **77**, 77.
54 J. P. Perdew, K. Burke and M. Ernzerhof, *Phys. Rev. Lett.*, 1998, **80**, 891.
55 B. Hammer, L. B. Hansen and J. K. Nørskov, *Phys. Rev. B: Condens. Matter Mater. Phys.*, 1999, **59**, 7413.
56 D. Vanderbilt, *Phys. Rev. B: Condens. Matter Mater. Phys.*, 1990, **41**, 7892.
57 H. J. Monkhorst and J. D. Pack, *Phys. Rev. B: Solid State*, 1976, **13**, 5188.
58 C. D. Taylor, R. G. Kelly and M. Neurock, First-principles calculations of the electrochemical reactions of water at an immersed Ni(111)/H_2O interface, *J. Electrochem. Soc.*, 2006, **153**, E207–E214.
59 C. D. Taylor, S. A. Wasileski, J. S. Filhol and M. Neurock, First principles reaction modeling of the electrochemical interface: Consideration and calculation of a tunable surface potential from atomic and electronic structure, *Phys. Rev. B: Condens. Matter Mater. Phys.*, 2006, **73**(16).
60 J. Rossmeisl, J. K. Norskov, C. D. Taylor, M. J. Janik and M. Neurock, Calculated phase diagrams for the electrochemical oxidation and reduction of water over Pt(111), *J. Phys. Chem. B*, 2006, **110**(43), 21833–21839.
61 J. S. Filhol and M. Neurock, Elucidation of the electrochemical activation of water over Pd by first principles, *Angew. Chem., Int. Ed.*, 2006, **45**(3), 402–406.
62 A. Cuesta, At least three contiguous atoms are necessary for CO formation during methanol electrooxidation on platinum, *J. Am. Chem. Soc.*, 2006, **128**(41), 13332–13333.
63 H. A. Gasteiger, N. Markovic, P. N. Ross and E. J. Cairns, Co Electrooxidation on Well-Characterized Pt-Ru Alloys, *J. Phys. Chem.*, 1994, **98**(2), 617–625.
64 W. Chrzanowski and A. Wieckowski, Surface structure effects in platinum/ruthenium methanol oxidation electrocatalysis, *Langmuir*, 1998, **14**(8), 1967–1970.
65 B. Hayden, Single Crystal Surfaces as Model Platinum-Based Hydrogen Fuel Cell Electrocatalysts, in *Catalysis and Electrocatalysis at Nanoparticle Surfaces*, ed. E. S. A. Wieckowski and C. Vayenas, Marcel Dekker, Inc, 2003, p. 171–210.
66 K. Sasaki, J. X. Wang, M. Balasubramanian, J. McBreen, F. Uribe and R. R. Adzic, Ultralow platinum content fuel cell anode electrocatalyst with a long term performance stability, *Electrochim. Acta*, 2004, **49**, 3873–3877.
67 Brankovic, J. X. Wang and R. R. Adzic, *J. Electrochem. Soc.*, 2001, **4**(12), A217–A220.
68 S. G. Sun, J. Clavilzer and A. Bewick, *J. Electroanal. Chem.*, 1988, **240**, 147–159.
69 M. D. Macia', E. Herrero and J. M. Feliu, *Electrochim. Acta*, 2002, **47**, 3653.

Surface structure effects on the electrochemical oxidation of ethanol on platinum single crystal electrodes

Flavio Colmati,[ab] Germano Tremiliosi-Filho,[a] Ernesto R. Gonzalez,[a] Antonio Berná,[b] Enrique Herrero*[b] and Juan M. Feliu[b]

Received 7th February 2008, Accepted 3rd April 2008
First published as an Advance Article on the web 7th August 2008
DOI: 10.1039/b802160k

Ethanol oxidation has been studied on Pt(111), Pt(100) and Pt(110) electrodes in order to investigate the effect of the surface structure and adsorbing anions using electrochemical and FTIR techniques. The results indicate that the surface structure and anion adsorption affect significantly the reactivity of the electrode. Thus, the main product of the oxidation of ethanol on the Pt(111) electrode is acetic acid, and acetaldehyde is formed as secondary product. Moreover, the amount of CO formed is very small, and probably associated with the defects present on the electrode surface. For that reason, the amount of CO_2 is also small. This electrode has the highest catalytic activity for the formation of acetic acid in perchloric acid. However, the formation of acetic acid is inhibited by the presence of specifically adsorbed anions, such as (bi)sulfate or acetate, which is the result of the formation of acetic acid. On the other hand, CO is readily formed at low potentials on the Pt(100) electrode, blocking completely the surface. Between 0.65 and 0.80 V, the CO layer is oxidized and the production of acetaldehyde and acetic acid is detected. The Pt(110) electrode displays the highest catalytic activity for the splitting of the C–C bond. Reactions giving rise to CO formation, from either ethanol or acetaldehyde, occur at high rate at any potential. On the other hand, the oxidation of acetaldehyde to acetic acid has probably the lower reaction rate of the three basal planes.

1. Introduction

Within the possible fuels for the fuel cell technology, ethanol has been the subject of many studies in the last decade. Ethanol presents some advantages with respect to methanol: it can be obtained directly from biomass after distillation, has a higher energy density, since 12 electrons are exchanged in the complete oxidation to CO_2, and is less toxic. However, the catalysis of ethanol oxidation presents additional challenges, since the catalytic activity of the electrode materials is normally lower than that observed for methanol oxidation.

The oxidation mechanism of ethanol on platinum electrodes can be decomposed into five different reactions as shown in eqn (1).

---

[a] *Instituto de Química de São, Carlos Universidade de São Paulo, Av. Trab. Sãocarlense 400, São Carlos, SP, 13560-970, Brazil*
[b] *Instituto de Electroquímica, Universidad de Alicante, Apdo. 99, Alicante, E-03080, Spain. E-mail: herrero@ua.es*

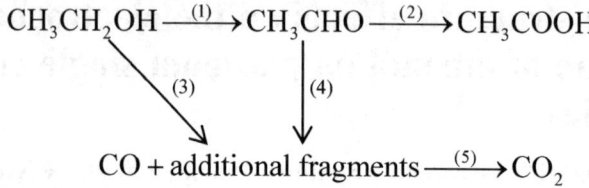

Reaction (1) is the oxidation of ethanol to acetaldehyde transferring two electrons. Acetaldehyde can, in turn, be oxidized to acetic acid, transferring two additional electrons. Acetic acid can be considered as a final product since its oxidation is very difficult. In fact, acetic acid is stable on platinum electrodes at potentials where oxygen evolves.[1,2] However, in the formation of acetic acid only 4 electrons are transferred, and, therefore, the energy density to yield acetic acid from ethanol is then lower than that obtained for the complete oxidation of methanol.

Reactions (3) and (4) represent the routes that allow the transfer of 12 e$^-$. Both reactions involve the cleavage of the C–C bond in the molecule to yield adsorbed CO and other fragments. These additional fragments, whose identification is not trivial,[3] are probably further oxidized to yield also adsorbed CO. The final oxidation of the adsorbed CO to CO_2 is difficult, since it requires the additional transfer of an oxygen atom to the adsorbed CO molecule. For platinum, the cleavage of the C–C bond and the oxidation to CO_2 seem to be favored in the presence of rhodium, ruthenium or osmium on the electrode surface.[4–9] For the modified electrodes, the activation of the C–C bond requires both platinum and osmium/ruthenium adjacent sites.

It is known that the oxidation of ethanol is a structure sensitive reaction.[10,11] Therefore, the final products of the oxidation will depend on the surface structure of the electrode. Additionally, the composition of the electrolyte can also change the reactivity of the electrode. Normally, the presence of strong adsorbing anions on the electrode surface hinders the interaction of the reacting molecules with the surface, resulting in a diminution of the current densities.[12] In this work, we have studied the effect of the structure of the electrode surface, as well as the effect of the competitive adsorption of the anion of the supporting electrolyte, in the catalytic activity of platinum electrodes for ethanol oxidation. Voltammetric and chronoamperometric experiments were used to evaluate the total reactivity of the surface under different conditions, whereas FTIR experiments allowed detection and determination of the different intermediate and final species involved in the reaction mechanism of ethanol oxidation on the different electrodes.

2. Experimental

Ethanol oxidation on platinum single crystal surfaces was studied in sulfuric and perchloric acid solutions. Single crystal platinum surfaces with basal orientations were used as working electrodes. Platinum single crystal surfaces were prepared from small single crystal beads, *ca.* 2 mm in diameter for voltammetric experiments and *ca.* 4.5 mm for infrared measurements, following the method developed by Clavilier.[13] Prior to any experiment, working electrodes were heated for around 10 s in a gas–oxygen flame, cooled down in a reductive atmosphere (H_2 + Ar)[14] and quenched in ultrapure water in equilibrium with this atmosphere. Electrode surfaces were protected with a water droplet during the transference to the electrochemical cell, to prevent the contamination of the surface.

Chronoamperometric and voltammetric curves were recorded with an Autolab Pgstat 30 potentiostat. The potential program for the chronoamperometric experiments is presented in Fig. 1. Prior to the final step, two conditioning potentials are applied. The purpose of these potential steps is to remove the maximum amount of adsorbates derived from the oxidation of ethanol from the surface so that the initial current recorded at the desired potential would be as close as possible to

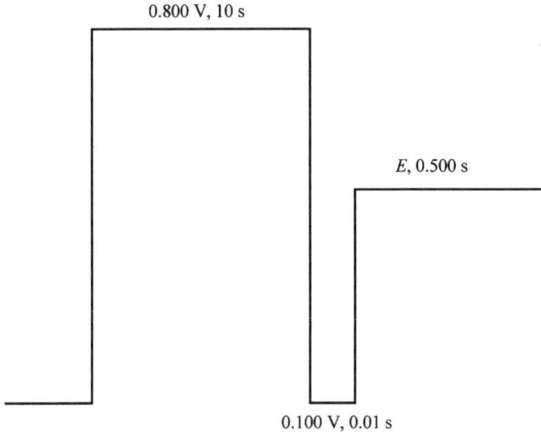

Fig. 1 Potential step program used in the chronoamperometric experiments.

the current measured on a clean surface. In this way, the electrode was kept for 10 seconds at 0.8 V to oxidize the adsorbed CO and then was polarized for 10 ms at 0.1 V to reduce the oxides and to desorb the acetate formed at the previous potential. The potential for this step coincides with the potential at which the reference spectra are collected in the FTIR experiments. For the Pt(100) and Pt(110) electrodes, potentials above 0.8 V should be avoided since they lead to changes in the surface structure. After that, the electrode is stepped to the desired potential. Data was collected for 500 ms with a time resolution of 0.1 ms. A voltammogram was recorded before and after a whole series of experiments to check that the potential steps have not changed the surface structure of the electrode.

Spectroelectrochemical experiments were carried out with a Nicolet Magna 850 spectrometer equipped with a narrow-band MCT detector. The spectroelectrochemical cell[15,16] had a prismatic CaF_2 window beveled at 60°. The spectra were collected with p-polarized light with a resolution of 8 cm^{-1}. The spectra are presented as the ratio $-\log(R_2/R_1)$, where R_2 and R_1 are the reflectance values corresponding to the single beam spectra recorded at the sample and reference potentials. Potential dependent series were collected according to the SPAIRS (single potential alteration infrared spectroscopy) technique.[17] In this way, the IR spectra collection was coupled with the voltammetric sweep at 2 mV s^{-1}. 260 interferograms, added to increase the signal to noise ratio, were recorded in a potential window of 50 mV and referred to that obtained between 0.075 and 0.125 mV. For clarity, the spectra were named using the mean potential value of the window.

Experiments were carried out at room temperature, 20 °C, in a classical two-compartment electrochemical cell de-aerated for 20 minutes by bubbling Ar (N50, Air Liquide for all gases used), including a platinum counter-electrode and a reversible hydrogen (N50) electrode (RHE) as reference. Solutions were prepared from sulfuric acid (Merck suprapur®), perchloric acid (Merck suprapur®), ethanol (Merck p.a.), ultrapure water from a Purelab Ultra (Elga-Vivendi) system and D_2O (99.95%, Sigma glass distilled). The cleanliness of the solutions was tested by the stability of the characteristic voltammetric features of well defined Pt(hkl) electrodes.

3. Results and discussion

3.1 Voltammetric results

Fig. 2 shows the voltammetric profiles (first cycle) of the three basal platinum electrodes in 0.2 M ethanol solutions. The comparison between the curves obtained

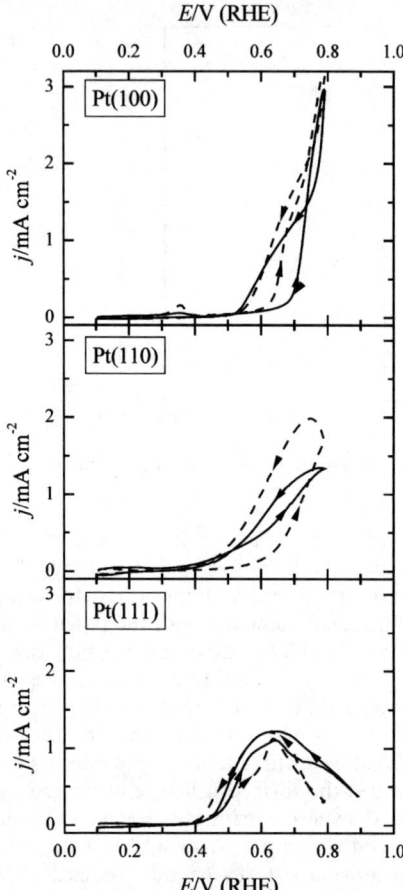

Fig. 2 Voltammograms (1st cycles) for ethanol oxidation in 0.1 M HClO$_4$ + 0.2 M EtOH (—) and 0.1 M H$_2$SO$_4$ + 0.2 M EtOH (- -) for Pt(111), Pt(110) and Pt(100) electrodes. Scan rate 50 mV/s.

in 0.1 M H$_2$SO$_4$ and HClO$_4$ will serve to understand the role played by the competitive adsorption of the anions in the oxidation mechanism, whereas the comparison between the three basal planes will help to clarify the role of the surface structure in the oxidation process. Probably, the most striking result is the effect of anions in the oxidation of ethanol, because the measured currents in perchloric acid are very similar to those observed in sulfuric acid. In fact, for the Pt(110) electrode, the observed behavior is the opposite to that anticipated, since the currents in perchloric acid are smaller than those in sulfuric acid. This fact seems to indicate that the adsorption of anions has no effect on the oxidation reaction, which is an unexpected result. Specific adsorption of anions is normally associated with a diminution in the oxidation currents, since the presence of anions adsorbed on the electrode surface interferes with the adsorption and oxidation of the reacting molecules. This can be observed in the voltammetric currents measured for methanol oxidation in 0.1 M sulfuric acid, which are between 1.5 (Pt(110)) and 2 times (Pt(111) and Pt(100)) smaller than those recorded in 0.1 M perchloric acid.[12]

The other important feature shown in Fig. 2 is that ethanol oxidation is a surface sensitive process, since the voltammetric profiles for the three electrodes are different, as has already been reported.[10,11] Additionally, the voltammograms show the typical behavior for a process in which the electrode is poisoned by CO formed during the positive going scan. CO is oxidized above 0.7 V and therefore currents in the negative going scan are higher, since the surface has been cleaned. For the Pt(111) electrode, the recorded voltammogram resembles that observed for the oxidation of methanol or formic acid.[12,18] The small hysteresis between the positive and negative going scans is, in all cases, linked to a low CO formation rate, which leads to a low blockage by CO in the positive scan. The diminution of the currents measured above 0.6 V has been linked to the formation of a strong bond between the anions and/or water and the surface at high potentials, which prevents the adsorption of reactant molecules on the electrode surface.[3,12] It is well known that the adsorption of (bi)sulfate on the Pt(111) electrode forms a well ordered structure[19,20] with higher coverage than that observed on Pt(100)[21] and this ordered structure has a strong influence on the reactions taking place at potentials above 0.6 V.[22] Another typical characteristic of the Pt(111) electrode is the low activity for the oxidation of small organic molecules, since maximum currents for this electrode are the smallest among the three basal planes. However, the onset potential for ethanol oxidation on Pt(111) electrodes is the lowest among the three electrodes, which is probably associated with the slow poisoning rate, *i.e.*, the low activity for the direct oxidation is also accompanied by a slow poisoning rate.

On the other hand, the Pt(100) electrode displays the highest peak current of the three electrodes and the higher poisoning rate as revealed by the absence of significant currents below 0.7 V in the positive-going scan, which indicates that CO has been formed at low potentials and completely blocks the electrode surface. This behavior has also been observed for formic acid and methanol oxidation.[12,23] At 0.7 V, CO oxidation starts and the currents immediately rise. In the negative scan, the currents are higher but the combination of a fast formation rate and a low oxidation rate for CO at potentials below 0.6 V leads to an accumulation of CO and to a sudden diminution of the currents.

Finally, the Pt(110) electrode shows a behavior that can be considered intermediate between the Pt(111) and Pt(100) electrodes, probably as a consequence of an activity and a poisoning rate which is in between those of Pt(111) and Pt(100) electrodes. In this case, a typical hysteresis loop is observed in the voltammetric profile, which is more pronounced in sulfuric acid solutions. Since the size of the hysteresis loop is linked to the poisoning of the surface, it can be concluded that the CO formation rate is smaller in sulfuric acid solutions.

3.2. Chronoamperometric results

In order to investigate the role of the different adsorbates in the ethanol oxidation on platinum electrodes, chronoamperometric experiments were carried out. The potential program (Fig. 1) was designed so that the surface is as clean as possible from any adsorbate at the measuring potential. A sequence of conditioning potentials was designed for that purpose, ensuring that this sequence does not alter the surface structure of the electrode. For that reason, step potentials above 0.8 V for Pt(100) and Pt(110) electrodes were avoided. It is known that repetitive and/or long steps above this potential lead to changes in the surface structure of the electrode. A typical series of chronoamperometric transients are shown in Fig. 3. As can be seen, currents decay with time at all potentials. Since the measured currents are much smaller than those expected for a diffusion controlled process, the decay has to be associated with the adsorption of intermediates on the electrode that hinder the overall oxidation. At potentials below 0.7 V, CO is not oxidized and its formation and accumulation on the surface can explain the decay in the currents. However, at potentials above 0.7 V, CO is readily oxidized and therefore the decay has to be associated either to a CO formation rate that is faster than the oxidation rate or to another intermediate adsorbed species. The only exception to this general

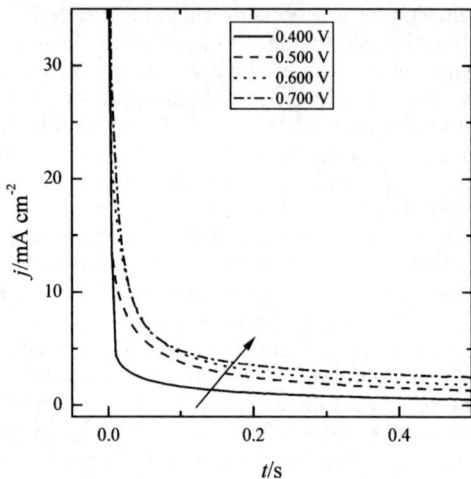

Fig. 3 Selected chronoamperometric transients obtained for the Pt(110) electrode in 0.1 M H_2SO_4 + 0.2 M EtOH. The arrow indicates the changes in the transients with increasing potential.

behavior is the Pt(111) electrode in 0.1 M H_2SO_4 + 0.2 M EtOH, for which the current decay between 0.015 and 0.500 s is negligible for potentials below 0.7 V.

In order to analyze the transient currents, 3 different sampling times were selected: 0.015, 0.200 and 0.500 s. Ideally, it would have been desirable to analyze the currents at zero time to obtain the "intrinsic" activity of the surfaces in the absence of strong adsorbing intermediates. However, owing to the double layer charging processes, the direct measurement is not possible and the current at 0.015 s was selected to represent that intrinsic activity (at this sampling time, all the transients recorded in the absence of ethanol have reached zero current). 0.200 and 0.500 s represent a situation where the surface has already been partially or completely blocked by the strongly adsorbed intermediates and close to the situation attained in the negative going scan of the voltammogram. The difference between the currents at 0.015 and 0.500 s then reflect the poisoning rate of the surface.

Fig. 4 shows the measured currents at the different sampling times for the three basal planes. For simplicity, only the currents at 0.015 s for the sulfuric acid solutions are shown in Fig. 4. The currents at 0.200 and 0.500 s for this medium (not shown) are very close to those obtained in perchloric acid solutions, in agreement with the voltammetric results.

The Pt(111) electrode in perchloric acid solutions exhibits the highest activity among the three basal planes at short times, with a maximum current value of 34 mA cm^{-2} at 0.8 V and the highest difference between the currents at 0.015 and 0.500 s. In contrast, the currents in sulfuric acid are the smallest at 0.015 s and the current difference between 0.015 and 0.500 s is very small. This indicates that the adsorbed intermediate, which highly affects the currents in perchloric acid solutions, is not being formed or does not affect the oxidation of ethanol in sulfuric acid to a large extent. This situation is very different from what is found in the voltammetry since the behavior of the Pt(111) electrode in both media is very similar. Since CO formation on this electrode should be small (the hysteresis between positive and negative scan is very small), the difference must be associated with some other intermediate species, which is formed on the electrode and remains adsorbed on the surface in perchloric acid. The formation or the adsorption of this species has to be very favorable in perchloric acid.

The qualitative behavior of the other two basal planes is very similar to that observed in the voltammetry. The maximum currents for the Pt(100) electrode are

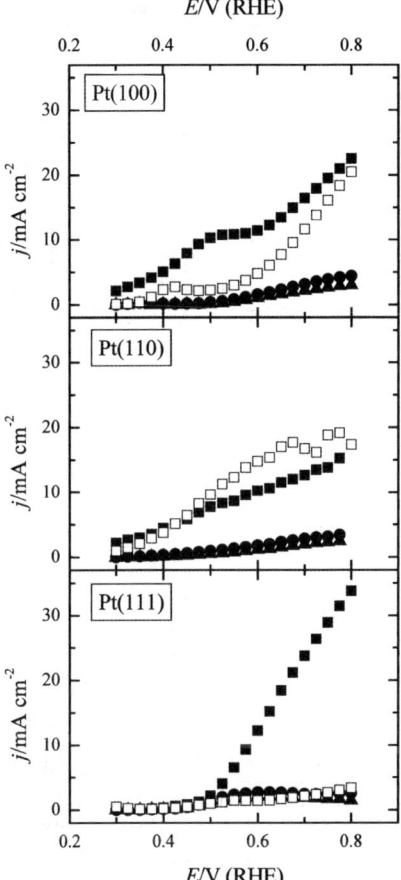

Fig. 4 Current densities measured in the chronoamperometric experiments at different sampling times. (■) 0.015 s, (●) 0.200 s, (▲) 0.500 s in 0.1 M $HClO_4$ + 0.2 M EtOH and (□) 0.015 s in 0.1 M H_2SO_4 + 0.2 M EtOH.

higher than those for the Pt(110) electrode. For this latter electrode, the currents in sulfuric acid are higher than those in perchloric acid.

3.3 FTIR experiments. Determination of the adsorbed intermediates

In order to determine which intermediates are being formed and trigger the current decay in the chronoamperometric transients, SPAIRs experiments in H_2O and D_2O were carried out. The use of D_2O presents some advantages for the detection of acetic acid and acetaldehyde since the frequency for the bending mode of water shifts from 1640 to 1200 cm^{-1}, eliminating the interference with the carbonyl group. The main disadvantage is that the stretching modes for D_2O overlap with the frequency for CO_2, the desired final product. An additional benefit of using D_2O solutions is that they allow the determination of acetic acid and acetaldehyde present in mixtures by using the carbonyl stretching band. As can be seen in Fig. 5, the transmission spectra of acetaldehyde presents a sharp band at 1713 cm^{-1} associated with the stretching mode of the C=O group. On the other hand, there is a split of this band for acetic acid, with a clear shoulder at lower wavenumbers. The curve fitting of this broad band using two Lorentzian functions yields two bands at 1685 and 1713 cm^{-1}, which have the same integrated intensities (within the experimental

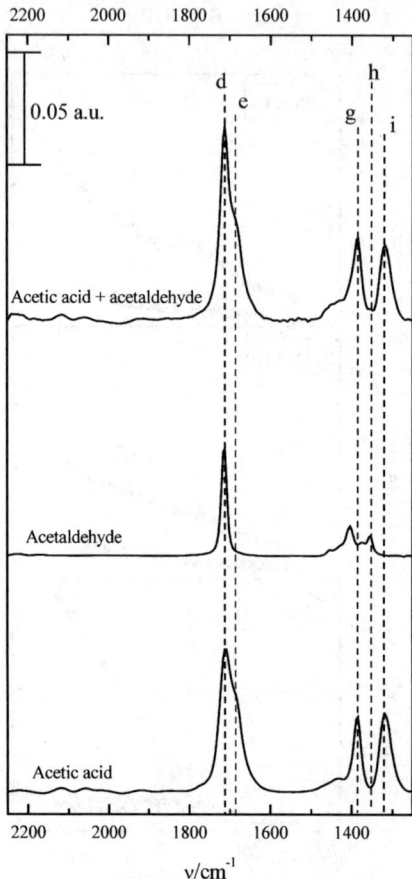

Fig. 5 Transmission spectra of 0.01 M acetic acid, 0.005 M acetaldehyde and a mixture of 0.01 M acetic acid + 0.005 M acetaldehyde in D_2O. The positions of the different bands are marked by the dashed lines and named according to Table 1.

error). A mixture of acetaldehyde and acetic acid will then have two bands at 1685 and 1713 cm^{-1} (Fig. 5); the first one comes from the acetic acid whereas the second one contains the contributions from both species. Since the integrated intensity for both acetic acid bands are the same, the contribution of acetaldehyde to the band at 1713 cm^{-1} can be easily calculated by subtraction of the total integrated intensity of the band at 1685 cm^{-1} from that of the band at 1713 cm^{-1}. This approach was experimentally confirmed by analyzing the transmission spectra of several mixtures with known concentrations of acetaldehyde and acetic acid in D_2O. For these mixtures, the molar absorption coefficient of acetic acid for the 1713 band is *ca.* 1.5 times larger than that of acetaldehyde.

Acetic acid and acetaldehyde also present other bands in the studied region of the spectrum. Acetic acid has two bands at 1380 and 1320 cm^{-1} which correspond to the CH$_3$ deformation and the coupling of the C–O stretching and OH deformation, respectively, whereas acetaldehyde has a weak band at 1350 cm^{-1}, corresponding to the CH$_3$ deformation. Due to the low intensity and the overlapping with the bands associated with ethanol, they are not suitable for the determination of both compounds. It should be mentioned that the small shoulder in the transmission spectra at *ca.* 1430 cm^{-1} corresponds to the formation of some HDO due to the large amount of acetic acid and acetaldehyde added.

Regarding the voltammetric profile of ethanol oxidation in D_2O solutions, it should be mentioned that the oxidation currents are slightly smaller than those recorded in H_2O. This behavior was also found for methanol oxidation under the same experimental conditions.[12] The diminution, much smaller than that expected for a true isotopic effect in the reaction kinetics, can be ascribed either to the presence of traces of impurities in the D_2O or to a solvent effect.

Pt(111) electrode. The IR spectra for the oxidation of ethanol in H_2O and D_2O on this electrode are presented in Fig. 6. Positive bands correspond to species formed or whose concentration has increased at the sampling potential whereas negative bands are associated with a diminution of the concentration of the species. Bipolar bands were found for adsorbed species which are present at the reference and sampling potential and whose frequency is dependent on the potential (Stark effect). Table 1 gives the frequencies of relevant bands that can be found in the spectra.

Significant changes in the spectra can only be observed at potentials above 0.3 V in the positive direction scan. As expected from the voltammetric experiments, the amount of CO formed and accumulated on this surface is very small. The formation of CO is associated with the reactions (3) and (4) in the general oxidation mechanism of eqn (1). Only a very small band for the linearly bonded CO (band b) can be detected. In H_2O, a small positive band for linearly bonded CO is only visible between 0.4 and 0.6 V in the positive going scan. This band is not bipolar, which indicates that the amount of CO formed at the reference potential is negligible. At 0.7 V, which is the onset potential for CO oxidation on this surface, the band disappears and, in parallel, the CO_2 band develops. The integrated intensity of this latter band is also very small, which indicates that the total CO_2 production is limited and associated with the CO formed on the surface. Due to the low intensity, the integrated band intensity has a significant error, but a general trend can be observed. A diminution of the intensity of this band is observed in the negative direction scan, which indicates that CO_2 is diffusing away from the thin layer and no additional CO_2 is being formed. In D_2O experiments, the observed behavior is slightly different.

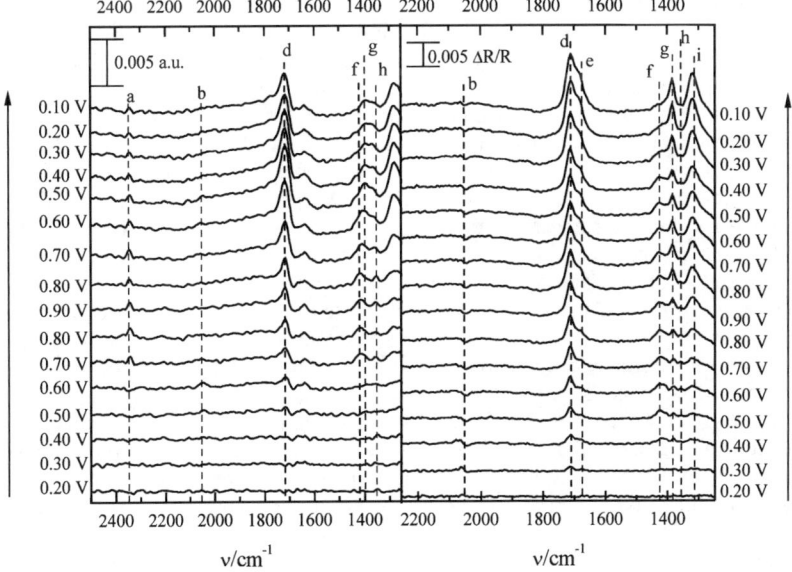

Fig. 6 SPAIR spectra for the oxidation of 0.2 M EtOH + 0.1 M $HClO_4$ in H_2O (left panel) and D_2O (right panel) for the Pt(111) electrode. The position of the different bands are marked by the dashed vertical lines and named according to Table 1.

Table 1 Assignment of the IR frequencies in the spectra for ethanol oxidation in H_2O and D_2O

Band	$\bar{\nu}H_2O/cm^{-1}$	$\bar{\nu}D_2O/cm^{-1}$	Functional group	Mode
A	2341	2341	CO_2	O–C–O asymmetric stretching
B	2030–2070	2030–2070	Adsorbed CO	Linearly bonded
C	1830–1860	1830–1860	Adsorbed CO	Bridge-bonded
D	1713	1713	COOH or CHO	C=O stretching
E	—	1685	COOH	C=O stretching
F	1410–1420	1410–1420	Adsorbed –COO–	C–O symmetric stretching
G	1385	1385	COOH	CH_3 deformation in acetic acid
H	1355	1355	–CH_3	CH_3 deformation in acetaldehyde
I	1280	1318	COOH	Coupling C–O stretching + OH deformation

The CO band between 0.3 and 0.6 V in the positive going scan is bipolar, which indicates that a small amount of CO was already present at the reference potential. From 0.7 V this band is always negative, suggesting that all the CO molecules adsorbed on the surface have been oxidized and no more CO is being formed. The small differences between both experiments are probably related to small changes in experimental conditions, namely, in small irreproducibilities in the surface preparation and electrode immersion in the cell and how the potential is controlled during the process.

The other bands observed during the experiment correspond to acetaldehyde, acetic acid and adsorbed acetate, which are formed through reactions (1) and (2). At potentials as low as 0.3 V, the formation of acetic acid and acetaldehyde can be detected in D_2O (bands d and e). As aforementioned, the detection of the carbonyl stretching band in D_2O is improved by the redshift of the bending modes of water, avoiding mutual interference. The voltammetric experiments for this electrode showed that the onset potential for ethanol oxidation was the lowest of the three basal planes, and IR experiments indicate that the oxidation currents have to be linked to the formation of acetic acid and acetaldehyde, *i.e.*, the main oxidation process in this potential range involves between 2 and 4 electrons per ethanol molecule. At 0.6 V, all the bands associated with the acetaldehyde/acetic acid system are already well developed: namely, the bands for the C=O stretching (d,e), for the coupling of C–O stretching and OH deformation of acetic acid (i), for the CH_3 deformation in acetaldehyde (h) and acetic acid (g) and for the adsorbed acetate (f). This latter band is already detected at 0.4 V in the positive going scan and only disappears at potentials below 0.3 V in the negative going scan. The acetate band observed in these experiments has the same characteristics than those observed in 10^{-2}–10^{-3} M acetic acid solutions in perchloric acid,[2] with two overlapping bands at 1410 and 1420 cm^{-1} and whose frequencies are potential dependent. Furthermore, acetate desorption takes place at the same potentials as in acetic acid solutions, between 0.4 and 0.3 V.[2] This fact indicates that the concentration of acetic acid in the interface is large enough to maintain a saturated adsorbed layer of acetate on the electrode surface. It should be stressed that the potential of zero total charge for the Pt(111) electrode is 0.33 V, and, therefore, anion adsorption is enhanced at potentials positive to that value.[24] If acetic acid is formed at those potentials and there is no other competing anion, the adsorption of acetate should take place. Therefore, the ethanol oxidation at those potentials is occurring in the presence of strongly adsorbed anions on the surface, although in the initial solution there were no adsorbing anions. To our knowledge, this is the first time that the adsorbed bands for acetate are so clearly visible in the presence of ethanol.

The presence of the acetate band explains why the voltammograms in perchloric and sulfuric acid solutions are very similar and the currents measured at 0.015 s are so different. Since the adsorption strength of the anion in 0.1 M H_2SO_4 and 10^{-3} M acetate on the platinum electrodes is very similar, which results in comparable voltammetric profiles for the Pt(111) electrode in both supporting electrolytes,[24] the oxidation of ethanol in the voltammetric experiments takes place under similar conditions for perchloric and sulfuric acid. In both cases, after the initial oxidation stages, in which some acetic acid is formed, the incoming ethanol molecules found adsorbed anions (either (bi)sulfate or acetate) on the electrode surface and therefore, its oxidation is hindered.

On the other hand, the conditions for the chronoamperometric experiments at short times are different. In the last conditioning step at 0.1 V, all the anions have been desorbed. After the step to the sampling potential, (bi)sulfate anions in 0.1 M H_2SO_4 immediately adsorb on the electrode surface and, thus, the oxidation of ethanol faces uniform conditions throughout the transient. In fact, the transient currents can be considered constant between 0.015 and 0.500 s for potentials below 0.7 V. This result is in agreement with the almost total absence of CO for these potentials, as the IR results indicate. On the other hand, the conditions for the chronoamperometric experiment in perchloric acid solutions change during the experiment. Just after the step (0.015 s), no anions are strongly adsorbed on the electrode surface and ethanol molecules can freely interact with the surface, facilitating a fast oxidation to yield acetaldehyde and acetic acid, which give rise to large currents. Once acetic acid is produced, the reversible adsorption of acetate begins and the currents start to decay, since the adsorbed anion hinders the oxidation. Therefore, the ethanol oxidation to acetic acid can be considered as a "self-poisoning reaction", since the product of this reaction, acetic acid, inhibits the reaction. At 0.500 s, the currents in sulfuric and perchloric acid are the same, since the conditions on the electrode surface are comparable. The current diminution by anion adsorption can be considered an example of non-specific poisoning, since it is independent of the nature of the adsorbing anion.

The poisoning role of acetic acid can be highlighted using a hanging meniscus rotating disk configuration (Fig. 7). For the measured currents, no effect of the rotation rate should be expected, since the currents are much smaller than those anticipated for a process controlled by diffusion. For instance, oxidation curves for 0.2 M formic acid on Pt(111) electrodes in 0.1 M $HClO_4$ (which have a maximum current density of *ca.* 2 mA cm^{-2}) are not affected by the rotation rate.[18] However, as

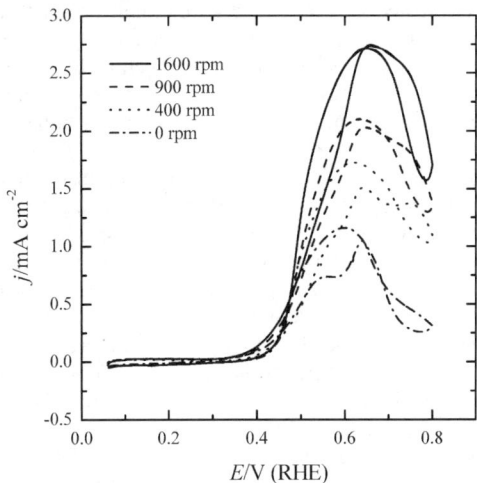

Fig. 7 Voltammetric profile for ethanol oxidation in 0.1 M $HClO_4$ + 0.2 M EtOH for the Pt(111) electrode in a hanging meniscus rotating disk configuration at different rotation rates.

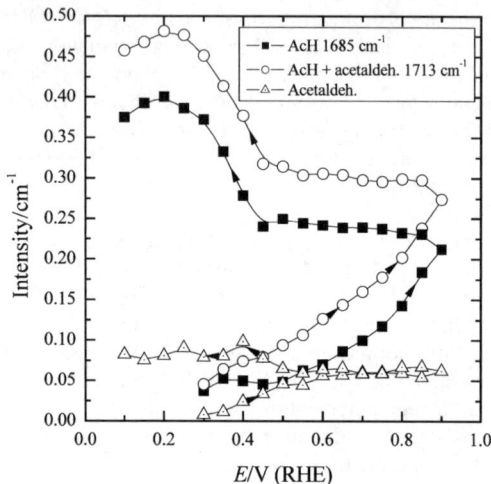

Fig. 8 Integrated intensities for the bands at 1713 (acetaldehyde + acetic acid) and 1685 cm^{-1} (acetic acid) and the subtraction of both bands (acetaldehyde) measured in SPAIRS experiments for the oxidation of 0.2 M EtOH + 0.1 M HClO$_4$ in D$_2$O for the Pt(111) electrode. Arrows indicate the scan direction.

Fig. 7 shows, the currents for ethanol oxidation increase with the rotation rate. Since the diffusion of ethanol to the electrode surface is not the limiting step, the increase in the currents with the rotation rate has to be associated with the diffusion of acetic acid away from the interface. Increasing rotation rates facilitate the diffusion of acetic acid formed at the electrode and the concentration in the interface diminishes. The diminution of the acetic acid concentration leads to a weaker acetate adsorption and to an increase in the oxidation currents.

The other important piece of information that can be gained from the IR spectra in D$_2$O is the ratio between produced acetaldehyde and acetic acid. The result of IR integrated intensities obtained for both species after the curve fitting of the bands associated with the carbonyl group are presented in Fig. 8. The ratio of the amount of products obtained for this electrode is in good agreement with previous results using other bands for the analysis, which validates the procedure used in this work.[3,25] For this electrode the acetic acid is the main product of the oxidation of ethanol. Only between 0.4 and 0.6 V is some acetaldehyde formed in both scan directions; for the other regions, the signal for acetaldehyde remains constant or diminishes due to the diffusion of acetaldehyde out of the thin layer. Therefore, the current measured in the voltammogram can be attributed almost exclusively to the oxidation of ethanol to acetic acid.

In the positive direction scan, the acetic acid band intensity increases with potential, indicating that acetic acid is being produced. In the negative scan, the intensity slightly increases between 0.9 and 0.5 V. The increase is much smaller than that observed in the positive direction scan, although the currents are similar in both scans. The difference is probably a consequence of the diffusion of acetic acid from the thin layer. At 0.9 V, the concentration in the thin layer is already high and acetic acid starts to diffuse at a high rate from the thin layer between the electrode and the prism. Therefore, although acetic acid is being formed in the negative direction scan, the total concentration in the thin layer is not increasing significantly, because acetic acid is diffusing away. The acetic acid signal increases again between 0.4 and 0.3 V in the negative direction scan. This increase is associated with the acetate desorption, which takes place at these potentials.[2]

Once the different species involved in the oxidation of ethanol on the Pt(111) electrode have been identified, the specific mechanism for this electrode can be discussed. Within the general mechanism of ethanol oxidation (eqn (1)), acetic acid comes from the consecutive oxidation of ethanol first to acetaldehyde and then to acetic acid. Each of these steps can have a side reaction to yield CO, that is, CO can be formed directly from ethanol (reaction (3)) or once acetaldehyde has been formed (reaction (4)). The low amounts of CO formed for the Pt(111) electrode implies that both paths to yield CO are strongly inhibited for this electrode. Probably, the small amounts of CO formed on this electrode arise from the acetaldehyde, since the formation of CO takes place only in the potential region where the interfacial concentration of acetaldehyde increases, *i.e.*, between 0.4 and 0.6 V. For the oxidation of pure acetaldehyde at high concentrations, adsorbed CO is readily formed and the measured coverages are much higher than those obtained here.[26] In fact, the electrode is completely poisoned during the positive direction scan under those conditions and the oxidation currents are negligible. The observed differences in the behavior can be attributed to the different concentration of acetaldehyde in both cases. It has been demonstrated that the distribution of products of ethanol oxidation depends on the concentration,[27] and it is possible that the formation of acetic acid is favored at low concentrations. Additionally, the role of surface defects in this process cannot be discarded. For instance, it has been demonstrated that CO formation from formic acid on the Pt(111) electrode only takes place on the surface defects.[28] In this respect, it should be mentioned that the amount of CO detected in these experiments is smaller than that reported in the literature for similar conditions[3,25] and also that the quality of the electrodes employed in this study is high. Experiments performed with stepped surfaces having (111) terraces showed that the presence of steps favors the formation of CO.[29] In addition, an experiment in which the defects have been blocked by small amounts of bismuth revealed that no CO was formed under these conditions. All these evidences suggest that defects play an important role in the formation of CO, and probably are the only sites responsible for the formation of CO in this surface.

Pt(100) electrode. The IR spectra obtained for this electrode are presented in Fig. 9. The observed behavior is significantly different from that of the Pt(111) electrode. As can be seen, CO bands are already visible at 0.3 V. The band for linear CO (at *ca.* 2050 cm^{-1}) is always positive, indicating that no linear CO is adsorbed at the reference potential, whereas the band for bridge-bonded CO is bipolar, implying that bridge-bonded CO was already present at the reference potential. The total integrated intensities for both bands are shown in Fig. 10. For the bridge-bonded CO band, the amount of adsorbed CO present at the reference potential should be taken into account. This can be easily measured at 0.8–0.75 V when the band is only negative. For the other potentials, the bipolar band has been fitted using two Lorentzian peaks, positive and negative, and the value for the integrated intensity of the negative one was set at the value obtained at 0.75–0.85 V. The total bridge-bonded CO at that potential corresponds to the integrated intensity of the positive band. As aforementioned, some bridge-bonded CO is adsorbed at the reference potential, and at only 0.3 V the band for linear CO appears, which means that the total amount of CO increases significantly. The increase coincides with a small bump observed in the voltammetric profile for this electrode (Fig. 2), which should be, then, linked to the cleavage of the C–C bond of ethanol to yield CO (reaction (3)). For higher potentials, the observed behavior for the CO bands is consistent with that expected for CO layers at high coverages.[30–32] Between 0.50 V and 0.65 V the total amount of CO does not change significantly, and the band frequencies have the typical values for a CO layer on the Pt(100) electrode with high coverage.[30] At 0.65 V the oxidation of CO begins, as revealed by the start of the CO_2 band and the sharp diminution of the integrated intensity of both types of CO. Interestingly, the CO coverage is finally zero at 0.75 V in the negative direction scan. It should be mentioned that CO should be already completely oxidized at 0.800 V. However, the band assigned to that potential contains the information

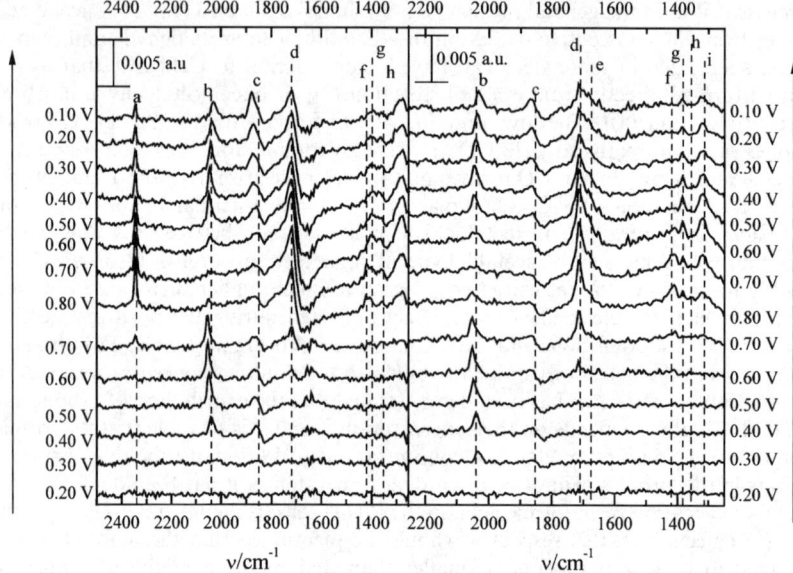

Fig. 9 SPAIR spectra for the oxidation of 0.2 M EtOH + 0.1 M HClO$_4$ in H$_2$O (left panel) and D$_2$O (right panel) for the Pt(100) electrode. The positions of the different bands are marked by the dashed lines and named according to Table 1.

acquired in a window between 0.775 V in the positive direction scan and 0.775 V in the negative direction one. The amount of CO measured at this potential should correspond to that adsorbed on the electrode surface at the initial part of the potential window. Thus, in the chronoamperometric experiments, CO is completely oxidized in the step at 0.8 V, and therefore the surface is free from CO at the beginning of the transient.

In the negative direction scan, the amount of CO increases between 0.65 and 0.50 V. From this latter potential, the intensity of both bands remains constant and the only significant change is the interconversion between both types of adsorbed CO, typical behavior of CO at high coverages and low potentials.[30]

Formation of acetaldehyde and acetic acid through reactions (1) and (2) is only observed at potentials above 0.60 V, when the oxidation of CO has started and some free sites are available on the surface. The band of adsorbed acetate can also been seen in the same potential region. The acetate band should always be observed at these potentials, since they are more positive than the pztc.[33] The only requirements are acetic acid in solution and some free sites available on the electrode surface. In acetic acid solutions the desorption of acetate on the Pt(100) electrode takes place between 0.3 and 0.4 V,.[2,33] However, in this case, this band disappears completely at 0.5 V in the negative going scan. Therefore, this desorption is not a consequence of the applied potential but the result of the formation of a CO adlayer that completely blocks the surface and displaces adsorbed acetate. The total disappearance of the band indicates that the CO layer has reached a coverage value high enough to fully block the surface. This CO coverage is able to displace completely the adsorbed acetate from the surface at potentials higher than the pztc, but it is lower than that obtained when CO is adsorbed from a CO containing solution, as has already been reported.[17]

The presence of adsorbed acetate on the electrode surface justifies the similar behavior of this electrode in the voltammetric experiments in sulfuric and perchloric acid, as in the case of the Pt(111) electrode. It also explains why the chronoamperometric currents at 0.015 s are different, since no strongly adsorbed anions are present

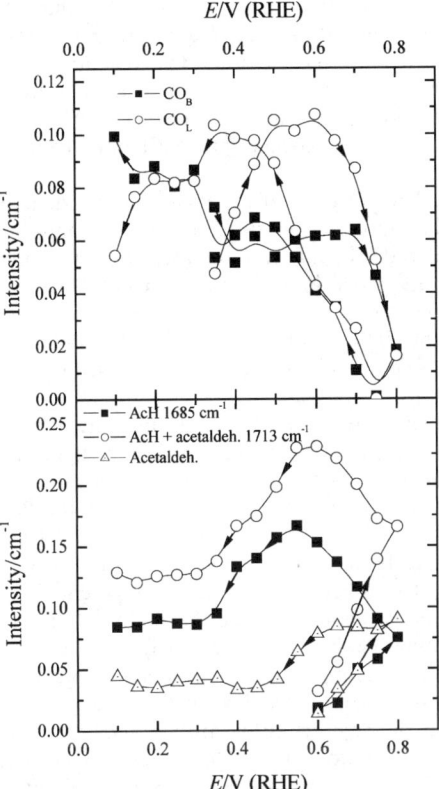

Fig. 10 Integrated intensities for the bands at 2050 (CO_L), 1840 (CO_B), 1713 (acetaldehyde + acetic acid), 1685 cm^{-1} (acetic acid) and the subtraction of both bands (acetaldehyde) measured in SPAIRS experiments for the oxidation of 0.2 M EtOH + 0.1 M $HClO_4$ in D_2O for the Pt(100) electrode. Arrows indicate the scan direction.

in the interface in perchloric acid solutions. It should be mentioned that the currents in the present case are smaller than those measured for the Pt(111) electrode, and thus, the total activity of this electrode is lower than that of the Pt(111) electrode in perchloric acid.

The onset potential for the oxidation of ethanol to acetaldehyde and acetic acid coincides with the onset for CO oxidation implying that acetaldehyde and acetic acid are only formed when some surface sites are not blocked by CO. The evolution of the integrated intensity for both species is shown in Fig. 10. In the positive going scan, the integrated intensities of acetaldehyde and acetic acid are similar but acetic acid is the major product in the negative going scan. Since the applied potentials are the same, the difference should be the result of the different conditions at the interface, namely, the different concentrations of the species. In the positive going scan, the concentrations of acetic acid and acetaldehyde are initially low, but they increase significantly at 0.8 V. For relatively high concentrations of acetaldehyde, the fall in the concentration can occur through three different mechanisms: (i) diffusion from the thin layer, (ii) oxidation to acetic acid (reaction (2)) and (iii) dissociation to yield CO (reaction (4)). Experiments with acetaldehyde reveal that CO is formed at potentials below 0.7 V, and the behavior of the CO band formed from acetaldehyde is very similar to that presented here.[26] This indicates that part of the formed acetaldehyde is being dissociated to CO, and that adsorbed CO at these potentials comes mainly from acetaldehyde (reaction (4)) and not from ethanol (reaction(3)). Within the general mechanism depicted in eqn (1), reactions (2) and (4) take place simultaneously at this potential and, thus, the acetaldehyde concentration in the thin layer does not increase. Since the CO oxidation rate is potential dependent,[34] the diminution of the potential in the negative going scan leads to a decrease in the oxidation rate and therefore to the accumulation of adsorbed CO on the electrode surface, until the surface is completely blocked (at 0.5 V) and ethanol oxidation stops.

Pt(110) electrode. The behavior of the Pt(110) electrode is similar to that of the Pt(100) electrode. The most visible difference is that only linear CO is observed on the Pt(110) electrode,[35] and in this case, the CO band is always bipolar (Fig. 11). The total integrated intensity for the CO band is shown in Fig. 12. The total CO remains constant at potentials below 0.5 V, that is, the surface is already completely blocked by CO even at low potentials. Since no acetaldehyde is present in the solution at those potentials, CO comes directly from ethanol through reaction (3).

From that potential, a decrease of the CO integrated intensity is observed, in parallel to the increase in the CO_2 band, which implies that CO is being oxidized to CO_2. However, the intensity of the CO band is never zero, even at 0.8 V. It should be stressed that the oxidation of a CO monolayer on the Pt(110) electrode takes place at potentials below 0.8 V. In fact, the potential for the CO stripping peak is the lowest of the three basal planes.[34] The observation of the adsorbed CO layer at 0.8 V implies that CO is being formed at those potentials at a high rate so that the CO formation rate, (reactions (3) and (4)) and the CO oxidation rate (reaction (5)) have similar values and a constant coverage is maintained. The non-zero coverage values at 0.8 V imply that the chronoamperometric transients were not actually recorded in the absence of CO. Probably, the amount of CO accumulated at 0.8 V depends on the supporting electrolyte and, for that reason, currents at 0.015 s in sulfuric acid solutions are slightly higher than those in perchloric acid. Total elimination of CO would have been observed at potentials above 0.80 V, but these potentials should be avoided because they lead to the disordering of the surface.

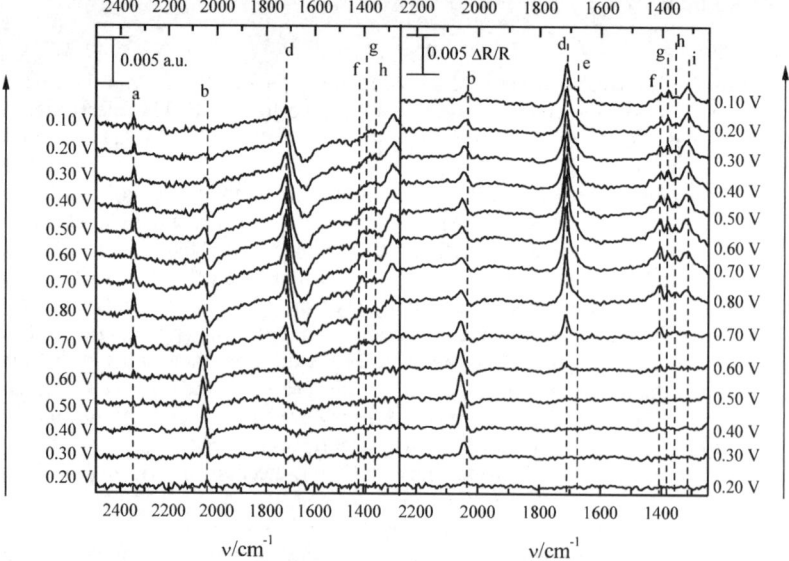

Fig. 11 SPAIR spectra for the oxidation of 0.2 M EtOH + 0.1 M $HClO_4$ in H_2O (left panel) and D_2O (right panel) for the Pt(110) electrode. The positions of the different bands are marked by the dashed lines and named according to Table 1.

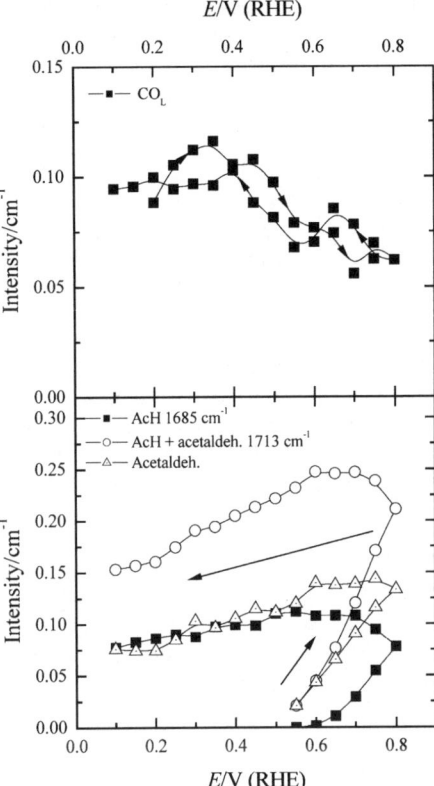

Fig. 12 Integrated intensities for the bands at 2050 (CO_L), 1713 (acetaldehyde + acetic acid), 1685 cm^{-1} (acetic acid) and the subtraction of both bands (acetaldehyde) measured in SPAIRS experiments for the oxidation of 0.2 M EtOH + 0.1 M HClO$_4$ in D$_2$O for the Pt(110) electrode. Arrows indicate the scan direction.

The total integrated intensity for the CO band is very similar in the positive and negative direction scans. The consequence of that is a voltammetric profile that shows a small hysteresis between the positive and the negative scans, as shown in Fig. 2 for perchloric acid solutions.

Formation of acetic acid and acetaldehyde occurs at potentials for which the CO layer does not block completely the surface, *i.e.*, at potentials above 0.5 V. At those potentials, adsorbed acetate is also detected. For this electrode, acetaldehyde is always the major product. Acetaldehyde is probably the species responsible for the formation of CO at high potentials, as observed with the Pt(100) electrode. The behavior of this electrode can be considered as being opposite to that of the Pt(111) electrode. For Pt(110) electrodes, the reactions that imply the C–C bond cleavage yielding adsorbing CO are very active at any potential and the oxidation of acetaldehyde to acetic acid is not favored.

4. Conclusions

The combined electrochemical and IR experiments have provided helpful insight into the oxidation mechanism of ethanol on platinum single crystal electrodes (eqn (1)). For Pt(111) electrodes, the main product of the oxidation is acetic acid and acetaldehyde is formed as secondary product, that is reactions (1) and (2) are the main path for this electrode. In fact, the amount of CO formed is very small, and probably associated with the defects present on the electrode surface. For that reason, the amount of CO_2 is also small. This electrode has the highest catalytic activity for the formation of acetic acid in perchloric acid. However, the formation of acetic acid is inhibited by the presence of specific adsorbed anions, such as (bi)sulfate or acetate. Since acetic acid is adsorbed above 0.3 V as acetate, the reaction is self-inhibited.

The reactivity of the Pt(100) electrode is different. Reaction (3) takes place at low potentials, especially between 0.3 and 0.4 V, where some oxidation currents are detected and the intensity for CO bands increase significantly. Between 0.65 and 0.80 V, the CO layer is oxidized (reaction (5)) and the production of acetaldehyde and acetic acid starts through reactions (1) and (2). In the negative scan, reactions (1) and (2) are still active, but the formation of CO through reaction (4) and the diminution of the CO oxidation rate leads to a progressive poisoning of the surface and the inhibition of the oxidation of ethanol.

The Pt(110) electrode displays the highest catalytic activity for the splitting of the C–C bond. Reactions giving rise to CO formation, from either ethanol or acetaldehyde, occur at high rate at any potential. On the other hand, the oxidation of acetaldehyde to acetic acid has probably the lower reaction rate of the three basal planes.

5. Acknowledgements

This work has been carried within the framework of a joint project supported by MEC (Spain) and CAPES (Brazil). Partial financial support through project CTQ2006–04071 (MEC-FEDER, Spain) is also acknowledged. GTF and ERG acknowledge the support of CNPq and FAPESP, Brazil.

6. References

1 A. Wieckowski, J. Sobrowski, P. Zelenay and K. Franaszczuk, *Electrochim. Acta*, 1981, **26**, 1111.
2 A. Rodes, E. Pastor and T. Iwasita, *J. Electroanal. Chem.*, 1994, **376**, 109.
3 X. H. Xia, H. D. Liess and T. Iwasita, *J. Electroanal. Chem.*, 1997, **437**, 233.
4 J. P. I. Souza, F. J. B. Rabelo, I. R. de Moraes and F. C. Nart, *J. Electroanal. Chem.*, 1997, **420**, 17.
5 V. Pacheco Santos and G. Tremiliosi-Filho, *J. Electroanal. Chem.*, 2003, **554**, 395.
6 V. Pacheco Santos, V. Del Colle, R. Batista de Lima and G. Temiliosi-Filho, *Langmuir*, 2004, **20**, 11064.
7 V. Pacheco Santos, V. Del Colle, R. Batista de Lima and G. Temiliosi-Filho, *Electrochim. Acta*, 2007, **52**, 2376.
8 J. P. I. de Souza, S. L. Queiroz, K. Bergamaski, E. R. Gonzalez and F. C. Nart, *J. Phys. Chem. B*, 2002, **106**, 9825.
9 K. Bergamaski, E. R. Gonzalez and F. C. Nart, *Electrochim. Acta*, 2008, , in press.
10 M.-C. Morin, C. Lamy, J. M. Léger, J.-L. Vazquez and A. Aldaz, *J. Electroanal. Chem.*, 1990, **283**, 287.
11 F. Cases, M. López-Atalaya, J. L. Vázquez, A. Aldaz and J. Clavilier, *J. Electroanal. Chem.*, 1990, **278**, 433.
12 E. Herrero, K. Franaszczuk and A. Wieckowski, *J. Electroanal. Chem.*, 1993, **361**, 269.
13 J. Clavilier, D. Armand, S. G. Sun and M. Petit, *J. Electroanal. Chem.*, 1986, **205**, 267.
14 J. Clavilier, K. E. Achii, M. Petit, A. Rodes and M. A. Zamakhchari, *J. Electroanal. Chem.*, 1990, **295**, 333.
15 T. Iwasita, F. C. Nart and W. Vielstich, *Ber. Bunsen-Ges. Phys. Chem.*, 1990, **94**, 1030.
16 A. Rodes, J. M. Pérez and A. Aldaz, in *Handbook of Fuel Cells. Fundamentals, Technology and Applications*, ed. W. Vielstich, H. A. Gasteiger and A. Lamm, John Wiley & Sons Ltd., Chichester, 2003, vol. 2, ch. 4.
17 L.-W. H. Leung and M. J. Weaver, *J. Electroanal. Chem.*, 1998, **240**, 341.
18 M. D. Macia, E. Herrero and J. M. Feliu, *J. Electroanal. Chem.*, 2003, **554**, 25.
19 K. Itaya, *Prog. Surf. Sci.*, 1998, **58**, 121.
20 A. M. Funtikov, U. Linke, U. Stimming and R. Vogel, *Surf. Sci.*, 1995, **324**, L343.
21 M. E. Gamboa-Aldeco, E. Herrero, P. S. Zelenay and A. Wieckowski, *J. Electroanal. Chem.*, 1993, **348**, 451.

22 M. D. Macia, J. M. Campiña, E. Herrero and J. M. Feliu, *J. Electroanal. Chem.*, 2004, **564**, 41.
23 E. Herrero, M. J. Llorca, J. M. Feliu and A. Aldaz, *J. Electroanal. Chem.*, 1995, **383**, 145.
24 J. M. Orts, R. Gómez, J. M. Feliu, A. Aldaz and J. Clavilier, *Electrochim. Acta*, 1994, **39**, 1519.
25 S. C. Chang, L.-W. H. Leung and M. J. Weaver, *J. Phys. Chem.*, 1990, **94**, 6013.
26 J. L. Rodríguez, E. Pastor, X. H. Xia and T. Iwasita, *Langmuir*, 2000, **16**, 5479.
27 G. A. Camara and T. Iwasita, *J. Electroanal. Chem.*, 2005, **578**, 315.
28 M. D. Macia, E. Herrero, J. M. Feliu and A. Aldaz, *J. Electroanal. Chem.*, 2001, **500**, 498.
29 D. J. Taranowski and C. Korzeniewski, *J. Phys. Chem. B*, 1997, **101**, 253.
30 F. Kitamura, M. Takahashi and M. Ito, *J. Phys. Chem.*, 1990, **92**, 4582.
31 X. H. Xia, T. Iwasita, F. Y. Ge and W. Vielstich, *Electrochim. Acta*, 1996, **41**, 711.
32 T. Iwasita, X. H. Xia, E. Herrero and H.-D. Liess, *Langmuir*, 1996, **12**, 426.
33 K. Domke, E. Herrero, A. Rodes and J. M. Feliu, *J. Electroanal. Chem.*, 2003, **552**, 115.
34 E. Herrero, B. Álvarez, J. M. Feliu, S. Blais, Z. Radovic-Hrapovic and G. Jerkiewicz, *J. Electroanal. Chem.*, 2004, **567**, 139.
35 S. C. Chang, L. W.-H. Leung and M. J. Weaver, *J. Phys. Chem.*, 1989, **93**, 5341.

PAPER

Electro-oxidation of ethanol and acetaldehyde on platinum single-crystal electrodes

Stanley C. S. Lai and Marc T. M. Koper

Received 3rd March 2008, Accepted 3rd April 2008
First published as an Advance Article on the web 20th August 2008
DOI: 10.1039/b803711f

The electrochemical oxidation of ethanol and acetaldehyde in sulfuric acid and perchloric acid were studied at Pt (111), Pt (110), and a number of Pt [n(111)×(111)] single-crystal electrodes. The oxidation of ethanol shows a marked dependence on the surface structure, roughly increasing as the surface step density increases. The oxidation of acetaldehyde shows a reversed correlation, the activity decreasing with increasing step density. Based on the results obtained here and reported earlier in electrochemical and ultrahigh vacuum literature, a detailed reaction scheme for the ethanol oxidation mechanism is suggested.

1. Introduction

The electrochemical oxidation of ethanol on platinum electrodes has been a subject of increasing interest.[1–5] From an environmental point of view, ethanol, which can be produced by the fermentation of bio-mass, is often mentioned as one of the main candidates for use in low temperature fuel cells.[6] Apart from practical advantages, such as the ease in storage and transportation and its non-toxicity, ethanol has a high theoretical energy content of 8.0 kWh kg^{-1}, corresponding to 12 electrons per molecule for its total oxidation to carbon dioxide.[3] Therefore, the main challenge in the electrochemical oxidation of ethanol is to achieve total conversion at a low overpotential. The main problem in achieving an efficient conversion is that the ethanol oxidation occurs by different reaction pathways. In some pathways, large amounts of partially oxidized products are produced due to the difficulty in breaking the C–C bond. These partially oxidized products, which include acetaldehyde[4,7–9] and acetic acid (or acetate),[1,2,4,7,9] do not only decrease the total efficiency of the system, but are also unwanted due to their polluting nature. In other pathways, strongly adsorbed species such as carbon monoxide[4,7,9] and carbohydrate (CH_x) fragments[5,10] are formed, which poison the platinum surface.

Despite the great number of papers dedicated to the electrochemical oxidation of ethanol, there is still a significant number of issues in the understanding in the mechanism of ethanol electro-oxidation. One of the main mechanistic issues lies in determining in which pathways carbon dioxide is produced, and therefore, in determining the molecular species in which the carbon–carbon bond is broken as well as the electrochemical nature of the resulting decomposition products. In this context, the present study will employ a series of well-defined platinum single-crystal electrodes to investigate the effect of surface structure on the oxidation of ethanol and acetaldehyde. We will compare our results to previous results, especially in relation to ethanol decomposition on similar (stepped) Pt single-crystals in ultrahigh

Leiden Institute of Chemistry, Leiden University, PO Box 9502, RA Leiden, 2300, The Netherlands

vacuum (UHV). In the discussion section, we will propose a reaction scheme that is able to rationalize most of the results obtained here and in previous work.

2. Experimental

Electrochemical measurements were carried out in a conventional single-compartment three-electrode glass cell. The cell and all other glassware were cleaned by boiling in a 1 : 1 mixture of concentrated nitric acid and sulfuric acid followed by repeated boiling with ultra-pure water (Millipore MilliQ A10 gradient, 18.2 MΩ cm, 2–4 ppb total organic content) before each experiment.

In all experiments, a platinum wire was used as counter electrode. For experiments with H_2SO_4 as electrolyte, a mercury–mercury sulfate electrode (MMSE: $Hg/Hg_2SO_4/K_2SO_4$) was employed as a reference electrode, while a reversible hydrogen electrode (RHE) was used for measurements in experiments with $HClO_4$ as electrolyte. In all cases, the reference electrode was connected *via* a Luggin capillary. All potentials reported here have been converted to the RHE scale in the same electrolyte.

The supporting electrolytes, 0.5 M H_2SO_4 and 0.1 M $HClO_4$, were prepared with concentrated sulfuric or perchloric acid (Merck, "Suprapur") and ultra-pure water. Electrolytes containing ethanol, acetaldehyde or acetic acid consisted of the blank electrolyte and 0.5 M ethanol (Merck, pro analyse), 0.1 M or 0.5 M acetaldehyde (Sigma–Aldrich, "ReagentPlus") or 0.1 M acetic acid (Sigma–Aldrich, "Reagent-Plus"), respectively. Argon (Air Products, "BIP Plus", 6.6) was used to deoxygenate all solutions. It should be noted that the concentration of 0.5 M ethanol corresponds to a value that would be a sensible concentration in a real fuel cell. According to the work of Camara and Iwasita,[4] this ethanol concentration leads to acetaldehyde as the primary reaction product. Similarly, the main reaction product of acetaldehyde oxidation in the concentrations employed in this study is acetic acid.[11]

The working electrodes used in this study for the voltammetric measurements were platinum bead-type single-crystal electrodes of Pt $[n(111)\times(111)]$ (or, equivalently, Pt $[(n - 1)(111)\times(110)]$) orientation. The surfaces studied were Pt (15 15 14) with $n = 30$, Pt (554) with $n = 10$, Pt (553) with $n = 5$ and the limiting cases Pt (111) and Pt (110), which were prepared according to Clavilier's method.[12] Both the terrace width n and the step density (defined as $\theta_{step} = 1/(n - 2/3)$) will be used in this study as a quantitative measure for the amount of steps.[13] Prior to each experiment, the electrodes were flame-annealed and cooled down to room temperature in an argon (Air Products, "BIP Plus", 6.6)–hydrogen mixture (Air Products, "Ultrapure Plus", 6.0) (*ca.* 3 : 1), after which they were transferred to the electrochemical cell under the protection of a droplet of deoxygenated ultrapure water.[14] Voltammetric measurements were performed at room temperature (20 °C) using a computer-controlled Autolab PGSTAT 12 potentiostat (Ecochemie).

3. Results and discussion

3.1 Ethanol

3.1.1 Continuous oxidation. Fig. 1 shows typical voltammograms obtained for the electro-oxidation of 0.5 M ethanol on Pt (111), Pt (110) and on the stepped surfaces Pt (15 15 14), Pt (554) and Pt (553) in 0.5 M H_2SO_4, sweeping initially positive from 0.07 V at 10 mV s^{-1}. The positive potential limits were chosen in each case before the onset of surface oxidation to avoid surface disordering due to irreversible oxide formation. A number of general features can be observed on all electrode surfaces: in the positive scan, the current is negligible up to 0.4 V. At this potential, a pronounced increase in the current occurs, forming a small shoulder at ~0.5 V followed by a large peak with a maximum between 0.6 and 0.8 V. During the return scan, the current increases again to produce a single peak. In the case of Pt

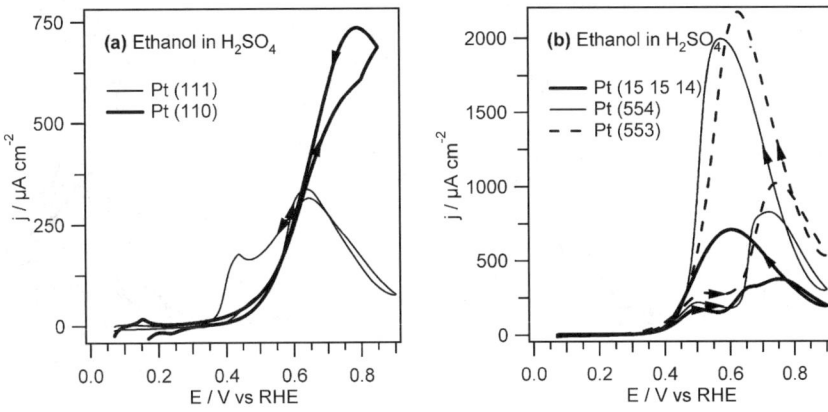

Fig. 1 Cyclic voltammograms for the electro-oxidation of 0.5 M ethanol in 0.5 M H_2SO_4 on (a) Pt (111) and Pt (110) and (b) Pt (15 15 14), Pt (554) and Pt (553), respectively. All voltammograms were recorded at a scan rate of 10 mV s^{-1}. Arrows indicate the scan directions.

(111), a small but reproducible increase in the current can also be observed at 0.44 V. The position of this current increase corresponds strongly with the location of the so-called Pt (111)-butterfly,[15] which is commonly ascribed to the order–disorder transition of the (bi)sulfate adlayer.[15,16] In addition, similar to the Pt (111)-butterfly, this feature was found to be very sensitive to the long-range crystalline order of the electrode, being, for example, more pronounced on a Pt (111) cooled in an H_2/Ar mixture after flame annealing and disappearing entirely on a Pt (111) electrode cooled in air after flame annealing, even though the blank voltammetry showed only small differences.[14] Therefore, we attribute this feature to the increased availability of surface sites due to a sharp decrease of the coverage of the strongly adsorbing (bi)sulfate anion.

By comparing the CV's of the different surfaces in Fig. 1, it can be observed that upon increasing the step density by going from Pt (111) to Pt (15 15 14), the maximum activity towards ethanol oxidation roughly doubles, while a further increase in the step density to Pt (554) yields another increase in the maximum current by a factor of three. As the step density is increased further from Pt (554) to Pt (553), the maximum activity increases slightly, before dropping by a factor of three in passing from Pt (553) to Pt (110). These results indicate that, similar to the oxidation of CO[17,18] and methanol,[19] defects enhance the ethanol oxidation rate in sulfuric acid. There are, however, significant differences in the quantitative effect of step density. In the case of CO, the oxidation rate constant was found to be linearly dependent on step density over the entire range of step densities.[18] In the present study, the peak current density appears to show a rough linear dependence for terraces of over 10 atoms wide (Fig. 3a). This is also the range in which the overpotential required for the peak activity is decreasing with step density (Fig. 3b). For narrower terraces, this trend is observed to reverse; it is especially noteworthy that the oxidation activity of Pt (110) in sulfuric acid is below that of Pt (554) and Pt (553), both in terms of the peak current density as well as in the location of the peak potential. This effect was also found in the case of methanol oxidation on the same surfaces,[19] indicating that the oxidation of ethanol requires some combination of steps and terrace sites. Considering the deviation from the linear trend when the terrace width is decreased below 10 atoms, it is likely that there is an optimum terrace width for ethanol oxidation.

To study the effects of anion adsorption, the same experiments were conducted using 0.1 M $HClO_4$ perchloric acid as supporting electrolyte. Typical voltammograms for the oxidation of ethanol on Pt [n(111)×(111)] type electrodes in perchloric

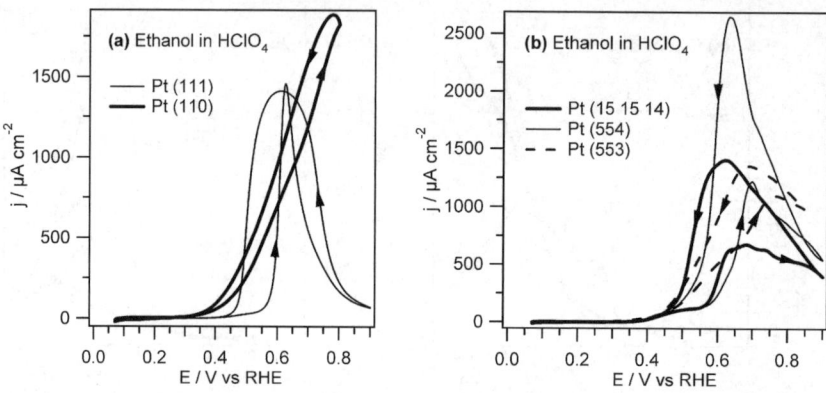

Fig. 2 Cyclic voltammograms for the electro-oxidation of 0.5 M ethanol in 0.5 M HClO$_4$ on (a) Pt(111) and Pt(110) and (b) Pt(15 15 14), Pt(554) and Pt(553), respectively. All voltammograms were recorded at a scan rate of 10 mV s^{-1}. Arrows indicate the scan directions.

acid are shown in Fig. 2. Similar to methanol electro-oxidation, the shape of the voltammetric profiles does not differ significantly in both acidic media.[19] Furthermore, there were no significant shifts in the peak potentials, although the trend is markedly different (Fig. 3b). Whereas the peak potential as a function of step density shows a minimum for oxidation in sulfuric acid, in perchloric acid it is

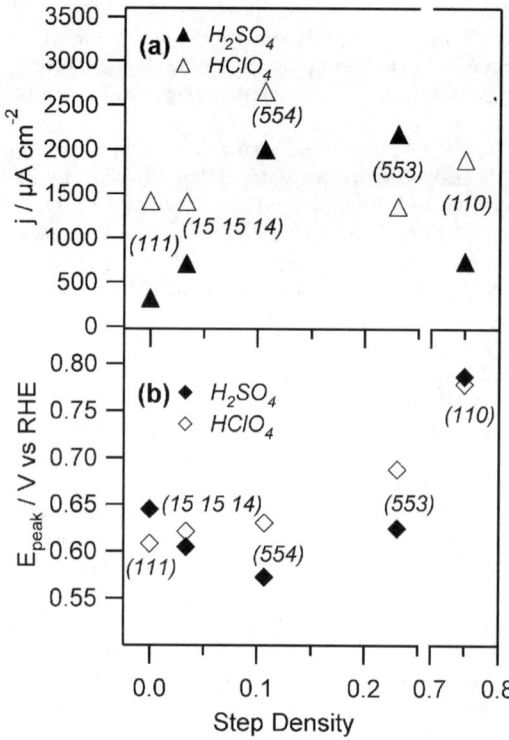

Fig. 3 Dependence of (a) the maximum current density and (b) the peak potential of bulk ethanol oxidation on the step density. Closed symbols denote 0.5 M H$_2$SO$_4$ as supporting electrolyte while open symbols indicate 0.1 M HClO$_4$ as supporting electrolyte.

monotonously increasing with increasing step density. With the exception of Pt (553), total ethanol oxidation activities were found to be higher in perchloric acid than in sulfuric acid. The relative increase, however, is inversely related with the step density, especially on surfaces with large terraces: on Pt (111), a four-fold increase in peak current can be observed compared to sulfuric acid. As steps are introduced, this increase is reduced to about 100% for Pt (15 15 14) and 25% for Pt (554). These results can be readily explained by the fact that (bi)sulfate adsorption plays a larger role on surfaces with large terraces due to a preferential adsorption of (bi)sulfate on terraces rather than on steps, in agreement with a study by Mostany et al.[20] Therefore, the effect of changing the anion is larger for more surfaces with wider terraces and decreases with step density. The divergence of this trend for the surfaces with a higher step density may well be the result of the markedly different interaction between (bi)sulfate and Pt (110) compared to the interaction with (111)-terraces.

Chronoamperometric measurements were also performed to study the ethanol oxidation activities over longer time-scales. The results of these experiments in perchloric acid are shown in Fig. 4, where the potential is stepped to a value of 0.7 V, close to the maximum current in the voltammetry. On all electrode surfaces, the current shows a monotonous decay, reaching a steady state after ~3 minutes. The steady state current is positively correlated with the step density, similar to the voltammetric measurements. An important difference between the two methods, however, is the activity on Pt (110), which shows the highest activity in the chronoamperometric measurements but which has a lower activity than the stepped surfaces in the voltammetric measurements. This discrepancy could be an indication that the stepped surfaces are blocked faster by decomposition products and/or partially oxidized products than Pt (110). Although Fig. 4 shows a marked

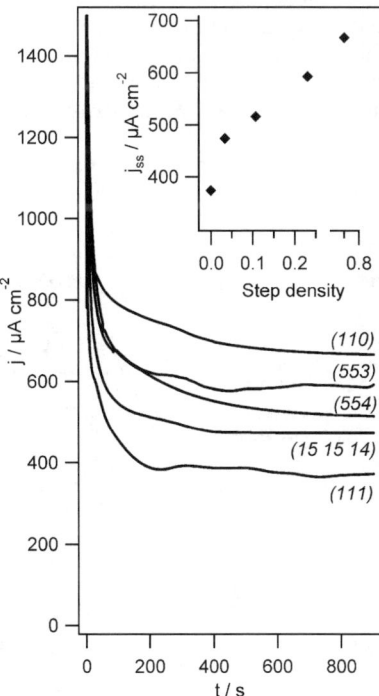

Fig. 4 Current–time transients for the oxidation of 0.5 M ethanol in 0.1 M $HClO_4$. The step potential was 0.7 V. The inset shows the current density after 15 minutes as a function of step density.

dependence of the ethanol oxidation current on the surface structure, the qualitative effect is rather small, the maximum current between the five electrodes varying by less than a factor of two. We believe this is related to the fact that, with 0.5 M ethanol, we are mainly probing the oxidation pathway to acetaldehyde,[4] and this will be discussed in more detail in section 3.4.

3.1.2 Adsorbate stripping. In order to determine the nature of the adsorbed intermediates involved in the oxidation of ethanol, stripping experiments were conducted. In these experiments, the working electrode was introduced in a solution of 0.5 M ethanol in the supporting electrolyte under potential control and kept at this potential (E_{ad}) for 15 minutes to adsorb ethanol decomposition products. Next, the working electrode was transferred to a cell containing only the supporting electrolyte. In this cell, voltammetric sweeps in the hydrogen underpotential deposition (H_{UPD}) region, which ranges from the hydrogen evolution onset potential to roughly 0.3 V, were recorded in order to determine the amount of adsorbates (or, more precisely, the amount of sites blocked for hydrogen adsorption). After determining the total amount of adsorbates, the potential was held at 0 V for five minutes to remove all reducible adsorbates (CH_x fragments, forming methane and ethane upon reduction),[10] followed by three voltammetric sweeps up to 0.85 V to oxidize the remainder of the adsorbates (CO-like intermediates and possible remaining CH_x, forming CO_2 upon oxidation). After both the reductive and oxidative stripping processes, a cyclic voltammogram of the H_{UPD} region was recorded in order to determine the initial amounts of reducible species and oxidizable species resulting from the adsorption of ethanol in terms of H_{UPD} sites made available in the stripping process. Finally, a voltammogram was recorded at the end of the experiment and compared with a blank voltammogram recorded prior to the experiment to ascertain that a clean electrode surface is recovered by the stripping processes. With this method, it is possible to circumvent the unknown composition of the adsorbates by relating the amount of adsorbates to the amount of charge it blocks in the hydrogen adsorption area rather than quantifying it directly through evaluation of the charge involved in the stripping process, which requires the knowledge of the number of electrons needed to reduce or oxidize the ethanol decomposition products.

The results of a typical experiment as described above with Pt (553) are shown in Fig. 5a. Curve '1' shows the voltammogram of the clean surface before adsorption.

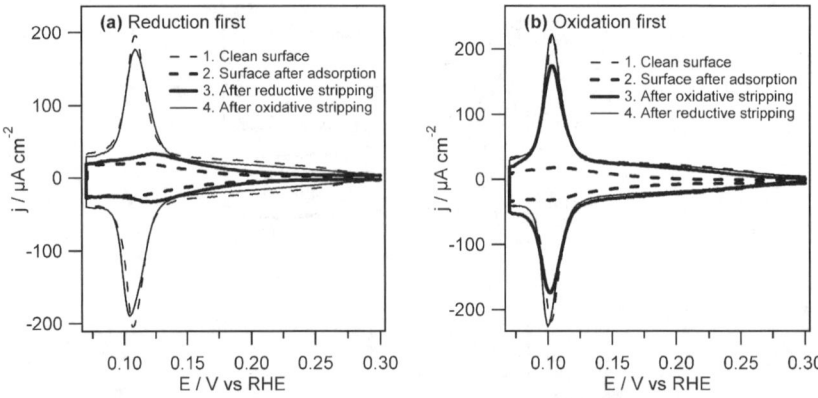

Fig. 5 Typical voltammetric profiles of the H_{UPD} region at various stages of the ethanol adsorbate stripping experiments on Pt (553) in 0.5 M H_2SO_4. After initial adsorption at 0.4 V, the surface was subjected to (a) reductive stripping first followed by oxidative stripping, and (b) oxidative stripping first followed by reductive stripping. The voltammograms were recorded at 50 mV s^{-1}.

Curve '2' is the voltammogram of the Pt (553) electrode after ethanol chemisorption. From comparison with curve '1', it is seen that the step sites are completely blocked and the terrace sites are partially blocked by decomposition products. After reductive stripping (curve '3'), it can be seen that mostly terrace sites are made available by the stripping process, while the step sites are still blocked completely by the remaining adsorbates. The clean surface is recovered almost entirely by the following oxidative stripping (curve '4'). Fig. 5b shows the effect of performing an oxidative rather than a reductive stripping process first after the initial adsorption. In this case, the first stripping procedure step liberates all terrace sites and part of the step sites (curve '3'). The remaining step sites are freed by the reductive stripping. Comparing Fig. 5(a) and (b), several conclusions can be drawn. Firstly, considering the amount of sites liberated by the reductive stripping treatment in both methods, part of the CH_x-species can be removed oxidatively within a few voltammetric sweeps. This is more clearly illustrated in Fig. 6, which shows the oxidative voltammetric sweeps of both experiments in Fig. 5. If a reductive stripping process is performed first after the initial adsorption, the oxidative stripping sweep gives a single peak which starts at 0.5 V and has a maximum around 0.7 V. If, however, the CH_x is not removed by reduction after the initial adsorption before the oxidative stripping, an additional feature appears in the stripping voltammetric sweep around 0.5 V, with the onset at 0.4 V. In addition, the feature around 0.8 V has a larger intensity than in the case where CH_x has been removed by a reductive stripping process. Based on these findings, we tentatively attribute the feature around 0.5 V and part of the feature around 0.8 V to the oxidation of CH_x species. Most likely, the low potential feature starting at 0.4 V would signify the oxidation of $CH_{x,ads}$ to CO_{ads}, and the high potential feature to further oxidation of CO_{ads} to CO_2. This is in good agreement with results of Hahn and Melendres,[21] who reported that CO_{ads} is an observed intermediate in the oxidation of methane to CO_2.

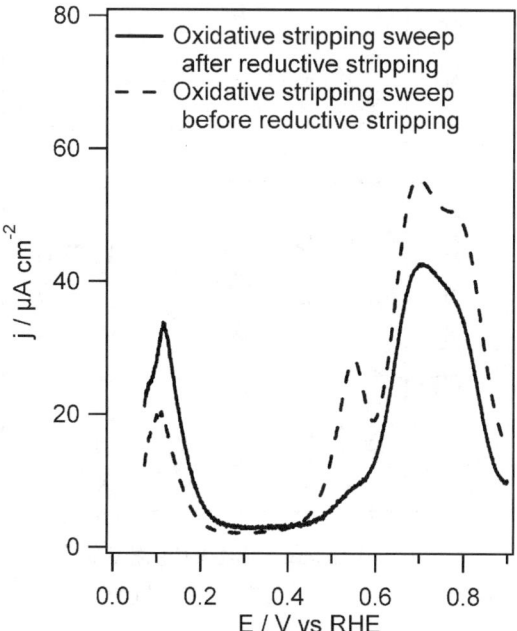

Fig. 6 Oxidative stripping of ethanol decomposition products at $E_{ad} = 0.3$ V on Pt (553) on 0.5 M H_2SO_4. The dashed line shows the stripping profile directly after adsorption, while the solid line shows the profile after the electrode surface was subjected to reductive stripping process first after the adsorption.

Furthermore, in both cases, the adsorbates remaining after the first stripping procedure are preferentially adsorbed at the step sites. This is especially the case when reductive stripping is applied first. This suggests that both types of adsorbed species have a preferential affinity for step sites as well as a significant surface mobility to migrate to the step sites once these are liberated by the removal of other adsorbates. Carbon monoxide has been known to exhibit a high surface diffusion rate on platinum in acidic media.[18,22,23] The diffusion rates of CH_x fragments under electrochemical conditions, on the other hand, have been much less studied. An alternative interpretation of Fig. 5 could be that CO primarily occupies the step sites, and the CH_x fragments primarily occupy the terrace sites. Comparing Fig. 7(a) and (b), it can be seen that changing the anion of the supporting electrolyte causes little change in the stripping results. This suggests that the anion effects observed earlier in Fig. 1 and Fig. 2 in the electro-oxidation of ethanol are not related to the C–C bond breaking leading to adsorbed intermediates. In both media, it is clear that the total amount of adsorbates increases with increasing adsorption potential, indicating that breaking of the ethanol carbon–carbon bond and the subsequent formation of strongly adsorbed intermediates occurs to a higher degree at higher potentials. Fig. 7 also shows that the amount of oxidizable adsorbed species increases with adsorption potential. For adsorption potentials above 0.2 V, the amount of reducible species decreases with increasing adsorption potential. This could be due to the oxidation of the CH_x fragments to CO (or COH/CHO), which, according to Hahn

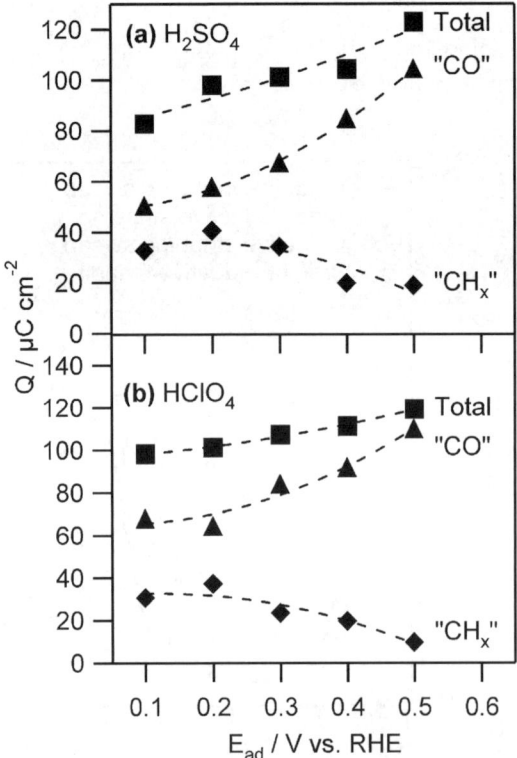

Fig. 7 The dependence of the amount of charge in the H_{UPD} region that is inhibited by the adsorbates resulting from ethanol decomposition on the ethanol decomposition potential on Pt (553). Triangles denote oxidizable species, while diamonds denote reducible species. Squares indicate the sum of the oxidizable and reducible species. The supporting electrolyte was (a) 0.5 M H_2SO_4 or (b) 0.1 $HClO_4$. The dashed lines are drawn to guide the eye.

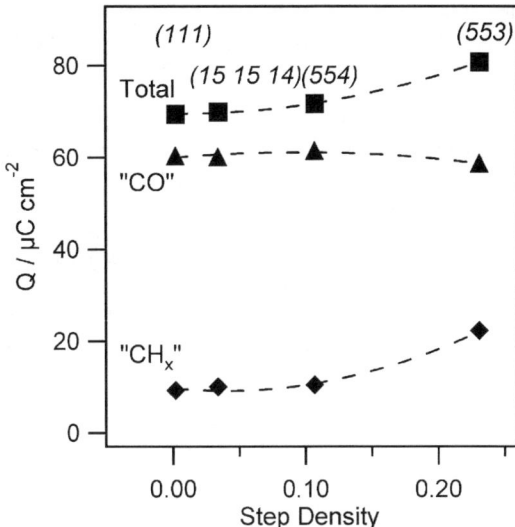

Fig. 8 Structure sensitivity of ethanol adsorbate stripping in 0.1 M HClO$_4$ as implied by the amount of charge in the H$_{UPD}$ region that is inhibited by the adsorbates. Triangles denote oxidizable species, while diamonds reducible species. Squares indicate the sum of the oxidizable and reducible species. The dashed lines are drawn to guide the eye.

and Melendres,[21] may happen already at potentials as low as 0.26 V. Decreasing the adsorption potential from 0.2 V to 0.1 V, however, yields a small decrease in reducible species. Since earlier DEMS experiments[9,10] reported the formation of small amounts of methane and ethane below 0.2 V, this finding can be attributed to the continuous reduction of the reducible species during the adsorption process.

The effect of the surface structure on the surface coverage of the different adsorbates was also investigated by employing several stepped platinum single-crystal electrodes. For these experiments, the Pt [n(111)×(111)] electrodes were kept at the 0.3 V for 30 minutes during the adsorption phase. The results of these experiments are shown in Fig. 8, which reveal several interesting features. First of all, by comparing Fig. 8 with Fig. 7b, the total amount of adsorbed species was found to be lower when the adsorption time was increased from 15 minutes to 30 minutes. In addition, the amount of oxidizable (CO-like) intermediates relative to reducible (CH$_x$) intermediates is greater for 30 minutes of adsorption than for 15 minutes of adsorption, which could indicate that slow reduction of the ethanol decomposition products already occurs during adsorption. The freed sites can accommodate new decomposition products, making the surface more and more 'CO'-rich with increasing adsorption time. An alternative explanation could be that slow conversion of CH$_x$ to CO already occurs during the adsorption.

Furthermore, although the total amount of adsorbates shows a slightly increasing trend with step density, the amount of CO-like intermediates is relatively constant over this range of surface step density. The amount of reducible intermediates, on the other hand, roughly doubles over the investigated range of step densities.

3.2 Acetaldehyde

3.2.1 Continuous oxidation. Since acetaldehyde has been reported repeatedly[4,7,9,24,25] as one of the main oxidation products of ethanol and the primary oxidation product at the 0.5 M ethanol concentration used in this work, it is clear that an improved understanding of the mechanism of acetaldehyde oxidation may assist in understanding the processes involved in the electro-oxidation of ethanol.

To this end, the oxidation of acetaldehyde was also investigated in this study. The results of the oxidation of 0.1 M acetaldehyde in 0.5 M H_2SO_4 on the different single-crystal electrodes are shown in Fig. 9. The general shapes of the cyclic voltammograms are similar to those of the oxidation of ethanol. However, a closer inspection reveals that the currents in the positive sweep compared to the currents in the negative sweep are much lower, suggesting different electrode-reactant interactions at low potentials for ethanol and acetaldehyde. Furthermore, the results suggest that increasing the step density of the electrode surface has a smaller effect on oxidation currents than is the case for ethanol. In addition, compared to Pt (111) and the stepped surfaces, the activity of Pt (110) in acetaldehyde electro-oxidation is very low. More interestingly, the effect of increasing step density on the oxidation of acetaldehyde is opposite to that on ethanol electro-oxidation in sulfuric acid: the oxidation activity is found to decrease with increasing step density. This is more

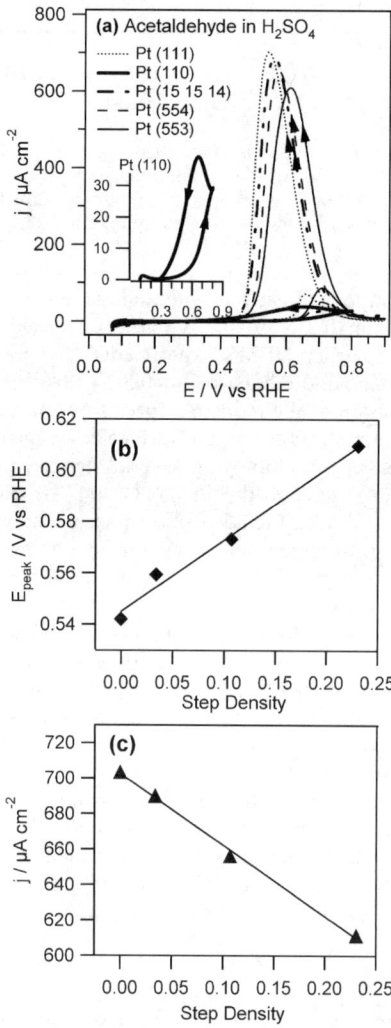

Fig. 9 (a) Cyclic voltammograms of the oxidation of 0.1 M acetaldehyde on platinum single-crystal electrodes in 0.5 M H_2SO_4 at a scan rate of 10 mV s^{-1}. (b) Dependence of the peak potential and (c) the maximum current density of acetaldehyde oxidation on the step density. The solid lines are the least-squares fit of the data.

clearly shown in Fig. 9b and c, in which the peak current and the peak potential, respectively, are shown as a function of step density. The correlation between both peak parameters and the step density is roughly linear, suggesting a quantitative and cumulative effect of steps in the deactivation of the oxidation of acetaldehyde.

The same experiments were also performed using perchloric acid as supporting electrolyte. Fig. 10 shows the results of these experiments. The voltammetric profiles correspond strongly to the results employing sulfuric acid as supporting electrolyte. In general, the oxidation currents are somewhat higher in perchloric acid (again, with the notable exception of Pt (553)), although the difference is quite small. Also, as is the case for sulfuric acid electrolyte, Pt (110) is very inactive for acetaldehyde oxidation. Moreover, the negative correlation between oxidation activity and step density found for oxidation in sulfuric acid is also observed, to a greater extent, in a perchloric medium. The trend of acetaldehyde oxidation activity decreasing with step density was also confirmed in potential step measurements (Fig. 11), at least for the stepped surfaces. The effect of the anion can again be attributed to the preferential adsorption of (bi)sulfate on terraces rather than on steps: if acetaldehyde oxidation and (bi)sulfate adsorption are both processes that are directly related to the number of terrace sites, it is clear that effect of switching from sulfuric acid to perchloric acid as supporting electrolyte has a larger effect on the oxidation of acetaldehyde on surfaces with wide terraces than on surfaces with narrow terraces.

3.2.2 Adsorbate stripping. Stripping experiments similar to those described in section 3.1.2 were also conducted with acetaldehyde. The results of these experiments are shown in Fig. 12. In contrast to ethanol, for acetaldehyde, increasing the adsorption potential decreases the total amount of adsorbates, although the absolute effect is rather small when using perchloric acid as supporting electrolyte. Apart from a more marked decrease of the amounts of adsorbates with adsorption potential in sulfuric acid, both electrolytes show the same behavior. From evaluating the redox behavior of the adsorbates, it can be seen that the quantity of reducible species decreases with adsorption potential in the potential range between 0.1 V and 0.5 V. Comparing this to the stripping results of the decomposition products of ethanol described in section 3.1.2, where the amount of reducible adsorbates increased when the adsorption potential was increased from 0.1 V to 0.2 V, this would imply that the reducible adsorbates resulting from ethanol adsorption are easier to reduce than those resulting from acetaldehyde adsorption. The amounts of oxidizable species increase strongly by stepping the adsorption potential from 0.1 V to 0.2 V, after which it remains essentially constant for adsorption potentials up to 0.5 V.

Results of acetaldehyde adsorbate stripping on different surfaces are shown in Fig. 13. For these experiments, acetaldehyde was adsorbed on the different platinum single-crystal electrodes at 0.3 V for 30 minutes. It can be seen that, similar to ethanol, the total amount of adsorbates increases with step density, indicating that step sites catalyze the carbon–carbon bond breaking. In addition, the composition of the adsorbates is strongly dependent on the surface structure: increasing the step density decreases the amount of oxidizable species while increasing the amounts of reducible species.

3.3 Acetic acid

Voltammetric profiles of acetic acid on Pt (111) in different electrolytes are shown in Fig. 14. It can be seen that, although the blank voltammetry of a well-ordered Pt (111) electrode shows large differences depending on the nature of the electrolyte, no significant differences are observed in the cyclic voltammograms recorded in 0.1 M acetic acid in all solutions. For all electrolytes, there is no oxidation activity within the potential range of interest. Furthermore, the voltammetric behavior of

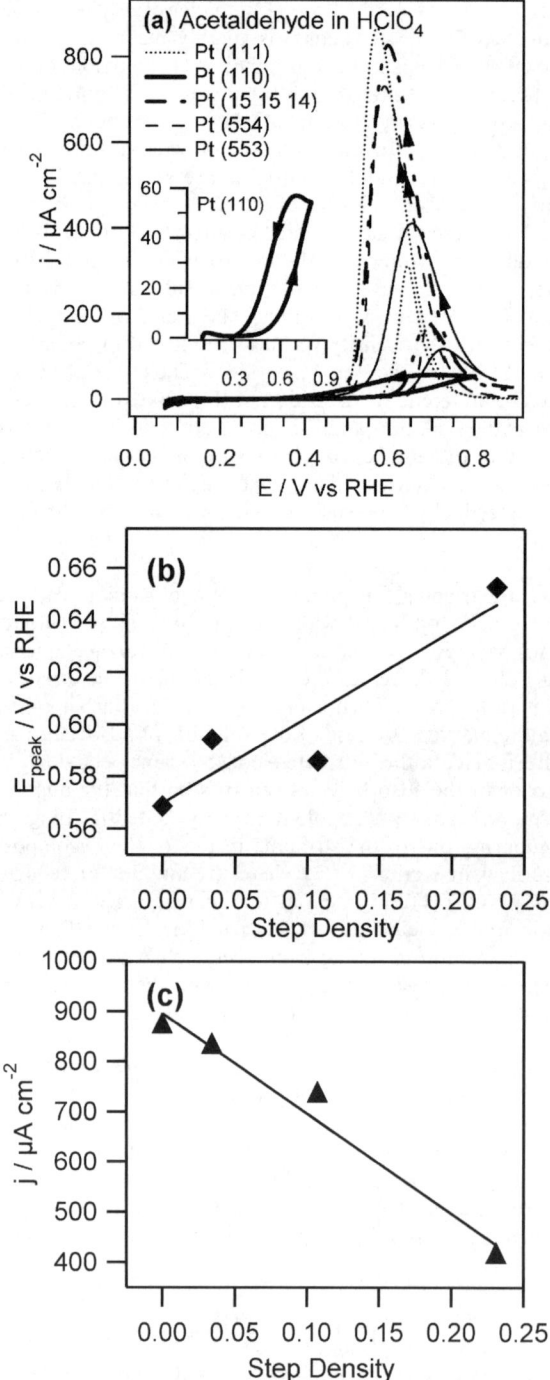

Fig. 10 (a) Cyclic voltammograms of the oxidation of 0.1 M acetaldehyde on platinum single-crystal electrodes in 0.1 M $HClO_4$ at a scan rate of 10 mV s^{-1}. (b) Dependence of the peak potential and (c) the maximum current density of acetaldehyde oxidation on the step density. The solid lines are the least-squares fit of the data.

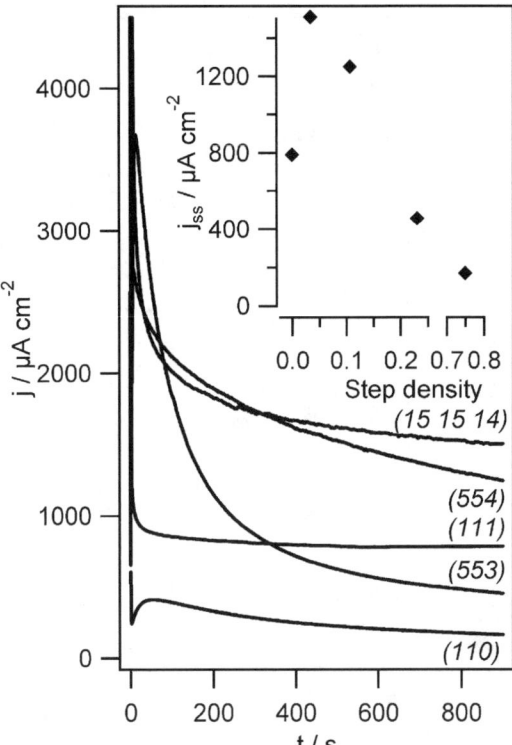

Fig. 11 Current-time transients of the oxidation of 0.5 M acetaldehyde in 0.1 M HClO$_4$. The step potential was 0.7 V. The inset shows the current density after 15 minutes as a function of step density.

acetic acid on Pt (111) in all electrolytes is very comparable to that of sulfuric acid. Similar results were found for the other electrodes of the Pt [$n(111)\times(111)$] series studied in this work. Since the voltammetric profile of acetic acid closely mirrors that of sulfuric acid, or, more generally, of (bi)sulfate species in the electrolyte solution, the voltammetric behavior of acetic acid can be considered in an analogous manner. The potential region between 0.0 V and roughly 0.3 V can be described as the hydrogen underpotential region. Starting at 0.3 V, a broad feature can be seen extending up to 0.8–0.9 V. The charge density derived from this feature was 100 ± 4 µC cm^{-2} for all electrolytes, which is somewhat higher than the charge density for the (bi)sulfate adsorption, which is around 80–90 µC cm^{-2}.[15,26] The charge involved in this feature can be ascribed to the adsorption of acetic acid, most likely as acetate, and correspond to a coverage of 0.41 ± 0.02 monolayers. The fact that the presence of the specifically adsorbing (bi)sulfate does not significantly alter the charge involved in the adsorption of acetic acid, indicates that acetic acid adsorbs more strongly than (bi)sulfate. Fig. 14 clearly shows that the formation of acetic acid should be considered a 'dead end' in the ethanol oxidation scheme.

3.4 General discussion

Based on the results presented here, as well as previously published studies, we propose a mechanism for the electrochemical oxidation of ethanol as shown in Scheme 1. This scheme includes the formation of strongly adsorbed intermediates and the formation and oxidation of acetaldehyde, as well as the non-reactivity of

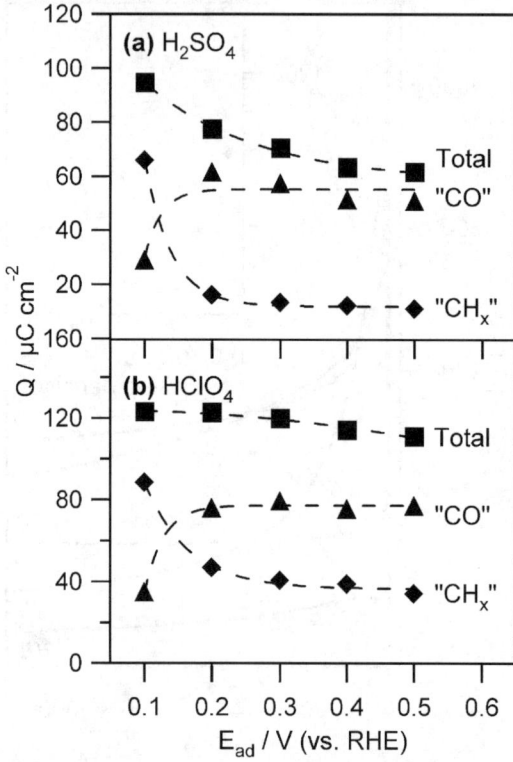

Fig. 12 The dependence of the amount of charge in the H_{UPD} region that is inhibited by the adsorbates resulting from acetaldehyde decomposition on the adsorption potential on Pt (553). Triangles denote oxidizable species, while diamonds denote reducible species. Squares indicate the sum of the oxidizable and reducible species. The supporting electrolyte was (a) 0.5 M H_2SO_4 or (b) 0.1 M $HClO_4$. The dashed lines are drawn to guide the eye.

acetic acid. Scheme 1 is similar to the reaction scheme proposed by Tremiliosi-Filho et al. for the oxidation of ethanol on gold,[27] with the difference that that scheme does not include C–C bond breaking. Our scheme relies heavily on the results obtained in ultrahigh vacuum (UHV) that the C–C bond primarily breaks after dehydrogenation to acetaldehyde.[28,29] Theoretical[30] and UHV[29] studies suggest that, initially, ethanol adsorbs weakly to the platinum surface through the lone pair electrons on the oxygen (reaction 1). Once adsorbed, ethanol can be oxidized to (weakly) adsorbed acetaldehyde, a reaction requiring two dehydrogenation steps (reaction 2). In UHV, the first dehydrogenation step is the cleavage of the O–H bond, giving rise to an adsorbed ethoxy,[29] but under electrochemical conditions it has been suggested that the first bond to be broken is the α-C–H bond.[8,31] Since, for the ethanol concentration investigated in this study, DEMS measurements have shown that the main ethanol oxidation product is acetaldehyde,[4] it can be safely assumed that the measurements shown in Fig. 1, Fig. 2 and Fig. 4 mainly show the current arising from the conversion of ethanol to acetaldehyde. Due to the relatively small anion effect illustrated in Fig. 3 and a small but significant effect of surface structure, it can be concluded that the oxidation of ethanol to acetaldehyde is a surface sensitive process, occurring preferentially at or near step sites on the electrode.

Since strongly adsorbed species are observed after adsorbing ethanol at potentials as low as 0.1 V, while no significant oxidation currents for ethanol are observed

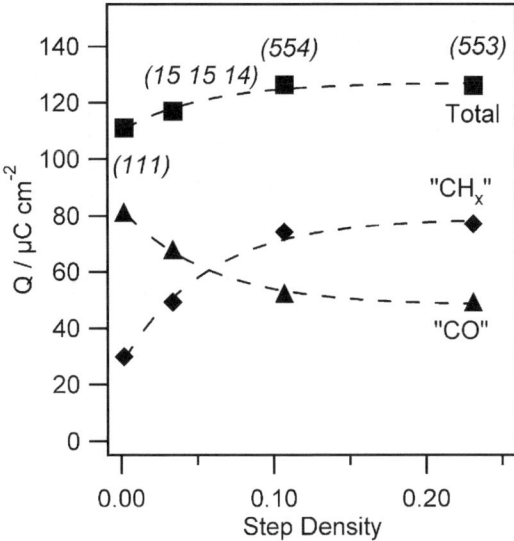

Fig. 13 Structure sensitivity of acetaldehyde adsorbate stripping in 0.1 M HClO$_4$ at $E = 0.3$ V, as implied by the amount of charge in the H$_{UPD}$ region that is inhibited by the adsorbates. Triangles denote oxidizable species, while diamonds denote reducible species. Squares indicate the sum of the oxidizable and reducible species. The dashed lines are drawn to guide the eye.

below 0.3 V in this study, we cannot exclude a direct pathway from adsorbed ethanol to decompose into adsorbed CH$_x$ and adsorbed CO (reaction 3). At the same time, acetaldehyde can also decompose into these species at these low potentials (reaction 4).[11] Comparing Fig. 8 and Fig. 13, it can be seen that the total coverage resulting from acetaldehyde decomposition is higher than that from ethanol decomposition, indicating that carbon–carbon bond breaking occurs to a greater extent in acetaldehyde. This is supported by most UHV studies,[28,29] which, as mentioned, often indicate acetaldehyde as a necessary intermediate in the decomposition of ethanol. On the other hand, a study by Cong and coworkers[28] hints indirectly at the ethanolic carbon–carbon bond partially being broken before the formation of acetaldehyde, especially at the (110) step sites on a Pt (331) surface. Therefore, we suggest that the main species in which the ethanol carbon–carbon bond is broken is acetaldehyde, especially in the potential range in which ethanol is readily oxidized to acetaldehyde. However, at low potentials (<0.3 V), C–C bond breaking may occur in ethanol or ethoxy, especially at step sites, in agreement with the suggestion made by Cong et al.[28] CO resulting from the ethanol and acetaldehyde decomposition can be oxidized at potentials above ~0.5 V (reaction 5). The adsorbed hydrocarbon fragment can be removed completely by reduction below 0.1 V (reaction 6). Our stripping results suggest that the adsorbed CH$_x$ fragments may be oxidized to CO$_{ads}$ at 0.4 V and subsequently to CO$_2$ at somewhat higher potential, and that therefore they are not more poisoning than adsorbed CO.

Apart from decomposition to yield strongly adsorbed carbon containing species, acetaldehyde can be further oxidized without the breaking of the C–C bond to yield acetic acid (acetate). Relatively little is known about the pathway in which acetic acid is generated. It has been suggested that acetaldehyde can react directly with a surface oxygenated species,[11] such as OH$_{ads}$, into adsorbed acetate (reaction 8). Alternatively, since acetaldehyde is readily hydrolyzed in an aqueous environment, acetic acid is formed by the oxidation of the resulting 1,1-ethane-diol (reactions 9–12),[32] similar to the suggested oxidation mechanism of ethanol on gold[27] and of formaldehyde on platinum to formic acid.[19]

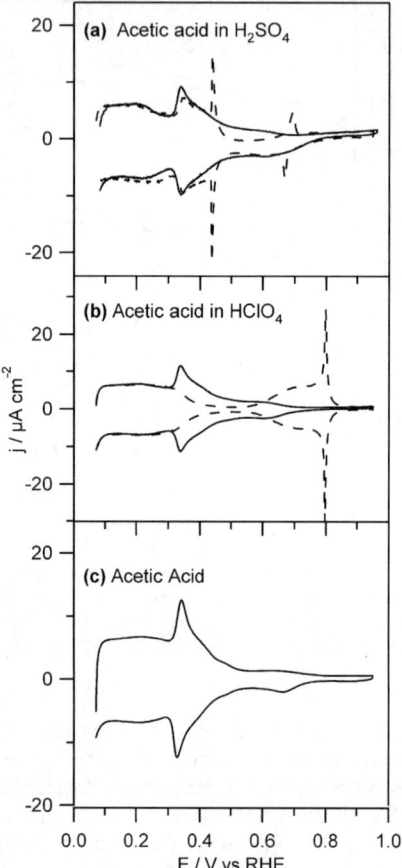

Fig. 14 Voltammetric profiles of 0.1 M acetic acid on Pt (111) in (a) 0.5 M H_2SO_4 (b) 0.1 M $HClO_4$ and (c) water. The dashed lines represent the voltammograms recorded in blank supporting electrolyte. The scan rates were 10 mV s^{-1}.

With Scheme 1 in mind and by combining the results of the adsorbate stripping experiments with the results of bulk oxidation experiments, we propose the following hypothesis for the (seemingly counter-intuitive) behavior of the effect of steps on the electro-oxidation activity for ethanol and acetaldehyde. For ethanol oxidation at a high ethanol concentration, the current is mainly the result of the oxidation of ethanol to acetaldehyde (reaction 2). This reaction has a (small) preference for step sites. Since direct C–C breaking in ethanol or ethoxy (reaction 3) is sluggish, and the concentration of acetaldehyde is still too low to lead to step poisoning, the overall activity increases with rapid step density. For acetaldehyde, the current is mainly due to the reaction of acetaldehyde to acetic acid (reactions 8–12) at potentials above 0.7 V.[11] However, C–C bond breaking in acetaldehyde (reaction 4) is fast, and therefore, step sites are rapidly blocked. If reaction 8 or reaction 11 prefers step sites, this would explain a decreasing overall activity with step density.

A remaining puzzling feature is the different stability of the CH_x fragments from ethanol and acetaldehyde, those from ethanol apparently being less stable, especially at low potentials (compare Fig. 7 and Fig. 12). This may be related to the oxidizability of the CH_x fragments, as Hahn and Melendres have reported the possibility of

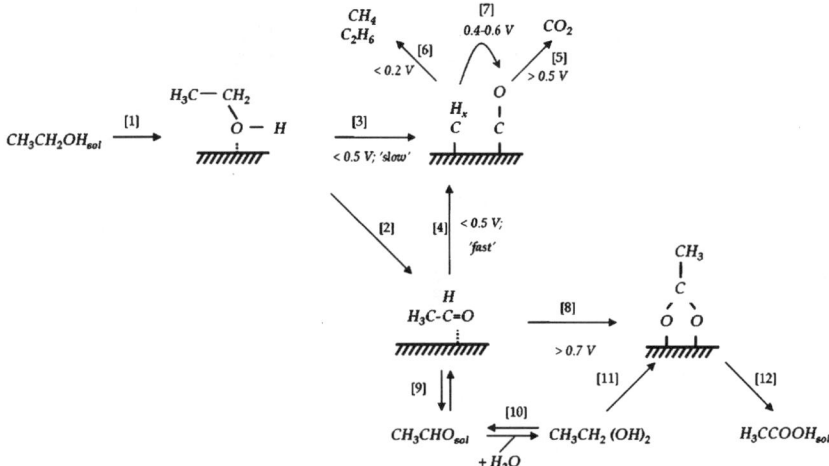

Scheme 1 Schematic representation of a suggested mechanism of the electrochemical oxidation of ethanol on platinum electrodes at the high concentrations of ethanol used in this study.

oxidizing methane adsorbates on platinum to COH_x fragments at potentials close to or in the hydrogen region.

4. Conclusion

In this paper, we have presented the results of a systematic electrochemical study of the oxidation of ethanol and acetaldehyde on well-defined platinum single-crystal surfaces for high concentrations of ethanol and acetaldehyde in solution. For ethanol, it was found that the continuous oxidation show a significant degree of surface sensitivity, with the total oxidation activity in general increasing with the concentration of defect sites (steps). In the case of acetaldehyde, however, this finding is reversed. To elucidate the effect of the surface structure on the breaking of the carbon–carbon bond, adsorbate stripping experiments were performed both with ethanol and acetaldehyde. Our results suggest that the C–C bond breaking occurs preferentially in acetaldehyde, suggesting that acetaldehyde is the main species in which the bond is broken in the ethanol reaction mechanism, although slow C–C bond breaking in ethanol and ethoxy cannot be excluded. Finally, based on our results and the existing literature, a detailed scheme of the ethanol oxidation mechanism was presented, which incorporates the formation of carbon dioxide, acetaldehyde and acetic acid, products of the electrochemical oxidation of ethanol, as well as the formation of adsorbed intermediates detected by spectroscopic methods. Based on this mechanism, we suggest that step poisoning is slow for ethanol oxidation, leading to a positive effect of the step density on the conversion of ethanol to acetaldehyde. On the other hand, the step poisoning is rapid for acetaldehyde because of the relative ease of C–C bond breaking, leading to an adverse effect of steps on the oxidation of acetaldehyde to acetic acid.

References

1 D. J. Tarnowski and C. Korzeniewski, *J. Phys. Chem. B*, 1997, **101**, 253.
2 J. Shin, W. J. Tornquist, C. Korzeniewski and C. S. Hoaglund, *Surf. Sci.*, 1996, **364**, 122.
3 F. Vigier, C. Coutanceau, F. Hahn, E. M. Belgsir and C. Lamy, *J. Electroanal. Chem.*, 2004, **563**, 81–89.
4 G. A. Camara and T. Iwasita, *J. Electroanal. Chem.*, 2005, **578**, 315–321.
5 T. Iwasita and E. Pastor, *Electrochim. Acta*, 1994, **39**, 531–537.

6 J. T. Wang, S. Wasmus and R. F. Savinell, *J. Electrochem. Soc.*, 1995, **142**, 4218–4224.
7 S. C. Chang, L. W. H. Leung and M. J. Weaver, *J. Phys. Chem.*, 1990, **94**, 6013–6021.
8 J. Willsau and J. Heitbaum, *J. Electroanal. Chem.*, 1985, **194**, 27–35.
9 H. Wang, Z. Jusys and R. J. Behm, *J. Phys. Chem. B*, 2004, **108**, 19413–19424.
10 U. Schmiemann, U. Muller and H. Baltruschat, *Electrochim. Acta*, 1995, **40**, 99–107.
11 M. J. S. Farias, G. A. Camara, A. A. Tanaka and T. Iwasita, *J. Electroanal. Chem.*, 2007, **600**, 236.
12 J. Clavilier, D. Armand, S. G. Sun and M. Petit, *J. Electroanal. Chem.*, 1986, **205**, 267–277.
13 J. Clavilier, K. Elachi and A. Rodes, *Chem. Phys.*, 1990, **141**, 1–14.
14 N. P. Lebedeva, M. T. M. Koper, J. M. Feliu and R. A. van Santen, *Electrochem. Commun.*, 2000, **2**, 487–490.
15 A. M. Funtikov, U. Stimming and R. Vogel, *J. Electroanal. Chem.*, 1997, **428**, 147–153.
16 C. G. M. Hermse, A. P. van Bavel, M. T. M. Koper, J. J. Lukkien, R. A. van Santen and A. P. J. Jansen, *Surf. Sci.*, 2004, **572**, 247–260.
17 N. P. Lebedeva, M. T. M. Koper, E. Herrero, J. M. Feliu and R. A. van Santen, *J. Electroanal. Chem.*, 2000, **487**, 37–44.
18 N. P. Lebedeva, M. T. M. Koper, J. M. Feliu and R. A. van Santen, *J. Phys. Chem. B*, 2002, **106**, 12938–12947.
19 T. H. M. Housmans, A. H. Wonders and M. T. M. Koper, *J. Phys. Chem. B*, 2006, **110**, 10021–10031.
20 J. Mostany, E. Herrero, J. M. Feliu and J. Lipkowski, *J. Phys. Chem. B*, 2002, **106**, 12787–12796.
21 F. Hahn and C. A. Melendres, *Electrochim. Acta*, 2001, **46**, 3525–3534.
22 M. Bergelin, E. Herrero, J. M. Feliu and M. Wasberg, *J. Electroanal. Chem.*, 1999, **467**, 74–84.
23 W. Akemann, K. A. Friedrich and U. Stimming, *J. Chem. Phys.*, 2000, **113**, 6864–6874.
24 J. F. Gomes, B. Busson and A. Tadjeddine, *J. Phys. Chem. B*, 2006, **110**, 5508–5514.
25 X. H. Xia, H. D. Liess and T. Iwasita, *J. Electroanal. Chem.*, 1997, **437**, 233–240.
26 A. Kolics and A. Wieckowski, *J. Phys. Chem. B*, 2001, **105**, 2588–2595.
27 G. Tremiliosi-Filho, E. R. Gonzalez, A. J. Motheo, E. M. Belgsir, J. M. Leger and C. Lamy, *J. Electroanal. Chem.*, 1998, **444**, 31–39.
28 Y. Cong, V. vanSpaendonk and R. I. Masel, *Surf. Sci.*, 1997, **385**, 246–258.
29 A. F. Lee, D. E. Gawthrope, N. J. Hart and K. Wilson, *Surf. Sci.*, 2004, **548**, 200–208.
30 R. Alcala, M. Mavrikakis and J. A. Dumesic, *J. Catal.*, 2003, **218**, 178–190.
31 E. Sokolova, *Electrochim. Acta*, 1975, **20**, 323–330.
32 R. P. Bell and B. D. Darwent, *Trans. Faraday Soc.*, 1950, **46**, 34–41.

General Discussion

Professor Gewirth opened the discussion of the paper by Christopher F. Blanford, Carina E. Foster, Rachel S. Heath and Fraser A. Armstrong*:

I am wondering about the characterisation of the adsorbed laccase. Have you performed any magnetic measurements, particularly susceptibility, to characterise the on-electrode enzyme. What is the degree of heterogeneity in the adsorbed material?

Dr Armstrong answered:

We have not carried out magnetic susceptibility experiments but we have started to run EPR experiments on microscopic graphite particles to which laccase molecules are attached. We are hoping to use advanced pulsed EPR methods to probe the interaction of laccase with the graphite surface and with neighbouring laccase molecules. There are two main sources of heterogeneity: the catalyst itself, and its orientation on the electrode. The laccase is chromatographically pure, although there is still a range of glycosylation sites that may or may not be occupied. Some degree of orientation heterogeneity—dispersion—is always likely in all enzymes we have studied. An interesting point is that the active sites of enzymes are usually buried and the close-up interaction with the electrode does not effect their properties; however, electrode–enzyme electron-transfer kinetics will be very prone to heterogeneity. We aren't sure if the trailing edge on the voltammograms is a result of different orientations (as modelled in *J. Phys. Chem. B*, 2002, **106**, 13058–13063) or if it is an inherent feature of the enzyme catalysis.

Miss Nestoridi asked:

What is the maximum number of laccase layers you could attract on a carbon surface and what would be the highest current density you could achieve in that case?

Dr Armstrong responded:

We believe that it is unlikely that we have a multilayer in our experiments. If we assume that each laccase molecule can turnover 300 molecules per second, that it has a footprint of 30 nm^2 (based on the crystal structure), and because it carries out a 4 e^- reduction, we calculate we can get 660 $\mu A\ cm^{-2}$, almost identical to our best films. This coverage is roughly 6 pmol cm^{-2}.

Professor Schmickler asked:

Could the strange shape of the Tafel curves that you observed be caused by transport, in particular by transport of the reactant through the pores?

Dr Armstrong replied:

This is a difficult question and it ties into our response to Miss Nestoridi previously. The electrode is not a porous system so transport shouldn't affect the Tafel plot. We are carrying out impedance experiments to comment on this more fully.

Professor Wieckowski commented:

From the fuel cell perspective I would appreciate your comments on three essential components, performance, cost and durability. After the lecture one can be optimistic about the performance but what about the remaining two?

Dr Armstrong answered:

We have estimated that in our laboratory laccase costs about 10 times as much per turnover frequency, and about 2.5 times as much as platinum, when one compares the cost of one laccase molecule *versus* one 5.2 nm platinum particle. Our platinum costs are based on the cost of 20% Pt black on carbon from Alfa Aesar; our laccase costs are based on the cost of purification columns, two days of technician's time, and one gram of commercial laccase from Fluka. Our best attempt at durability is shown in the manuscript, with nearly 2000 hours of continual use.

Professor Wittstock said:
You emphasised the attempts to design the electrode surface in order to obtain a "correct" orientation of the T1 centre. The enzyme has, however, several hydrophobic regions. Do you know which fraction of the bound enzyme have the correct orientation? If you also have enzymes in less favourable orientation, could they be switched on at higher overpotentials leading to "strange" slopes on Tafel plots? How do you see the enzymes in less favoured orientation, assuming that the ones in "correct" orientation can seen by T1 redox chemistry? Would the use of mediators be an option to make a kind of different measurement?

Dr Armstrong responded:
Regarding the fraction of enzyme molcules in the "slow" orientation: this may tie into the "trailing edge" of the voltammogram, but I don't think we can comment.

Relating to the Tafel plot: this is an interesting idea. A range (dispersion) of ET rates could give rise to curvature in the Tafel plot. This is similar to modelling of the catalytic waveform that we reported in *J. Phys. Chem. B*, 2002, **106**, 13058–13063. Relating to resolving the question of orientation in a way that would 'by-pass' the T1 Cu: we have attempted to resolve this by producing a T1-depleted or T1-substituted version of our high-potential laccase. So far we have not produced the desired product, perhaps due to the high stability of T1 in the Cu(I) state, which makes it very difficult to remove using cyanide.

Professor Ren asked:
Since the exact active part of the enzyme is a small metal centre, is there any attempt to synthesize the smallest possible metal centre that can still retain the function of catalytic activity?

Dr Armstrong replied:
We know of many attempts to do this, but so far the overpotentials observed with simple models are quite a lot greater than observed with laccase.

Professor Janik opened the discussion of the paper by Jan Rossmeisl*, Gustav S Karlberg, Thomas Jaramillo and Jens K Nørskov:
The reaction mechanism, as shown in Fig. 3 in your paper, includes the reductive dissociation of OOH* to form a hydroxyl adsorbed and a water molecule. What is the justification for writing this, rather than a dissociative chemical step followed by reduction? Further, how would this impact the rate expressions and subsequent volcano plot analysis? Specifically, in eqn (14) in the paper, a steady state approximation applied to the hydroxyl coverage, there would then be a "2" in front of the first term ion the right hand side of the equation, as then reaction 1 (eqn (10)) would derive 2 hydroxyls *versus* 1. On a potential energy surface, *i.e.* your Fig. 3, this would then make a limiting barrier of a $2e^-$ process, though I am unsure how this further alters the rate expressions.

Dr Rossmeisl answered:
We have found that at the ORR relevant potentials 0.8–0.9V the associative reaction is favored over the dissociative for metals at the top of the volcano.[1] For very weak binding metals the associative is favored even at higher potentials. The kinetics with a dissociative mechanism would change, first: since the barrier is directly dependent on potential and second: since $(1 - \theta)^2$ enters in the dissociative mechanism

instead of just $(1 - \theta)$, and as Prof. Janik points out you would then make two OH, this will change the following equations. In eqn (14) the $(1 - \theta)^2$ would also enter. I am not sure that Prof. Janik is right that it would look like a 2-electron process. The rate expressions become quite complicated due to the squared term.

1. G. S. Karlberg, J. Rossmeisl and J. K. Nørskov, *Phys. Chem. Chem. Phys.*, 2007, **9**, 5158–5161

Professor Schmickler addressed Dr Rossmeisl and Dr Wang:
I noticed that you assumed rather different values for the transfer coefficient: Dr Wang used the normal Marcus-type value of 1/2, Dr. Rossmeisl a value of unity corresponding to an activationless reaction. Could you please comment on your different approaches?

Dr Rossmeisl replied:
In our paper we have used a transfer coefficient of one, because we still do not have a reliable simulation for the proton transfer barrier based on *ab initio* calculations. In our model no fitting parameters are used, we try to base every thing directly on the DFT-calculations. Many of the equations and parameters that are input in Dr Wang's approach are output in ours. In that way the two studies are complementary to each other.

Dr Wang answered:
We tested different assumptions on the transfer coefficients in fitting measured polarization curves. The results suggest a value of 1/2 because this yields the 118 mV dec^{-1} Tafel slope at large overpotentials observed in rotating electrode measurements; the smaller Tafel slope at low overpotentials can be well explained by the effect of adsorption isotherms (Fig. 4e and 4d in our paper). A unity transfer coefficient, corresponding to the slope of 70 mV dec^{-1} at 80 °C, is used commonly in analyzing the polarization curves measured in fuel cells. The discrepancies with data at large overpotentials are often attributed to non-kinetic complications that might exist in fuel cells. Using the intrinsic kinetic equation with the electron transfer coefficient being 1/2, we reproduced a typical PEMFC polarization curve (Fig. 7 in the paper). Further, we proposed two criteria for distinguishing between the possible causes for the failure of the B–V equation with the unity transfer coefficient. One is by changing the Pt loading (Fig. 8 in the paper) and another is by varying the partial pressure of oxygen (Fig. 9 in the paper). Published data support predictions based on the intrinsic kinetic equation with the transfer coefficient being 1/2.

Dr Jinnouchi said:
Do you have any idea how we can predict the pre-exponential factor, which is assumed at 200 site^{-1} s^{-1}, by using a first principles method?

Dr Rossmeisl responded:
It would take a simulations of the proton transfer reaction. From previous calculations[1] it seems that the low prefactor cannot be understood from the harmonic frequencies of H$^+$ in the waterlayer. My feeling is that prefactor may be more related proton transport between water molecules.

1. E. Skulason, G. S. Karlberg, J. Rossmeisl, T. Bligaard, J.Greeley, H. Jonsson, and J. K. Nørskov, *Phys. Chem. Chem. Phys.*, 2007, **9**, 3441–3250

Professor Behm asked:
What would change in the predicted calculated volcano plot if one would include O$_2$ reduction *via* H$_2$O$_2$ formation and subsequent decomposition as an important pathway, as it was discussed yesterday?

Dr Rossmeisl replied:

Nothing would change, since, according to DFT-calculations and table values for the free energy of H_2O_2, reduction of this is downhill in free energy at the relevant ORR potentials. The first reaction step $O_2 \rightarrow OOH$ or the last step $OH \rightarrow H_2O$ are, according to the DFT-calculations, rate determining. We haven't included transport limitations as you discussed in the model.

Dr Jinnouchi opened the discussion of the paper by Jia X Wang*, Francisco A. Uribe, Thomas E. Springer, Junliang Zhang and Radoslav R. Adzic:

The point of this work is that the potential where the change in the Tafel slope occurs corresponds to the reversible potential for getting O from OH, and the reversible potential was set at 0.77 V (RHE). Is the value of 0.77 V (RHE) lower than previous experimental measurements, which showed that the reversible potential may be around 1.0 V (RHE)?

Dr Wang answered:

Although the Tafel slope continuously changes with potential, we showed in Fig. 4f in our paper that the potential separating the regions of the low and high Tafel slopes is close to the reversible potential for the O and OH transition (*i.e.*, where the activation barriers are equal for both reaction directions). The difference in adsorption free energies for the O and OH determines its value: it may vary with catalysts, temperature, or oxygen pressure. For Pt(111) with 1 atm pressure of oxygen at 23 °C, this value is 0.87 V (Fig. 4f in the paper), while for Pt/C with 0.42 atm oxygen partial pressure at 80 °C, it is 0.77 V (Fig. 7 in the paper). Voltammetry curves measured on Pt(111) in the absence of oxygen show a rising current above 0.95 V, related to the $OH \rightarrow O$ transition. This might explain why the questioner suggested 1 V as the reversible potential for O and OH transition. However, the cathodic and anodic current peaks are not symmetric when the positive limit rises to 1.2 V. Accordingly, the reversible potential determined by averaging the transitions in both directions should be lower than 1 V. The point we made is that the reversible potential of 0.77 V for the Pt/C in fuel cells should not change with Pt loading; therefore, the range of measured polarization curves that can be fitted with a single Tafel slope would be determined by the same potential, not at a constant current (Fig. 8 in the paper).

Professor Gewirth addressed Dr Wang and Dr Rossmeisl :

I am confused about the proposed mechanisms of oxygen reduction developed by each of you during the session. In one paper, an associatiive mechanism for oxygen reduction was asserted, while in the other a dissociative mechanism proposed. And then, additionally, we heard other talks where a peroxide intermediate was suggested.

How much does the particular mechanism matter in your analysis? If you were incorrect on the mechanism, what would that do to your overall conclusions? And what do we really know—on the basis of experiment—about the real mechanism(s) of oxygen reduction on Pt in acid?

Dr Rossmeisl responded:

The dissociative and associative mechanisms are close in energy at the relevant ORR potentials. When including the field, the associative mechanism seems to be favored.[1] The volcanos obtained for the two mechanism are very similar.

Concerning the peroxide intermediate the model is robust concerning including that. Also I would like to stress that the model is concerned with trends and there are still things to be done in order to fully understand the details in the kinetics.

1. G. S. Karlberg, J. Rossmeisl and J. K. Nørskov, *Phys. Chem. Chem. Phys.*, 2007, **9**, 5158–5161

Dr Wang responded:

Our double-trap kinetic model includes both associative and dissociative pathways. The latter often is neglected due to the preference for using the simplest model possible, an approach that can lead to semi-quantitative agreement with the experimental results. Rossmeisl showed that for both adsorption isotherms and kinetic currents, simulated curves for the Pt(111) and Pt$_3$Ni(111) largely reproduced the potential shift between them, but not the shape of the measured curves. We developed the intrinsic kinetic equations from models with and without the dissociative pathway. Better fits and, more importantly, reasonable adsorption isotherms were found using the dual-path model (Fig. 4a and 4c compared to Fig. 3a and 3b in our paper). In our previous publication [Wang *et al.*, J. Phys. Chem. A, 2007, **111**, 12702], we detailed how adsorption isotherms can be used to distinguish different models and explained why the existing data support the model with both pathways.

The double-trap model was proposed for the 4e-ORR on Pt with water as the only product. Disk-ring electrode experiments revealed the presence of H_2O_2 at low potentials. However, the yield often is less than 1% above 0.4 V. Thus, neglecting the 2e-reduction product, H_2O_2 should not significantly affect our analyses of the Pt polarization curves. At this conference, Behm (and the microelectrode work shown in the discussion) demonstrated that some peroxide molecules could be re-adsorbed and further reduced to water, and thus acted as an intermediate of the 4e-reduction. The question then arises; do we need to add this reaction pathway in our model? We think probably not because the re-adsorbed H_2O_2 still needs to undergo an O–O bond breaking by forming either O_{ad} or OH_{ad}. In the double-trap model, we have omitted the detailed steps leading to the formation of O_{ad} and OH_{ad}. That is, we considered only how the effective barriers for all reaction steps before the formation of OH_{ad} and O_{ad} affect the overall kinetic behavior. Details about the various routes to the adsorbed O and OH do not alter the values of effective barriers we obtained from analyzing the measured polarization curves.

Concerning the last question, we considered the following are the experimentally established facts for the ORR on Pt-like catalysts in acid: (1) 4e-reduction dominates under common conditions for PEMFC applications (*i.e.*, with sufficient catalyst loading and at potentials positive of 0.4 V); (2) catalytic activity is desorption limited because O and OH adsorption saturate near the reversible potential, so causing a large overpotential for the onset of ORR reaction currents; (3) the Tafel slope increases with increasing overpotential over a wide potential region in most cases. In contrast, the HOR is adsorption limited because the coverage of the intermediate at the reversible potential is much less than 0.5 (see the θ_{IR} curve for atop-site-adsorbed H in Fig. 1b in our paper). Therefore, while the dual-path for the H adsorption process plays a major role in determining the HOR activity, the effect of the deep energy traps for O and OH is the key for understanding the ORR kinetic behavior. Thus, the double-trap kinetic model successfully reproduced measured polarization curves and explained the change of Tafel slope over a wide potential region. However, due to the simplification we made for the adsorption steps, our analysis offers little on how the O–O bond breaks on Pt among the many possible reaction steps.

Professor Wieckowski commented:

There is a possibility that with the complicated equation used, containing several exponents, one can fit practically everything. So what are the constraints (parameters, experimental entries) of the fit?

Dr Wang answered:

In general, the uniqueness of fit declines as the number of fitting parameters rises. However, complicated equations do not necessarily produce good fits. Using relevant principles and well-justified assumptions in modeling are of primary importance for ensuring meaningful results from fitting. In fitting the ORR polarization curves, we fixed two, and allowed the other four parameters to vary. Through systematically changing the two preset values, we obtained good fits with several sets of parameters. In Table 1 of a previous publication, we listed five sets of parameters with nearly perfect fits [Wang *et al.*, *J. Phys. Chem. A*, 2007, **111**, 12702]. The first three sets reveal that we could not distinguish whether the activation barrier for the reductive desorption (RD) is slightly higher or lower than that for the reductive adsorption (RA). However, different sets of parameters produced distinctly different adsorption isotherms. If we can experimentally determine the adsorption isotherms, even only the OH coverage at large overpotentials (see the last column of our paper), we shall be able to find which set of parameters reflects the reality. We also found uncertainties in the adsorption free energies of the intermediates. Only the difference between ΔG_O and ΔG_{OH} can be determined from fitting (i.e., $\Delta G_O - \Delta G_{OH}$ is constant for different ΔG_{OH}). Therefore, to reduce the uncertainties in fitted kinetic parameters, we need measured adsorption isotherms to identify constraints in singling out the right set of parameters.

Professor Tryk asked:

I would like to go back for a moment to the HOR for a moment, I would like to ask whether it is possible that the dual pathway arises naturally from the fact that on a polycrystalline electrode or nanoparticle surface, you have intrinsically different pathways being followed on the different faces. For example, in Markovic's 1997 paper (*J. Phys. Chem. B*, 1997, **101**, 5405), there is a 28mV decade^{-1} slope on Pt(110) and a 74mV decade^{-1} slope on Pt(111), indicating a clear difference in pathway, with a homolytic (Tafel) step on Pt(110) and possibly a heterolytic (Heyrovsky) step on Pt(111).

Dr Wang responded:

By lowering the temperature to 1 °C, Markovic *et al.*,[*J. Phys. Chem. B*, 1997, **101**, 5405] revealed distinctly different polarization curves for the HOR on Pt(111), Pt(100), and Pt(110) surfaces. We obtained good fits to their data using the dual-pathway kinetic equation with coverage-dependent repulsion (see the answer to Professor Behm's question later in the discussion). As we show in Fig. 1 below, the activation barrier for dissociative adsorption or the Tafel reaction, ΔG_T, decreases in the order of (111) > (100) > (110). The kinetic current from this pathway (dashed lines in the right panel) is negligible on (111), quite significant on (100), and

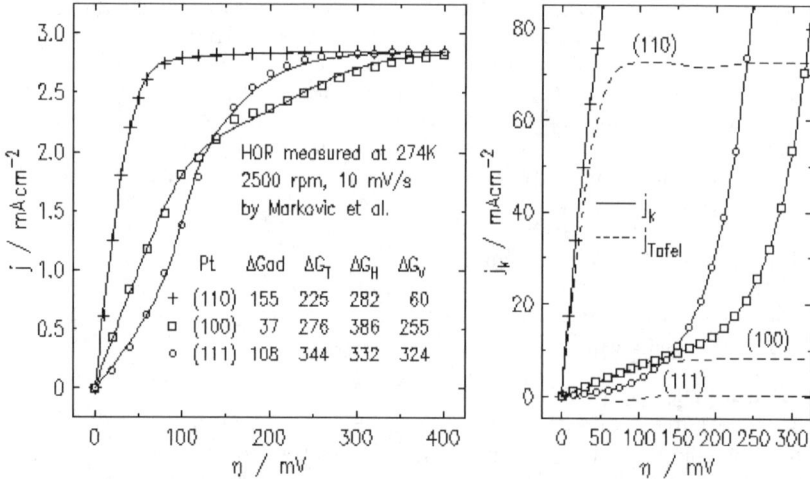

Fig. 1 The activation barrier for dissociative adsorption or the Tafel reaction, ΔG_T.

almost dominant on (110) at low overpotentials. We attribute this trend to the increase of reactivity with more open structures that promotes the dissociative adsorption, rather than to selectivity based on different geometries. The latter is unlikely because the HOR activity on polycrystalline Pt is even higher than that on Pt(110).

Professor Savinova returned the discussion of the paper by Jan Rossmeisl*, Gustav S. Karlberg, Thomas Jaramillo and Jens K. Nørskov:

I must admit that I am rather unhappy with the overall discussion of the oxygen reduction reaction (ORR) and with the model utilized to describe it. In the paper it is written that the model "predicts both the potential dependence of the OH coverage and the measured current densities seen in experiments, and that it offers an understanding of the oxygen reduction reaction at the atomic level." However, if we look at the model and its predictions (see *e.g.* Fig. 4 in the paper), we will see that it predicts that on Pt(111), removal of OH from the surface is the rate determining step (rds). Meanwhile, all the experimental evidence gathered until now (note *e.g.* the first order in O_2) tells us that this is rather unlikely. Moreover, recent investigations from the groups of A. Kucernak (Imperial Colledge, London) and R. J. Behm (University of Ulm) strongly suggest that the ORR on Pt proceeds *via* formation of the H_2O_2 intermediate. It is likely that the rds on Pt is concerned with the transfer of either the first or the second electron, rather than with the electrochemical desorption of $O(H)_{ads}$.

Thereby, I conclude that if the model is based on the wrong assumptions concerning the reaction mechanism, and still predicts the experimentally observed trends, it cannot be realistic.

Dr Rossmeisl answered:

I am unhappy that Prof. Savinova is unhappy with the paper.

The model presented in the paper is a necessary but not sufficient analysis for understanding trends in ORR catalysis. Including barriers for all steps is still needed to get the full understanding of ORR. Concerning if the first or last step is rate determining: many experiments suggest that a surface with an oxygen surface bound that is slightly weaker than platinum is optimal for ORR, this was predicted by the original Sabatier model in 2004. The model actually also predicts first order in oxygen, see Fig 5 in the paper. First order in O_2 is not the same as saying the first step is the rds, again this is an output not an input in the model. So I do not agree with the first statement from Prof. Savinova. Concerning the H_2O_2 path I have a few comments. Taking into account an H_2O_2 step will not change the predictions, since the reduction of H_2O_2 is easy at 0.8–0.9 V. Concerning wrong assumptions, the rds predicted is an output and not an input in the simulations. In general, as little as possible is assumed in this kind of modeling. In other words we do not speculate about the rds we calculate.

Professor Schmickler asked:

You claim that the steps involving peroxide are not important. Why does the oxygen reduction then stop at peroxide on some metals, and why does this even depend on the face of single crystals?

Dr Rossmeisl responded:

Peroxide steps are of cause important in order to fully understand the details in ORR, however it seems that it is not very important to understand to difference in catalytic ORR activity between Pt and Pt_3Ni. (and other modified Pt surfaces) In this study, we have focused on the most active 111 surface. The question by Prof. Schmickler is very interesting and something that I would like to investigate, also in connection with the other discussion and Prof. Behm's paper.

The different facets of gold has been investigated with the same method, see ref. 1.

1. P. Vassilev and Marc T. M. Koper, *J. Phys. Chem. C*, 2007, **111**, 2607

Professor Behm addressed Dr Wang and Dr Rossmeisl:
I would like to comment on the previous discussion on the selectivity and different reaction pathways in the ORR. In general, the selectivity is not a question of 'either-or' but a competition between different pathways which occur at the same time, which results in certain probabilities for the different products. A question, mainly to Dr Wang—I realise that using coverage dependant adsorption energies would add even more to the many parameters. Nevertheless they are known to exist. Wouldn't that affect the fitted other parameter values or the fits?

Dr Wang replied:
We agree that generally there are two ORR products: peroxide and water. The selectivity mainly is determined by the strength of the O–Metal bond relative to the O–O bond in peroxide. Au and Pt are the commonly used examples for one of the products being dominant. However, the 4e-reduction might occur on small Au clusters due to the enhanced O–Au bonding; conversely, the yield of peroxide might increase on Pt-alloy catalysts that weaken the M–O bond. Ideally, the model should encompass both products. We chose not to include peroxide in the model to keep the equation simple for studying the cases where peroxide formation is negligible, which is most relevant to fuel cell applications.

Treating coverage-dependent repulsion in kinetic analysis is cumbersome and requires an independently determined adsorption isotherm. We studied this issue in our previous work on the HOR on Pt [Wang *et al.*, J. Phys. Chem. C, 2007, **111**, 12425], and presented the results in the supporting information available at the following link, http://pubs.acs.org/subscribe/journals/jpccck/suppinfo/jp073400i/jp073400isi20070503_021339.pdf

Briefly, we found that adding a coverage-dependent repulsion in the model (Fig. S3d compared to Fig. S3a in the supporting information) can improve the agreement between calculated (dashed lines) and measured adsorption isotherm for the HOR intermediate (plus signs). However, the impact on the kinetic current is marginal, as revealed by the nearly identical curves in Fig. S3c and S3f. In our new analyses of low-temperature polarization curves measured on Pt (111), (100), and (110) surfaces, we obtained more significant improvements by adding the coverage-dependent lateral repulsion. These results suggest that the effect does exist, but explicit treatment is not essential.

Dr Kucernak reopened the discussion of the paper by Jia X Wang*, Francisco A Uribe, Thomas E Springer, Junliang Zhang and Radoslav R Adzic:
It would seem that in order to fully understand the intermediates in oxygen reduction we need to include in modelling studies, we need to develop methods for measuring the surface peroxide coverage. This is a challenge for experimentalists but is an important target.

Dr Wang answered:
The adsorption of peroxide probably is similar to that of water, *i.e.*, much weaker than OH and O, having low barriers for desorption and re-adsorption. Therefore, the H_2O_2 yield varies with the density of the Pt sites, and the H_2O_2 diffusion rate that depends on measuring conditions. Better understanding on how H_2O_2 yield varies with catalysts, pH values, catalyst loading, and other conditions can provide valuable insights for designing catalysts.

Professor Gewirth asked
I'm wondering whether the model of the ORR you utilized, featuring a dissociative mechanism, has any basis in experiment? In particular, the results of Behm, featured at this conference, suggest that peroxide is produced, at least some of the time. More

generally, do you think there is a single mechanism for the ORR on Pt? Or can both associative and dissociative mechanisms be active at the same time?

Dr Wang responded

We think both associative and dissociative mechanisms are active for the ORR on Pt. The former involves electron transfer in forming adsorbed intermediates, which we termed as the reductive adsorption (RA) to indicate that its reaction rate increases with decreasing potential. In contrast, dissociative adsorption (DA) does not involve electron transfer, and thus, can occur at open circuit. At room temperature, STM studies of oxygen adsorption find atomic O, not molecular O_2 on Pt. Electrochemically, one finds that exposure of Pt to oxygen leads to a larger reduction current peak in the initial potential sweeps from the open circuit potential. These observations provide the experimental basis for oxygen dissociative adsorption on Pt.

Professor Markovic said:

Currently there is no spectroelectrochemical method capable of detecting reaction intermediates formed in the course of the ORR. The surface coverage by intermediates is rather small.

Dr Wang replied:

Some progress has been made for determining of the nature and coverage of adsorbed ORR intermediates in the past ten year, after Adzic's review that pointed out its importance in studying the ORR kinetics [R. Adzic, in: *Electrocatalysis*, ed. J. Lipkowski and P. N. Ross, Wiley/VCH, New York, 1998, ch. 5]. However, we agree that the adsorption isotherm for the OH and O under the ORR conditions remain to be clearly determined.

The second sentence in the comment implies that the coverage of O_{ad} and OH_{ad} changes little with oxygen pressure, and that the majority of adsorbates are not actively involved in the ORR. These assumptions are questionable. Experimentally, we know that CO oxidation on Pt occurs at lower potentials after adding a small amount of oxygen (air bleeding) because oxygen interacts with Pt forming additional OH_{ad} or O_{ad}. Naturally, we expect the OH coverage to increase considerably under ORR conditions (1 atm oxygen pressure) compared to that measured by voltammetry in the absence of oxygen, especially at potentials below 0.8 V (see Fig. 2a in *J. Phys. Chem. C*, 2007, **111**, 12425). To see whether OH/O act purely as a site blocker, we introduced a trace amount of CO in the ORR measurements. At potentials between 0.6 and 0.9 V, OH/O coverage should be reduced through the surface reaction with CO. It would enhance the ORR activity if the majority of OH/O adsorbates act as idled site-blockers. We observed small oscillation in the ORR currents; but not an increase on average. This finding suggests that significant amounts OH_{ad} and O_{ad} are the ORR intermediates. Reducing their coverage opens up more available sites for reaction steps involving adsorption, but lowers the reactant coverage for further reduction steps in desorption process. Thus, there is no net enhancement to the overall ORR currents. We are hopeful that experimentally determined adsorption isotherms in future will resolve directly how oxygen pressure modulates the coverage of OH and O.

Conceptually, the comments also imply that the ORR intermediates, O_{ad} and OH_{ad}, adsorb more weakly than those formed from the oxidation of water do. The idea probably arose from analogy to the HOR on Pt. The H_{ad} formed from the HOR on Pt exhibits lower coverage and higher HOR activity than those from UPD (see Fig. 1 in our paper); the reason is that they are adsorbed, respectively, at atop and in hollow/bridge sites. However, we note that site preference is much stronger for OH_{ad} and O_{ad}, with the former at atop sites and the latter in hollow ones based on DFT calculations. Since adsorption on the other type of site is unlikely for either one of them, there is no apparent reason for some of them being more weakly bound or more active in the ORR than others. Therefore, we consider that OH_{ad} and O_{ad} formed from oxygen reduction are indistinguishable from those generated from water oxidation.

Professor Gewirth returned to the discussion of the paper by Christopher F. Blanford, Carina E. Foster, Rachel S. Heath and Fraser A. Armstrong*:

The mechanism of oxygen reduction by laccase remains the focus of intensive study by a number of groups. I note that some of these proposed mechanisms feature a stabilized Cu(I) hydroxyl intermediate, and wonder whether a connection between this and the ORR on surfaces can be made?

Dr Armstrong replied:

I agree. There is a very good similarity between the Cu–Cu spacing plus orientation in the trinuclear cluster of laccase and the structure and distances present on the (110) face of Pt. In the case of laccase, Solomon (Stanford) believes the third copper, and the orientation of its orbitals are crucial for breaking the O–O bond. The active hydroxide intermediate (termed the Native Intermediate, NI), which is the first detectable product of peroxide cleavage, contains a bridging hydroxido and a tri-bridging oxido in close proximity on the trinuclear site.

Professor Wittstock asked:

1. From where comes the evidence that laccase does not release H_2O_2?
2. Does the evidence obtained from dissolved enzymes also hold for enzymes on surfaces, where adsorption may induce conformational strain and could make them work less optimal than in solution?

Dr Armstrong answered:

For evidence that laccase does not release H_2O_2, we refer to the article by Palmer et al. (*J. Am. Chem. Soc.*, 2001, **123**, 6591–6599). This paper describes the formation and decay of the peroxy-intermediate which is stable until a second pair of electrons is able to be transferred.

The main evidence is from a laccase in which the T1Cu is substituted by Hg. Because it is unable to supply one of the second pair of electrons to the peroxide bound at the [peroxo-Cu(II),Cu(II),Cu(I)] the T1Hg derivative stabilises the peroxy intermediate; the derivative reacts about a million times more slowly than the native form. We cannot comment on whether there is any change in conformation upon binding laccase to a surface, but don't forget that many (most?) redox enzymes are associated naturally with membranes. Who says that the conformation of the enzyme in dilute aqueous solution, or in a crystal, is the 'correct' one ?

Professor Savinova remarked:

I would like to point out that considering the complex mechanism of the ORR, in my opinion, in order to substantially reduce the overpotential, we must turn towards hybrid multifunctional materials rather than to metals and alloys. Until now these are the natural catalysts—enzymes—which provide the highest turnover numbers for the ORR. This may be attributed to the fact that enzymes are complex entities, comprising multiple sites and functions: recognition, binding, charge transfer, proton transfer, *etc*.

Dr Armstrong responded:

The active site of an enzyme is often extensively buried, and all the participating functional groups are in the correct position, essentially where they need to be to facilitate different pre-equilibria in the electrocatalytic cycle as well as enforcing geometries close to that of the transition state. This is one reason why enzymes are so large. A surface provides only part of the organisation that is possible in an enzyme. We need '3D electrodes'—conducting porous materials with complex cavities.

Dr Wang returned to the discussion of the paper by Jan Rossmeisl*, Gustav S. Karlberg, Thomas Jaramillo and Jens K. Nørskov:

The onset potentials for OH/O adsorption measured by CV in $HClO_4$ solutions, E_{onset}, often are used to indicate the strength of the OH/O adsorption. However, experimental trends apparently do not always agree with the DFT-calculated oxygen-adsorption energy, E_O. For example, the E_{onset} increases in the order Ag(111) < Pt(111) < Pt$_3$Ni (111) < Au(111), [Markovic *et al.*, *Faraday Discuss.*, 2008, **140**, DOI: 10.1039/b803714k, Rossmeisl *et al.*, *Faraday Discuss.*, 2008, **140**, DOI: 10.1039/b802129e], while in the DFT-calculated E_O the order is: Pt(111) < Pt$_3$Ni(111) < Ag(111) < Au(111). Can you explain why the position of Ag differs between the two? Markovic in *Faraday Discuss.*, 2008, **140**, DOI: 10.1039/b803714k correlated the PZC with the E_{onset}, and the ORR activity for Cu, Ag, and Au. If E_O is not the only factor determining the E_{onset}, can you comment on other factors that can determine the stability of a metal or alloy in acidic solutions? In particular, what makes Pt$_3$Ni much more stable in acid solutions than Ag while their E_O's are similar?

Dr Rossmeisl responded:

The first oxidation peak for Ag cannot be due to OH or O binding. It is dependent on the electrolyte. In surface science experiments the order of E_O is the same. We discuss this in ref. 1. Concerning stability, the dissolution potentials for Ag and Pt$_3$Ni are very different and not directly related to the oxygen binding.

(1) H.A. Hansen, J. Rossmeisl and J.K. Nørskov. *Phys. Chem. Chem. Phys.*, 2008, DOI: 10.1039/b803956a.

Professor Markovic reopened the discussion of the paper by Jia X. Wang*, Francisco A. Uribe, Thomas E. Springer, Junliang Zhang and Radoslav R. Adzic:

We used IB group metal to discuss the importance of the $(1 - \theta)$ and ΔG terms. For Au we found that ΔG is determining the reaction mechanism in acid solution. However activities of Ag and Cu are determined usually through $(1 - \theta)$ term

Dr Wang answered:

These statements seemingly suggest that Ag and Au are on different sides of the volcano plot for the ORR in acid. This would be inconsistent with the volcano plot produced from DFT calculations that shows Pt on one side and Ag with Au on the other. The ORR activity on Ag probably is not limited by strong O and OH adsorption, *i.e.*, the $(1 - \theta)$ term, but the low Ag dissolution-potential in acid

Professor Schmickler returned to the discussion of the paper by Jan Rossmeisl*, Gustav S Karlberg, Thomas Jaramillo and Jens K, Nørskov:

On your volcano plot there are a number of metals that are covered by oxide or by hydroxyl films such as iron or nickel. Obviously, such surfaces will behave quite differently from pure metals.

Why have you included them in this plot?

Dr Rossmeisl answered:

This was a comment to a previous paper,[1] and not directly relevant for the analysis of Pt and Pt$_3$Ni both close to the top of the volcano. In that previous paper we investigated trends and doing so it is sometimes a good idea to include both very noble and very reactive metals. Any trend volcano like this is not to be quantitatively trusted, especially not for the elements far away from the top. Prof. Schmickler is therefore quite right that every surface too reactive will be oxidized. To do this correctly one have to find the stable surface at the relevant conditions. We have tried to do this for Ni, Pt and Ag, see ref. 2.

(1) J. K. Nørskov, J. Rossmeisl, A. Logadottir, L. Lindqvist, J. Kitchin, T. Bligaard and H. Jnsson, *J. Phys. Chem. B*, 2004, 108, 17886
(2) H. A. Hansen, J. Rossmeisl and J. K. Nørskov. *Phys. Chem. Chem. Phys.*, 2008, DOI: 10.1039/b803956a

Professor Behm said:
If I understand it correctly the activity for the O_2 reduction is determined by the first step, by addition of H or O_2(ad) to OOH(ad), while the selectivity is at least partly determined by the competition between desorption acid decomposition of adsorbed H_2O_2. Assuming for simplicity that desorption is potential independent, while decomposition depends on the potential, measuring the potential dependant decomposition of H_2O_2, in the electrolytic should describe the change in selectivity with potential. Would you agree with this description?

Dr Rossmeisl replied:
I think an experiment like that would be very interesting. I would like to look at this if you are planning to do this experiment in the future. The desorption might, *via* the surface coverage, depend on the potential.

Professor Wittstock addressed Professor Behm:
H_2O_2 desorption is likely also potential dependant because the adsorption of ions must be considered as a competitive process. The energy balance must include this. Ion adsorption takes place on the free place left by H_2O_2 adsorbate, and this would be potential dependent.

Professor Behm responded:
I agree in the final result that (potential dependent) anion adsorption may play a role. But I think it important for the re-adsorption of H_2O_2, while for desorption of adsorbed H_2O_2 species this should be of little relevance.

Professor Strasser remarked:
What is the activity of ORR at top of volcano curve? How much room is there on top?

Dr Rossmeisl answered:
For a metal alloy that follows the trends for scaling between the different intermediates[1] there is not much room at the top, Pt_3Ni is very close to the optimal. However, other kinds of materials, *e.g.* oxides show a different scaling relation and thereby a different volcano. Also enzymes have a performance that is beyond the volcano for metals.

1. F. Abild-Pedersen, J. Greeley, F. Studt, J. Rossmeisl, T.R. Munter, P.G. Moses, E. Skulason, T. Bligaard and J.K. Nørskov, *Phys. Rev. Lett.*, 2007, **99**, 016105
2. J. Rossmeisl, Z.-W. Qu, H. Zhu, G.-J. Kroes and J. K. Nørskov, *J. Electroanal. Chem.*, 2007, **607**, 83–89

Professor Gross opened the discussion of the paper by Matthew Neurock*, Michael Janik and Andrzej Wieckowski:
We recently performed a detailed computational study of the methanol oxidation over Cu(110)[1] which is one of the relevant substrates for the understanding of this reaction in heterogeneous catalysis. On Cu, the methanol oxidation only proceeds *via* the initial activation of the O–H bond, which corresponds to the secondary path over Pt. In our calculations, we found that along this path, because of the volatile nature of formaldehyde, methanol becomes only partially oxidised unless additional oxygen is offered, in agreement with the experiment.[2] Have you looked at the details of the further oxidation of methanol over Pt along the secondary path after formaldehyde is formed? Does the oxidation of formaldehyde require specific oxidants?

1 Sung Sakong and Axel Gross, *J. Phys. Chem. A*, 2007, **111**, 8814.
2 I. E. Wachs and R. J. Madix, *J. Catal.* 1978, **53**, 208.

Professor Neurock replied:
We have recently published the results on the dehydrogenation of formaldehyde to CO [ref. 8 in the manuscript] and the subsequent oxidation of CO [ref. 39 in the manuscript]. In addition, we have also examined the oxidation of various CH_xO intermediates over Pt but have not yet published the results. As you correctly point out, the oxidation is limited due to the fact that formaldehyde is bound very weakly to the surface and that the supply of a strong oxidant is limited. Formaldehyde likely desorbs before undergoing further oxidation steps. This is seen experimentally under electrocatalytic conditions as appreciable amounts of formaldehyde are formed. We are just beginning to look at hydroxyl coverage effects which are likely important in aiding the activation of the O–H bond and carrying out any further oxidation of formaldehyde.

Dr Jinnouchi asked:
Did you check the dependence of calculated work function on the number of water layers? If you did check, could you tell me how big the dependence is?

Professor Neurock responded:
We have analyzed the changes in the work function with respect to bilayer thickness over the Cu(111) surface. This is the subject of a recent manuscript that we have submitted [C. D. Taylor, M. Neurock and R. G. Kelly, *Phys. Rev. B*, 2008]. The change in work function in moving from a bilayer to a double bilayer of water is only 0.1 eV. The addition of subsequent water layers lead to changes that are less than 0.05 eV. The more critical changes are those that result from changes in the orientation of the water molecules that can occur at the interface. The two extremes are when the water is oriented with its hydrogens tilted upwards and those where the hydrogens are oriented toward the surface. The difference in the work function for these two systems is more than 1 eV. Significant changes in the work function and the potential are also noted experimentally between liquid water and ice on well defined metal surfaces due to the differences in the orientations of the dipoles of the water molecules with respect to the surface. *Ab initio* molecular dynamics simulations of water on Cu also reveal systematic fluctuations in the potential due to changes in the orientation of the water layer with respect to the surface.

Dr Herrero commented:
Experimentally, it is found that the dehydration step of the formic acid oxidation mechanism takes place only on the defects of the Pt(111) surface, that is, the dehydration would be completely inhibited on a perfect Pt(111) electrode at any potential.[1] However, your computational results show that this step is the most favorable at low potentials. How can you explain this significant difference?

1. M. D. Maciá, E. Herrero and J. M. Feliu, *Electrochim. Acta,* 2002, **47**, 3653.

Professor Neurock replied:
This is a very important point. We did not specifically calculate the barrier for the activation of the C–O bond in formic acid in the presence of solution and an applied potential. However, based on previous gas phase calculations that we have carried out, we estimate this barrier to be >1 eV and too high to occur on the ideal (111) surface. We suggest that this step occurs instead at defect sites present in the surface. The CO that forms builds up and ultimately inhibits the surface at lower potentials. We have included the reference now in the revised manuscript.

Professor Korzeniewski asked:

What is the role of the methoxy as an intermediate in formaldehyde formation? For the low index Pt surface planes, do your calculations show a difference in the energetics of methoxy formation?

Professor Neurock answered:

Our calculations show that methoxy is the precursor to the formation of formaldehyde as you suggested in your early work [ref. 46–48 in our paper]. The primary dehydrogenation path for methanol is to CO and occurs *via* the initial C–H bond activation to the form hydroxyl methyl intermediate and then on to hydroxyl methylene. The activation of the O–H bond requires step sites and only appears to occur at higher potentials. The methoxy surface intermediate can then readily react to form formaldehyde. While we have not explicitly explored the reaction of methanol to methoxy over Pt(100) or Pt(110) surfaces, we have looked at the reaction of methanol to methoxy over the Pt(211) surface and show that there is significant structure sensitivity as the reaction becomes 70 kJ mol^{-1} more favorable on the (211) surface [ref. 8]. I would also expect some enhancement on the Pt(100) and Pt(110) surfaces.

Professor Koper remarked:

From Fig. 9 plotted in your paper, it cannot be concluded that the pathway through formate (HCOO) is less likely than through COOH. In such a simple energy diagram, the preferred pathway is that with lowest absolute transition state energy, and in Figure 9 this is still the pathway through formate. In chemical terms: although the activation energy for formate oxidation is high, its concentration is high enough to lead to an appreciable overall rate. According to Fig. 9, the barrier for the formation of COOH is prohibitive. If the formate pathway is to be excluded on the basis of Fig. 9, other effects not incorporated in Fig. 9, such as the existence of a saturation coverage, must be taken on board.

Professor Neurock replied:

The energetics shown in Fig. 9, as well as the experimental results reported by Behm [ref. 9, 11 and 49] and Owsawa [ref. 21, 25 and 26] clearly indicate that the surface is covered with adsorbed formate. As such, the apparent activation barrier should be taken from the adsorbed state and not from the gas phase. This is what we have reported in the manuscript. From these results we can safely conclude that the COOH pathway is much more favored than the path through formate. This falls out of a simple Langmuir–Hinshelwood analysis where the two paths can be written as:

HCOOH* = HCOO* + H$^+$ + e$^-$ (K1) HCOO* → CO_2(g) + H$^+$ + e$^-$ (k_1) and HCOOH* = COOH* + H$^+$ + e$^-$ (K2) COOH* → CO_2(g) + H$_+$ + e$^-$ (k_2)

The rate which proceeds through the formate intermediate (r_1) can be written as: $r_1 = k_1 K_1'[\text{HCOOH}]/[1 + (K_1' + K_2')[\text{HCOOH}]]$

At high formate coverages, the second term in the denominator dominates and the rate of reaction is then only dependent on the rate constant from the adsorbed state, $r_1 \sim k_1$. The apparent activation energy is then $E_{1\text{app}} = E1^* = 1.0$ eV (at 1 V), which is the barrier we have reported.

The rate for the second reaction can be written as:
$r_2 = k_2 K_2'[\text{HCOOH}]/[1 + (K_1' + K_2')[\text{HCOOH}]]$

At high formate coverages, this reduces to the following. $r_2 \sim k_2 K_2/K_1$ The apparent barrier to proceed through the hydroxycarbonyl is then:

$E_{2\text{app}} = E2^* + D_{\text{Hads2}} - D_{\text{Hads1}}$

Since $D_{H2} < D_{H1}$ the value of $E_{2\text{app}} < E_2^*$ The value for $E_{1\text{app}} \gg E_{2\text{app}}$ and thus the route through the hydroxyl carbonyl is favored.

Professor Sun asked:

The experimental data concerning the mechanism of HCOOH oxidation in your paper were cited from Professor Osawa and Professor Behm's work. Both used

SEIRA which involved film electrodes that present a lot of defects in comparison with Pt(111) sample crystal electrode.

We have published a paper about the HCOOH oxidation mechanism an Pt(111) and Pt (100) electrodes, which may be more suitable for you taking into account your theoretical consideration. The paper reference is S. G. Sun, J. Clavilzer and A Bewick, *J. Electroanal. Chem.*, 1988, **240**, 147–159

Professor Neurock replied:
Thank you. This is very helpful. We will cite this in the revised manuscript.

Professor Wieckowski said:
One needs to emphasise electronic effects of Ru or Pt for the direct path of formic acid oxidation to CO_2.

Professor Neurock replied:
This indeed is an important point. We have focused herein predominantly on the ensemble and geometric effects. There are clearly important electronic effects as well that take place upon alloying. We have examined both the geometric and electronic effects for CO oxidation and CO desorption over PtRu alloys previously in detail [ref. 40 in the paper and M. Janik and M. Neurock, *J. Am. Chem. Soc.*, 2008, submitted]. The addition of Ru in the surface or subsurface decreases the d-band center which ultimately enhances the desorption of CO and the reaction of CO* with OH* on the surface. Neighboring Ru, however, increases the barriers for the activation of the C–H and O–H bonds over Pt. Overall the electronic effect is likely to aid both methanol and formic acid.

Professor Wittstock opened the discussion of the paper by Flavio Colmati, Germano Tremiliosi-Filho, Ernesto R Gonzalez, Antonio Berna, Enrique Herrero*:
Could you explain the experiment in Fig. 1, 3 and 4 in the paper. If you pretreat the electrode at 0.8 V, do you oxidise ethanol and form a diffusion layer? The diffusion layer would not relax at the short 0.01 s step at 0.1 V.

Dr Herrero answered:
The stationary currents measured at 0.8 V are much smaller than the diffusion limiting current for the ethanol concentration used, that is, the currents measured for the oxidation of 0.2 M ethanol are completely controlled by the kinetics and therefore a negligible diffusion layer is formed at 0.8 V. For instance, it has been shown that voltammetric profile for the oxidation of 0.2 M formic acid solution, whose currents are very similar to those reported in the manuscript, is independent of the rotation rate of the electrode and equal to that obtained under quiescent conditions.[1]

1. M. D. Maciá, E. Herrero and J.M. Feliu, *J. Electroanal. Chem.*, 2003, **554**, 25.

Professor Behm remarked:
The data very nicely explains the short term behaviour of the clean Pt surface. For longer reaction times on a more or less adsorbate covered surface, slow but irreversible processes such as CO(ad) formation may dominate the reaction kinetics. Could you make predictions for the dominant reaction processes on a longer time scale?

Dr Herrero answered:
For the long term scale, the reactivity is probably dominated by the ability of the surface to oxidise the CO formed when the C–C bond is cleaved. The final product of the oxidation will be a mixture of acetaldehyde, acetic acid and CO_2, whose ratios will depend on the experimental conditions, *i.e.*, surface structure, ethanol concentration, supporting electrolyte and electrolyte flow. The only exception is the

Pt(111) electrode since the formation of CO can be considered negligible. For this electrode, the long term currents will be not very different from those measured in the voltammetric profile.

Professor Wieckowski commented:
Acetic acid desorbs from Pt at low potentials, of fuel cell relevance. Like (bi) sulfate, the acid is adsorbed reversibly.

Dr Herrero responded:
Acetic acid adsorption on Pt electrodes is stronger than sulphuric acid adsorption for the same concentration. As shown in figure 6, the band for adsorbed acetate only disappears below 0.3 V vs. RHE. Although it would be desirable to have an anode working at potentials more negative than 0.3 V, for the existing electrocatalysts, the anode potential is higher and, therefore, adsorption of acetate may hinder further reaction.

Professor Korzeniewski remarked:
In the Pt(111) results, how can acetaldehyde be ruled out as a potential product based on infrared measurements? The two species have overlapping bands. Also, formation of acetic acid is more complicated than acetaldehyde, because water activation is required for acetic acid production from ethanol.

Dr Herrero answered:
This is true in H_2O solutions. D_2O presents some advantages for the detection of acetic acid and acetaldehyde since the frequency for the bending mode of water shifts from 1640 to 1200 cm^{-1}, eliminating the interference with the carbonyl group. In D_2O, the C=O stretching band for acetic acid shows two peaks, at 1713 and 1685 cm^{-1}, whereas only one peak is observed for acetaldehyde at 1713 cm^{-1}. As explained in the paper, the analysis of the integrated intensities of the peaks at 1713 and 1685 cm^{-1} can be used to determine the relative amounts of acetic acid and acetaldehyde produced. This approach has been validated analyzing the spectra of several mixtures of acetic acid and acetaldehyde in different concentrations.

Professor Ren opened the discussion of the paper by Stanley C. S. Lai* and Marc T M Koper:
You assigned two bands in the low frequency region (400–420, 490–500 cm^{-1}) related to Pt–CH_x and Pt–CO species. Actually, Weaver's and our group have done a lot of work on CO adsorption on Pt. Indeed, we detected both bands in systems with only pure CO, and we assigned the bands to bridge-bonded CO and linearly-bonded CO. I would recommend you check your system carefully with pure CO gas.

Mr Lai responded:
We have extensively measured signals of pure CO in our SERS setup. Under the conditions used in the ethanol experiments (0.1 M $HClO_4$, Pt on Au working electrode), no bridge bonded CO has been observed both from the decomposition of ethanol nor from the direct adsorption of CO (at any coverage), as witnessed by lack of bands at 1800–2000 cm^{-1} and ~400–450 cm^{-1}. In addition, experiments employing deuterated ethanol shifts the band at 425 cm^{-1} while the bands at 500 cm^{-1} remains at the same place,[1] strongly suggesting that the adsorbate is a protonated fragment rather than (bridge-bonded) CO. Finally, the intensity of the feature at 425 cm^{-1} shows a strong correlation with features at 2880 cm^{-1},[1] which is in the typical wavenumber region of a C–H bond.

[1] S. C. S. Lai, S. E. F. Kleyn, V. Rosca and M. T. M. Koper, 2008, in preparation.

Professor Sun said:

You reported that when introducing steps on Pt single crystal plane, the activity for ethanol oxidation is increasing, but is decreasing for acetaldehyde oxidation. What is the origin of the completely different effects, and how do you explain the surface processes of the two kinds of molecules on different Pt single crystal planes?

Mr Lai responded:

As described in the paper, we assume that in the continuous oxidation experiments, we are mainly probing partial oxidation pathways (*i.e.* oxidation to acetaldehyde in the case of ethanol and oxidation to acetic acid in the case of acetaldehyde). We believe that both bulk processes are enhanced by low coordination sites, such as steps. The voltammetric response of these systems, however, are not only the response of the partial oxidation, but is also influenced by other processes, such as the formation of strongly adsorbed fragments through carbon–carbon bond breaking. From the stripping experiments in our paper, it is clear that the amount of adsorbates resulting from the decomposition of acetaldehyde is larger than the amount of adsorbates from ethanol decomposition, suggesting that surface poisoning is relatively slow for ethanol compared to acetaldehyde. Therefore, the steps become blocked more rapidly (or more completely) in the case of acetaldehyde, causing a net negative correlation between the activity and the step density.

Professor Wieckowski said:

CH_x species present on platinum from ethanol decomposition may be speculative at this moment.

Mr Lai responded:

Although the results in the paper do not show any conclusive proof for the existence of CH_x species, there is a clear similarity between our stripping results shown in Fig. 6 and the results of Hahn and Melendres for the oxidation of methane on platinum.[1] In addition, we have a very strong indication for the existence of CH_x from our recent SERS experiments.[2]

1. F. Hahn and C. A. Melendres, *Electrochim. Acta*, 2001, **46**, 3525-3534.
2. S.C.S. Lai, S.E.F. Kleyn, V. Rosca and M.T.M. Koper, 2008, in preparation.

Professor Ren commented:

When you are doing deuterated experiments, you have to be very careful about the possible change in surface coverage, as both the high frequency and low frequency bands depends very much on the surface coverage and environment.

Mr Lai responded:

While it is certainly true that the vibrational frequencies of certain bands are strongly dependent on a number of factors, such as surface coverage, deuteration shows only a significant shift in a pair of bands (425 cm^{-1} to 400 cm^{-1} and 2880 cm^{-1} to 2200 cm^{-1}), while the bands we attribute to adsorbed CO do not shift significantly.[1] This suggests that the surface coverage and electrode environment in the experiments with deuterated compounds are very similar as those obtained with non-deuterated compounds.

1. S. C. S. Lai, S. E. F. Kleyn, V. Rosca and M. T. M. Koper, 2008, in preparation.

Professor Koper remarked:

In my mind there is no serious doubt about the involvement of a CH_x adsorbed species in ethanol oxidation on platinum. To summarize some of the evidence:

(1) Our paper, as well as many earlier papers cited in our paper, show the existence of a reducible surface-bonded species on platinum. DEMS experiments cited in our

paper detect methane or ethane during such a reduction experiment. (2) Voltammetry also clearly shows two different kinds of surface-bonded species after ethanol and acetaldehyde dissociation. One of these species may be oxidized irreversibly to carbon monoxide (see Fig. 6 in the paper).

(2) This is very similar to an observation made by Hahn and Melendres in the oxidation of methane adsorbates (ref. 21 in our paper). Together with point 1, this points to the coexistence a CHx and a CO surface-bonded species.

(3) The SERS results with deuterated ethanol and acetaldehyde, as discussed by Mr. Lai, are most consistently explained by, and fully in agreement with, a CH_x surface-bonded species. According to DFT calculations, the most stable CH_x species on Pt(111) is CH, *i.e.* with 1 hydrogen.[1]

1. J. Kua and W. A. Goddard, *J. Phys. Chem. B*, 1998, **102**, 9492

Professor Wieckowski responded:

Data in Fig. 6 in the Lai and Koper paper are quite convincing about formation of a yet another (other than CO) reducible surface product. This could indeed be CH_x. In terms of the spectroscopic assignment, the paper by F. Hahn and C. A. Melendres, *Electrochim. Acta*, 2001, **46**, 3525–3534 is not that clear as the authors do not provide evidence for CH_x formation. Instead, they believe they identified CO and –CHO (or –COOH) species, similar to those involved in methanol oxidation. (The discussion on page 3529 of the Hahn and Melendres paper could be read against any of the C–H species formation, but the issue is open for further study.)

Professor Wittstock addressed Mr Lai and Dr Herrero :

Apart from some discussion on some intermediates, given that the data for single crystal faces are correct, can you predict the behaviour of a polycrystalline electrode?

Mr Lai replied:

It is clear that the activity of polycrystalline electrodes does not simply equal the sum of the activities of (low index) single crystal planes, and it remains a question to what degree the low coordination sites of polycrystalline electrodes can be modeled by controlling the step density of single crystal electrodes. More specifically, it is often found that polycrystalline electrodes are more active than even single crystal electrodes with a high 'defect' density. Since we have found that low coordination sites are the active sites for C–C bond breaking, as well as for the bulk oxidation of ethanol, and, to a smaller degree, acetaldehyde, we would expect the (onset) potentials for oxidation to be lower for polycrystalline electrodes. In addition, it is expected that the amount of CO_2 as the oxidation product will be higher than for the single-crystal electrodes, due to the enhanced bond breaking, at least at higher potentials were adsorbed CO may be oxidized.

Furthermore, it should be noted that, although controlling the amount of step sites might not be the best model for polycrystalline electrodes, it does offer the introduction of low coordinated sites in a controlled fashion to make the study of surface structural effects possible.

Dr Herrero replied:

The response of the Pt(111), Pt(110) and Pt(100) electrodes cannot be directly extrapolated to that of polycrystalline electrodes since they represent ideal surfaces with only one type of surface site and ordered domains of infinite size. On a general polycrystalline surface, ordered domains with variably size and low coordination sites can be found. In order to fully predict its behaviour, stepped surfaces have to be used. Stepped surfaces have terraces with different width and steps. The behaviour of the ordered domains in the polycrystalline sample can be assimilated to that observed for the terraces whereas the low coordinated surface atoms will have a response close to that of the steps. Our preliminary studies with stepped surfaces indicate that the steps play a key role in the cleavage of the C–C bond. Probably, for a typical polycrystalline surface, the behaviour will not be very far from that obtained for the Pt(110) electrode.

Professor Behm commented:
The last contributions showed very nice data on the oxidation of small organic molecules. For comparison with realistic situations it would be very important to expand this work, both in theory and experiment, to higher reaction temperature and to include electrolyte transport, to warrant defined transport conditions. This is a plea also for the use/development of *in-situ* spectroscopic methods that can be employed for single crystal studies.

Professor Wieckowski answered:
About: "This is a plea also for the use/development of in-situ spectroscopic methods that can be employed for single crystal studies," I may quote the paper: A. Lagutchev, G.Q. Lu, T. Takeshita, D.D. Dlott and A. Wieckowski, Vibrational Sum Frequency Generation Studies of the $(2 \times 2) \rightarrow (\sqrt{19} \times \sqrt{19})$ Phase Transition of CO on Pt(111) Electrodes, *J. Chem. Phys.*, 2006, **125**, 154705/1–10, where we responded to this call. The BB-SFG method we use allows for infrared and simultaneous voltammetry measurements of the CO type adsorbates on platinum single crystal electrodes. Other adsorbates are or can be examined.

Professor Sun said:
Using single crystal micro electrode and design flow IR cell, the *in situ* IR study of small organic molecules on single crystal surface can be done. In this case we will use IR microscope together with step-scan facility to do fast IR measurements.

Dr Fernandes Gomes returned to the discussion of the paper by Stanley C S.Lai* and Marc T M Koper:
During the investigation of the electro-oxidation of ethanol on platinum by Raman spectroscopy, you observed some bands as function of the applied potential. One of them, you interpreted as to be related to CH vibration of the Pt–CH. Once Pt–CH$_3$ is possibly a precursor for Pt–CH, why don't you see CH$_3$ vibration of Pt–CH$_3$? In addition, previous works based on the conventional infrared spectroscopy suggest that CH$_x$ and CO are not the only adsorbed intermediates of the ethanol electro-oxidation on Pt. Did you have any evidence of adsorbed species other than CH and CO?

Mr Lai answered:
Additional adsorbed intermediates for ethanol electro-oxidation observed in infrared spectroscopy usually include acetaldehyde and acetate, species formed from the oxidation of ethanol without the breaking of the C–C bond. Previous FTIR measurements, however, are mainly performed with the presence of ethanol in solution. Our SERS experiments, on the other hand, are performed after ethanol adsorption followed by an exchange of the electrolyte, so no bulk ethanol remains. An early DEMS study by Willsau and Heitbaum[1] found only CO$_2$ as the ethanol oxidation product after electrolyte exchange, suggesting that only C$_1$-species are chemisorbed and remain on the electrode surface after the exchange. Similarly, no features for molecularly adsorbed species were observed in our SERS experiments.

In the SERS experiments,[2] there are two features which we have assigned for the CH$_x$ species. These features are located at ~2880 cm^{-1} and ~425 cm^{-1}. The feature at 2880 cm^{-1} is typical for the C–H stretch vibration in organic molecules and lies close to the value observed for the C–H stretch vibration of ethanol on Pt(111) under UHV conditions.[3] The feature at 425 cm^{-1} is typical for a stretch vibration between a metal atom and a carbon atom. The assignment of these features to CH$_x$ species was based on the correlation between the intensities of these features. In addition, both bands shifted upon deuteration of the ethanol. Based on (the magnitude of) these shifts, the assignment was made for CH rather than CH_3, which might be due to the fact that CH_3 is not stable enough to have a sufficient lifetime to be observed by SERS under electrochemical conditions. Alternatively, DFT calculations have shown that the most favoured pathways for C–C breaking lies in the dehydrogenation of ethanol to a CHCO species which decomposes to give adsorbed CH and adsorbed CO.[4]

1. J. Willsau and J. Heitbaum, *J. Electroanal. Chem.*, 1985, **194**, 27–35.
2. S. C. S. Lai, S. E. F. Kleyn, V. Rosca and M. T. M. Koper, 2008, in preparation.
3. P. Gao, C. H. Lin, C. Shannon, G. N. Salaita, J. H. White, S. A. Chaffins and A. T. Hubbard, *Langmuir*, 1991, **7**, 1515–1524.
4. R. Alcala, M. Mavrikakis and J. A. Dumesic, *J. Catal.*, 2003, **218**, 178–190.

Professor Sun returned to the discussion of the paper by Flavio Colmati, Germano Tremiliosi-Filho, Ernesto R Gonzalez, Antonio Berna, Enrique Herrero*:

The catalysis of small organic fuels depends strongly on surface structure. Using single crystals provides model catalytic surface and gives us knowledge of surface at atomic scale. However, we still know little about the surface processes and the detail of mechanism of molecule interacting with single crystal planes. For example, in the case of well defined Pt(111), no CO is produced, but on defected Pt(111) it is. What are your opinions to investigate the surface processes? How far can we go?

Dr Herrero responded:

If we want to fully understand the process in detail, we have to use all the available techniques and well defined and complex surfaces. Studies with single crystal surfaces are essential for surface sensitive processes since they have a well defined surface structure which helps disentangling the role of the different factors affecting the reactivity. Probably, the most difficult problem is the identification of the reaction intermediates (other than CO), since their concentration is very low and their spectroscopic signals overlap to a great extent with those coming from reactants or final products. Some techniques, such as ATR-FTIR or Raman spectroscopies can help to identify those species, but single crystal electrodes cannot be used, right now. For that reason, we have to design strategies to overcome these difficulties and be able to identify the whole mechanism and the role of the different sites.

Professor Behm commented:

I would like to speculate on what would be a good catalyst for oxidation of C2 and larger molecules. If a catalyst is good C–C bond breaking, it will result in CO(ad) formation, and we are limited by the activity for CO(ad) oxidation. Otherwise, the reaction will be dominated by formation of incomplete oxidation products. So, one could speculate, the best low potential activity would be observed on materials with zero activity for C–C bond breaking, which would be in contrast to all discussion. The question is, therefore, what do we expect of these catalysts and could we predict a similar kind of volcano plot as we saw today for O_2 reduction?

Dr Herrero responded:

In fact, the voltamograms show that the Pt(111) electrode, on which the cleavage of the C–C bond does not take place, has the highest currents of the three basal planes below 0.5 V. However, in this case, the final product is acetic acid, and only 4 electrons are exchanged. The maximum theoretical energy density for the oxidation of ethanol is obtained when CO_2 is the final product and 12 electrons are exchanged. A good electrocatalyst for that process has to be able to effectively break the C–C bond and also to oxidise all the residues completely to CO_2 at low potentials. In this latter step, adsorbed CO molecules will be formed as intermediate products. Since it is going to be very difficult to find an electrocatalyst for which both properties are optimized, a bi-functional or tri-functional electrocatalyst will be required. For instance, we have shown that the oxidation of ethanol is enhanced when small amounts of ruthenium are deposited on the step of (111) vicinal surface.[1] The step is effective for the C–C bond cleavage and the ruthenium adatom helps to oxidise the residues to CO_2. Probably, adding a third adatom that enhanced the rate for the C–C bond cleavage will improve the performance of the electrode material.

Regarding the prediction of volcano curves for the oxidation of ethanol, I do not expect a typical volcano curve in this case. The volcano curves have been obtained for reactions in which only one type of adsorbed species is involved, *i.e.*, hydrogen evolution reaction or oxygen reduction reaction. In this latter case, probably several adsorbed species are involved but all of them are different oxygen groups whose interactions with the surfaces follow the same trends. For ethanol, several and different adsorbed species are involved and the final oxidation rate is not determined by only one parameter. As aforementioned, the performance of the electrocatalyst depends, at least, on the ability to break the C–C bond and on the oxidation of the residues to CO_2. In this case, the theoretical curve will be much more complex than the typical volcano curve. However, depending on the balance of the different contributions, curves resembling volcano plots may be experimentally observed.

1. V. Del Colle, A. Berná, G. Tremiliosi-Filho, E. Herrero and J. M. Feliu, *Phys. Chem. Chem. Phys.*, 2008, **10**, 3766

Dr Thompsett added:

From a practical viewpoint (*i.e.* oxidising EtOH in an MEA environment), it is much more preferable to encourage the breaking of the C–C bond to form CO as the formation of CH_3CHO and CH_2CO_2H is undesirable. Partly because only 4 electrons are yielded from EtOH(from a maximum of 12e) and more importantly desorbed CH_3CHO and CH_2CO_2H can cross the membrane, adsorbing on the cathode catalyst greatly inhibiting the catalyst for oxygen reduction activity.

All dressed up, but where to go? Concluding remarks for FD 140

David J. Schiffrin

Received 19th September 2008
First published as an Advance Article on the web 2nd October 2008
DOI: 10.1039/b816481a

The aim of this meeting was to bring together experimentalists and theoreticians to discuss the interplay between recent developments in theoretical and computational tools with experimental results in electrocatalysis. Intense and rewarding discussions in aspects of very topical electrochemical research arose during the three days of this Faraday Discussion. A closer collaboration between experimentalists and theoreticians is a prerequisite for successful development of the field and this meeting definitely aided mutual understanding of these areas of research.

Why was this meeting important and timely? It is an interesting observation that, since we are immersed in rapid fundamental changes in the way scientific research is carried out, we sometimes cease to notice how much this has changed. Clear examples of this are the better availability of advanced instrumental and computational techniques and the instantaneous access to scientific information through the Internet. Thus, we are in the middle of revolutionary changes in scientific research resulting from three new fundamental advances. The first refers to the possibility of carrying out quantum chemical calculations at a high level of theory and using these for the modelling of complex electrochemical reactions to give an insight into interfacial structure and reaction mechanisms. In some cases, this is providing quantitative information of preferred reaction channels that it would have not been possible to obtain a decade ago. The second refers to the new instrumental techniques that have become available. The third is developments in materials science.

Quantum chemical calculations, theory and modelling

Some of the calculation tools that have been developed are very powerful for the prediction of reactivity but, as the work of Schmickler and Santos shows, advanced computational methods by themselves are not sufficient to analyse electrocatalytic reactions and a better fundamental understanding of electron transfer theory is essential. An example of this, although not discussed in this meeting, is the development of the theory of simultaneous electron and proton transfer reaction (the "diagonal" reaction channel in the scheme of squares reaction model). The complexity of the calculation methods employed can be a deterrent, however, for their widespread use and a very important task that became clear during this Discussion is to ensure the training of the next generation of scientists in *both* instrumental and computational techniques.

Another aspect that should be considered is the need to have a better communication between people doing calculations and experimentation or, in very plain terms, you need two to tango. The questions that experimentalists need to solve must be well defined and very clearly stated, with an understanding of the problems that can be solved with different types of calculations, and importantly, to be aware of the level of theory that is required to obtain meaningful results. Also, in order to be able to compare experimental results with quantum chemical calculations, it would be advantageous, when possible, to have values of the relevant

Chemistry Department, University of Liverpool, UK L69 7ZD

thermodynamic functions *i.e.*, Gibbs energies at 298 K. Electronic energy calculations at 0 K alone are very useful to map the course of reactions but may give an incomplete picture; zero point energy contributions corrected to room temperature and entropic contributions should ideally be taken into account. These can be very large as is the case for reactions involving a change in the total number of species or involving large bond reorganisations. In addition, solvation contributions are important, since these can determine the course of reaction. Improvements from simple polarized continuum models[1] that take into account the discrete nature of the solvent molecules would be expected to become common. Calculations following this wish list may prove to be computationally very expensive at present but the rapid developments in computers will make this type of calculations more feasible.

DFT calculations were extensively discussed in relation to interfacial structure and reaction mechanisms. Interesting examples of the importance of quantum chemical calculations used for the prediction of interfacial properties and reactivity were presented during the Discussion. For example, the work by Gross *et al.* demonstrated that the properties of the Pt(111) surface are hardly affected by a water layer attached to it. A general problem that was highlighted during the discussions was how to include the electrode potential in these calculations. Another example of the usefulness of these calculations was the comprehensive paper by Neurock, Janik and Wieckowski that demonstrated very clearly the relevance of quantum chemical calculations to the analysis of mechanisms for the oxidation of methanol and formic acid. It was possible to establish which reaction would follow a "direct" or an "indirect" pathway and in addition, interesting predictions on the likely reactivity of bifunctional catalysts were derived.

New instrumentation techniques

The second aspect that is a fundamental driver for new developments in electrocatalysis is the availability of new instrumental techniques that are providing an unprecedented access to structural information. Examples of this were clearly shown during this Discussion, for example, in the use of X-ray scattering techniques to analyse the surface properties of the Pt_3Ni alloys or the brave attempts by Dan Scherson to measure transient SHG signals on microfacetted Pt(111) surfaces. The latter experiments would have been regarded as extremely difficult some 10 years ago. Although this is still at the edge of what is possible, these results are very encouraging. The discussion on the beautiful work presented by Lucas *et al.* on the layered structure of single crystal Pt_3Ni is an example of the profound influence of the surface on the interfacial alloy composition.

The development of SERS in flow cells employing core–shell nanoparticles is an interesting method for probing reactions. The basic ideas were originally proposed by Mike Weaver but their realisation using core–shell particles in a well-controlled environment is a novel approach that has great potential due to the large surface sensitivity that can be achieved. Sum frequency generation (SFG) is providing a wealth of information on the potential dependence of orientation of interfacial water and thickness of the double layer, as was shown by Noguchi *et al.* Importantly, the spectral resolution achieved with modern instrumentation allows discrimination between highly structured and disordered interfacial water.

Another exciting development that was presented was the use of anomalous small-angle X-ray scattering (ASAXS) to determine scattering profiles of the two components of a PtCu alloy. It is hoped that future developments and refinements of this technique could provide surface and interfacial composition of alloys in an electrochemical environment. This technique could determine the separate scattering profiles of Pt and Cu and from this information, the removal of surface copper could be detected as well as changes in the structure of the nanoalloys leading to the formation of core-shell materials. The determination of metal distribution in PtRu nanoparticles has also been assessed by Tong *et al.* employing ^{195}Pt electrochemical

NMR. Significantly, this technique provides information on the Fermi level local density of states.

A fundamental issue in electrocatalytic materials is the development of models that can be used for analysing the processes that occur in a fuel cell. In this respect, the paper by Behm attracted a lively discussion, since the microelectrode arrays prepared gave well-defined structures that can be used to map the fate of intermediates. For example, the decay of hydrogen peroxide during the course of the reduction of oxygen is of particular importance in practical applications and indeed, for establishing the conditions under which significant concentrations of this intermediate would be observed. The simple approach proposed in this paper gives an experimental handle for the analysis of electrochemical reactions for which the small dimensions of the electrodes makes it necessary to develop a new modelling approach not determined by macroscopic diffusion and reaction laws. This type of modelling still needs to be investigated.

Materials science and reactivity

The third aspect that has to be considered refers to developments in materials science. The materials employed as electrocatalysts and their properties were extensively discussed. Geometry of nanocrystals can determine their properties and although there is a considerable body of work published, we still do not know how to predict morphology. The paper by Sun *et al.* showed that it is possible to change geometry by oxidation–reduction cycles to produce, in the case described by these authors, hexocathedral nanocrystals. Although the mechanism by which these changes occur is still unclear, it is very interesting that the simple procedure described can lead to a reproducible evolution of shape.

An example of the importance of combining theory and experiment to investigate the modification of surface morphology was the elegant studies of microfacetting of Ir(210) surfaces by Timo Jacob *et al.*, who was able to construct surface phase diagrams using DFT calculations and then relate these to the structures formed. Theoretical calculations were first made for facetting under UHV conditions and these results were then transposed to an electrochemical environment. The relationship between geometry and reactivity still requires investigation. In this respect, the Alicante group has made significant advances in demonstrating that it is possible to transpose results obtained with single crystals to nanoparticles.

Oxygen reduction on substrates other than Pt is of interest for its replacement in fuel cells and the work on ruthenium selenide represents an interesting effort in this direction. As was discussed, there are practical obstacles for the use of this interesting material. The simultaneous formation of hydrogen peroxide in ~10% yield is an additional difficulty. These results beg the question of which surface features are responsible for the formation of peroxide.

The study of the properties of alloy nanoparticles presents some serious challenges for example, for predicting the conditions for the appearance of a miscibility gap in nanoparticles of single-phase bulk alloys. This is an important question since it would be very useful to be able to predict the surface composition of alloy nanoparticles. The distribution of components in bimetallic catalysts, for example, for the industrially relevant Pt:Ru catalyst, is of great interest. It is clear that bulk phase diagrams do not provide reliable information on the properties of nanoparticles. A clear example of this is the appearance of a miscibility gap in Au:Ag nanoparticles; these two metals form solid solutions in the bulk in the whole compositional range but metal segregation occurs in the nanometre range.

A different approach to electrocatalytic materials was presented by Armstrong *et al.*, who described the use of lacasse for oxygen reduction. This work shows the interest in the development of bioinspired electrocatalytic materials. Another example of this approach was the paper by Chorkendorff *et al.* who attempted to mimic the active sites of nitrogenases and hydrogenases using nanoparticulate transition metal sulfides. This work related the kinetics of the hydrogen evolution reaction to DFT calculations of the Gibbs energy of adsorption of hydrogen to the sulfur edge for the material and differences in reactivity could be predicted. This is another example of how the combination of calculations and experiments can offer a powerful approach for predicting reactivity.

The use of biological materials to produce electricity offers other interesting possibilities that are mentioned below.

The reaction mechanisms of oxygen reduction and of methanol, ethanol, acetaldehyde and formic acid oxidation on single crystal electrodes were vigorously discussed during the last session. The delicate surface sensitivity of some of these reactions and the influence of anion adsorption on reactivity were highlighted by Herrero *et al.* and Koper gave a good example of the relationship between UHV and electrochemical measurements. These discussions showed that although work on these systems has been going on for a long-time, mechanistic details are still lacking. Single crystals are very useful not only in providing well defined surfaces but also, they offer the possibility of changing the step density and hence allow a better understanding of reaction mechanisms.

What next?

We have three approaches to investigate electrocatalytic reactions, experiments, simulations of experimental data and theoretical models. Theoretical models do not aim to reproduce experimental results as such but to explain reactivity at a molecular level. This Faraday Discussion presented examples of these methods and showed that the tools available to investigate electrocatalytic reactions are experiencing a fundamental change, in particular in computational and advanced instrumental methods. In parallel with these developments, questions concerning the utilisation of science to solve societal issues are being actively discussed and, indeed, sections of the large current Framework 7 programme of the European Union are targeting these issues. Examples of this are concerns regarding energy, transport, food, health, clean (or "green") chemical syntheses and general environmental problems. As is common when distribution of scarce resources are being discussed, the scientific community is presented with a dilemma: do we encourage people to pursue their ideas regardless of their application or do we direct scientific efforts to solve problems that have been clearly identified to be of importance to society in general?

These questions are, of course, not new and they are not as contradictory as they would appear at first sight. The topic of this Discussion, electrocatalysis, is a good example of this, since progress in applications such as fuel cells, that could have a profound impact in society, would be slow without a fundamental understanding of the electron transfer processes to catalytic centres. My view is that we cannot abstract ourselves from the problems that surround us. This is not just simply a question of following the funding, but whether scientists have a social responsibility. Besides doing research because of our desire (and need) for understanding Nature, and also because of the joy and satisfaction of advancing knowledge, we should be aware of the consequences of our activity, which can affect many people, and that we should not ignore.

This Discussion has highlighted two other questions. The first is the international and interdisciplinary character of modern science. This is a consequence of the highly interdisciplinary character of research that requires the mobilisation of a great variety of resources. The study of problems in electrocatalysis necessitates a combination of expertise in theory, modelling, interfacial physical chemistry, the use of advanced instrumental facilities and materials science. These new requirements have been recognised in European research and for this reason, the need for collaborative research is at the core of the EU Framework programmes. The second issue is the need to train a new generation of young researchers with a more general outlook of the subject and in particular, the need to impart knowledge in both theoretical and experimental aspects in PhD training programmes.

It is difficult to predict where the next fundamental developments in electrocatalysis will occur. The importance of the general area of electrocatalysis must be stressed and the topics where this research will have a societal impact can be approximately mapped. Fuel cells are likely to be important in the future since the use of fossil fuels for transport is not sustainable in the long term and from what is known at present, these represent a reasonable alternative. The developments discussed in this meeting demonstrate that the combination of quantum chemical calculations and advanced instrumental techniques can represent a powerful approach for the rational search for new materials to replace Pt as an electrocatalytic material.

The study of the electrochemical properties of materials at the nanoscale is essential for applications requiring large area electrodes. As mentioned before, although we know now how to synthesise a great variety of metal nanoparticles, we do not know yet how to control the morphology and reactivity of these new materials. In addition, practical applications require good contact between the nanoparticles and the support and there is still a need to have appropriate techniques to study nanoparticle–electrode interactions or indeed, the properties of single nanoparticles.

Biologically inspired reaction centres can provide new avenues of inquiry for reactions of practical importance. Oxygen reduction is a case in point that was discussed during this meeting. The synthesis of artificial enzymatic centres mimicking the reactivity of biological materials represents an intriguing challenge, which, if properly developed, can bring different scientific communities to a common objective. In addition, the use of electrocatalysts in sensors in bioanalytical chemistry, although not a topic of this Discussion, will have a significant impact on health issues.

Another significant development that can be foreseen is the transformation of carbon dioxide into useful products. Work on the electrocatalytic reduction of CO_2 has been carried out for many years with limited success. The electrochemical reduction on metals such as copper or using macrocyclic complexes gives a very wide variety of products. Without deploying the techniques discussed in this meeting, progress in this area is bound to be slow. This reaction is a prime candidate for using strategies derived from the known dark reactions of photosynthesis. Although it is not possible at present to consider all the reactions in the Calvin cycle, it is reasonable to study the feasibility of considering sections of it. In this respect, it is interesting to notice that plants have evolved a reduction strategy that does not attempt the direct reduction of CO_2 but rather, the reduction step is the reduction of the COOH function to an aldehyde.[2] In the Calvin cycle, this is the key intermediate for the various reactions that regenerate the ribulose-1,5-diphosphate that acts as the trapping agent for CO_2 at the start of the cycle. A bioinspired approach may provide a route for this reduction.

The use of biological materials as electrocatalysts discussed in the meeting offers interesting possibilities: (1) the syntheses of catalytic materials inspired by the structure of the redox centre of enzymes can provide novel electrocatalysts, and (2) microbial fuel cells employing bacteria that have cytochromes in their external membranes are attracting renewed interest.[3] Although the power that can be obtained for the latter is small compared with a high energy density H_2/O_2 fuel cell, there are many applications in water treatment where combining remediation with energy production could be of economical and great practical use in effluent control.

In conclusion, this Discussion has highlighted some very topical contemporary issues and has helped to overcome the very common difficulties in communicating across disciplines. Like its predecessors, it will serve as a reference point for new developments.

References

1 M. Cossi, G. Scalmani, N. Rega and V. Barone, *J. Chem. Phys.*, 2002, **117**, 43.
2 M. Calvin and J. A. Bassham, in *The Photosynthesis of Carbon Compounds*, W. A. Benjamin, Inc., New York, 1962.
3 See for example: D. R. Bond and D. R. Lovley, Electricity production by *Geobacter sulfurreducens* attached to electrodes, *Appl. Environ. Microbiol.*, 2003, **69**, 1548–55.

Poster titles

Influence of the structure of the carbon support on the oxidation of carbon monoxide and methanol at Pt/C and PtX/C, **J. J. Quintana, J. Flórez, L. Calvillo, M. J. Lázaro, P. L. Cabot, I. Esparbé and E. Pastor**, *Universidad de La Laguna, Spain*

Interactions between H_2O and preadsorbed H or O on Pt(533), **M. J. T. C. van der Niet, I. Dominicus, O. T. Berg, L. B. F. Juurlink and M. T. M. Koper**, *Leiden University, The Netherlands*

Electrocatalytic reduction of nitrite on various metals. A preliminary mechanistic study, **M Duca, V. T. Kavvadia and M. T. M. Koper**, *University of Leiden, The Netherlands*

On the elucidation of the mechanism of the CO oxidation on platinum vicinal surfaces in alkaline media, **Gonzalo García, Paramaconi Rodríguez and Marc T. M. Koper**, *Leiden University, The Netherlands*

Measuring fuel cell electrocatalysis under fuel cell conditions in the absence of mass transport, electronic and ionic losses, **Eishiro Toyoda and Anthony R. Kucernak**, *Imperial College London, UK*

The effect of anode degradation on MEA components in direct methanol fuel cells, **N. Cabello-Moreno, E. M. Crabb, J. M. Fisher, A. E. Russell and D. Thompsett**, *Johnson Matthey Technology Centre, UK*

Probing operating fuel cell catalysts using X-ray absorption spectroscopy, **S. C. Ball, S. L. Hudson, A. E. Russell, B. Theobald and D. Thompsett**, *Johnson Matthey Technology Centre, UK*

Spectroelectrochemical detection of proton-coupled redox processes in novel tatpp-based photocatalysts, **Reynaldo O Lezna, Norma R de Tacconi and Frederick M MacDonnell**, *The University of Texas at Arlington, USA*

A first principles study of direct electrooxidation of aqueous borohydride on Au and Pt surfaces, **Michael J Janik and Gholamreza Rostamikia**, *Pennsylvania State University, USA*

X-ray absorption spectroscopy: A versatile tool for the in-situ and operando study of fuel cell catalysts, **V Croze, F Ettingshausen, J Melke, D E Ramaker and C Roth**, *Technische Universität Darmstadt, Germany*

Modelling platinum surface area loss in PEM fuel cell cathodes, **Edward F Holby, Yang Shao-Horn, Wenchao Sheng and Dane Morgan**, *University of Wisconsin – Madison, USA*

Tuning the activity of Pt-based electrocatalysis by shape control and by surface decoration, **Ceren Susut, Bingchen Du and YuYe Tong**, *Georgetown University, USA*

Gold based electrocatalysts for fuel cells, **Prabalini Sivasubramaniam and Andrea E. Russell**, *University of Southampton, UK*

Pt_{2ML}/Pd/C core-shell electrocatalysts for the ORR in PEMFCs, **B. C. Tessier*, A. E. Russell, B. Theobald and D. Thompsett**, *University of Southampton, UK*

H_2 electrooxidation on PdAu alloy catalyst, **Faisal Al Odail, Alexandros Anastasopoulos, Jens P Suchsland and Brian E Hayden**, *University of Southampton, UK*

Oxygen reduction at core–shell Pt/Me/C electrocatalysts, **Gael Chouchelamane*, Andrea E Russell, Dave Thompsett and Brian Theobald**, *University of Southampton, UK*

Layer-by-layer grown Pd(Core)-Pt(shell) nanoparticle electrocatalysts for the oxygen reduction reaction, **J. X. Wang, H. Inada, L. Wu, Y. Zhu and R. R. Adzic**, *Brookhaven National Laboratory, USA*

Dissociation of H_2 on water covered Pt(111), **Sebastian Schnur and Axel Groß**, *Ulm University, Germany*

Model studies of electrocatalytic processes using spectroelectrochemical techniques, **M. Heinen, Z. Jusys and R. J. Behm**, *Ulm University, Germany*

High resolution electrochemical imaging with the Scanning Micropipette Contact Method (SMCM), **Hollie V. Patten, Agnieszka J. Rutokowska, Martin A. Edwards, Cara G. Williams, Julie V. Macpherson and Patrick R. Unwin**, *University of Warwick, UK*

Single-walled carbon nanotube-gold nanoparticle composites, **Petr V. Dudin*, Julie V. Macpherson and Patrick R. Unwin**, *University of Warwick, UK*

Study of the activity of copper, gold and copper-gold alloys for the electroreduction of carbon dioxide, **Jennifer Christophe and Claudine Buess-Herman**, *Université Libre de Bruxelles (ULB), Belgium*

Electrochemical responses of Te upd layers at Au-Pd core-shell nanostructures, **Maria Montes de Oca and David J. Fermín**, *University of Bristol, UK*

Oxygen reduction catalysis by supported transition metal compounds, **Dawid Wodka, Aleksandra Pacula and Pawel Nowak**, *Polish Academy of Sciences, Poland*

On the mechanism of the ethanol electro-oxidation on Pt as probed by sum frequency generation spectroscopy, **Janaina F. Gomes, Kleber Bergamaski, Hilton B. de Aguiar, L. Jay Deiner, Paulo B. Miranda and Francisco C. Nart**, *Fritz-Haber-Institut der Max-Planck-Gesellschaft, Germany*

Platinum nanoparticles supported on conducting polymers using synthetic conditions, **Carlos Sanchis, Horacio J. Salavagione and Emilia Morallón**, *Instituto Universitario de Materiales de Alicante (IUMA), Spain*

Characterization of fuel cell cathode catalysts based on metal oxide supports, **Hideo Notsu, Takuya Kitamura and Ichizo Yago**, *National Institute of Advanced Industrial Science and Technology (AIST), Japan*

Development on nano-structured catalyst supports and in situ electrochemical SERS-active substrates for ORR investigation, **Ichizo Yagi, Akari Hayashi, Ken'ichi

Kimijima and **Narumi Ohta**, *Polymer Electrolyte Fuel Cell Cutting-edge Research Centre (FC-Cubic), Japan*

First principles molecular dynamics simulation of water/Pt interface under negative potential, **Minoru Otani, Ikutaro Hamada, Osamu Sugino Yoshitada Morikawa, Yasuharu Okamoto and Tamio Ikeshoji**, *National Institute of Advanced Industrial Science and Technology (AIST), Japan*

Hydrogen adsorption, evolution and oxidation on the Pt/111 electrode in aqueous solution: a critical evaluation based on *ab initio* MD/DFT studies of HOR and UPD H states on the Pt(111) surface of nanoparticles, **Yasuyuki Ishikawa, Juan J. Mateo, Carlos R. Cabrera and Donald A. Tryk**, *University of Yamanashi, Japan*

Density functional theory/modified Poisson-Boltzmann theory approach to interfacial electrochemistry, **Ryosuke Jinnouchi and Alfred B. Anderson**, *Case Western Reserve University, USA*

Controlled modification of carbon supported Pt cathode electrocatalysts for the PEM fuel cell, **Y. Qian, P. P. Wells, E. M. Crabb, L. E. Smart, A. E. Russell and D Thompsett**, *The Open University, UK*

Particle size influence on Ru and Sn modified Pt/C electrocatalysts for direct methanol fuel cells, **X. Liu, E. M. Crabb, L. E. Smart, D. Thompsett, N. Cabello-Moreno and J. Fisher**, *The Open University, UK*

Heat treated $PtCO_3$ nanoparticles as catalyst for oxygen reduction, **H. Schulenburg, E. Miller, T. Roser, G. Khelashvili, H. Bönnemann, A. Wokaun and G. G. Scherer**, *Paul Scherrer Institut, Switzerland*

The role of surface strain on the particle size effect of Pt in electrocatalysis, **Fan Yang and Shengli Chen**, *Wuhan University, China*

Soft landed proteins on MWCNTs surface: electron transfer kinetic characterization of immobilized MP-11, **Marco Frasconi, Gabriele Favero, Federico Pepi, Alessandra Tata and Franco Mazzei**, *Sapienza Università di Roma, Italy*

In Situ IR and surface voltage spectroscopy of WC electrocatalyst surfaces, **David S. Warren, Jan Scholz and A. James McQuillan**, *University of Otago, New Zealand*

The Skinner Prize for the best poster was awarded to M. J. T. C. van der Niet of Leiden University, The Netherlands, for her poster on Interactions between H_2O and preadsorbed H or O on Pt(533).

List of participants

Professor E. Ahlberg, *Gothenburg University, Sweden*
Mr F. Alodail, *University of Southampton, United Kingdom*
Mr A. Anastasopoulos, *University of Southampton, United Kingdom*
Dr F. Armstrong, *University of Oxford, United Kingdom*
Mr A. Baars, *Ivium Technologies, United Kingdom*
Dr S. Ball, *Johnson Matthey Technology Centre, United Kingdom*
Professor P. Bartlett, *University of Southampton, United Kingdom*
Professor J. Behm, *University of Ulm, Germany*
Dr P. Biedermann, *Max-Planck Institut Fuer Eisenforschung GmbH, Germany*
Dr I. Buder, *Centre of Fuel Cell Technology, Germany*
Professor C. Buess-Herman, *Universite Libre De Bruxelles Servie Chani, Belgium*
Miss N. Cabello, *Johnson Matthey Technology Centre, United Kingdom*
Dr I. Cerri, *Toyota Motor Europe, Belgium*
Dr K. Chandler, *Johnson Matthey PLC, United Kingdom*
Professor S. Chen, *Wuhan University, China*
Professor I. Chorkendorff, *Technical University of Denmark, Denmark*
Mr G. Chouchelamane, *University of Southampton, United Kingdom*
Mr F. Ciucci, *Caltech, U.S.A.*
Dr M. Coward, *AWE PLC, United Kingdom*
Dr E. Crabb, *The Open University, United Kingdom*
Dr C. Cremers, *Fraunhofer ICT, Germany*
Dr K. Dawes, *Windsor Scientific Ltd, United Kingdom*
Dr D. Dawson, *Alphasense Ltd., United Kingdom*
Mr M. Duca, *Leiden Institute of Chemistry, The Netherlands*
Mr P. Dudzin, *University of Warwick, United Kingdom*
Mr T. Esterle, *University of Southampton, United Kingdom*
Professor J. Evans, *University of Southampton, United Kingdom*
Dr D. Fermin, *University of Bristol, United Kingdom*
Dr J. Fernandes Gomes, *Fritz-Haber-Institut, Germany*
Dr J. Fisher, *Johnson Matthey Technology Centre, United Kingdom*
Dr S. Frank, *Leiden University, The Netherlands*
Dr M. Frasconi, *Sapienza Universita Di Roma, Italy*
S. Fryatt, *Alvatek Ltd, United Kingdom*
Miss M. Gamero, *University of Southampton, United Kingdom*
Dr G. Garcia, *Leiden University, The Netherlands*
Professor A. Gewirth, *Univeristy of Illinois, U.S.A.*
Mr A. Gieske, *Leiden University, The Netherlands*
Ms M. Gilbert, *Royal Society of Chemistry, United Kingdom*
Professor A. Gross, *University of Ulm, Germany*
Professor B. Hayden, *University of Southampton, United Kingdom*
Mr M. Heinen, *Ulm University, Germany*
Dr E. Herrero, *Universidad De Alicante, Spain*
Mr J. Hodge, *Royal Society of Chemistry, United Kingdom*
Dr S. Horswell, *University of Birmingham, United Kingdom*
Miss S. Hudson, *Johnson Matthey Technology Centre, United Kingdom*
Mr L. Hussein, *Institute for Mikrosystem Technology, Germany*
Dr T. Ikeshoji, *AIST, Japan*
Dr T. Jacob, *University of Ulm, Germany*
Professor M. Janik, *Pennsylvania State University, U.S.A.*
Dr R. Jinnouchi, *Toyota Central R&D Labs, Japan*

Mr L. Johnson, *University of Nottingham, United Kingdom*
Miss A. Juskowiak, *University of Southampton, United Kingdom*
Dr P. Keil, *Max-Planck-Institut Fuer Eisenforschung, Germany*
Dr L. Kibler, *Universitat Ulm, Germany*
Dr J. Kim, *Massachusetts Institute of Technology, U.S.A.*
A. King, *Pacer International Limited, United Kingdom*
Dr C. King, *Windsor Scientific Ltd., United Kingdom*
Mr P. Kleszyk, *University of Southampton, United Kingdom*
Professor M. Koper, *Leiden University, The Netherlands*
Professor C. Korzeniewski, *Texas Tech University, U.S.A.*
Dr C. Kriek, *North-West University, South Africa*
Dr A. Kucernak, *Imperial College London, United Kingdom*
Mr S. Lai, *Leiden University, The Netherlands*
Mr J. Lapinski, *University of Southampton, United Kingdom*
Miss S. Latham, *Royal Society of Chemistry, United Kingdom*
Mr S. Lee, *Massachesetts Institute of Technology, U.S.A.*
Dr R. Lezna, *INIFTA-CONICET, Argentina*
Dr F. Liu, *CMR Fuel Cells (UK) Ltd, United Kingdom*
Mr X. Liu, *The Open University, United Kingdom*
Dr J. Low, *University of Southampton, United Kingdom*
Dr C. Lucas, *University of Liverpool, United Kingdom*
Professor N. Markovic, *Argonne National Laboratory, U.S.A.*
Mr H. Matej, *Solvay S.A., Belgium*
Miss H. Mohd, *University of Southampton, United Kingdom*
Miss M. Montes-de Oca, *University of Bristol, United Kingdom*
Professor D. Morgan, *University of Wisonsin - Madison, U.S.A.*
Mr E. Morgan, *Royal Society of Chemistry, United Kingdom*
Dr H. Naohara, *Toyota Motor Corporation, Japan*
Miss M. Nestoridi, *University of Southampton, United Kingdom*
Professor M. Neurock, *University of Virginia, U.S.A.*
S. Newport, *Pacer International Limited, United Kingdom*
Professor H. Noguchi, *Hokkaido University, Japan*
Dr H. Notsu, *FC-Cubic, AIST, Japan*
Dr N. Nugent, *Royal Society of Chemistry, United Kingdom*
Dr G. Offer, *Imperial College London, United Kingdom*
Mr M. Otani, *AIST, Japan*
Professor R. Parsons, *United Kingdom*
Professor E. Pastor, *University of La Laguna, Spain*
Miss H. Patten, *University of Leicester, United Kingdom*
Mrs M. Perdjon-Abel, *University of Southampton, United Kingdom*
Miss D. Plana, *University of Manchester, United Kingdom*
Professor D. Pletcher, *University of Southampton, United Kingdom*
Prof Dr B. Ren, *Xiamen University, China*
Dr J. Rossmeisl, *CAMD_Technical University of Denmark, Denmark*
Dr C. Roth, *TU Darmstadt, Germany*
Professor A. Russell, *University of Southampton, United Kingdom*
Mr P. Ruvinskiy, *Universite Louis Pasteur, France*
Mr C. Sanchis, *Universidad De Alicante, Spain*
Dr E. Santos, *Ulm University, Germany*
Dr M. Sarwar, *Johnson Matthey Technology Centre, United Kingdom*
Professor E. Savinova, *LMSPC-UMR 7115 Du CNRS ULP ECPM, France*
Professor D. Scherson, *Case Western Reserve University, U.S.A.*
Professor D. Schiffrin, *University of Liverpool, United Kingdom*

Professor W. Schmickler, *Univeritat Ulm, Germany*
Mr S. Schnur, *University of Ulm, Germany*
Mr J. Scholz, *Otago University, New Zealand*
Mr H. Schulenburg, *Paul Scherrer Institut, Switzerland*
Mr A. Scott, *Royal Society of Chemistry, United Kingdom*
Professor Y. Shao-Horn, *Massachusetts Institute of Technology, U.S.A.*
Ms P. Sivasubramanian, *University of Southampton, United Kingdom*
Dr A. Slipszenko, *SRA Developments, United Kingdom*
Professor P. Strasser, *University of Houston, U.S.A.*
Prof Dr S. Sun, *Xiamen University, China*
Ms C. Susut, *Georgetown University, U.S.A.*
Dr J. Talbot, *AWE Plc, United Kingdom*
Miss B. Tessier, *Johnson Matthey Technology Centre, United Kingdom*
Mr V. Thomas, *Caltech, U.S.A.*
Dr D. Thompsett, *Johnson Matthey Technology Centre, United Kingdom*
Miss G. Tomba, *King's College London, United Kingdom*
Dr M. Toney, *Stanford Synchrotron Radiation Lab, U.S.A.*
Professor Y. Tong, *Georgetown University, U.S.A.*
Professor D. Tryk, *University of Yamanashi, Japan*
Mr N. Tsiouvaras, *University of Southampton, United Kingdom*
Professor G. Tsirlina, *M.V. Lomonosov Moscow State University, Russia*
Ms J. Van Der Niet, *Leiden University, The Netherlands*
Mr S. Veltze, *Syddansk Universitet, Denmark*
Dr D. Walsh, *University of Nottingham, United Kingdom*
Dr J. Wang, *Brookhaven National Laboratory, U.S.A.*
Dr P. Wells, *University of Southampton, United Kingdom*
Professor A. Wieckowski, *University of Illinois, U.S.A.*
Professor G. Wittstock, *University of Oldenburg, Germany*
D. Wodka, *Polish Academy of Sciences, Poland*
Dr R. Wreland Lindstrom, *Royal Institute of Technology, Sweden*
Dr I. Yagi, *FC-Cubic, AIST, Japan*
Mr A. Yanson, *Leiden University, The Netherlands*

Index of contributors*

Ahlberg, E., 303, 311
Armstrong, F., **319**, 417, 418, 429
Behm, J., 94, **167**, 200, 202, 203, 204, 205, 206, 314, 317, 419, 424, 431, 435, 436
Biedermann, P., 191
Buder, I., 108, 313
Buess-Herman, C., 189
Chen, S., 193, 317
Chorkendorff, I., 99, 109, **219**, 301, 302, 309
Cremers, C., 313
Duca, M., 110
Dudzin, T., 110
Fermin, D., 205, 311, 317
Fernandes Gomes, J., 193, 435
Gewirth, A., 93, 102, 104, **113**, 185, 186, 187, 189, 190, 200, 309, 310, 314, 317, 417, 419, 424, 426
Gross, A., 194, **233**, 303, 304, 305, 306, 309, 428
Hayden, B., 107
Herrero, E., 96, 202, **379**, 429, 431, 432, 434, 436
Hudson, S., 316
Ikeshoji, T., 191, 297, 305
Jacob, T., **69**, 101, 102, 103, 104, 106, 108
Janik, M., 302, 418
Jinnouchi, R., 429
Kibler, L., 101
Koper, M., 185, 430
Korzeniewski, C., 192, 432
Kucernak, A., 203, 206, 424
Lai, S., **399**, 432, 433, 434, 435
Lucas, C., **41**, 93, 94, 98
Markovic, N., **25**, 94, 95, 96, 97, 98, 99, 100, 101, 302, 310, 425, 427
Morgan, D., 204, 301

Nestoridi, M., 417
Neurock, M., 96, 193, 299, 302, 304, **363**, 429, 430, 431
Noguchi, H., **125**, 190, 191, 192, 193, 194
Parsons, R., 297
Pletcher, D., 187, 190
Ren, B., 111, **155**, 197, 198, 199, 418, 432
Rossmeisl, J., 306, **337**, 418, 419, 420, 423, 427, 428
Russell, A., 101, 195, 301, 316
Santos, E., 100, 101, 300
Savinova, E., 95, 104, 108, 196, 200, 301, 423, 426
Scherson, D., **59**, 100, 101, 102, 111, 197, 310, 310
Schiffrin, D., 94, 98, 106, 185, 186, 188, 198, 206, 299
Schmickler, W., 95, 103, **209**, 297, 298, 299, 300, 309, 310, 417, 419, 423, 427
Strasser, P., 106, 204, **283**, 316, 317, 428
Sun, S., **81**, 93, 99, 104, 105, 106, 107, 109, 110, 111, 190, 199, 314, 430, 432, 435, 436
Thompsett, D., 196, 437,
Tong, Y., 97, 139, 194, 195, 196, 197
Tryk, D., 98, 206, 299, 305, 306, 422
Tsirlina, G., 96, 102, 190, **245**, 300, 310, 311, 312, 313
Wang, J., 97, 109, 200, 305, **347**, 420, 421, 422, 424, 425, 427
Wieckowski, A., 94, 100, 102, 103, 105, 185, 190, 192, 195, 197, **269**, 298, 302, 312, 314, 315, 316, 417, 431, 432, 433, 434, 435
Wittstock, G., 199, 418, 434

* The page numbers in **bold** type indicate papers submitted for discussions.